Lecture Notes in Civil Engineering

Volume 33

Lecture Notes in Civil Engineering (LNCE) publishes the latest developments in Civil Engineering - quickly, informally and in top quality. Though original research reported in proceedings and post-proceedings represents the core of LNCE, edited volumes of exceptionally high quality and interest may also be considered for publication. Volumes published in LNCE embrace all aspects and subfields of, as well as new challenges in, Civil Engineering. Topics in the series include:

- Construction and Structural Mechanics
- Building Materials
- Concrete, Steel and Timber Structures
- Geotechnical Engineering
- Earthquake Engineering
- Coastal Engineering
- Hydraulics, Hydrology and Water Resources Engineering
- Environmental Engineering and Sustainability
- Structural Health and Monitoring
- Surveying and Geographical Information Systems
- Heating, Ventilation and Air Conditioning (HVAC)
- Transportation and Traffic
- Risk Analysis
- Safety and Security

To submit a proposal or request further information, please contact the appropriate Springer Editor:

- Mr. Pierpaolo Riva at pierpaolo.riva@springer.com (Europe and Americas);
- Ms. Swati Meherishi at swati.meherishi@springer.com (India);
- Ms. Li Shen at li.shen@springer.com (China);
- Dr. Loyola D'Silva at loyola.dsilva@springer.com (S-E Asia and Australia/NZ).

Indexed by Scopus

More information about this series at http://www.springer.com/series/15087

Jayanta Kumar Ghosh ·
Irineu da Silva
Editors

Applications of Geomatics in Civil Engineering

Select Proceedings of ICGCE 2018

Springer

Editors
Jayanta Kumar Ghosh
Civil Engineering Department
Indian Institute of Technology Roorkee
Roorkee, India

Irineu da Silva
Sao Carlos School of Engineering
University of Sao Paulo
Sao Carlos, São Paulo, Brazil

ISSN 2366-2557 ISSN 2366-2565 (electronic)
Lecture Notes in Civil Engineering
ISBN 978-981-13-7066-3 ISBN 978-981-13-7067-0 (eBook)
https://doi.org/10.1007/978-981-13-7067-0

Library of Congress Control Number: 2019933715

This Springer imprint is published by the registered company Springer Nature Singapore Pte Ltd.
The registered company address is: 152 Beach Road, #21-01/04 Gateway East, Singapore 189721, Singapore

Dedicated
to
The Civil Engineering Department
Indian Institute of Technology (IIT) Roorkee
[Erstwhile: University of Roorkee
(2001–1949), Thomason College of Civil
Engineering (1949–1854) and Roorkee
College of Civil Engineering (1854–1847)]

Preface

An International Conference on Geomatics in Civil Engineering (ICGCE 2018) was organised by the Department of Civil Engineering, IIT Roorkee, during 05–06 April 2018. The conference had six technical themes based on applications of geomatics engineering for six limbs of civil engineering, i.e. applications of geomatics in structural engineering; applications of geomatics in geotechnical engineering; applications of geomatics in hydraulics and water resources engineering; applications of geomatics in environmental engineering; applications of geomatics in transportation engineering and miscellaneous aspects and application of geomatics engineering. In these technical sessions, keynote speakers, invited speakers and authors deliberated various applications of geomatics engineering in different aspects of civil engineering.

Part I of the proceedings contains the technical papers presented by two keynote speakers. One of the papers deliberated harnessing of remote sensing data and technology for different domains of civil engineering with special emphasis on monitoring and control of quality of civil engineering materials and constructions of structure. And the other discusses on why, how and when geomatics engineering is relevant to civil engineering, followed by possible course contents of geomatics engineering for civil engineering practices. Part II contains papers on applications of geomatics engineering on different aspects of structural engineering with special emphasis on structural damage assessment, estimation of concrete strength during its early age, 3D reconstruction of civil structure, hydration monitoring of concrete during construction, etc. Part III contains papers on applications of geomatics engineering on geotechnical engineering with special emphasis on estimation of spatial variability in depth of weathered rock, soil moisture, landslide study, etc. Part IV contain papers on applications of geomatics engineering on hydraulics and water resources engineering with special emphasis on assessment and mapping of flood hazard, estimation of groundwater recharge as well as river discharge, vulnerability assessment of glacier as well as of droughts, reservoir sedimentation and soil erosion study, etc. Part V discusses on applications of geomatics engineering on environmental engineering with special emphasis on estimation of air pollution parameters; assessment of ground and surface water quality; monitoring, management and disposal of solid wastes, etc.

Part VI contains papers on applications of geomatics engineering on transportation engineering with special emphasis on the development of road information and management system; mass transit system; accessibility analysis; study of pedestrian movement; planning and acquiring land for road; vulnerability analysis of road networks and accident-prone locations, etc. Finally, in Part VII contains papers on different aspects and applications of geomatics engineering like digital land record; dynamics of mangrove forest; integration of GCPs with terrestrial laser scanner and close-range photogrammetry, assessment of backscattering, snow-cover analysis, built-up and bare soil indices, Web application for urban planning; land surface temperature study, GPR-based sub-surface study, digital elevation/terrain model, deep convolution neural network for satellite images; extraction of building footprint; Public Health SDI; glacier hazard study, biomass estimation; cellular network planning, mapping of cultural heritage sites and drainage network, among others.

As an overview, it may be found that geomatics engineering is marching towards indispensable presence towards civil engineering. Thus, to cope up with the future needs, geomatics engineering should be incorporated more and more in civil engineering education.

Roorkee, India
Sao Carlos, Brazil

Jayanta Kumar Ghosh
Irineu da Silva

Contents

Editors and Contributors

About the Editors

Dr. Jayanta Kumar Ghosh is Associate Professor in the Civil Engineering Department, Geomatics engineering group at the Indian Institute of Technology, Roorkee. He holds a Ph.D. in Civil Engineering (Remote Sensing) from IIT Roorkee. He has supervised 38 Masters and 14 Ph.D. students, and has published more than 100 papers in scientific conferences and reputed journals, as well as two scientific books in the area of Geomatics. He has been engaged in teaching, research and consultancy works in geomatics engineering for more than 33 years. His areas of interest include geomatics engineering, remote sensing, GPS, geodesy, AI, soft computations, automated system development and cognition.

Dr. Irineu da Silva is Associate Professor of civil engineering in the Department of Transportation Engineering at the University of Sao Paulo, Brazil. He received his Ph.D. in Geodetic sciences from the Federal University of Lausanne. He has guided several Masters and Ph.D. students over the years, and has published more than 80 papers in reputed journals and scientific conferences. He has also written two scientific books on Geomatics. Dr. da Silva's field of research is related to geomatics quality and geodetic structural monitoring. He has been invited as a keynote speaker in more than 20 scientific conferences, and he has been on the editorial board of several international journals. In 2007, he received the Order of Cartographic Merit from the Brazilian Society of Cartography.

Contributors

S. Agrawal Photogrammetry and Remote Sensing Division, Indian Institute of Remote Sensing, ISRO, Dehradun, India

Sonam Agrawal GIS Cell, Motilal Nehru National Institute of Technology Allahabad, Prayagraj, Uttar Pradesh, India

Ishtiyaq Ahmad Department of Civil Engineering, NIT Raipur, Raipur, India

Odai Ibrahim Mohammed Al Balasmeh Department of Civil Engineering, Thapar Institute of Engineering and Technology, Patiala, Punjab, India

B. Aneesha Satya Department of Civil Engineering, NIT Warangal, Warangal, India

N. P. Anoona Transportation Engineering, Visvesvaraya National Institute of Technology, Nagpur, India

R. Aravinth Department of Civil Engineering, Chitkara University, Baddi, Himachal Pradesh, India

Aman Kumar Balihar Punjab Remote Sensing Centre, Ludhiana, India

Balwan Punjab Remote Sensing Centre, Ludhiana, India

Tapti Baskey Punjab Remote Sensing Centre, Ludhiana, India

Alolika Basu Atur Centre, Symbiosis Institute of Geoinformatics, SIG Pune, Pune, Maharashtra, India

Suresh Bhalla Civil Engineering Department, Indian Institute of Technology (IIT) Delhi, Hauz Khas, New Delhi, India

Narayana Rao Bhogapurapu NIT Warangal, Warangal, India

Maisnam Bipinchandra Department of Civil Engineering, NIT-Manipur, Imphal, India

Sujata Biswas Department of Civil Engineering, IIEST Shibpur, Howrah, West Bengal, India

Subha Chakraborty Department of Architecture, Town and Regional Planning, IIEST, Shibpur, India

Swarnadeepa Chakraborty Department of Civil Engineering, IIEST Shibpur, Howrah, West Bengal, India

Rajat Chatterjee Department of Civil Engineering, Visvesvaraya National Institute of Technology, Nagpur, Maharashtra, India

Devarshi Chaurasia School of Planning and Architecture, Bhopal, India

Sasmita Chaurasia Space Applications Centre, ISRO, Ahmedabad, India

Rajat Chhabra Civil Engineering Department, Indian Institute of Technology (IIT) Delhi, Hauz Khas, New Delhi, India

Sandeep Kumar Chouksey Department of Civil Engineering, NIT Raipur, Raipur, India

Josodhir Das Department of Earthquake Engineering, Indian Institute of Technology Roorkee, Roorkee, India

Debajit Datta Department of Geography, Jadavpur University, Kolkata, India

Karun Reuel Dayal Photogrammetry and Remote Sensing Division, Indian Institute of Remote Sensing, ISRO, Dehradun, India

Manisha Desai Civil Engineering Department, Sardar Vallabhbhai National Institute of Technology, Surat, Gujarat, India

Champat Lal Dewangan Department of Civil Engineering, NIT Raipur, Raipur, India

B. Divya Priya Computational Geomechanics Lab, Department of Civil Engineering, Indian Institute of Technology Madras, Chennai, India

G. R. Dodagoudar Computational Geomechanics Lab, Department of Civil Engineering, Indian Institute of Technology Madras, Chennai, India

Saket Dubey Department of Civil Engineering, NIT Raipur, Raipur, India

K. H. V. Durga Rao NRSC, Hyderabad, India

P. K. Garg Department of Civil Engineering, IIT Roorkee, Roorkee, Uttarakhand, India

Rahul Dev Garg Indian Institute of Technology Roorkee, Roorkee, Uttarakhand, India

L. Gnanappazham Department of Earth and Space Sciences, Indian Institute of Space Science and Technology, Thiruvananthapuram, Kerala, India

V. V. Govind Kumar Indian Institute of Technology (Indian School of Mines) Dhanbad, Dhanbad, Jharkand, India

R. D. Gupta Civil Engineering Department and Member in GIS Cell, Motilal Nehru National Institute of Technology Allahabad, Prayagraj, Uttar Pradesh, India

Harmanpreet Punjab Remote Sensing Centre, Ludhiana, India

Vinayak Huggannavar NITK, Surathkal, Mangalore, India

Gaurav V. Jain Space Application Centre (ISRO), Ahmedabad, Gujarat, India

Varinder S. Kanwar Department of Civil Engineering, Chitkara University, Baddi, Himachal Pradesh, India

Pramodkumar Kappadi National Institute of Technology Karnataka, Surathkal, Mangalore, India

Tapas Karmaker Department of Civil Engineering, Thapar Institute of Engineering and Technology, Patiala, Punjab, India

Preeti Kashmira Punjab Remote Sensing Centre, Ludhiana, India

Y. B. Katpatal Civil Engineering Department, Visvesvaraya National Institute of Technology, Nagpur, India

Nivedita Kaul Department of Civil Engineering, MNIT Jaipur, Jaipur, India

Naveet Kaur CSIR-CRRI, New Delhi, India

Simerjit Kaur Punjab Remote Sensing Centre, Ludhiana, India

S. L. Kesav Unnithan Department of Earth and Space Sciences, Indian Institute of Space Science and Technology, Thiruvananthapuram, Kerala, India

Sumit Khandelwal Department of Civil Engineering, MNIT Jaipur, Jaipur, India

Tripti Khanduri Civil Engineering Department, Indraprastha Engineering College, Ghaziabad, Uttar Pradesh, India

Deepak Khare Department of Water Resource Development and Management, Indian Institute of Technology, Roorkee, Roorkee, India

Rupali Khare Department of Mining Engineering, Indian Institute of Technology (Indian School of Mines) Dhanbad, Dhanbad, Jharkhand, India

Radhakanta Koner Department of Mining Engineering, Indian Institute of Technology (Indian School of Mines) Dhanbad, Dhanbad, India

Kaushal Kumar GIS Cell, MNNIT, Allahabad, UP, India

Roshan Kumar Geomatics Engineering, Indian Institute of Technology Roorkee, Roorkee, Uttarakhand, India

Rupendra Kumar Geomatics Engineering, Indian Institute of Technology Roorkee, Roorkee, Uttarakhand, India

Kumar Kumarapu Department of Civil Engineering, NIT Warangal, Warangal, India

Supriya Kumari Central University of Jharkhand, Ranchi, India

S. K. P. Kushwaha Photogrammetry and Remote Sensing Division, Indian Institute of Remote Sensing, ISRO, Dehradun, India

Divya Lad Department of Environmental Engineering, Sarvajanik College of Engineering and Technology, Surat, India

Debra F. Laefer Center for Urban Science and Progress and Department of Civil and Urban Engineering, Tandon School of Engineering, New York University, Brooklyn, NY, USA

Dilip H. Lataye Department of Civil Engineering, Visvesvaraya National Institute of Technology, Nagpur, Maharashtra, India

Shirshendu Layek Department of Mining Engineering, Indian Institute of Technology (Indian School of Mines) Dhanbad, Dhanbad, India

K. N. Loukika NIT Warangal, Warangal, India

Chandramani Loukrakpam Department of Civil Engineering, NIT-Manipur, Imphal, India

Debaleena Majumdar Department of Geography, Jadavpur University, Kolkata, India

Santanu Mallik Department of Civil Engineering, National Institution of Technology, Barjala, Tripura (W), India

Sahoo Biswa Manaschintan Department of Civil Engineering, NIT Rourkela, Rourkela, Odisha, India

Ajay Mathur Punjab Remote Sensing Centre, Ludhiana, India

Mehali Mehta Department of Environmental Engineering, Sarvajanik College of Engineering and Technology, Surat, India

Rashmi A. Mestri Walchand College of Engineering, Sangli, Maharashtra, India

Umesh Mishra Department of Civil Engineering, National Institution of Technology, Barjala, Tripura (W), India

M. S. Mukesh Research Scholar, Transportation Engineering, Visvesvaraya National Institute of Technology, Nagpur, India

Nandita Mukherjee Atur Centre, Symbiosis Institute of Geoinformatics, SIG Pune, Pune, Maharashtra, India

M. K. Nagaraj National Institute of Technology Karnataka, Surathkal, Mangalore, India

B. Nagarajan Department of Civil Engineering, Indian Institute of Technology Kanpur, Kanpur, Uttar Pradesh, India

Sahoo Sanat Nalini Department of Civil Engineering, NIT Rourkela, Rourkela, Odisha, India

Y. Navatha NIT Warangal, Warangal, India

Prateek Negi Civil Engineering Department, Indian Institute of Technology (IIT) Delhi, Hauz Khas, New Delhi, India

Rahul Nijhawan Department of Earthquake Engineering, Indian Institute of Technology Roorkee, Roorkee, India

Mahesh Pal Department of Civil, NIT Kurukshetra, Kurukshetra, India

Sudipta Pal RITES Ltd., Kolkata, West Bengal, India

Hina Pande Photogrammetry and Remote Sensing Division, Indian Institute of Remote Sensing, ISRO, Dehradun, India

Bipin Chand Pandey Indian Institute of Surveying and Mapping, Survey of India, Hyderabad, India

Dharmendra Kumar Pandey Space Applications Centre, ISRO, Ahmedabad, India

P. K. Parida Odisha Space Applications Centre, Bhubaneswar, India

Ajay Kumar Patel Geomatics Engineering Group, Indian Institute of Technology Roorkee, Roorkee, Uttarakhand, India

Jayantilal Patel Civil Engineering Department, Sardar Vallabhbhai National Institute of Technology, Surat, Gujarat, India

Brijendra Pateriya Punjab Remote Sensing Centre, Ludhiana, India

B. Pavan Kumar Civil Engineering Department, NIT Warangal, Warangal, India

C. Prakasam Department of Civil Engineering, Chitkara University, Baddi, Himachal Pradesh, India

Deva Pratap Department of Civil Engineering, NIT Warangal, Warangal, India

A. P. Prathiba Indian Institute of Technology (Indian School of Mines) Dhanbad, Dhanbad, Jharkand, India

Umesh Pruthviraj NITK, Surathkal, Mangalore, India

G. Punithraj NITK, Surathkal, Mangalore, India

Deepak Putrevu Space Applications Centre (ISRO), Ahmedabad, India

S. Raghavendra Photogrammetry and Remote Sensing Division, Indian Institute of Remote Sensing, ISRO, Dehradun, India

Prajjwal Singh Rahar Department of Civil, NIT Kurukshetra, Kurukshetra, India

Sharwan Ram Geomatics Engineering Group, Indian Institute of Technology Roorkee, Roorkee, Uttarakhand, India

Ramandeep Punjab Remote Sensing Centre, Ludhiana, India

Kriti Rastogi Space Application Centre (ISRO), Ahmedabad, Gujarat, India

Ravindra R. Rathod Walchand College of Engineering, Sangli, Maharashtra, India

Dhanashree Raut Atur Centre, Symbiosis Institute of Geoinformatics, SIG Pune, Pune, Maharashtra, India

Keesara Venkata Reddy Department of Civil Engineering, NIT Warangal, Warangal, India

Ngangbam Romeji Department of Civil Engineering, NIT-Manipur, Imphal, India

Sachchidanand Remote Sensing Applications Centre – RSAC, Lucknow, UP, India

Swati Saha Department of Architecture, Town and Regional Planning, IIEST, Shibpur, India

Rakesh Sahu GIS Cell, Motilal Nehru National Institute of Technology Allahabad, Prayagraj, India

B. Sai Teja NIT Warangal, Warangal, India

P. Sampath Kumar NRSC, Hyderabad, India

Sana Punjab Remote Sensing Centre, Ludhiana, India

M. K. Sanabada Odisha Space Applications Centre, Bhubaneswar, India

Suraj Sawant Geomatics Engineering Group, Indian Institute of Technology Roorkee, Roorkee, Uttarakhand, India

Kul Vaibhav Sharma Department of Civil Engineering, MNIT Jaipur, Jaipur, India

Reenu Sharma Punjab Remote Sensing Centre, Ludhiana, India

M. Shashi Department of Civil Engineering, NIT Warangal, Warangal, India

Zoheb Sheikh Atur Centre, Symbiosis Institute of Geoinformatics, SIG Pune, Pune, Maharashtra, India

Amba Shetty NITK, Surathkal, Mangalore, India

Shivam Department of Civil Engineering, National Institution of Technology, Barjala, Tripura (W), India

Reedhi Shukla NRSC, Hyderabad, India

Irineu da Silva Department of Transportation Engineering, São Carlos School of Engineering (EESC), University of São Paulo (USP), São Carlos, SP, Brazil

Amanpreet Singh NRSC, Hyderabad, India

Lakhvir Singh Punjab Remote Sensing Centre, Ludhiana, India

Lakhwinder Singh Department of Water Resource Development and Management, Indian Institute of Technology, Roorkee, Roorkee, India

Raj Mohan Singh Department of Civil Engineering, MNNIT, Allahabad, UP, India

Rajat Kumar Singh Department of Mining Engineering, Indian Institute of Technology (Indian School of Mines) Dhanbad, Dhanbad, India

Ujjwal Singh Institute of Environment and Sustainable Development, Banaras Hindu University, Varanasi, India

Ashish Soni Department of Mining Engineering, Indian Institute of Technology (Indian School of Mines) Dhanbad, Dhanbad, India

Prashant K. Srivastava Institute of Environment and Sustainable Development, Banaras Hindu University, Varanasi, India

S. K. Srivastava Geospatial Technology and Outreach Programme Group, Indian Institute of Remote Sensing, ISRO, Dehradun, India

Shikha Tayal Uttarakhand Technical University, Dehradun, Uttarakhand, India

Praveen K. Thakur Indian Institute of Remote Sensing, Dehradun, Uttarakhand, India

Poonam S. Tiwari Geoweb Services, IT & Distance Learning Division, Indian Institute of Remote Sensing, ISRO, Dehradun, India

Ashutosh Kumar Tripathi GIS Cell, Motilal Nehru National Institute of Technology Allahabad, Prayagraj, Uttar Pradesh, India

Sandeep Tripathi Odisha Space Applications Centre, Bhubaneswar, India

Manisha Vashi Department of Environmental Engineering, Sarvajanik College of Engineering and Technology, Surat, India

K. Venugopal Rao NRSC, Hyderabad, India

Sandip Vijay Shivalik College of Engineering, Dehradun, Uttarakhand, India

Vasanta Govind Kumar Villuri Department of Mining Engineering, Indian Institute of Technology (Indian School of Mines) Dhanbad, Dhanbad, Jharkhand, India

Yogender Department of Civil Engineering, National Institute of Technology Kurukshetra, Kurukshetra, India

Abbreviations

ABC	Artificial bee colony optimisation
ACP	Auxiliary control points
ACS	Aerial camera systems
ADAM	Adaptive moment estimation
AGB	Above-ground biomass
AHP	Analytic hierarchy process
AML	Arc Macro Language
AMSL	Above mean sea level
ANN	Artificial neural network
AOD	Aerosol optical depth
AOI	Area of interest
API	Application programming interface
AQI	Air quality index
ASF	Alaska Satellite Facility
ASTER	Advanced Spaceborne Thermal Emission and Reflection Radiometer
AWS	Automatic weather station
BCI	Biophysical composition index
BI	Bare soil index
BIM	Building information modelling
BOD	Biochemical oxygen demand
BRTS	Bus rapid transit system
BS	Base station
CDO	Climate Data Operator
CEL	Central Electronics Limited
CI	Consistency index
CLDP	Cellular layout design problem
CLR	Computerisation of land records
CLU	Change in land use
CMM	Concrete maturity method
CMOS	Complementary metal–oxide–semiconductor

CMS	Carbon monitoring system
COD	Chemical oxygen demand
CPCB	Central Pollution Control Board
CR	Consistency ratio
CRP	Close-range photogrammetry
CRU	Climate Research Unit
CSI	Consortium for Spatial Information
CSS	Cascading Style Sheets
CSW	Catalogue Service for Web
CTM	Chemistry transport models
CUDA	Compute Unified Device Architecture
CVS	Concrete vibration sensor
CWC	Central Water Commission
DBH	Diameter at breast height
DCNN	Deep convolution neural network
DEA	Data envelopment analysis
DEM	Digital elevation model
DGPS	Differential global position system
DHRU	Development and Health Research Unit
DILRMP	Digital Land Record Modernisation Program
DisTrad	Disaggregation of radiometric surface temperature
DO	Dissolved oxygen
DOLR	Department of Land Resources
DOP	Dilution of precision
DPAP	Drought Prone Area Programme
DRDO	Defence Research and Development Organisation
DSLR	Digital single-lens reflex camera
DSM	Digital Surface Model
DSS	Decision support system
EBBI	Enhanced built-up and bareness index
ELU	Existing land use
EMI	Electro-mechanical impedance
ESA	European Space Agency
ETS	Electronic Total Station
EVI	Enhanced vegetation index
FAO	Food and Agriculture Organisation of United Nations
FCC	False-colour composite
FCM	Fuzzy C-means clustering
FDTD	Finite-difference time-domain
FLIR	Forward-looking infrared
FNEA	Fractal net evolution approach
GA	Genetic algorithm
GBD	Global Burden of Diseases
GCMs	General climate models
GCPs	Ground control points

GDA	Geocentric Datum of Australia
GDEM	Global Digital Elevation Map
GEOS	Global atmospheric chemistry model
GIS	Geographical Information System
GLCF	Global Land Cover Facility
GLCM	Grey level co-occurrence matrix
GNSS	Global navigation satellite system
GPR	Ground-penetrating radar
GPU	Graphical processing unit
GSI	Geographical Survey Institute
GUI	Graphical user interface
HDF-EOS	Hierarchical Data Format Earth Observing System
HDR	High dynamic range
HEC-RAS	Hydrologic Engineering Center's River Analysis System
HOG	Histogram of oriented gradient
HRSI	High-resolution image
HRU	Hydrologic response unit
HTML	Hypertext Markup Language
IBI	Index-based built-up index
ICP	Iterative closest point
IDW	Inverse distance weighted
IGP	Indo-Gangetic plain
IGS	International GNSS Service
IIT(ISM)	Indian Institute of Technology (Indian School of Mines)
ILSVRC	ImageNet Large-Scale Visual Recognition Challenge
IMD	India Meteorological Department
IMU	Inertial Navigation Unit
ISO	International Organisation for Standardisation
ISODATA	Iterative Self-Organizing Data Analysis Technique Algorithm
ISRO	Indian Space Research Organisation
ITCZ	Intertropical convergence zone
JSP	Java Server Pages
KDE	Kernel density estimation
LAI	Leaf area index
LaSRC	Landsat surface reflectance corrected
LiDAR	Light detection and ranging
LISS	Linear imaging self-scanning sensor
LOS	Level of service
LPA	Local planning area
LPM	Land parcel map
LS	Length and steepness
LST	Land surface temperature
LULC	Land use/land cover
LWIR	Large-wave infrared
MASW	Multichannel analysis of surface wave

MCDA	Multi-criteria decision analysis
MCGM	Municipal Corporation of Greater Mumbai
MEM	Mixed effect model
MGA	Map Grid of Australia
MIHAN	Multi-model International Cargo Hub and Airport
MLR	Multiple linear regression analysis
MM	Mathematical morphology
MMS	Mobile mapping system
MODIS	Moderate Resolution Imaging Spectroradiometer
MOSDAC	Meteorological and Oceanographic Satellite Data Archival Centre
MRV	Measurement, reporting and verification
MSL	Mean sea level
MSS	Multispectral scanner system
MSW	Municipal solid waste
MSWM	Municipal solid waste management
MTSO	Mobile Telephone Switching Office
MUAVs	Micro-unmanned aerial vehicles
MWIR	Middle-wave infrared
MXL	Maximum likelihood classification
NAAQS	National Ambient Air Quality Standard
NAG	Nesterov's accelerated gradient
NAMP	National Air Quality Monitoring Programme
NASA	National Administrative Space Application
NCDC	National Climate Data Centre
NDBaI	Normalised difference bareness index
NDBI	Normalised difference built-up index
NDE	Non-destructive evaluation
NDISI	Normalised difference impervious surface index
NDSI	Normalised difference soil index
NDVI	Normalised differential vegetation index
NE	Normalised energy
NED	National Elevation Dataset
NEHRP	National Earthquake Hazards Reduction Programme
NIC	National Informatics Centre
NIR	Near infrared
NLRMP	National Land Records Modernisation Programme
NMC	Nagpur Municipal Corporation
NMT	Non-motorised transport
NRDMS	National Resource Data Management System
NSDI	National Spatial Data Infrastructure
NSE	Nash–Sutcliffe efficiency
NSGA-II	Non-dominated sorting genetic algorithm II
OBIA	Object-based image analysis
OGC	Open geospatial consortium
OK	Ordinary kriging

OLI	Operational land imager
ORSAC	Odisha Space Applications Centre
OWA	Overlay weighted average
PCP	Primary control points
PDAF	Phase detection auto focus
PHIS	Public health information system
PHP	Hypertext Preprocessor
PML	Perfectly matched layers
POP	Plaster of Paris
PSO	Particle swarm optimisation technique
PUDA	Punjab Urban Development Authority
PZT	Lead zirconate titanate
QC	Quality checking
R&DM	Revenue and Disaster Management Department
RADAR	Radio detection and ranging
RCMES	Regional climate model evaluation system
RDBMS	Relational database management system
RELU	Rectified linear units
RIS	Road information system
RJGC	Royal Jordanian Geographic Center
RJP	Resin Jacketed Piezo
RMS	Root mean square
RMSD	Root mean square deviation
RMSE	Root mean square error
RNDS	Ratio normalised difference soil index
ROC	Receiver operating characteristics
ROI	Region of interest
RTK	Real-time kinematic
RVI	Ratio vegetation index
SA	Simulated annealing
SAR	Synthetic aperture radar
SASW	Spectral analysis of surface waves
SCA	Snow-cover area
SCP	Secondary control points
SCR	Signal-to-clutter ratio
SCS	Soil Conservation Services
SDI	Spatial data infrastructure
SERB	Science and Engineering Research Board
SFM	Structure from motion
SGOS	Single geode operating software
SHAW	Simultaneous heat and water
SHM	Structural health monitoring
SIFT	Scale invariant feature transforms
SMC	Surat Municipal Corporation
SNAP	Sentinel Application Platform

SNN	Standard nearest neighbour
SNR	Signal-to-noise ratio
SNTHERM	Snow THERmal Model
SOI	Survey of India
SOP	Standard operating procedure
SOR	Statistical outlier removal
SOS	Strategic Option Study
SRA&ULR	Strengthening of Revenue Administration and Updating of Land Records
SRTM	Shuttle Radar Topography Mission
STD	Standard deviation error
STI-FM	Spatio-temporal image fusion model
SUFI	Sequential uncertainty fitting
SVM	Support vector machines
SWAT	Soil and Water Assessment Tool
TCP	Tertiary control points
TDS	Total dissolved solids
TGA	Total geographic area
TIN	Triangulated irregular network
TIR	Thermal infrared
TIRS	Thermal infrared sensor
TLS	Terrestrial laser scanning
TM	Thematic mapper
TOA	Top of atmosphere
TOD	Globally transit-oriented development
TOPSIS	Technique for Order of Preference by Similarity to Ideal Solution
TS	Tabu search
TsHARP	Sharpening thermal imagery
UBA	Unnat Bharat Abhiyan
UEB	Utah Energy Balance Model
UMMS	Unmanned mobile mapping system
UMSL	Under mean sea level
USB	Universal serial bus
USDA	United States Department of Agriculture
USGS	United States Geological Service
USLE	Universal Soil Loss Equation
UTM	Universal Transverse Mercator
UTs	Union territories
VHR	Very high spatial resolution
VIC	Variable infiltration capacity
VNIR	Visible and near infrared
WBHDC	West Bengal Highway Development Corporation
WCM	Water cloud model
WCS	Web Coverage Service
WFS	Web Feature Service

WGS	World Geodetic System
WLC	Sum and Weighted Linear Combination
WMS	Web Map Service
WPS	Web Processing Service
WRIS	Water Resources Information System

Part I
Papers of Key Note Speakers

Harnessing Remote Sensing for Civil Engineering: Then, Now, and Tomorrow

Debra F. Laefer

Abstract Despite enormous advances in remote sensing data over the past 20 years, harnessing and exploiting that data by the Civil Engineering community has been relatively limited. To understand the full potential of such data, first this paper briefly recaps the Civil Engineering community's engagement with remote sensing for dike monitoring and post-earthquake damage assessment. Next, the state of the art is introduced with special considerations for recent advances in quality, affordability, accessibility, and equipment size; the role of national aerial laser scanning data collection programs; and the increasing applicability of remote sensing to a wide range Civil Engineering applications. Finally, the paper concludes with a vision of how Civil Engineering can better benefit from existing technologies not regularly exploited today, as well as the logistical challenges of storing and integrating such data in a computationally meaningful manner.

Keywords Remote sensing · LiDAR · Hyperspectral imaging · Terahertz scanning · Spatial databases · Data indexing · Distributed computing

1 Introduction

Today is an amazing time to be part of the remote sensing community. Over the past two decades, the industry has undergone nothing short of a revolution in terms of capabilities, accessibility, and affordability. In some ways, this is all the more surprising given the discontinuation of surveying classes in many civil engineering programs and the closing of many departments of geography at major universities through much of the 1980s and 1990s. However, with an estimated 80% of all data

D. F. Laefer (✉)
Center for Urban Science and Progress and Department of Civil and Urban Engineering, Tandon School of Engineering, New York University,
370 Jay Street, 12th Floor, Brooklyn, NY 11201, USA
e-mail: debra.laefer@nyu.edu

J. K. Ghosh and I. da Silva (eds.), *Applications of Geomatics in Civil Engineering*, Lecture Notes in Civil Engineering 33, https://doi.org/10.1007/978-981-13-7067-0_1

being generated today having some form of spatial component [1], there is a new set of opportunities and challenges for harnessing such data in a Civil Engineering context.

The aims of this paper are threefold. The first is to give examples showing the historical development of remote sensing technologies in Civil Engineering. The second is to highlight recent advances in remote sensing equipment, platforms, and data processing with respect to improved functionalities and usability in a Civil Engineering context. The third and final goal is to present a set of challenges that the Civil Engineering community is currently facing that are impeding the full exploitation of the current generation of remote sensing capabilities and are likely to do so even more acutely in coming decades.

2 Civil Engineering's Use of Remote Sensing

Arguably, the most common forms of remote sensing in Civil Engineering are imagery and light detection and ranging (LiDAR) [also known as laser scanning]. While the growth of photogrammetry from as early as World War I is well documented [2], the evolution of LiDAR in a Civil Engineering context is less well known. Thus, to help illustrate the evolution of that relationship, a brief historical overview of two application areas is provided herein. The first is in dike monitoring, and the second is in response to the threat of urban earthquakes.

2.1 LiDAR Usage for Dike Monitoring

As early as 1987 there were Dutch-based publications providing a clear and cogent vision for the opportunities presented by remote sensing in the form of satellite imagery, orthophotos, and soundings for national flood risk management, although usage was still some years off [3, 4]. By the late 1990s, initial experiments were being undertaken to use LiDAR for dike monitoring [5], while its deployment for evaluating flood plain mapping and flood risk estimation were already well underway in Europe [6] and the United States [7]. Eventually, dike deformation monitoring was tried with terrestrial and aerial LiDAR, photogrammetry, passive microwave radiometry, and Interferometric Synthetic Aperture Radar (InSAR) [8], all in an attempt to facilitate faster dike inspection [9–12]. Today, both LiDAR (airborne) and InSAR (satellite borne) are common commercial approaches to overall monitoring, as well as detection and prediction of shallow surficial failures [13]. More recent experimental efforts have considered airborne hyperspectral imagery for model generation and slide detection [14]. Even passive microwave radiometry [15], polarimetric SAR [16], and thermal imaging [17] have been investigated for soil moisture estimation in earthen dams as precursors to failure events; see Cundill [18] for a more in-depth overview.

2.2 *LiDAR Usage in Urban Earthquakes*

Dating back to 1987, laser altimetry (an early form of LiDAR) was proposed as a tool for monitoring faults and predicting earthquakes [19]. A decade later, examples appear of attempting to use aerial LiDAR to map and monitor tectonic movement in seismically active areas [20] and to compare urban areas before and after an earthquake [21]. These initial efforts and the use of aerial LiDAR in support of the post-911 disaster recovery in New York City in 2001 [22] lead to the National Geospatial-Intelligence Agency conducting high resolution laser scans over core areas of major cities including both Los Angeles and San Francisco to create a baseline to enable post-disaster evaluation in case of a major man-made, storm, or seismic event [23, 24]. By 2002, processing of that type of data was already underway by Steinle and Bahr [25], whose goal was to find a reliable way to measure seismic related damage without falsely identifying generic urban change caused by the data's temporal gap across the multiple scans. In 2008, Rehor and Voegtle suggested that, for this application and others, the integration of spectral data with LiDAR could be advantageous in achieving higher reliability rates [26].

Generally, the earthquake community has embraced the idea of using LiDAR for a wide range of applications related to earthquakes. Prominent examples include the 2010 work by Shen et al. to identify inclined buildings after the Haiti earthquake [27]; the research by Borfecchia et al. also in 2010 to pair multispectral data with LiDAR to predict seismic vulnerability of buildings [28], the efforts by Moya et al. to automatically identify collapsed buildings after a 2016 Japanese earthquake using a combination of before and after LiDAR scans and digital surface models [29], and Toprak et al.'s 2018 research efforts to consider seismic related pipeline damage, in which those authors concluded that LiDAR data correlated more strongly with pipeline damage than satellite data [30]. Finally, the National Science Foundation of the United States (US) has recently funded the acquisition of multiple LiDAR units that can be used on a combination of terrestrial, mobile, and aerial platforms for rapid post-disaster deployment in both the US and abroad [31].

3 Recent Advancements

Today, remote sensing is regularly used in various aspects of the creation, monitoring, and assessment of civil infrastructure. Applications include planning and checking road and tunnel alignments before and during construction [32, 33], providing a daily record of progress at quarries and construction sites [34, 35], conducting periodic bridge inspection [36], documenting historic structures [37], monitoring slope failures and coastal erosion [38, 39], and many more.

The widespread adoption of such techniques over the past two decades has been born from a wide range of advancements that have occurred in remote sensing. These advances can be summarized in the following 10 categories: (1) substantial unit miniaturization, (2) major price reductions, (3) applications to new domains, (4) widespread open-source data, (5) improved data resolution, (6) enhanced acquisition speed, (7) better vertical data capture, (8) greater data yields, (9) more automation in post-processing, and (10) commercialization of emerging capabilities. Each factor has contributed to the accelerated adoption of remote sensing in the daily work of Civil Engineering. An example of each of these developments is described in the following subsections with respect to either LiDAR or some form of imagery (standard, multispectral, or hyperspectral). Importantly these are provided anecdotally, as they reflect only a sampling of the exciting advances happening in remote sensing today.

3.1 Substantial Unit Miniaturization

Following the consumer electronics trend, extensive remote sensing capabilities are now available in extremely small, lightweight units. This trend towards remote sensing miniaturization was first driven by efforts to make the technology more easily portable, followed by efforts to make products viable as unmanned aerial vehicle (UAV) payloads. The progression of this trend is easily seen in the hyperspectral imagery market. Spectrometers were initially lab-based devices dating back more than half a century [40]. Field units able to capture a large number of continuous bands only began to emerge in 1995 [41]. These hyperspectral cameras initially resembled survey units and were characterized by their significant weight, bulk, and cost (in excess of $150,000). Then in 2015, Cubert introduced a 500 g hyperspectral camera designed for small UAVs [42], although still with a substantial price tag of $54,000. While this was not the first UAV-mountable hyperspectral camera, it was less than one-tenth the weight of anything else available commercially, thereby enabling it to benefit from the explosion of small, low-cost UAVs that have flooded the market over the past few years. Since then, Consumer Physics has introduced the SCiO, a handheld sensor with a preprogrammed library of detectable materials, enabling usage with no data processing [43] for less than $4,000. While still fairly limited in detection capabilities, this type of unit definitely points in the direction of further miniaturization and possible development of a unit that could be attached to a smartphone; something that has already occurred in the 3D mapping arena with low-cost structured light sensors [44]. These hyperspectral instruments are shown in Fig. 1.

3.2 Major Price Reductions

Like the trend in miniaturization, remote sensing has also followed the consumer electronics trend towards major price reductions coupled with increased functionality. As an example, a state-of-the-art, medium range, terrestrial LiDAR scanner in 2002 cost $250,000. By 2005 a superior unit with the same range was on the market

(a) Lab-based spectrometer (b) Typical field hyperspectral unit (c) UAV-mountable unit (only 500g) (d) Hand-held unit

Fig. 1 Array of hyperspectral units now on the market

for $175,000, and today a lighter, faster, and much easier to use commercial grade terrestrial scanner with a broader range begins at only $25,000, although many of the top end units still cost in excess of $75,000. Additionally, lower functionality units are now available in form of the Velodyne puck at less than $10,000 and even as an attachment to an iPad in the form of the Occipital Structure Sensor at only $350 [44].

3.3 Applications to New Domain

The portability and affordability of the latest generation of remote sensing equipment have enabled many more researchers and practitioners to have access to these technologies, thereby creating the possibility of their application to new domains. Figure 2 shows the potential usefulness of hyperspectral imagery to facade assessment for historical documentation of interventions, identification of weak patches in what should be a homogeneous material, and ultimately characterization of materials (once a reference library of material values is established). Figure 2b readily highlights differences in materials that are not easily identifiable with the naked eye, especially from street level.

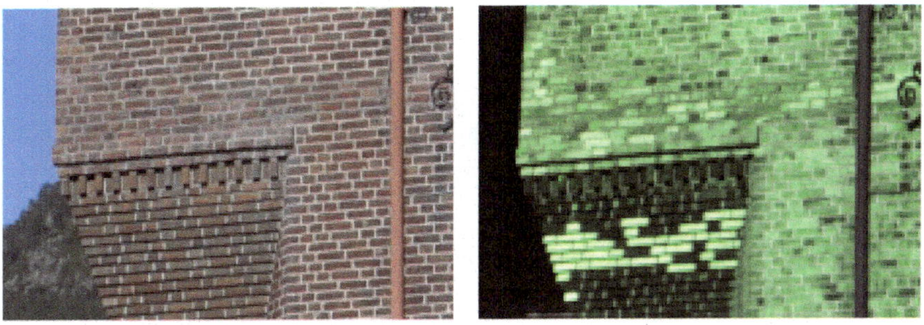

(a) Photograph (b) Hyperspectral Image

Fig. 2 Comparative Image of Masonry Wall

3.4 *Widespread Open-Source Data*

Because the cost of data collection can still be high (e.g., aerial LiDAR), the creation of national LiDAR surveys has become a rich new resource for practitioners and researchers alike, who otherwise could not obtain such data. At least seven countries have already completed country-wide ALS projects (i.e., Denmark, England, Finland, the Netherlands, Slovenia, Switzerland, and the United States), with many more in progress (e.g., Italy, Poland, Spain, and Sweden) and countries such as Japan nearly complete [45]. Notably, The Netherlands is now on its third generation of a national survey (at ever increasing densities). These data sets are supplemented by more limited data releases from individual researchers and specific cities. For example, New York City undertook one in 2011 at 10–12 pts/m^2 available as a 1 ft Digital Elevation Model [46] and is now considering an updated collection at a much higher resolution.

3.5 *Improved Data Resolution*

While some of the most exciting early work in remote sensing came in the form of INSAR, in recent years many large landmass considerations have given way to more localized concerns, as the resolution of other equipment and technologies have evolved. A good example of this is in laser scanning. The current generation of terrestrial scanners is capable of collecting tens of thousands of points in less than a minute. Data density is in part a function of the angle of incidence and the offset distance for terrestrial units [47, 48], minimum flight height for large aerial units, and flight speed for UAVs [49]. However, current commercial aerial scans are typically being delivered at less than 20 pts/m^2, even though the equipment can achieve 50–60 pts/m^2 within Federal Aviation Authority and local flight restrictions (e.g., 300 m) in urban areas. When such capability is coupled with innovative flight planning [50], densities of several hundred points per square meter are achievable. Figure 3 shows a comparison of 50 pts/m^2 (single pass) versus 335 pts/m^2 (multipass) from a 2015 flyover of Dublin, Ireland. The recent introduction of photon-based Geiger LiDAR holds the promise for even denser datasets [51]. These unprecedented outputs offer the ability to engage in data mining of objects that were not readily detectable in previous scans (e.g., utility lines, curbs).

3.6 *Enhanced Acquisition Speed*

There has also been a radical improvement in acquisition speed. For example, a 2013 comparison of a GS200 Trimble [52] purchased in 2005 and a Leica P20 [53] purchased in 2013 demonstrated an acceleration of more than two orders of

(a) Single pass scan with 50pts/m^2 (b) Multi-pass scan with 335pts/m^2

Fig. 3 Images showing single versus multi-pass LiDAR scan data

(a) Trimble GS200 (b) Leica P20 (c) Leica BLK360

Fig. 4 Examples of LiDAR Units from 2003 to 2017

magnitude in data collection capabilities using the same offset and quality settings (≈500 pts/s vs. ≈50,000 pts/s) (Fig. 4) [54]. Notably, this refers only to the data acquisition phase.

Other small gains were obtained in the setup stage, while leveling the unit and registering the targets. Furthermore, some of the newest terrestrial units that are designed primarily for interior uses are intended to sit on a table or desk [55]. Thus, they have an inherent assumption that the unit is already on a leveled plane and require no leveling in the setup.

3.7 Better Vertical Data Coverage

The nadir orientation of most aerial remote sensing devices results in excellent coverage and high data yields on horizontal surfaces. However, the construction of fully three-dimensional (3D) models of urban areas, large monuments, and even steep rock faces requires a rethinking of flightpath planning to obtain adequate vertical coverage with nadir-facing sensors as, under standard deployment, vertical coverage is only about 10% of the corresponding horizontal coverage. Furthermore, as most remote sensing approaches are line-of-sight technologies, a building will always have at least one backside that is non-visible (with respect to the sensor). In addition to this "self-shadowing", attempting to achieve 100% coverage in urban areas is further complicated by the phenomenon of "street shadowing" where one building presents a visual barrier to seeing the face of another (Fig. 5) and, therefore, prevents data capture on that building face, as previously described by Hinks et al. [50]. To minimize these occlusions, an evolutionary algorithm [56] can be devised to select the elevation and orientation of the flight path (Fig. 6). As the equipment's field of view and the minimum flight elevation are usually fixed, the variables relate to the orientation of the flight path to the prevailing grid of the community being flown. Where there is a fairly regular grid, this can be estimated as 45° to the grid [50].

Fig. 5 Schematic of self-shadowing

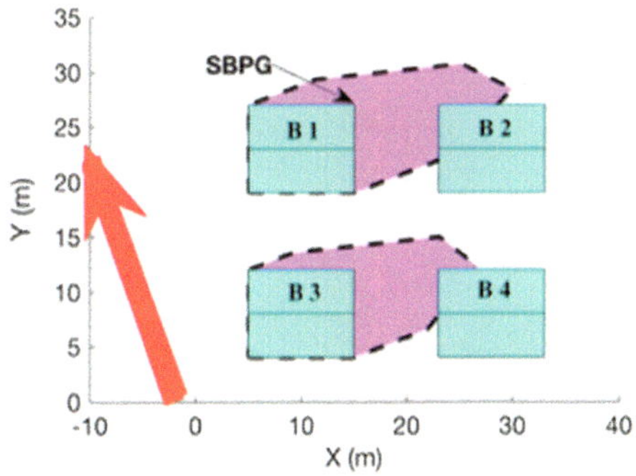

Fig. 6 Schematic setup for evolutionary algorithm calculation of shadowed areas to inform flightpath planning

3.8 Better Data Yields

When a laser beam is emitted from an aerial unit, the beam may only be 1 cm in diameter, but by the time it encounters a ground-based object, the beam may have spread to more than 10 times that, which often results in the beam encountering multiple objects and producing multiple returns to the sensor. The most common occurrence of this is with vegetation, as the beam may hit a leaf, and then a branch, and then the ground as it travels downward. Consequently, researchers have used the presence of multiple returns as an indication of vegetation and, therefore, an easy way to segregate that data from constructed objects (e.g., road, building) [57]. However, many returns are lost, because the current sensor reading systems have relatively long intervals between their successive reading capture abilities. This dead zone between readings precludes successive data echoes from being captured at intervals of less than several meters [58]. This inability to register closely spaced data limits vertical surface documentation on buildings.

In contrast, the full waveform (FW) version of LiDAR has the potential to be processed at much smaller intervals [59], which could vastly improve the quality of the resulting point cloud, especially with respect to small façade feature identification. As previously noted by Parrish [60], FW scanners record the full performance of the backscattered laser energy with the expected, ideal signal being a group of spikes—each representing a true object (true point) hit by the laser beam. In actuality, what is recorded is a blurred, noisy interpretation of the ideal signal. To manage this challenging data and extract the maximum number of true points from FW records, a deconvolution based approach is being developed to localize the true points of reflection. Work to date enables discrimination of 2 objects separated by a distance equal to the FW system sampling distance (in this case 1 ns, which equals 15 cm at a 2 cm accuracy) for LiDAR data acquired at elevations of more than 300 m (Fig. 7). The work demonstrates a new ability to extract an unprecedented level of vertical geometric façade data from existing data, thus, exploiting urban FW data in a heretofore unexplored manner, which is highly valuable for a wide range of 3D urban modeling applications.

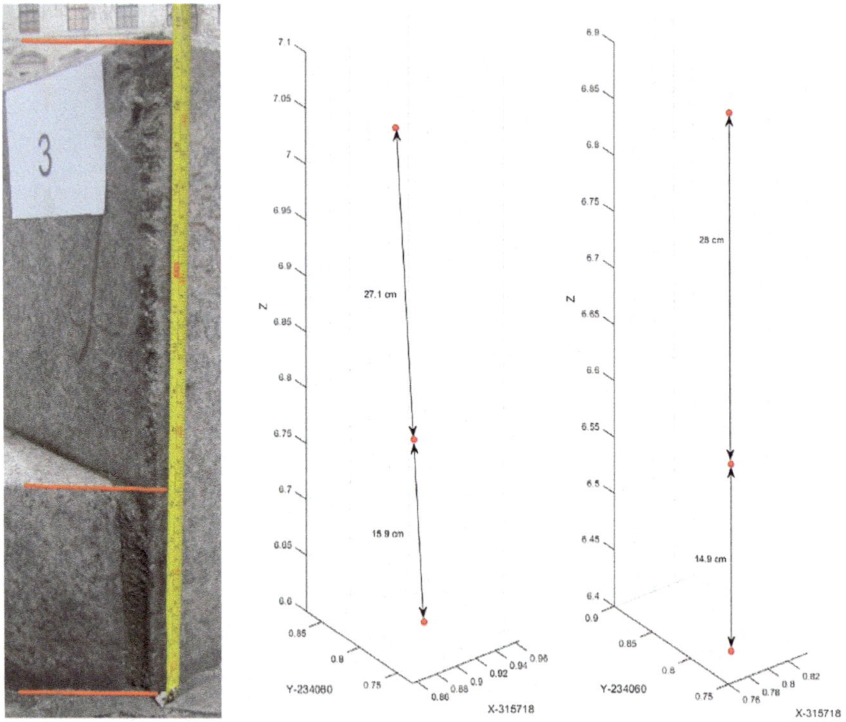

Fig. 7 Deconvolution-based outputs for 2 full waveform signals for 2 steps and the pavement

3.9 *More Automation in Post-processing*

As datasets have become more numerous, as well as larger in size, there is an increasing push for more automation both of processes previously partially automated and of processes not previously addressed. An example of this is seen in the documentation of steel bridges, where efforts have been made to auto-identify specific steel sections and then ultimately populate a BIM type of management system. My group's efforts in this area involve creating a set of surfaces for a 3D solid model by extruding cross-sections along the sweeping profile. The method starts by extracting the point cloud of the cross-section. This can be done by selecting the point cloud within the interval thickness along the longitudinal direction of a given component [61]. From this, a polygon can be fitted through the point cloud of the cross-section, in which the Euclidean distance between the data points and the fitted polygon is used to determine the best polygon. Next, the algorithm may need to determine the trajectory of the component to allow the cross-section sweeping. Another method involves generating multiple cross-sections along the longitudinal direction, after which, the surfaces of the components can be extruded through the cross-sections [61] (Fig. 8).

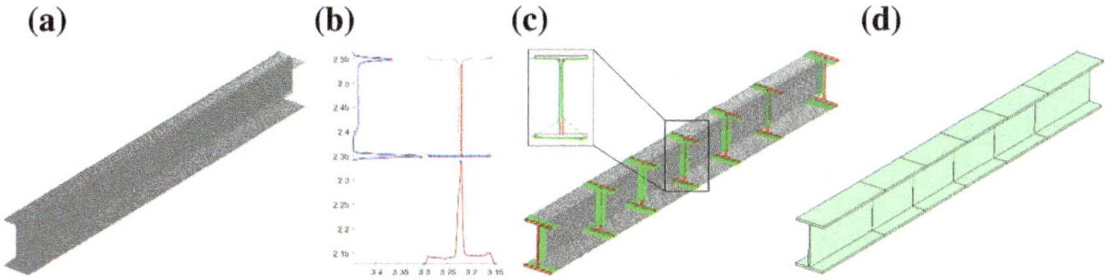

Fig. 8 Generating a 3D model of a steel beam; **a** Point cloud; **b** Estimation of section dimensions; **c** Cross-sections identification; **d** 3D model generation by sweeping through cross-sections

3.10 Commercialization of Emerging Capabilities

While the past 20 years have seen the emergence of a lot of new capabilities in the shape of delivery platforms (e.g., UAVs, autonomous vehicles, robots), there has also been the development of new forms of remote sensing. Perhaps, the most exciting one that has yet to be integrated into Civil Engineering workflows is terahertz scanning. In 2004, the US Department of Energy co-hosted a workshop on terahertz scanning as a powerful new remote sensing technology cross applicable to fields as divergent as astronomy and medicine [62]. Terahertz scanning (also known as sub-millimeter radiation) operates between 300 GHz and 10 THz depending upon the device energy input level and is able to detect unique material signatures at distances of several meters, even through solid materials such as asphalt and concrete [63]. This is possible because radiation wavelengths in the terahertz band range from 1 to 0.1 mm (or 100 μm), and so the photon energy in the THz regime is less than the band-gap energy of nonmetallic materials. On the spectrum, THz scanning is between microwaves and infrared light waves. THz beams transmitted through barrier materials can be used for material characterization, layer inspection [64], finding buried explosives [65], and as an alternative to X-rays for producing high-resolution images of the interior of solid objects [66] (Fig. 9). Large-scale THz scanning [67] is starting to become more viable and there are significant efforts to make the equipment lighter, more compact, and more versatile, including wearable units for a near-range investigation [68]. An unpublished report by Mott MacDonald [69] demonstrated the viability of the THz scanning as an alternative to ground penetrating radar for establishing soil layers and identifying subsurface objects in a civil engineering context to soil depths of 6 m, even below paved surfaces.

Subsurface object detection is one of these areas that remains a major challenge for communities, with unintended encounters between excavation equipment and utility lines responsible for dozens of injuries and multiple deaths each year [70], as well as extensive power and water outages, while current technologies struggle to achieve reliable and cost-effective detection (Fig. 10). With its ability to identify subsurface objects [69], THz scanning may provide a critical opportunity for breakthroughs in this area.

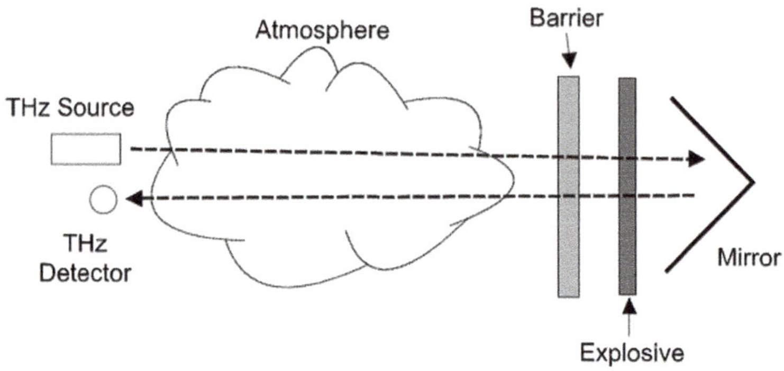

Fig. 9 Schematic of explosives detection with THz scanning as per Federici et al. [65]

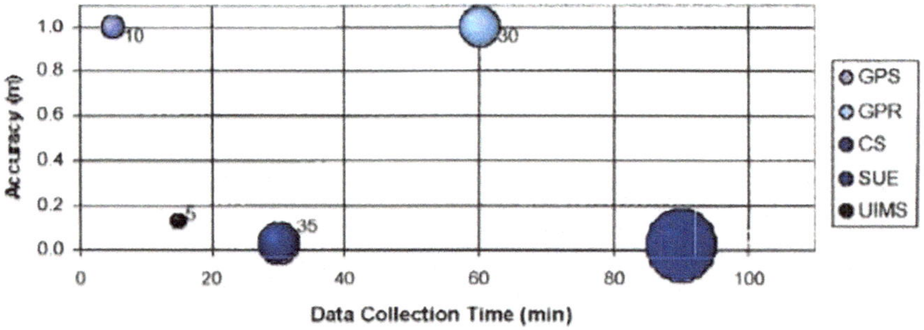

Fig. 10 Analysis of subsurface detection technologies [71]

4 Challenges

There are many challenges to harnessing the usefulness of remote sensing data in a Civil Engineering context. The major ones include the following: (1) data representations being incompatible with existing workflows; (2) data integration mechanisms having significant limitations; and (3) data sets exceeding the capacities of standalone solutions, while current distributed systems do not yet provide adequate 3D functionality. Each is described below.

4.1 Data Representations Being Incompatible with Existing Work Flows

Remote sensing data appear in representations that are not readily compatible with existing engineering workflows [e.g., Finite Element Modeling (FEM), Building Information Modeling (BIM)]. Points, pixels, or other representations common to remote sensing data are fundamentally, uninformed, partial portrayals of larger objects. Furthermore, these smaller elements do not know that they are part of larger objects or which other data have a relationship to them, with respect to a specific object.

In the current mindset, this has meant that such data must be processed to make it usable. The processing is typically an attempt to affiliate specific subsets of the data with some particular geometric shape or combination of shapes. Common means often involve plane fitting through RANSAC [72], gridding [73], machine learning [74], or other approaches that require training sets or libraries. Strategies for extracting these geometries may relate to a road surface [75] or its features [76], a building [77, 78], vegetation [79], or even utility poles [80]. To this end, thousands of papers have been written related to the processing of such data sometimes for its own sake [81, 82], sometimes for creating a solid model to populate an FEM (Fig. 11) [37, 83], and sometimes in an attempt to support a BIM model [35, 84]. As buildings and built environments are highly variable and can differ significantly from neighborhood to neighborhood, as well as city to city, developing a robust algorithm that is equally effective for cities of different sizes and in different parts of the world is particularly challenging. Thus, the question arises as to whether or not a better approach can be devised; one that is wholly data driven and is more holistic in its data conversion. To begin to answer this question, a broader look at what is happening in digital representation is perhaps warranted, specifically with regard to the digital twinning movement (Fig. 12).

In 1994, Renaudin et al. [85] introduced the term "digital twin" to describe a 3D anthropomorphic phantom on which testing could be done—in this case for the purpose of coronary vessel reconstruction. Since then, the digital twin concept has been extended to mean "a comprehensive physical and functional description of a component, product or system, which includes more or less all information which

Fig. 11 Voxelization of laser scanning points shows an automated strategy for point conversion into a solid model

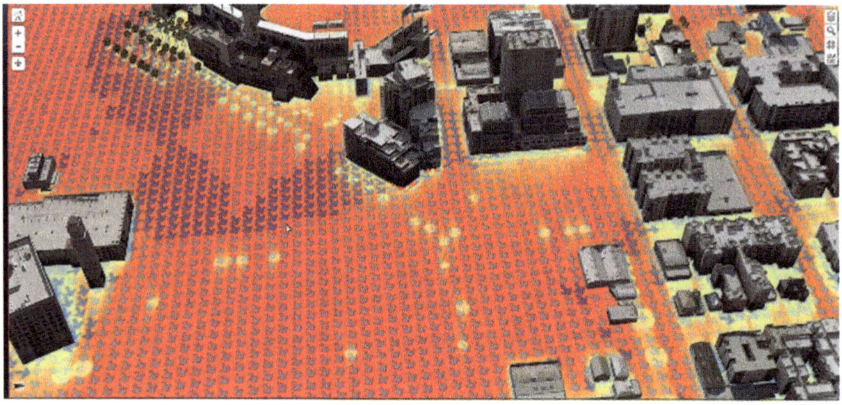

Fig. 12 Preliminary efforts by ESRI toward Singapore's digital twin

could be useful in all—the current and subsequent—lifecycle phases" [86]. The concept has been applied widely to areas as disparate as fuel manufacturing [87], NASA vehicle design [88], and vehicle fleet management [89].

In the context of the built environment, the concept has been defined as "a representation of a system which mimics its real-world behavior (and, in some cases, the surrounding environment). This is typically a real-time updated collection of data, models, algorithms or analysis." This definition, outlined in the landmark British document "Data for the Public Good" [90], is now the basis for a major national British investment involving the creation of a digital twin for the entirety of the United Kingdom [90]. Their motivation for this major undertaking relates to expected improvements in energy and water usage, traffic flow, anti-terrorism resiliency, natural disaster response, and general quality of life. Presently, a pilot effort for the city of Exeter is being led by Sir Alan Wilson, who heads the Alan Turing Institute.

Similar efforts are being pursued elsewhere. For example, later this year, Singapore has been building a platform for a dynamic, 3D virtual model of its urban areas incorporating static and real-time data on climate, demographics, terrain attributes, and energy consumption. Funded by its National Research Foundation in conjunction with Dassault Systèmes, the system is expected to contain semantic information such as building composition and materials but with a focus on dynamic, human behavior [91].

To date, most city-scale representations have not been developed with engineering functionality [92]. As such, the concept of a true digital twin differs quite significantly from most of the existing representations of the built environment. This is notable in the following three aspects for which remote sensing data are pivotal, as will be described below: functionality, accuracy, and completeness.

4.1.1 Functionality

The vast majority of current 3D models were created for visual representation, with limited (if any) functionality beyond visual exploration. While there are notable exceptions such as the Energy Atlas Berlin in Germany with 500,000 buildings [93, 94] and SEMANCO in Spain with 150,000 buildings across 3 urban areas [93, 95], both models only address one topic and are not presently usable for other, even closely related, topics such as urban heat island modeling. Furthermore, both projects predict only rudimentary annual energy usage for each building within the study area. More detailed energy modeling is not possible because the building representations are only defined to a Level of Detail (LOD) 2 [96], which involves only a bounding box representation of a given structure and its general roof shape. A detailed energy model of a building requires LOD 3 representation, which includes detailed exterior representation, where the major external features of a building are present including the exact roof shape. The limited functionality of existing models is due in large part to a combination of the data input stream and the platforms used to store the data. With respect to data storage, available options begin at the most local level with BIMs, which largely serve as a means for achieving an accurate historical record and as a tool for asset management (e.g., knowing the type and quantity needed for replacement materials). BIMs rely on a highly prescribed set of industry foundation classes [97] to store the relevant information, thereby making integration, to say nothing of interoperability with other data sets, extremely difficult, though a limited amount has been done with four-dimensional virtual reality [98], structural analysis [99], and geographic information system (GIS) integration in mind [100]. The state of the art is graphically summarized in Fig. 13, along with a possible form of a new digital twin approach.

Today's city-scale systems tend to be outgrowths of two-dimensional (2D) GIS layers traditionally capable of storing only vector or raster data and struggle to support fully 3D models (Fig. 13). While much has been achieved through the introduction of CityGML (i.e., an open standardized data model and exchange format devised for exchanging 3D building model in virtual models) [101] functionality in both GIS and virtual models is of very limited extent and generally achieved only through predefined querying (e.g., finding the shortest route from point A to B or quantifying object frequency like the number of intersections or amount of a land cover type). These systems cannot support predictive modeling due to their relative inflexibility with respect to dynamic (real time) mutability, their accuracy level, their completeness in representation, and their inability to introduce uncertainty.

4.1.2 Accuracy

Current 3D models of the built environment are derived from a variety of sources that cover a wide spectrum of accuracies (Fig. 14) from measured surveys (as the historical gold standard at a sub-centimeter level accuracy), through low-density laser scanning (<20 pt/m^2) to 3D reconstructions from imagery, to nonexpert community initiatives

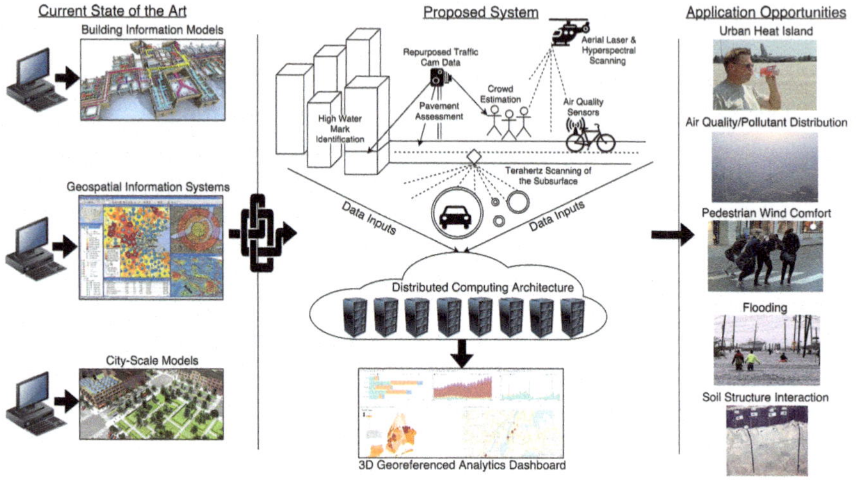

Fig. 13 Schematic representation of the current state-of-the-art and a proposed digital twinning approach

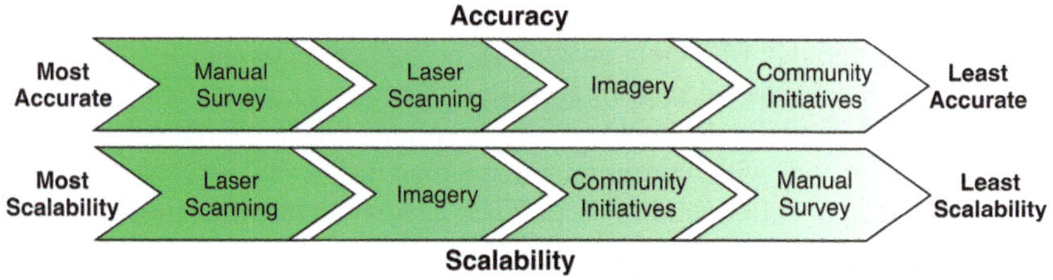

Fig. 14 Data gathering techniques compared by accuracy and scalability

at a meter-level accuracy (e.g., OpenStreetMap). Today, aerial laser scanning can rival ground-based manual surveying and offer a scalable alternative. While not typically needed for models for virtual tourism, the film industry, and gaming, having such accuracy can be critical for a wide variety of engineering modeling and analysis problems. This is shown in the context of shadow modeling with a state-of-the-art imagery based approach for NYC (Fig. 15) versus that from a high-density laser scan for Dublin, Ireland (Fig. 16). The small features that are captured in laser scanning can strongly influence not only shadow generation and solar potential, but also pedestrian wind comfort, roof-integrated wind turbines, pollution dispersion, urban heat island effects, blast vulnerability, and noise propagation. Identifying above ground utility lines and ground level curbs and steps can also create opportunities for asset management with respect to tree growth and accessibility options for those with mobility limitations. As most 3D models have not been used for such comprehensive and predictive modeling (because of system limitations), having such accuracy has not been a factor.

Fig. 15 Typical shadow information (NYC) provided for this paper by the authors of [102]

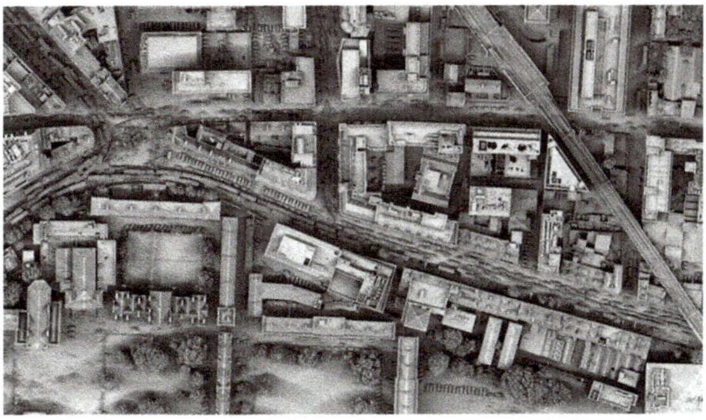

Fig. 16 Unpublished shadow work by Laefer

4.1.3 Completeness

Current 3D models typically lack many of the components that make cities feel like cities including vegetation, street furniture, building ornamentation, etc. While some aboveground features can be captured in imagery and are used as textures for selective 3D models, these textures do not actually occupy space and, thus, do not contribute to the actual modeling capabilities. Though street-level laser scanning data from vehicles circumvents some of these geometric shortcomings with their resulting dense, geo-positioned point clouds, these data sets are filled with occlusions at the roof level, in back alleys (and other non-vehicle accessible areas), and even the back sides of street furniture (e.g., utility poles). Thus, exclusive use of these data streams is problematic as a source of comprehensive data input, as well as for material identification and surface roughness calculations. While individualized solutions can overcome such occlusions either through dedicated algorithms [103] or through machine learning [104], these solutions are complicated as they may: (1) produce only inferred results, (2) entail extra processing time, (3) require high levels of computational resources and/or (4) demand extensive ground-truthing, all of which impede scalability.

4.2 *Data Integration Mechanisms Having Significant Limitations*

Typically, remote sensing data are collected in an ad hoc manner. This means that data sets cover different spatial extents at disparate granularities and at irregular intervals. This presents significant difficulties when trying to use such data sets together. The problem is exacerbated when multiple remote sensing modalities are included, especially imagery where parallax problems may emerge. These complications impede data storage, querying, and visualization.

The difficulty can be understood using a database concept with the one to many versus the many to one approach. Namely that when a high-resolution dataset is mapped to a low resolution one, the two standard ways to address this are downsampling and upsampling (Fig. 17). In downsampling, many pieces of data (from the high-resolution dataset) are assigned to a single low-resolution piece of data that covers the same spatial extent. In upsampling, some form of interpolation is required to affiliate one low-resolution piece of data to multiple high-resolution ones. Both approaches have significant downsides.

In downsampling the data to the lowest spatial and temporal resolutions, the user does not benefit from advances that regularly occur in sensing equipment and delivery platforms. There is also the quandary of whether information from the denser data set should be amalgamated or whether the closest data point should be used. The upsampling strategy is equally problematic, as it typically relies on some form of interpolation, which may not be justified. This is particularly problematic for the end user, as current systems do not contain easy to use mechanisms to manually explore such decisions in an interactive way (e.g., with an on-screen slider). Furthermore, there is no option for introducing cutting-edge probabilistic means for interpolation. Adding such functionality is typically precluded by the current generation of storage solutions.

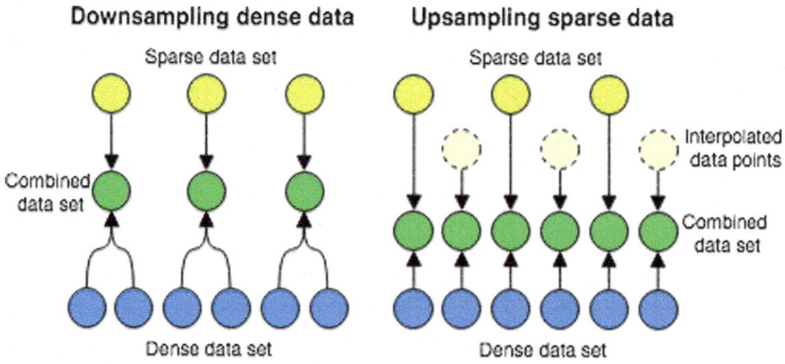

Fig. 17 The process of downsampling dense data versus upsampling sparse data

4.3 Datasets Exceeding the Capacities of Standalone Solutions, While Current Distributed Systems Are Not Providing Adequate 3D Functionalities

Arguably, data storage now poses the greatest threat to having usable data, as the size of today's data demand a distributed computing solution. For example, a single remote sensing data collection mission can harvest multiple terabytes of data per square kilometer [105]. When extrapolated across a single city, the result is a petabyte or more of data, thereby making standalone solutions obsolete for many use cases.

The landscape on the distributed computing side is not much more encouraging, as the current generation of distributed-computing enabled spatial database solutions have offered only minimal support with respect to three-dimensional (3D) spatial data, especially with regard to querying, visualization, and change detection capabilities (Table 1). Notably, while some of these abilities are available in standalone solutions, those systems cannot be easily transitioned into a distributed computing environment, because of the data storage structures that they employ.

What is needed is a new data storage architecture approach that has three main characteristics. First, it must fully benefit from the 50+ years of research and investment that has been made in distributed computing by devising and employing data structures and indexes that thoroughly exploit the scalability of that arrangement without introducing data storage redundancies and data query uncertainties. The problem is not simple, as spatial features are often continuous (e.g., a road), while in contrast, data sets and hosting capabilities are finite. This means that multiple data sets that belong to a single feature may be present on multiple cloud-based machines. Thus, an effective distributed system must have an inherent awareness as to the location of specific sets of data with respect to a global positioning schema. This capability does not presently exist.

Second, the system must support vertical data integration, hosting a wide range of spatial and nonspatial data without the loss of metadata. This is not simple, as

Table 1 Analysis of available spatiotemporal storage systems with respect to 3D data

System	3D Storage	Distributive enabled	3D Indexing	Temporal indexing	3D Querying	3D Visualization	3D Change analysis	Prediction
PAIRS	✘	✓	✘	✓	✘	✘	✘	✓
Spatial-Hadoop	✘	✓	✘	✘	✘	✘	✘	✘
GeoSpark	✘	✓	✘	✘	✘	✘	✘	✘
MongoDB	✘	✓	✘	✘	✘	✘	✘	✘
EarthDB	✘	✓	✘	✓	✘	✘	✘	✘
EarthServer	✓	✓	✘	✓	✓	✓	✘	✘
Geowave	✓	✓	✓	✓	✓	✓	✘	✘

different data sets have unique data formats and characteristics. To achieve this, a level of flexibility not presently available in the storage schema is required.

Third, the storage approach must support the retention of the raw data along with its derived objects to provide the next generation of input for supervised and unsupervised machine learning, as well as to provide support for advanced forms of predictive modeling (i.e., both those that are probability based and those that are rooted in finite element analysis). Specifically, such an arrangement would allow 3D geometric characteristics and multi-scale features, as well as mixed data sets (e.g., raw and processed; thermal and LiDAR) to be used as input data for machine learning in a way that is wholly unsupported today. In the proposed arrangement, supervised and unsupervised learning would be further supported by being able to store (and thus use) all aspects of a data set's original raw components. For example, commercial and open-source solutions and standards for laser scan storage do not enable the complete retention of the pulse and wave portions of full waveform data, which is becoming increasingly available in aerial, terrestrial, and mobile units. Only the most recent research has addressed this comprehensively in a standalone environment [106–108]. A comprehensive data storage scheme in a distributed environment would allow exploration of such data individually, but even more importantly, as part of a richer co-registered data resource. Furthermore, given the rapid acceleration of the autonomous vehicle industry and their use of laser scanners, the management and exploitation of that data stream is likely to play an increasingly critical role in asset management activities.

A possible solution is an HBase structure [109] to facilitate MapReduce [110] data intake in combination with a hybrid indexing structure [45], in which a piece of the data's global position is considered in a 2D context, but all other functionality occurs in a localized 3D arrangement. This configuration exploits the fact that spatial data have a strong horizontal bias. That is, the horizontal spatial extent is much larger than the vertical component. Even for cities with skyscrapers, cities are much wider than they are high (Fig. 18).

Selecting a local indexing mechanism poses an additional set of challenges as there are hundreds of options [111] and no easy way to rapidly screen for ones that are well-suited for various remote sensing data types. In an effort to provide rapid guidance for data structure selection, a gifi-based approach was pioneered by my

Fig. 18 Horizontal bias of data versus its vertical spatial extent shown for Dublin, Ireland's city center (1 km in width shown)

group. This approach appears to offer some assistance, as shown by the data flowers in Fig. 18. My research group implemented these "data flowers" for three of the most common data indexes. In Fig. 19a, the octree is shown as a highly efficient storage means for heterogeneous data. The raw data are in the top left-hand corner. Their storage is shown directly below, where space is occupied only where data exist. The heterogeneous nature of the storage is shown by the lack of homogeneity in the data flower.

The R-tree flower illustrated in Fig. 19b shows a less efficient storage mechanism both in terms of the solid representation of the bounding box of the building in the lower left and in terms of a much more uniform representation in each of the petals of the data flower. The representation of the k-d tree in Fig. 19c is slightly misleading. While the building representation in the lower left is fully accurate showing the repeated divisions (i.e., every data point is individually stored), the flower itself only contains 1/100th of the data, as their full representation would be too crowded to see any of the petals, thereby directly demonstrating the storage inefficiency of the approach for this type of highly heterogeneous data.

5 Conclusions

The remote sensing community has witnessed tremendous advancements in data collection capabilities over the past few decades, especially with respect to the speed, ease, and cost of acquisition, as well as the expansion of equipment acquisition platforms in the form of mobile and UAV-based options. These advancements have made data available at quantities and qualities that would have been hard to imagine only a few years ago. As such, the major challenge for the community today is how to harness this stream of data that is rapidly evolving into a torrent. This data proliferation will especially challenge existing storage solutions. New storage mechanisms are desperately needed to harness the 3D aspects of the data and other attributes that characterize their tremendous richness and heterogeneity. Without them, the Civil Engineering community will struggle to profit from the data at the levels that would otherwise be expected, especially with respect to asset management, inspection, and monitoring where the automated nature of the data capture opportunities could be of inestimable benefit.

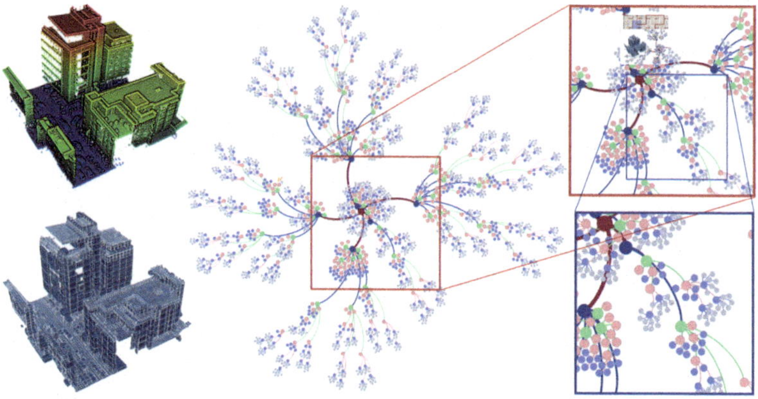

(a) Octree storage with accompanying data flower

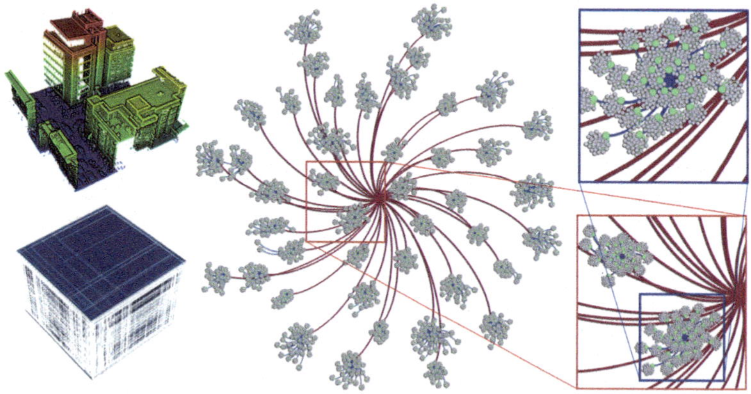

(b) R-tree storage with accompanying data flower

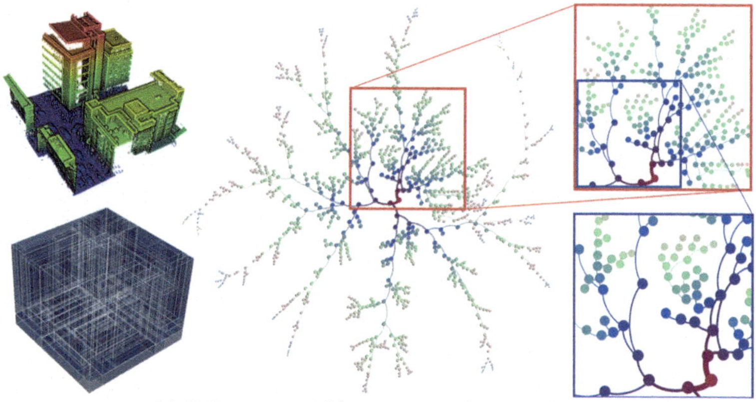

(c) K-D storage with accompanying data flower

Fig. 19 Data flower representations of three common remote sensing data indexes

References

1. Shekhar S, Chawla S (2003) Spatial databases: a tour. Pearson, London
2. Alspaugh D (2004) A brief history of photogrammetry. In: McGlone JC (ed) Manual of photogrammetry. American Society for Photogrammetry and Remote Sensing, Bethesda, MD, pp 1–12
3. van Heuvel T, Hillen R (1995) Coastal management with GIS in the Netherlands. EARSeL Adv Remote Sens 4:27–34
4. Misdorp R, Steyaert F, de Ronde J, Hallie F (1989) Monitoring in the western part of the Dutch Wadden: sea—sea level and morphology. Helgoländer Meeresuntersuchungen 43:333–345
5. Eleveld MA (1999) Exploring coastal morphodynamics of Ameland (the Netherlands) with remote sensing monitoring techniques and dynamic modelling in GIS. PhD thesis. Universiteit van Amsterdam
6. Tockner K, Schiemer F, Ward JV (1998) Conservation by restoration: the management concept for a river-floodplain system on the Danube River in Austria. Aquat Conserv Mar Freshw Ecosyst 8:71–86
7. Hodgson ME, Bresnahan P (2004) Accuracy of airborne lidar-derived elevation. Photogramm Eng Remote Sens 70:331–339
8. Hanssen RF, Van Leijen FJ (2008) Monitoring deformation of water defense structures using satellite radar inferometry. In: Measuring the changes: joint symposia—13th FIG symposium on deformation measurements and analysis and 4th IAG symposium on geodesy for geotechnical and structural engineering, Lisbon. https://doi.org/10.1109/RADAR.2008.4720874
9. Bishop MJ, Dunbar JB, Peyman-Dove LP (2003) Integration of remote sensing (LiDAR, electromagnetic conductivity) and geologic data toward the condition assessment of levee systems. In: Ehlers M (ed) Proceedings SPIE 4886, remote sensing for environmental monitoring, GIS applications, and geology II, pp 400–407. SPIE
10. Givehchi M, Vrijling JK, Hartmann A, Van Gelder PHAJM, Van Baars S (2002) Application of remotely sensed data for detection of seepage in dikes. In: Navalgund RR, Nayak SR, Sudarshana R, Nagaraja R, Ravindran S (eds) Proceedings of the ISPRS commission VII symposium: resource and environmental monitoring, Hyderabad, India
11. Kühn F, Brose F (1998) Die Auswertung von Fernerkundungsdaten zur Deichzustandseinschätzung [The analysis of remote sensing data for dike condition assessment]. Brand Geowissenschaftliche Beiträge 5:59–63
12. van der Schrier JS, Slob S, Straeter J, Hack HRGK, Rupke J (2004) Slim kijken naar dijken: patroon en anomalie herkenning door combinatie van verschillende air-borne remote sensing technieken [Looking intelligently at dikes: pattern and anomaly detection by combining various air-borne remote sensing techniques]. In: Kennisdag Waterkeringsbeheer en Remote Sensing - STOWA
13. Hossain A, Easson G, Hasan K (2006) Detection of levee slides using commercially available remotely sensed data. Environ Eng Geosci 12:235–246
14. Azad Hossain AKM, Easson G (2012) Predicting shallow surficial failures in the Mississippi River levee system using airborne hyperspectral imagery. Geomat Nat Hazards Risk 3:55–78
15. Swart LMT (2007) Remote sensing voor inspectie van waterkeringen [Remote sensing for inspection of flood defences]. Swartvast, Nieuw-Vennep, The Netherlands
16. Mahrooghy M, Aanstoos J, Hasan K, Prasad S, Younan NH (2011) Effect of vegetation height and volume scattering on soil moisture classification using synthetic aperture radar (SAR) images. In: 2011 IEEE Applied imagery pattern recognition workshop (AIPR). IEEE, pp 1–5
17. Moser GM, Zomer WS (2006) Inspectie van Waterkeringen, een Overzicht van Meettechnieken [Inspection of water barriers, an overview of measurement techniques]. Stowa, Utrecht, The Netherlands
18. Cundill SL (2016) Investigation of remote sensing for dike inspection. http://purl.org/utwente/doi/10.3990/1.9789036540360
19. Cohen S, Degnan J, Bufton J, Garvin J, Abshire J (1987) The geoscience laser altimetry/ranging system. IEEE Trans Geosci Remote Sens GE-25:581–592

20. Ridgway JR, Minster JB, Williams N, Bufton JL, Krabill WB (1997) Airborne laser altimeter survey of Long Valley, California. Geophys J Int 131:267–280
21. Murakami H, Nakagawa K, Hasegawa H, Shibata T, Iwanami E (1999) Change detection of buildings using an airborne laser scanner. ISPRS J Photogramm Remote Sens 54:148–152
22. NOAA (2001) NOAA Conducts more flights over World Trade Center site. http://www.noaanews.noaa.gov/stories/s798b.htm
23. FEMA (2008) Geospatial data coordination Appendix B: FEMA data collection in U.S. urban areas
24. National Geospatial-Intelligence Agency. National Geospatial-Intelligence Agency—LiDAR data over core areas of major cities
25. Steinle E, Bähr H (2002) Detectability of urban changes from airborne laserscanning data. In: Navalgund RR, Nayak SR, Sudarshana R, Nagaraja R, Ravindran S (eds) Proceedings of the ISPRS commission VII symposium: resource and environmental monitoring, Hyderabad, India
26. Rehor M, Voegtle T (2008) Improvement of building damage detection and classification based on laser scanning data by integrating spectral information. Int Arch Photogramm Remote Sens Spat Inf Sci XXXVII:1599–1605
27. Shen Y, Wu L, Wang Z (2010) Identification of inclined buildings from aerial LIDAR Data for disaster management. In: 2010 18th international conference on geoinformatics. IEEE, pp 1–5
28. Borfecchia F, Pollino M, De Cecco L, Lugari A, Martini S, La Porta L, Ristoratore E, Pascale C (2010) Active and passive remote sensing for supporting the evaluation of the urban seismic vulnerability. Ital J Remote Sens 42:129–141
29. Moya L, Yamazaki F, Liu W, Yamada M (2018) Detection of collapsed buildings from lidar data due to the 2016 Kumamoto earthquake in Japan. Nat Hazards Earth Syst Sci 18:65–78
30. Toprak S, Nacaroglu E, Koc AC, O'Rourke TD, Hamada M, Cubrinovski M, Van Ballegooy S (2018) Comparison of horizontal ground displacements in Avonside area, Christchurch from air photo, LiDAR and satellite measurements regarding pipeline damage assessment. Bull Earthq Eng 1–18
31. University of Washington Rapid Experimental Facility. https://rapid.designsafe-ci.org/
32. Aruga K, Sessions J, Akay AE (2005) Application of an airborne laser scanner to forest road design with accurate earthwork volumes. J For Res 10:113–123
33. Fekete S, Diederichs M, Lato M (2010) Geotechnical and operational applications for 3-dimensional laser scanning in drill and blast tunnels. Tunn Undergr Sp Technol 25:614–628
34. Tong X, Liu X, Chen P, Liu S, Luan K, Li L, Liu S, Liu X, Xie H, Jin Y, Hong Z (2015) Integration of UAV-based photogrammetry and terrestrial laser scanning for the three-dimensional mapping and monitoring of open-pit mine areas. Remote Sens 7:6635–6662
35. Wang J, Sun W, Shou W, Wang X, Wu C, Chong H-Y, Liu Y, Sun C (2015) Integrating BIM and LiDAR for real-time construction quality control. J Intell Robot Syst 79:417–432
36. Liu W, Chen S, Hauser E (2011) LiDAR-based bridge structure defect detection. Exp Tech 35:27–34
37. Truong-Hong L, Laefer DF, Hinks T, Carr H (2013) Combining an angle criterion with voxelization and the flying voxel method in reconstructing building models from LiDAR data. Comput Civ Infrastruct Eng 28:112–129
38. Lingua A, Piatti D, Rinaudo F (2008) Remote monitoring of a landslide using an integration of GB-INSAR and LIDAR techniques. Int Arch Photogramm Remote Sens Spat Inf Sci XXXVII:361–366
39. Hobbs PRN, Gibson A, Jones L, Pennington C, Jenkins G, Pearson S, Freeborough K (2010) Monitoring coastal change using terrestrial LiDAR. Geol Soc Lond Spec Publ 345:117–127
40. Wiley WC, McLaren IH (1955) Time-of-flight mass spectrometer with improved resolution. Rev Sci Instrum 26:1150–1157
41. Brelstaff GJ, Parraga A, Troscianko T, Carr D (1995) Hyperspectral camera system: acquisition and analysis. In: Lurie JB, Pearson JJ, Zilioli E (eds) Proceedings Volume 2587, geographic information systems, photogrammetry, and geological/geophysical remote sensing, pp 150–159

42. CUBERT. www.cubert-gmbh.com/&#x200E
43. Consumer Physics. SCiO for customers. https://www.consumerphysics.com/scio-for-consumers/
44. Occipital: Structure sensor. https://structure.io/
45. Vo A-V, Laefer DF, Bertolotto M (2016) Airborne laser scanning data storage and indexing: state-of-the-art review. Int J Remote Sens 37:6187–6204
46. 1 foot Digital Elevation Model (DEM). https://data.cityofnewyork.us/City-Government/1-foot-Digital-Elevation-Model-DEM-/dpc8-z3jc/data
47. Laefer DF, Fitzgerald M, Maloney EM, Coyne D, Lennon D, Morrish SW (2009) Lateral image degradation in terrestrial laser scanning. Struct Eng Int 19:184–189
48. Chen S, Laefer DF, Byrne J, Natanzi AS (2017) The effect of angles and distance on image-based, three-dimensional re-constructions. In: 27th Annual European safety and reliability conference (ESREL 2017), Portoroz, Slovenia
49. Chen S, Laefer DF, Mangina E (2016) State of technology review of civilian UAVs. Recent Patents Eng 10:160–174
50. Hinks T, Carr H, Laefer DF (2009) Flight optimization algorithms for aerial LiDAR capture for urban infrastructure model generation. J Comput Civ Eng 23:330–339
51. Geiger AR (1992) Mid-infrared light hydrocarbon DIAL LIDAR
52. Trimble Navigation Limited (2005) Trimble GS Series 3D Scanner [Datasheet]
53. Leica Geosystems (2013) Leica ScanStation P20 industry's best performing ultra-high speed scanner [Datasheet]
54. Truong-Hong L, Gharibi H, Garg H, Lennon D (2014) Equipment considerations for terrestrial laser scanning for civil engineering in urban areas. J Sci Res Rep 3:2002–2014
55. Schrock G BLK360 + ReCap Pro: collaboration for reality capture. http://www.xyht.com/lidarimaging/blk360-recap-pro/, (2017)
56. O'Neill M, Swafford JM, McDermott J, Byrne J, Brabazon A, Shotton E, McNally C, Hemberg M (2009) Shape grammars and grammatical evolution for evolutionary design. In: Proceedings of the 11th annual conference on genetic and evolutionary computation—GECCO '09, p 1035. ACM Press, New York, New York, USA
57. Höfle B, Hollaus M, Hagenauer J (2012) Urban vegetation detection using radiometrically calibrated small-footprint full-waveform airborne LiDAR data. ISPRS J Photogramm Remote Sens 67:134–147
58. Nayegandhi A, Brock JC, Wright CW, O'Connell MJ (2006) Evaluating a small footprint, waveform-resolving lidar over coastal vegetation communities. Photogramm Eng Remote Sens 72:1407–1417
59. Parrish CE, Jeong I, Nowak RD, Smith RB (2011) Empirical comparison of full-waveform lidar algorithms. Photogramm Eng Remote Sens 77:825–838
60. Parrish CE (2007) Vertical object extraction from full-waveform lidar data using a 3D wavelet based approach. PhD thesis. University of Wisconsin–Madison
61. Laefer DF, Truong-Hong L (2017) Toward automatic generation of 3D steel structures for building information modelling. Autom Constr 74:66–77
62. Sherwin MA, Bucksbaum PH, Schmuttenmaer CA, Allen J, Biedron S, Carr L, Chamberlain M, Crowe T, DeLucia F, Hu Q, Jones B, Noordham B, Norris T, Orenstein J, Unterrainer K, Van der Meer L, Wilke I, Williams G, Zhang X-C, Cheville A, Markelz A, Parks B, Plancken P, Shan J, Austin B, Basov D, Citrin D, Grundfest W, Heinz T, Kono J, Mittleman D, Siegel P, Taylor T, Jones B, Markelz A, Martin M, Nelson K, Smith T, Williams G, Allen M, Averitt R, Brunel L, Heilweil T, Heyman J, Jepsen P, Kaind R, Leemans W, Mihaly L, Rangan C, Tom H, Wallace V, Zimdars D (2004) DOE-NSF-NIH workshop on opportunities in THz science, 12–14 Feb 2004, Washington D.C.
63. DOE/Argonne National Laboratory (2007) New T-ray source could improve airport security, cancer detection. https://www.sciencedaily.com/releases/2007/11/071126121732.htm
64. Ho L, Müller R, Römer M, Gordon KC, Heinämäki J, Kleinebudde P, Pepper M, Rades T, Shen YC, Strachan CJ, Taday PF, Zeitler JA (2007) Analysis of sustained-release tablet film coats using terahertz pulsed imaging. J Control Release 119:253–261

65. Federici JF, Gary D, Barat R, Zimdars D (2005) THz standoff detection and imaging of explosives and weapons. In: Saito TT (ed) Proceedings Volume 5781, optics and photonics in global homeland security, pp 75–84. SPIE
66. Ahi K, Anwar M (2016) Advanced terahertz techniques for quality control and counterfeit detection. In: Anwar MF, Crowe TW, Manzur T (eds) Proceedings SPIE 9856, Terahertz physics, devices, and systems X: advanced applications in industry and defense, p 98560G
67. Zimdars D, White JS, Stuk G, Chernovsky A, Fichter G, Williamson S (2006) Large area terahertz imaging and non-destructive evaluation applications. Insight—Non-Destruct Test Cond Monit 48:537–539
68. Suzuki D, Oda S, Kawano Y (2017) A flexible terahertz scanner for omnidirectional imaging. In: 2017 42nd International conference on infrared, millimeter, and terahertz waves (IRMMW-THz), pp 1–2. IEEE
69. Mott MacDonald (2017) Unpublished report, Boulder Colorado
70. Anderson AR (2015) Health effects of cut gas lines and other petroleum product release incidents—Seven States, 2010–2012. Morb Mortal Wkly Rep 64:601–605
71. Tulloch M, Chapman M (2005) Mapping Ontario's underground utilities. Ontario Prof Surv 24–27
72. Hamid-Lakzaeian F, Laefer DF (2016) An integrated octree-RANSAC technique for automated LiDAR building data segmentation for decorative buildings. In: Bebis G, Boyle R, Parvin B, Koracin D, Porikli F, Skaff S, Entezari A, Min J, Iwai D, Sadagic A, Scheidegger C, Isenberg T (eds) Advances in visual computing, pp 454–463. Springer International Publishing, Cham
73. Iman Zolanvari SM, Laefer DF (2016) Slicing method for curved façade and window extraction from point clouds. ISPRS J Photogramm Remote Sens 119:334–346
74. Rajagopal A, Chellappan K, Chandrasekaran S, Brown AP (2017) A machine learning pipeline for automated registration and classification of 3D lidar data. In: Palaniappan K, Doucette PJ, Seetharaman G, Stefanidis A (eds) Proceedings Volume 10199, Geospatial informatics, fusion, and motion video analytics VII, p 101990D
75. Vo A-V, Truong-Hong L, Laefer DF, Tiede D, D'Oleire-Oltmanns S, Baraldi A, Shimoni M, Moser G, Tuia D (2016) Processing of extremely high resolution LiDAR and RGB data: outcome of the 2015 IEEE GRSS data fusion contest—Part B: 3-D contest. IEEE J Sel Top Appl Earth Obs Remote Sens 9:5560–5575
76. Soilán M, Truong-Hong L, Riveiro B, Laefer D (2018) Automatic extraction of road features in urban environments using dense ALS data. Int J Appl Earth Obs Geoinf 64:226–236
77. Aljumaily H, Laefer DF, Cuadra D (2016) Big-data approach for three-dimensional building extraction from aerial laser scanning. J Comput Civ Eng 30:4015049
78. Aljumaily H, Laefer DF, Cuadra D (2017) Urban point cloud mining based on density clustering and MapReduce. J Comput Civ Eng 31:4017021
79. Yang H, Chen W, Qian T, Shen D, Wang J (2015) The extraction of vegetation points from LiDAR using 3D fractal dimension analyses. Remote Sens 7:10815–10831
80. Yan L, Li Z, Liu H, Tan J, Zhao S, Chen C (2017) Detection and classification of pole-like road objects from mobile LiDAR data in motorway environment. Opt Laser Technol 97:272–283
81. Vo A-V, Truong-Hong L, Laefer DF, Bertolotto M (2015) Octree-based region growing for point cloud segmentation. ISPRS J. Photogramm Remote Sens 104:88–100
82. Truong-Hong L, Laefer DF (2014) Octree-based, automatic building façade generation from LiDAR data. Comput Des 53:46–61
83. Hinks T, Carr H, Truong-Hong L, Laefer DF (2013) Point cloud data conversion into solid models via point-based voxelization. J Surv Eng 139:72–83
84. Wang X, Chong H-Y (2015) Setting new trends of integrated Building Information Modelling (BIM) for construction industry. Constr Innov 15:2–6
85. Renaudin CP, Barbier B, Roriz R, Revel D, Amiel M (1994) Coronary arteries: new design for three-dimensional arterial phantoms. Radiology 190:579–582
86. Boschert S, Rosen R (2016) Digital twin—the simulation aspect. In: Hehenberger P, Bradley D (eds) Mechatronic futures: challenges and solutions for mechatronic systems and their designers, pp 59–74. Springer International Publishing, Cham

87. Schleich B, Anwer N, Mathieu L, Wartzack S (2017) Shaping the digital twin for design and production engineering. CIRP Ann 66:141–144
88. Glaessgen E, Stargel D (2012) The digital twin paradigm for future NASA and U.S. air force vehicles. In: 53rd AIAA/ASME/ASCE/AHS/ASC structures, structural dynamics and materials conference, pp 1–14
89. Reifsnider K, Majumdar P (2013) Multiphysics stimulated simulation digital twin methods for fleet management. In: 54th AIAA/ASME/ASCE/AHS/ASC structures, structural dynamics, and materials conference. American Institute of Aeronautics and Astronautics, Reston, Virginia. https://doi.org/10.2514/6.2013-1578
90. National Infrastructure Commission (2017) Data for the public good
91. Holstein WJ (2015) Virtual Singapore: creating an intelligent 3D model to improve experiences of residents, business and government. https://compassmag.3ds.com/en/8#/en/8/Cover-Story/VIRTUAL-SINGAPORE
92. Laefer D (2016) Emerging city-scale damage prediction options for urban tunnelling. In: Structural analysis of historical constructions. CRC Press, Taylor & Francis Group, 6000 Broken Sound Parkway NW, Suite 300, Boca Raton, FL 33487–2742, pp 15–22
93. Sicilia A, Madrazo L, Pleguezuelos J (2014) Integrating multiple data sources, domains and tools in urban energy models using semantic technologies. In: 5th eeBDM workshop in ECPPM, Vienna, Austria
94. Technical University Berlin. Energy Atlas Berlin. http://energyatlas.energie.tu-berlin.de/
95. Madrazo L. SEMANCO. http://semanco-project.eu
96. Open Geospatial Consortium. CityGML. https://www.opengeospatial.org/standards/citygml
97. Zhang J, Yu F, Li D, Hu Z (2014) Development and implementation of an industry foundation classes-based graphic information model for virtual construction. Comput Civ Infrastruct Eng 29:60–74
98. Hallberg D, Tarandi V (2011) On the use of open Bim and 4D visualisation in a predictive life cycle management system for construction works. J Inf Technol Constr 16:445–466
99. Qin L, Deng X, Liu X (2011) Industry foundation classes based integration of architectural design and structural analysis. J Shanghai Jiaotong Univ 16:83–90
100. de Laat R, van Berlo L (2011) Integration of BIM and GIS: the development of the CityGML GeoBIM extension. In: Kolbe TH, König G, Nagel C (eds) Advances in 3D geo-information sciences, pp. 211–225. Springer, Berlin, Heidelberg
101. Kolbe TH, Gröger G, Plümer L (2005) CityGML: Interoperable access to 3D city models. In: Geo-information for disaster management, pp 883–899. Springer, Berlin, Heidelberg
102. Miranda F, Doraiswamy H, Lage M, Wilson L, Hsieh M, Silva CT (2018) Shadow accrual maps: efficient accumulation of city-scale shadows over time. IEEE Trans Vis Comput Graph 1–1
103. Yang B, Fang L, Li Q, Li J (2012) Automated extraction of road markings from mobile lidar point clouds. Photogramm Eng Remote Sens 78:331–338
104. Yu Y, Li J, Guan H, Jia F, Wang C (2015) Learning hierarchical features for automated extraction of road markings from 3-D mobile LiDAR point clouds. IEEE J Sel Top Appl Earth Obs Remote Sens 8:709–726
105. Laefer DF, Abuwarda S, Vo A-V, Truong-Hong L, Gharibi H (2015) Aerial laser and photogrammetry survey of Dublin city collection record. https://geo.nyu.edu/catalog/nyu_2451_38684. https://doi.org/10.17609/N8MQ0N
106. Laefer D, Vo V, Bertolotto M (2018) A spatio-temporal index for aerial full waveform laser scanning data. ISPRS J Photogramm Remote Sens 138:232–251
107. Vo AV, Chauhan N, Laefer DF, Bertolotto M (2018) A 6-dimensional hilbert approach to index full waveform LiDAR data in a distributed computing environment. ISPRS Int Arch Photogramm Remote Sens Spatial Inf Sci XLII-4:671–678
108. Vo AV, Konda N, Chauhan N, Aljumaily H, Laefer DF (2018) Lessons learned with laser scanning point cloud management in Hadoop hbase. In: Smith I., Domer B. (eds) Advanced computing strategies for engineering. EG-ICE 2018. Lecture notes in computer science, vol 10863. Springer, Cham, pp 231–253

109. Carstoiu D, Lepadatu E, Gaspar M (2010) Hbase—non SQL database, performances evaluation. Int J Adv Comput Technol 2:42–52
110. Dean J, Ghemawat S (2008) MapReduce. Commun ACM 51:107
111. Samet H (1990) The design and analysis of spatial data structures. Addison-Wesley Longman Publishing Co, Boston

Geomatics Applied to Civil Engineering State of the Art

Irineu da Silva

Abstract Civil engineers, as well as other engineering careers, are facing new challenges in their professions due to new technologies and market forces that have been redefining engineer's roles toward a new productive vision (ASCE, American Society of Civil Engineers in The vision for civil engineering in 2025. Reston, Virginia) [1]. As one of the oldest civil engineering related activity and to this day an indispensable tool for any civil engineering work, Geomatics Engineering shares the same concerns, especially when related to the synergy between Geomatics and Civil Engineering. As all technical segments, Geomatics has been making important technological advances over the years, incorporating new instruments, new technologies, and new working methods that must be understood by civil engineers so that both continue to develop complementarily and work together seamlessly. Considering this, the present article discusses new technologies available in Geomatics and their applications in civil engineering projects and the level of teaching that must be passed on to civil engineers so that both can work together making efficient use of the new technologies at their disposal.

Keywords Geomatics engineering · Civil engineering · Topography · Surveying

1 Introduction

Discussing Geomatics applied to Civil Engineering is currently a challenge that needs to be carefully addressed due to the great innovations that have occurred in both professional disciplines and due to the high synergy between them. To be able to discuss this, it is essential, first and foremost, to understand the meaning of the term Geomatics in the engineering field. Taking this into account, like so many other new disciplines, scholars have proposed various definitions for the term, which, in

I. da Silva (✉)
Department of Transportation Engineering, São Carlos School of Engineering (EESC), University of São Paulo (USP), Av. Trabalhador são-carlense, 400 – Parque Arnold Schimdt, São Carlos, SP 13566-590, Brazil
e-mail: irineu@sc.usp.br

J. K. Ghosh and I. da Silva (eds.), *Applications of Geomatics in Civil Engineering*, Lecture Notes in Civil Engineering 33, https://doi.org/10.1007/978-981-13-7067-0_2

the end, converge to the same point of view, which is to consider it as an extension of Surveying Engineering to reflect new measurement technologies and data management capabilities. At this point, Geomatics should be understood as a modern engineering discipline that encompasses the sciences, techniques and methods that deal with acquisition, storage, processing, modelling, georeferencing, analysis, display, distribution, setting out, and management of spatially referenced data, to group them into a coherent discipline consistent with new technologies and engineering needs. Based on the scientific framework of Geodesy, it uses terrestrial, marine, airborne and satellite-based sensors to acquire spatial and image data, while relying on database management, topometric computation and computer vision to handle them and make them available to the users. In conjunction with Civil Engineering, it can be said that Geomatics is concerned with measuring existing features of the natural and built environment and presenting data in a format suitable for architects and engineers to use in designing construction projects. Providing the positioning, shape and nature of geographical features through topographic technologies, Geomatics lies between Civil Engineering and the real world, playing an important role from the early stages of designing the project to the final as-built mapping.

Needless to say, Geomatics is not only a data provider for civil engineering projects, although this is the subject of this article. Despite being closely related professional activities, Geomatics and Civil Engineering have had a controversial coexistence, mainly due to the level of knowledge that one has of the other. In many countries, Geomatics is a discipline on its own, and in its entirety, it is almost as broad in scope as Civil Engineering. In others, it is just a research group, most often linked to transportation engineering departments. In the first scenario, despite its broad scope, Civil Engineering is only one of the topics studied, and not the most important. In the second scenario, Geomatics is only a 6-month course (sometimes 1 year) offered to students in the third semester. In these scenarios, both sides have difficulty communicating with each other.

The point here is how to minimize this conflict. And this can only be done by having a complete understanding of civil engineering needs and the tools made available by Geomatics. New technologies and processing methods have destroyed old paradigms and have created new opportunities, which have prompted the need for a new approach in teaching civil and geomatics engineering, as discussed in the next pages of this article.

2 When Geomatics Matters to Civil Engineering

The importance of Geomatics for civil engineering projects can be described by Fig. 1. As depicted in the figure, Geomatics procedures are involved in almost all civil engineering workflow, i.e. planning, design, construction and management.

Planning. No significant civil engineering project can be developed without detailed information on the construction site. Moreover, no civil engineering project can be

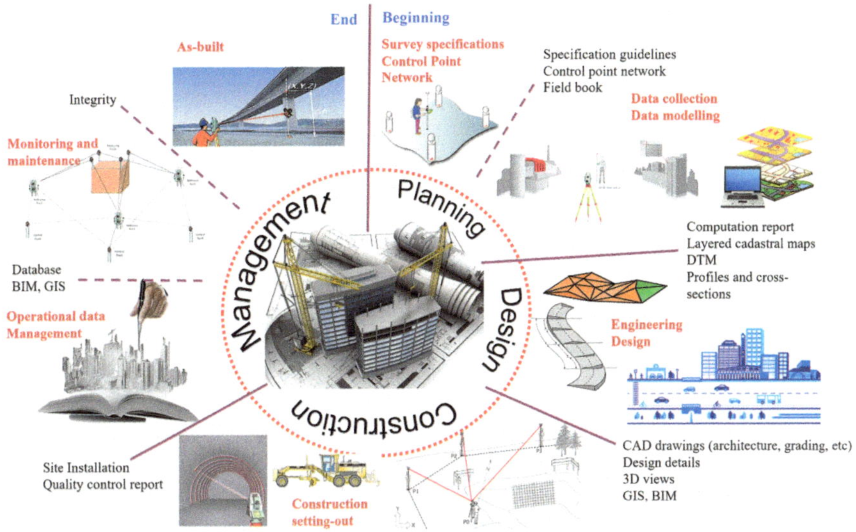

Fig. 1 Geomatics workflow in the civil engineering production chain

executed without a network of geodetic control points. Activities in this working phase basically include the preparation of surveying specifications, mapping specifications, database standardization and measurement tolerances; field measurements design and data collection; data modelling, referencing and digital mapping production to use in the engineering design phase. Geomatics products made available at this stage are: specification guidelines; provision of horizontal and vertical control point networks; measurement field books; computation reports and layered cadastral plans including digital orthophotos, land boundaries, point meshes, terrain profiles, cross-sections, Digital Terrain Modelling (DTM) or Digital Elevation Modelling (DEM) and levelling benchmarks. It can also include quantities for inventory, database management, economic assessment and cost accounting.

Design. Upon completion of the planning phase, engineers must have at their disposal a complete set of data required to develop engineering projects. Most often, this means Computer-Aided Design (CAD) drawings of surveying measurements on specified mapping projections and digital terrain models, complemented by specific geomatics products as described in the planning phase. As a result, civil engineers will gather all available information from a variety of sources to develop and manage infrastructure projects, including architectural and construction designs, parcelling, intermediate construction, 3D visualization, grading, earth movement, volume computation, mass haul plans, machine control layout, environmental mitigation plans, and many others, depending on the type of project and its lifetime. Geographic Information System (GIS) and Building Information Modelling (BIM) methodologies are also initiated at this stage. At this point, the greater the interaction between Geomatics and Civil Engineering, the greater the success of the project.

Construction. In the construction stage, information needs to be enlarged considerably and topographic information must be fine enough to control the progress of the construction on a daily basis. Measurement of spatial data in real-time and automated procedures are needed in order to manage the difference between the daily progress of the construction and its scheduled value. At the same time, setting out processes are implemented to enable pegs, profiles or other marks to be set out to control construction work, and ensure that each element of the work is constructed in the correct position and to the correct level. Execution and certification of quality control and production reports are also imperative tasks performed at this stage of the project.

Management. The management phase encompasses managing building elements, monitoring terrain and structural stability and executing as-built measurements. GIS and BIM formatting are currently essential for data management and effectively replace relational databases previously used. Real-time measurement sensors, digital cameras, drones and web-based data transmission are technologies currently available to ensure the quality and effectiveness of construction and structural monitoring. Needless to say, metadata, encoding, and standardization of communication protocol become indispensable in carrying out information integration at this stage. Finally, the completion of the construction is mapped by as-built survey in order to present the construction status-quo.

3 Geomatics Engineering Assignments

To properly apply geomatics engineering technologies to the civil engineering production workflow as previously outlined, civil engineers must have a clear understanding of geomatics engineering assignments [3]. To help clarify this understanding, the following is a brief description of them, from the civil engineering point of view. Please refer to Fig. 2 for a conjectural view of the geomatics production chain and subsequent assignments.

3.1 Surveying Specifications, Norms, Standards and Database Structuring

Structuring a database, whether geographic or not, aims to regulate how the dataset will represent an environment in a digital way. This structuring process analyses and determines which elements and characteristics should be represented in order to make the standardization of data and the contents of the database possible, including measurement tolerances and measurement methods. The objective, in this case, is

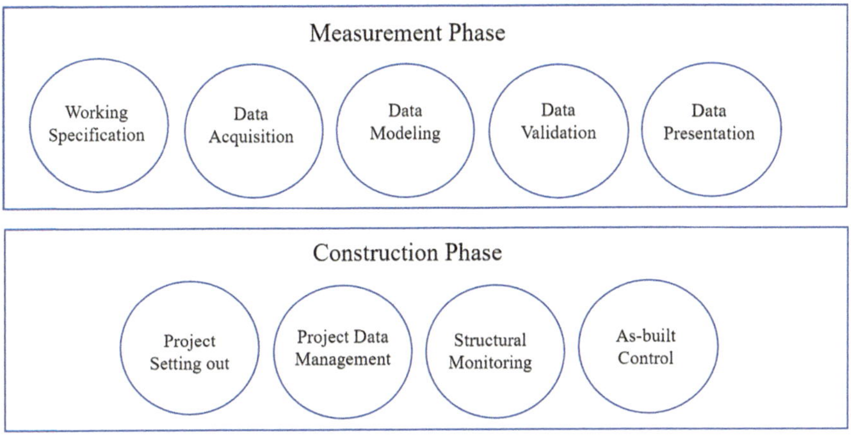

Fig. 2 Geomatics engineering production chain and assignments

to comply with international norms and standards to ensure that the same database has the potential to be used for different applications and for data exchange in the same project. Failure to comply with these standards entails risks of rework, data loss and wasted time in manual data formatting. Preventing this from happening is the responsibility of project managers who should include both Geomatics and Civil Engineers. To overcome this problem, Geomatics Engineers are requested to determine the specification guidelines on collecting, archiving and maintaining spatial data. The scope and details of these guidelines will depend on the type and lifecycle of the project, so that no summed-up suggestion can be presented here without challenging quality management principles.

3.2 Field Measurement Design and Data Acquisition

The second stage in a civil engineering project is the field data acquisition, which includes measurement planning and spatial data collection. Measurement planning entails a detailed study of the surveying area, choosing measurement instruments and measurement methods. Spatial data collection, on the other hand, means field data acquisition execution and data handling. For these purposes, measurements can be currently categorized into three groups, as shown in Fig. 3.

The **single point measurement** category is related to the instruments and measurement methods in which the surveying data is obtained by individual assessment. Included in this category are measurements performed with total stations, Global Navigation Satellite System (GNSS) sensors and levelling instruments, among others. The data acquisition, in this case, is obviously time-consuming, however, mea-

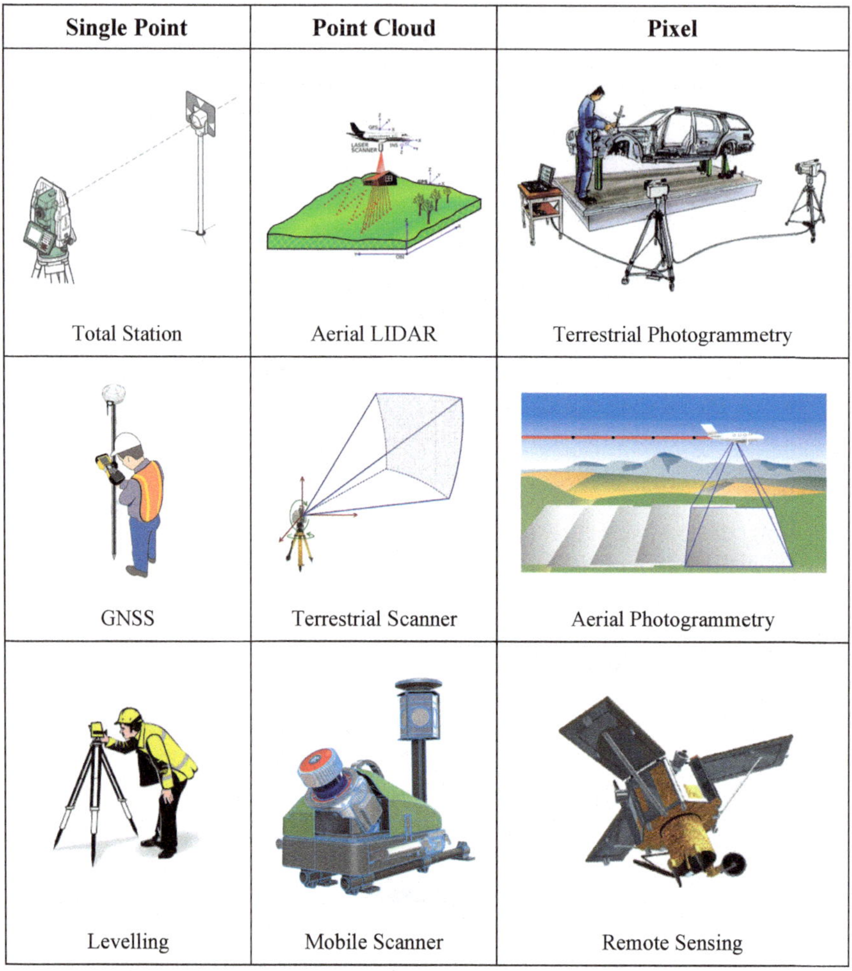

Fig. 3 Measurement categories

sured values tend to have better quality than the other measurement categories, reaching millimetre accuracy level.

The **point cloud measurement** category consists of instruments and measurement methods performed to produce huge amounts of surveying points—called point cloud—with precise (X, Y, Z) values together with their information on the ground reflected signal. It corresponds to a category of measurements based on Light Detection and Ranging (LiDAR) which can be airborne systems (Airborne Laser Scanning) or ground-based systems (Terrestrial and Mobile Laser Scanning) mounted on a tripod or a vehicle. The measurement methodology, in this case, is based on an active optical remote sensing system that measures the distance from the sensor to the object by calculating the time taken by a laser pulse from the release to the reception.

LiDAR is an upcoming technology which has enormous applications in civil and construction engineering. It has an accuracy of centimetre level in vertical as well as horizontal positioning, and hence its application ranges from precise topographic survey to a wide range of geotechnical, coastal, transportation, structural engineering, 3D modelling, planning and as-built applications, including computation of cut and fill quantities, detailed surveys of road and other construction projects [2].

The **pixel measurement** category consists of instruments and measurement methods based on digital images with the purpose of making geometric or radiometric inferences about the objects appearing in the images. Such images can be acquired at close range on the ground, from aircraft or from satellites, defining the so-called terrestrial photogrammetry, aerial photogrammetry and remote sensing. Measurement sensors, in this case, are digital cameras based on the combination of Charge-Coupled Devices (CCD), ranging from small formats of around 10 MP to large formats of around 400 MP, with a pixel size ranging from 4 to 15 micrometres. The main contributions of photogrammetry and remote sensing to Civil Engineering include topographic mapping, digital orthophoto production, 3D modelling, environmental monitoring, database development for GIS and BIM, resource inventory and monitoring, as well as deformation analysis.

Another point to be highlighted in the data acquisition phase is the fact that raw data can be collected through different types of sensors in the same project, and each sensor presents its own data typology and format. It is, therefore, important to consider using a data management system that would be able to properly manage these different data, enabling the user to retrieve and apply them conveniently, at any time, simply and quickly.

In order to properly use the measurement technologies presented, it is essential that civil engineers know the characteristics of the use and applications of each instrument, its accuracy and its limitations. It is also important to ensure that the selected instruments are properly calibrated and in good working order and that the specified procedures are rigorously followed.

3.3 Data Modelling and Referencing

After collecting information from the field, it must be modelled and referenced according to mathematical models and appropriate coordinate systems so that geographic data is available. All major civil engineering projects are currently modelled according to geodetic models. The altitudes are modelled according to the geoid model and the planimetric values according to the ellipsoidal models, as shown in Fig. 4.

Data modelling is performed through topometric calculations depending on the measurement method applied in the field. The most primary data modelling is based on plane surveying because it is so widely used in engineering and surveying practice. Based on the fundamentals of measuring distance, angle, direction and elevation, data modelling will provide point position, slope, area and volume. GNSS positioning is

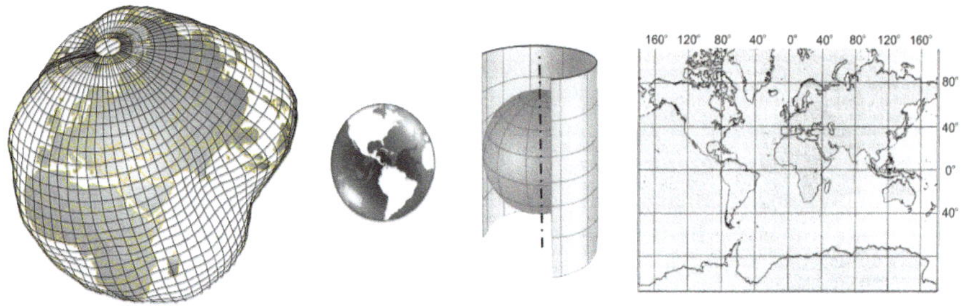

Geoidal model

Ellipsoidal model

Fig. 4 Geodetic models

another source of data modelling which can provide referenced data in real-time or post-processing positioning techniques. The interaction of the user, in that case, is simplified by using processing software, which does not mean less responsibility.

Another point to be considered in the data modelling concept is the increasing use of redundant measurements which leads to using adjustment techniques in order to resolve the inconsistency between the observations and the model. It is therefore recommended for civil engineers to have at least a slight idea of adjustment techniques in order to avoid misuse.

During data modelling procedures, a data referencing procedure is also followed, which means connecting the modelled data with geodetic reference systems. This can be done through map projections or local ground-based plane coordinates. In both cases, it is imperative to understand that distortion will arise from that geometric assumption. Not understanding their effects or disregarding them can be fatal to the development of many engineering projects.

3.4 Data Validation

After the data modelling phase, it is important to perform a data validation analysis, which consists of a series of data quality control to align the conceptual model (mathematical model) to the real world. This can be done through quality controls and quality management concepts that aim at total control of product quality, as well as its means of production. It is recommended to apply this management concept at every stage of production.

3.5 *Configuration and Data Presentation*

To be useful for civil engineering designs, geomatics data must be properly configured and rationally presented. For that purpose, a series of procedures are indicated as depicted in Fig. 5.

The primary method of topographic data presentation is its digital graphic representation in CAD format. In this type of data representation, the user has access to any planimetric information from the dataset, including the coordinates (X, Y) of any point represented in the drawing. However, the user has little information about the third dimension, which makes it necessary to consider using Digital Terrain Modelling (DTM). DTM consists of a mathematical model of the terrain surface, based on the application of one or more mathematical functions to describe the variation of its altimetry. The surface represented is called a modelled surface, and the name Digital Terrain Model is given to the set of functions, combined with the practical use of the modelled surface. This numerical modelling can identify the altitude value (H) of any point on the terrain surface, according to its planimetric coordinates (X, Y) represented in the drawing. This is an invaluable resource for many civil engineering projects.

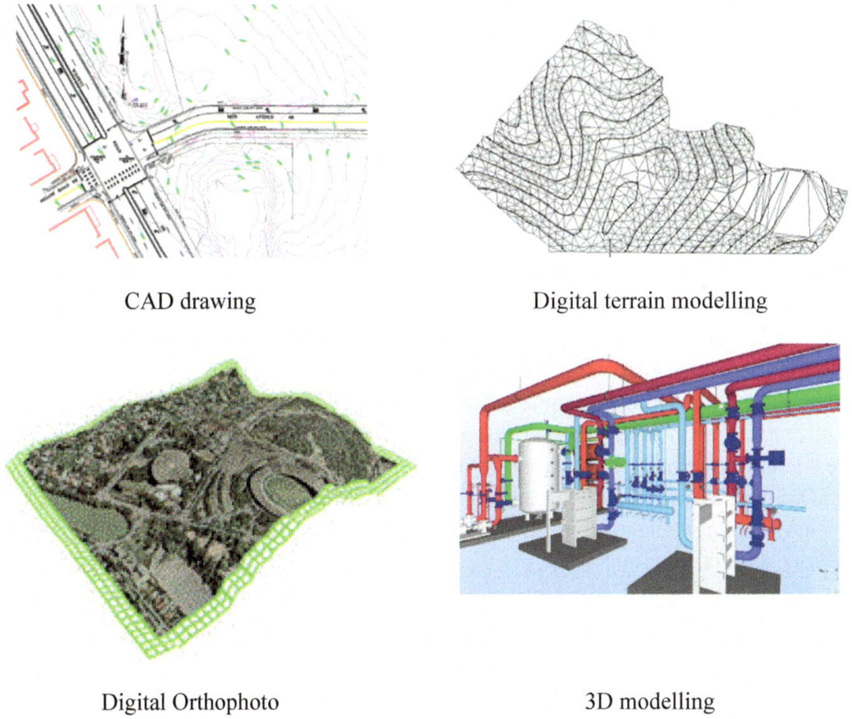

Fig. 5 Data presentation

Fig. 6 Example of orthorectified digital image superposed with engineering design

Another form of topographic data presentation that has gained space in Civil Engineering in recent years is the graphic representation from orthorectified digital images, that is, digital images represented in orthogonal projection, georeferenced, in which the deformations inherent to a photographic image are suppressed. This is the well-known 'digital orthophoto'. The user, in this case, has a digital map at his/her disposal with all its topographic elements in colour and referenced to the coordinate system of the project, as shown in Fig. 6. As it is an orthorectified digital image, the orthophoto allows the superimposition of designed features over the image itself, which evidently makes the graphic visualization clearer.

A digital orthophoto is produced from a pair of oriented aerial images and the numerical model of the corresponding terrain. The production process is automated requiring little operator interaction. Its quality depends, fundamentally, on the quality of the numerical terrain model. It has mainly been used for specific civil engineering designs in small terrain areas, as highway intersections, for instance. It should be noted that for specific cases, the designer can also use the orthorectified mapping technology through Unmanned Aerial Vehicles (UAV), which have proved to be suitable for generating small-area photogrammetric data, including generating orthophotos and 3D views.

In recent years the presentation of spatial data in the form of 3D models has contributed enormously to the development of a new kind of data presentation. It is the so-called scanning data, represented by a point cloud data format with millions of 3D points. How to handle this huge amount of points, however, is still a matter of discussion. Only a handful of civil engineering design software is currently enabled to work with point clouds. The practice has been to use these 3D values to generate flat views from which engineering designs are drawn. For that reason, few civil engineering projects are currently designed completely in 3D format. Most of the time, 3D point clouds or 3D models are just an auxiliary design tool. Expectation, however, is for a complete 3D design very soon.

3.6 Setting Out

The setting out procedures are generally performed using total stations or GNSS instrument operating in Real-Time Kinematic (RTK) mode. These instruments already have internal application programs for the work to be implemented, which indicate the direction and distance to be measured for the positioning of points on the ground.

For earthworks, however, the tendency is to use the so-called machine control systems, which allow dimensional control of earthwork movements in civil engineering construction areas, based on using stakeout techniques based on machinery automation systems. Using this type of system increases productivity and decrease costs, mainly in large earthworks, slopes and road construction. Depending on the type of work, it is considered that an increase in productivity of the order of 30–100% compared to traditional methods can be achieved. The most often automated machines are scrapers, dozers, bulldozers, motor graders, profilers and pavers. The control systems for such machines are based on using attitude sensors, hydraulic valves, onboard computers, and surveying instruments. The machine automation takes place, internally, by the convenient installation of attitude sensors in its different implements to determine the relative movements and, externally, by using surveying instruments, which can control the vertical positioning of its implements or its spatial coordinates (X, Y, Z), in relation to a specific coordinate system. For this purpose, three types of surveying instruments are used, which are: laser levels, robotic total stations, and GNSS receivers, as shown in Fig. 7. The whole set is managed by an onboard computer, whose interaction with the sensors is carried out through hydraulic valves and connection cables and, with the operator, using a Liquid Crystal Display (LCD) control panel.

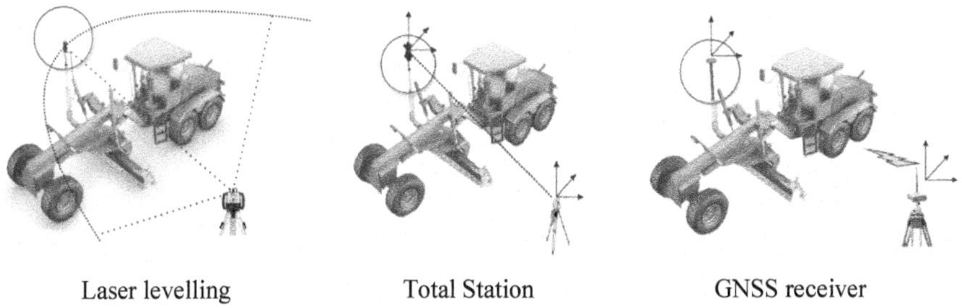

Fig. 7 Surveying instruments for machine control positioning

3.7 Data Management

As shown in precedent sessions, as in many disciplines, geomatics professionals face the problem of big data management and the extensive use of engineering software requiring high rates of velocity, variety, variability, and complexity in database management and result analysis. As a result, it becomes clearly necessary to standardize the storage structures and data management of this kind of dataset in order to provide mass storage and easy access to the users, mainly in big databases. The complexity of the problem is even greater when the geomatics data is connected to already deployed engineering structures on the ground, connecting the geomatics databases with engineering design databases. The solution to these cases has been to use GIS structures for civil engineering projects and BIM methodology for building data management. Whatever the solution, it is imperative that in this stage of the work, there is a complete connection between both Geomatics and Civil Engineer professionals to ensure the integrity of the databases.

3.8 Monitoring and Construction Safety

Depending on the type of project and its magnitude, there may be a need to control the safety of the work or monitor its structural behaviour after its final deployment. For this type of quality control, there are several monitoring methods available, and geodetic monitoring is one of the most important, depending on the type of project. Geodetic monitoring, due to its range of applications, measurement techniques, data processing methods, measuring instruments and accessories may be considered as a data management system comprising several components that together can perform structural health monitoring, as shown in Fig. 8. The ultimate goal of such monitoring is the periodic determination of spatial coordinates of specific points of the structure, from which structural deformations or displacements are calculated. For this purpose, current geodetic monitoring systems operate through robotic total stations or GNSS instruments operating in static or RTK measurement mode.

The physical behaviour of a civil engineering structure is a characteristic of its structural concept and of the load to which it will be subjected. Designer engineers are responsible for analysing these parameters. On the other hand, requirements of measurement accuracies, and consequently measurement methods and result analysis are the responsibility of Geomatics Engineers, as well as several other professionals involved in the project. For these reasons, structural geodetic monitoring should not be considered a simple measurement task and must be performed only by experienced professionals with a high knowledge of Geomatics.

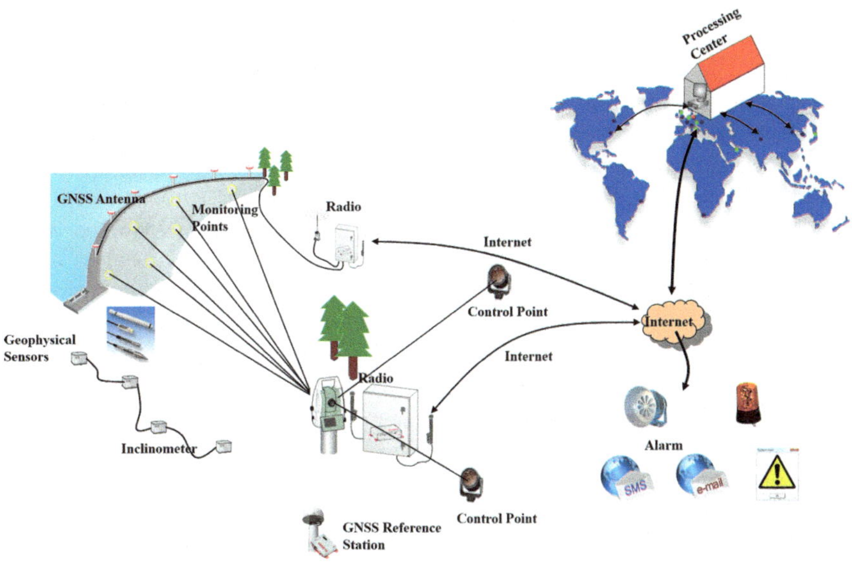

Fig. 8 Geodetic monitoring system

3.9 As-Built

The purpose of an as-built survey is to prove to the project owner or to government agencies that the designed construction details have been performed in accordance with the specifications established during the planning phase and shown in the work plan, or eventually to show exactly what has been completed to date or modified during the operational work. Considering this, as-built work is mostly performed in the same way as any surveying mapping work. Considering the amount of data to be measured, the crucial point here is the level of measurement automation and dataset management to allow easy connection between field books and processing/mapping software. RTK GNSS and terrestrial scanning instruments are of interest for this type of application.

4 Geomatics Teaching for Civil Engineers

The level of education in geomatics for civil engineering can currently be considered as ascending to the degree of attribution confirmations. Civil engineers know the available technologies and understand the measurement techniques, but they are still rooted in classic paradigms. In most developing countries and even in developing ones, there are islands of innovation, but knowledge has not been distributed properly. To solve this problem, the following topics should be highlighted in teaching Geomatics to civil engineers.

Topographic instruments. Civil engineers must understand the main constructive concept of each instrument presented in Fig. 3: its calibration methods; accuracy levels; their sources of errors and their corresponding operating techniques. It is important to consider, in this case, the studies related to the automation of measurement processes and instrument internal routines to perform easy and online communication with surveying mapping software.

Regarding instrument and surveying accuracies, they have been defined for a long time according to the dimension of the smallest detail to be measured and of the scale of the final graphical representation. Nowadays, however, with the computational resources available, it is possible to plot a drawing on any graphical scale, that is, with the 'zoom' features available in computer programs, the scale of a digital map is no longer fixed. For this reason, graphical representation becomes an important component for defining the surveying accuracy only when it is specified that the project delivery will be done by a graphic process printed on paper. For the cases in which surveying will be manipulated in digital and numerical means, the required accuracies should be indicated according to the technical specifications of the project or in function of technical standards. For example, the level of accuracy required for a tunnelling project is different from the accuracy level of a highway, water dam or a parcelling project. The accuracy specification must, therefore, be what ensures that the positioning of the constructive elements is within the limits accepted by the designing standards and by the construction techniques. In addition, engineers must also consider the accuracy of the instruments available in the market in order to choose the most appropriate ones.

Measurement technologies. In addition to conventional measurements of angles, distances, directions and levelling, civil engineers must understand the measurement fundaments of GNSS receivers, terrestrial and aerial laser scanners, a basic concept of digital photogrammetry and machine control technologies.

It is important to consider that the topographical information of the future will be three-dimensional, that is, more and more work will be done with point clouds and relational database, which suggests the gradual change of conventional teaching based on horizontal and vertical planes to 3D visualization and operation.

Geodetic and topographic references. Although considered a tough subject, civil engineers must have a substantial knowledge about plane and geodetic reference systems and their transformations. GNSS technologies have brought geodesy to the world of Civil Engineering, requiring appropriate knowledge of coordinate transformations and mapping projections for such instruments to be properly operated.

Data presentation. Since all topographic maps are presented in CAD format and many surveying software are CAD-based, it is important that civil engineers have adequate knowledge of the use and facilities of this technology. On the other hand, in view of the increased use of orthophotos for engineering projects, it is recommended that this mapping technology should be understood well by civil engineers.

Digital terrain modelling. Civil engineers must take into account that the numerical terrain models have totally supplanted the old altimetric representations by means of contour lines. Currently, all projects related to altimetry are made based on digital terrain modelling, including contour line representation. It no longer makes sense to study manual contour lines drawing methods. Instead, civil engineers must understand what affects DTM quality, which is its field of application and how it can be used in civil engineering projects.

Geographic information system and building information modelling. The detailed study of GIS and BIM methodologies, naturally demands the study of a series of complementary disciplines, as well as basic Geomatics teaching for civil engineers. For this reason, in a basic civil engineering course, it is not possible to cover all the topics needed for teaching GIS and BIM. Even so, it is important that civil engineers are taught the basics on this matter, which enables them to decide when and where to use them.

Data evaluation methods. It is essential that civil engineers have basic knowledge of error theory, error propagation and to some extent, basic concepts of observation adjustments. Besides that, Geomatics teaching for Civil Engineering should also include the fundamentals of surveying data control and big data management.

Geodetic monitoring. The increasing amount of bold structures requires civil engineers to know the principles of structural monitoring, comprising geodetic monitoring. Teaching, in this case, should include studies related to: the basic principles of geodetic monitoring; measurement techniques; available instruments, accuracies and the subsystems involved in the geodetic monitoring system.

5 The Future of Geomatics Engineering

In the foreseeable future, the field of Geomatics Engineering will continue to undergo significant changes due to technological development in digital image sensors, attitude and positioning sensors, laser sensing, artificial intelligence, computer vision, Big Data, Machine Learning and database management, as well as other related technologies. This ongoing and other emerging technologies are expected not only to change and increase the field of Geomatics Engineering applications but also to have an impact on the synergy between Geomatics and Civil Engineering. On the other hand, automation and 3D environment projects will be of special interest and should be part of future Geomatics course syllabuses for civil engineers.

References

1. ASCE, American Society of Civil Engineers (2010) The vision for civil engineering in 2025. Reston, Virginia
2. Jeganathan C, Pramod K, Kshama G, Rahul DC, Anand KrS, Kirti A, Ramesh H (2017) Remote sensing and GIS for civil engineering applications and human development. Int J Adv Remote Sens GIS Geogr 5(1):1–18
3. Silva I, Segantine PCL (2015) Topografia para Engenharia – Teoria e Prática de Geomática. Editora Elsevier, Rio de Janeiro

Part II
Geomatics in Structural Engineering

3D Digital Documentation of a Cultural Heritage Site Using Terrestrial Laser Scanner—A Case Study

S. K. P. Kushwaha, Karun Reuel Dayal, Sachchidanand, S. Raghavendra, Hina Pande, Poonam S. Tiwari, S. Agrawal and S. K. Srivastava

Abstract Cultural heritage sites are the important sites which are given more priority in terms of preservation and conservation, so that they last for a long duration of time. Although modern structures are designed to be resilient to several events, heritage structures usually undergo considerable damages. Irreparable damages have been inflicted on some old structures in such cases, documentation proved to be a very useful tool for the reconstruction of the structure and preserving it. Remote sensing techniques using Terrestrial Laser Scanner (TLS) and Photogrammetry are

S. K. P. Kushwaha (✉) · K. R. Dayal · S. Raghavendra · H. Pande · S. Agrawal
Photogrammetry and Remote Sensing Division, Indian Institute of Remote Sensing, ISRO, Dehradun, India
e-mail: s.k.p.kushwaha92@gmail.com

K. R. Dayal
e-mail: karunrdayal@gmail.com

S. Raghavendra
e-mail: raghav@iirs.gov.in

H. Pande
e-mail: hina@iirs.gov.in

S. Agrawal
e-mail: shefali_a@iirs.gov.in

Sachchidanand
Remote Sensing Applications Centre – RSAC, Lucknow, UP, India
e-mail: sachchidanandd11@gmail.com

P. S. Tiwari
Geoweb Services, IT & Distance Learning Division, Indian Institute of Remote Sensing, ISRO, Dehradun, India
e-mail: poonam@iirs.gov.in

S. K. Srivastava
Geospatial Technology and Outreach Programme Group, Indian Institute of Remote Sensing, ISRO, Dehradun, India
e-mail: sksrivastav@iirs.gov.in

J. K. Ghosh and I. da Silva (eds.), *Applications of Geomatics in Civil Engineering*, Lecture Notes in Civil Engineering 33, https://doi.org/10.1007/978-981-13-7067-0_3

very effective methods in acquiring 3D information and texture of the structure with least interaction with the structure. The point cloud data from TLS is textured using high-quality photographs acquired from a Digital Single-Lens Reflex camera (DSLR). The colored point cloud data was used to create different sections like top, front, back, left, and right and drawings were made in AutoCAD software. 3D digital documentation is necessarily sufficient to reconstruct the structure in case any damage occurs. The main advantage of using remote sensing technique is that it does not need any physical contact with the surface. Remotely accessed data are very vital in case of cultural heritage site because the present strength of the structure is not known.

Keywords Cultural heritage site · 3D digital documentation · Terrestrial laser scanner (TLS) · Digital single lens reflex (DSLR) camera · Photogrammetry · Point cloud · AutoCAD · Section drawings

1 Introduction

Cultural heritage sites are to be preserved and taken good care. Up-to-date information plays a major role which can be done with proper documentation of the structure. Documentation of these cultural heritage sites are important to reconstruct in case of any damage occurs to the heritage site. Documentation plays a major role in extracting the geometry and monitoring its health. Accurate measurements and reconstruction is possible with the help of 3D modeling [1].

Remote sensing technique plays a major role in data acquisition when the study area is preserved like heritage site which needs less interaction. In this research, terrestrial laser scanning and close-range photogrammetry are used to document heritage structures. Terrestrial laser scanning is a very effective technology for generating a detailed 3D model of the structure. Dense point cloud is generated with the pre-specified parameters, all the points are sampled over the surface of the structure. A particular location of a point is located on the basis of time taken by the lidar signal to hit the target and return back to the scanner [2]. The point cloud generated can be used for 3D modeling and to extract any other valuable information [3]. The scan parameters like horizontal and vertical scan area, horizontal and vertical resolutions are pre-entered before the data acquisition by the user. Similar kind of research has been carried out to extract information such as damages, structural analysis, deformities, etc., from the generated point cloud [4–6].

Fig. 1 St. Mary's Church, Lansdowne

2 Study Area

Study area chosen for this case study is St. Mary's Church located. Which is a cultural heritage site located in Lansdowne, Uttarakhand, India. The church is a masonry structure made on a hilly terrain. The church was built in 1896 by Lieutenant A. H. B. Hume of Royal Engineers. Pre-independence the church was in regular use but post-independence health started to deteriorate gradually. Its health is the main concern to select this site as a study area. Church has a sufficiently large area for carrying out the data acquisition process completely from all the sides (Fig. 1).

3 Data Acquisition

Data acquisition was carried out with the help of Terrestrial Laser Scanner (TLS) Riegl VZ 400. A total of 11 scans were carried out around the study area from different scan positions. Height of the instrument is measured ground to the bottom

Table 1 Different scan positions with their scan parameters

	Height of instrument (cm)	Horizontal coverage	Horizontal resolution	Vertical coverage	Vertical resolution
Scan position 1	173.2	80	0.05	100	0.05
Scan position 2	181.0	100	0.05	100	0.05
Scan position 3	180.9	120	0.05	100	0.05
Scan position 4	178.0	120	0.05	100	0.05
Scan position 5	200.4	100	0.05	100	0.05
Scan position 6	200.9	75	0.05	100	0.05
Scan position 7	173.3	90	0.05	100	0.05
Scan position 8	165.2	120	0.05	100	0.05
Scan position 9	175.3	140	0.05	100	0.05
Scan position 10	185.0	80	0.04	100	0.04
Scan position 11	179.5	100	0.04	100	0.04

of the laser scanner. Horizontal coverage is the range of horizontal angle that has to be covered in the scan (between 0° and 360°). Vertical coverage is the range of vertical angle that has to be covered in the scan (between 30° and 130°). Horizontal and Vertical resolution is the angle of difference with which the next scan line takes place (Table 1).

4 Methodology

Initially, proper field planning has to be carried out to estimate the number of scans that have to be acquired in the study area. So that the whole target is visible from all sides. The scans were acquired and co-registered and a dense point cloud was generated in RiScan PRO software. Noise was removed from the point cloud using Statistical Outlier Removal (SOR) filter. With the help of the point cloud, different sections were extracted and drawings were generated in AutoCAD (Fig. 2).

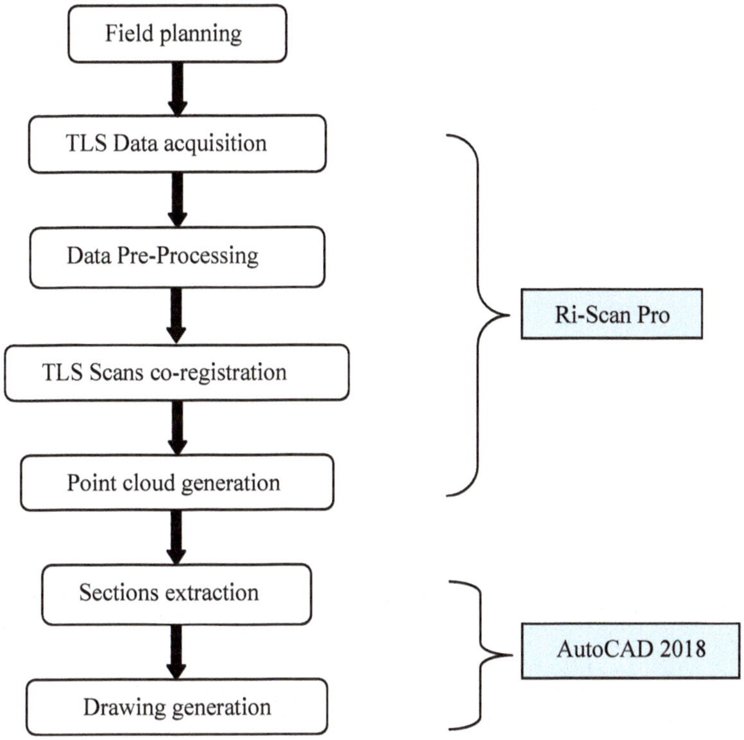

Fig. 2 Methodological workflow adopted for the study

5 Results and Discussions

To draw different views like floor plan, front, back, left, and right point cloud was exported into AutoCAD software and different views were drawn. Precise thickness of the wall and all the dimensions can be obtained with the help of Point cloud through dense laser scanner (Figs. 3, 4, 5, 6, 7, 8, and 9).

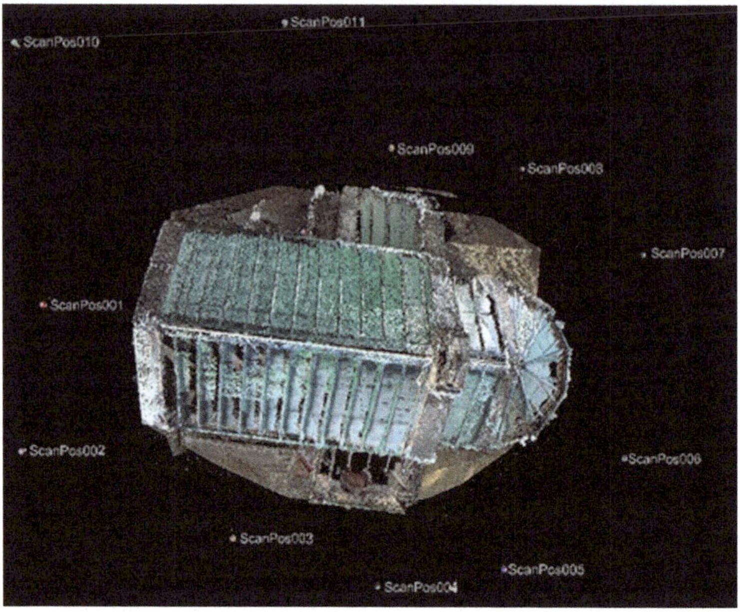

Fig. 3 Generated point cloud with scan positions

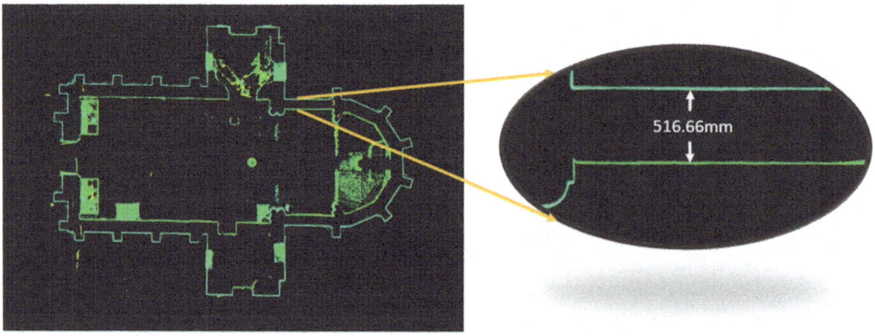

Fig. 4 Precise dimensions from the dense point cloud of the floor

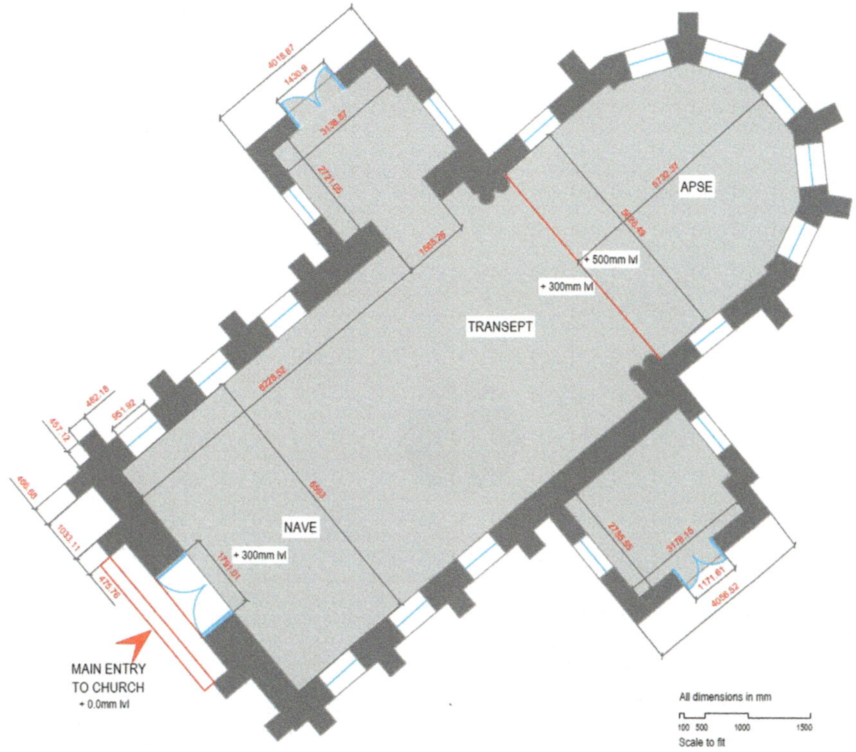

Fig. 5 Floor plan drawing

6 Conclusions

All the drawings that were extracted were of millimeter accuracy. These drawings will have a very great impact in reconstruction of the structure if it is damaged due to any reason. With the help of these documentations, complete geometry of the cultural heritage site can be stored and regenerated. One can never predict when the cultural heritage sites can get damaged due to any natural calamity or man-made disasters. So, 3D documentation of the heritage sites is very helpful.

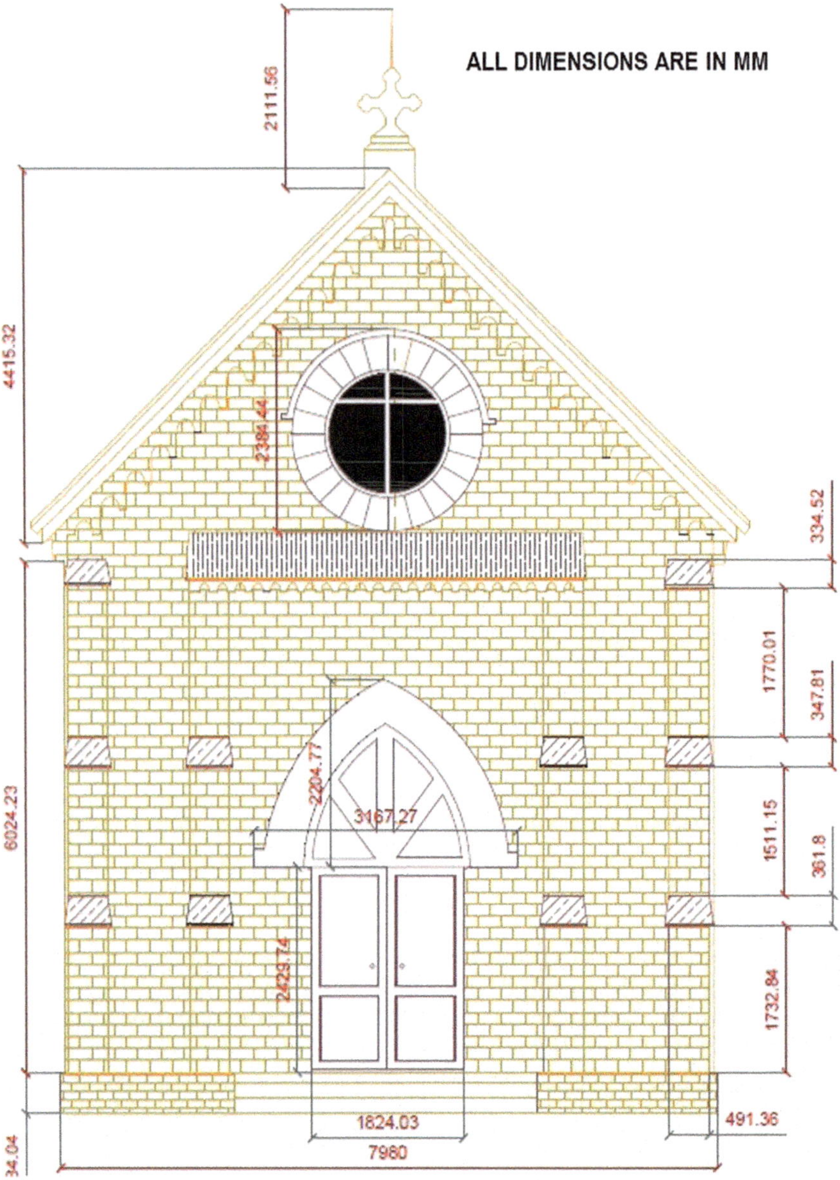

Fig. 6 Front elevation drawing

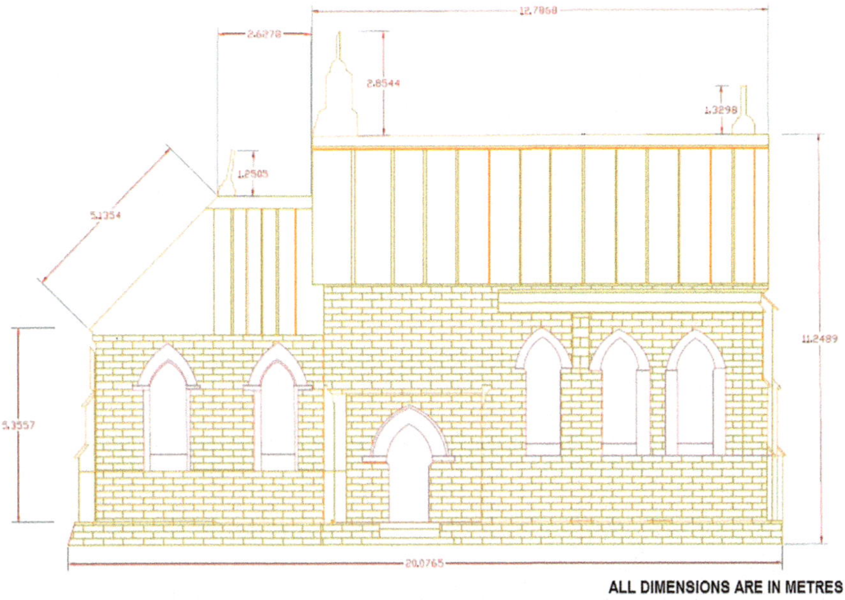

Fig. 7 Left elevation drawing

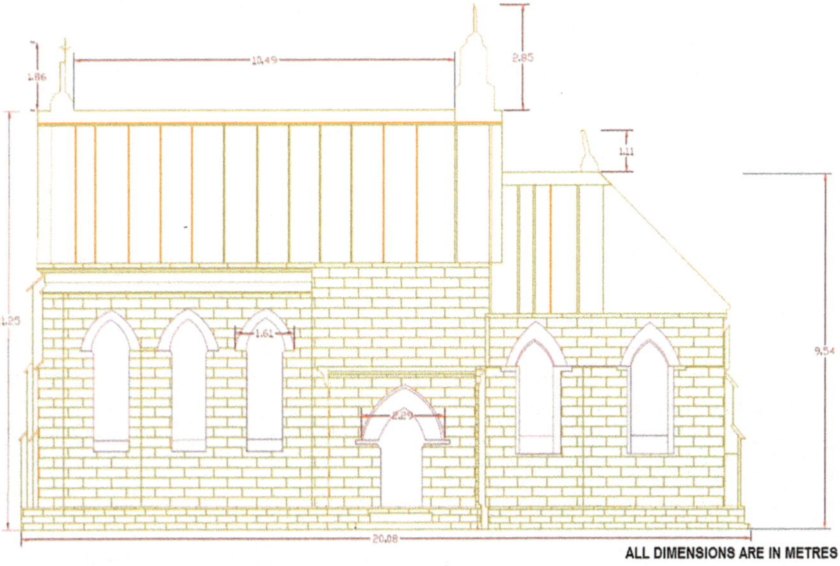

Fig. 8 Right elevation drawing

Fig. 9 Back elevation drawing

References

1. Remondino F (2011) Heritage recording and 3D modeling with photogrammetry and 3D scanning. Remote Sens 3(6):1104–1138. https://doi.org/10.3390/rs3061104
2. Burton G (2007) Terrestrial laser scanner, no 4, pp 45–48
3. Briese C, Pfeifer N, Haring A (2003) Laserscanning and photogrammetry for the modelling of the statue Marc Anton
4. Girardeau-Montaut D, Roux M, Marc R, Thibault G (2005) Change detection on points cloud data acquired with a ground laser scanner. Int Arch Photogramm Remote Sens Spat Inf Sci 36(3):W19. https://doi.org/10.1.1.221.8313
5. Guldur B, Hajjar J (2014) Laser-based structural sensing and surface damage detection
6. Lerma JL, Navarro S, Cabrelles M, Villaverde V (2010) Terrestrial laser scanning and close-range photogrammetry for 3D archaeological documentation: the Upper Palaeolithic Cave of Parpall? as a case study. J Archaeol Sci 37(3):499–507. https://doi.org/10.1016/j.jas.2009.10.011

Thermal Remote Sensing in Early Age Concrete Strength Estimation

Kumar Kumarapu, M. Shashi and K. Venkata Reddy

Abstract The construction industry is the most prominent sector which needs continuous evaluation and monitoring for structural stability and reliability. Monitoring the concrete at early ages can reduce structural failures which may result in fatal accidents. Maturity method is one such NDT method, particularly used for predicting the early age strength by heat generated from concrete. The temperature generated from the heat of hydration is considered as a key parameter in evaluating the maturity method. Conventionally, the maturity method is evaluated by installing temperature sensors inside the concrete and plotting the temperature graphs. NDT, in conjunction with remote sensing thermal imaging sensors, is a motivating alternative for strength estimation in early stages of construction. This study is aimed to replace the thermocouples with well-calibrated thermal infrared imaging sensor. Two different mix proportions, i.e., M20 and M40 are adopted for conducting the study. This study is conducted under a controlled environment without interacting with the external climatic temperature by placing the concrete cubes in thermocol box. Thermal images are obtained in specific time intervals like 15, 30, 60, and 90 min. Concrete cubes are tested for compressive strength simultaneously at the age of 3, 5, and 7 days developing calibration curve. Thermal images of concrete specimens are processed in FLIR SMART VIEW software for recording the surface temperature variations. Time-temperature graphs are plotted from the observed surface temperatures for calculating the maturity indices of concrete. From the developed graphs it is observed that drastic change in surface temperatures has occurred only in the first 24 h of the casting. Nurse-Saul calibration curve is generated by the observed temperatures and compressive strengths of concrete specimens. This calibration curve can be used for concrete specimens under controlled climatic conditions. The hybridization of thermography, photogrammetric, and computer vision techniques like image analysis serves in interpreting the early age strength gain of concrete.

K. Kumarapu · M. Shashi (✉) · K. Venkata Reddy
Department of Civil Engineering, NIT Warangal, Warangal 506004, India
e-mail: mshashi@nitw.ac.in

K. Kumarapu
e-mail: kkumarapu@student.nitw.ac.in

K. Venkata Reddy
e-mail: kvreddy229@gmail.com

J. K. Ghosh and I. da Silva (eds.), *Applications of Geomatics in Civil Engineering*,
Lecture Notes in Civil Engineering 33, https://doi.org/10.1007/978-981-13-7067-0_4

Keywords NDT · Heat of hydration · Thermography · Photogrammetry · Maturity index

1 Introduction

Infrastructure sector is a key driver for the Indian economy. The sector is highly responsible for propelling India's overall development. Construction industry plays a major role in infrastructure development. Technological advancements toward monitoring the construction industry have a huge potential for this sector. Monitoring of construction sites through the automated process than the manual process is in lagging. This can be achieved by advanced tools, machinery, and computer vision. Cost-effective real-time monitoring tools are required for construction site monitoring. Many technological advances in infra industries provide great benefits to one of the most important aspects of monitoring construction operations [1]. All over the world, there are at least 60,000 fatal accidents every year on construction sites due to underestimating early strength which leads to structural failure [2]. Monitoring the concrete at early ages can reduce structural failures which may result in fatal accidents. Each and every aspect of construction and assembly can be inspected for quality control of materials and structure and can be monitored and managed. The mechanical properties of concrete in the initial stages should be assessed for perceiving the structural durability. The estimation of in situ compressive strength by crushing of cubes is time-consuming and debris generating. It is also widely accepted that the concrete cube samples do not exactly replicate the exact strength of in situ casted concrete. The problem is that strength estimation is done by crushing only three specimens. This may not be necessarily accurate. There is a necessity of sophisticated technologies, which can replace the cube crushing for better understanding the curing requirements and behavior of concrete in the early stages.

1.1 Maturity Method

An advanced practical approach is required to deal with more challenges like the ease in handling, limiting manpower, reduction in debris and money for examination of early age concrete strength. The suitable method to address this issue and predict the strength gain at an early age is maturity method, with the concept that the concrete strength is directly proportional to the heat of hydration. The heat emission from the concrete is recorded and plotted for predicting the characteristics strength of concrete. There is ample amount of literature available for concrete maturity method. Concrete maturity method (CMM) is a practical and alternative method to test the cube strength. It is developed to estimate the mechanical properties of concrete by relating the historical temperatures during the hydration process, in which the maturity index is generated by the function of time and temperature [3].

Thermal emission in the hydration process is called as "apparent active energy" that is necessary to speed up the reaction in strength gain of cement paste. CMM is evaluated right from the initial time of water contacting with the cement, this historical data is very important for calibrating the whole setup [4]. Thermal sensitivity is the fundamental and crucial parameter in acknowledging the strength achieved by the concrete [5]. This study is aimed to replace the temperature sensors with well-calibrated thermal infrared imaging sensor. The concrete maturity method is considered as an appropriate and reliable method to estimate the strength gain of concrete at an early age [6] (ACI).

1.2 Thermal Remote Sensing

Due to the enhancements in imaging sensors like Near Infrared (NIR) and thermal bands of electromagnetic spectrum led to many new applications. The imaging sensor is well calibrated for sensor distortions, interior orientations, and exterior orientation for obtaining distortion-free images which can be used after minimal preprocessing. These well-calibrated sensors are very sensitive to take thermal variations of 0.1 °C. The image acquisition by thermal imaging sensors for obtaining the temperature of the concrete sample by processing and analyzing the image is called thermography. The spectral variations recorded by Baldridge [7] states that for concrete the emissivity is more than 0.9 which can be recorded at Middle Wave Infrared (MWIR) and Large Wave Infrared (LWIR) ranges with wavelengths of 3–5 μm and 8–12 μm, respectively. The thermal emissivity of concrete can also be recorded by well-calibrated thermal imaging sensors. The potential applications for thermography in the non-destructive testing focus on the tasks structural condition detection, damage analysis, and condition monitoring. Currently, thermography principles are used in various applications like detecting subsurface delamination in bridge decks [8] (ASTM 1997), moisture ingress detection in a masonry wall and thermal performance and energy efficiency of structures [9]. As describes in literature, the use of non-contact sensors has shown better results in various applications. The main advantage of thermography is observations are made without in contact, quicker, damage-free while dislocating, easy to operate than the conventional thermal probes. These principles are also used in the strength estimation at the early age of concrete, for improving the performance of structural elements. Thermography principles are used to account heat of hydration, which is produced by the chemical reaction of calcium aluminate (C_3A) with water in concrete [10]. The heat of hydration is directly proportional to the strength of concrete, which is released enormously in the initial period of 7 days.

2 Methodology

Thermography for strength assessment can be achieved by calibrating the setup. The set up can be calibrated by laboratory investigations on fresh concrete and the variation in temperature emissions at controlled climatic conditions. The investigations are carried based on the concrete maturity method code [11] (ASTM C1074-11). Figure 1 demonstrates a detailed methodology flow chart followed for conducting the study.

2.1 *Cube Casting for Calibrating the Setup*

To perform the study two different mix designs like M20 and M40 are selected. These mix proportions of concrete are selected for grasping temperature variations in the 15 * 15 cms cubes at different periods. The number of specimens is calculated in such a way that the variation in temperature and strength can be plotted accurately. The concrete cubes are kept in a thermocol box to create a controlled climatic condition and not interacting with the external climate. As per standards, three test cubes of each mix design are kept under a controlled environment for observing changes in temperature. These concrete cubes are kept in a water bath within the box for undergoing curing. Every sample is immersed in water in such a way that top 1 cm of concrete cube is left for thermal imaging. Under each mix designs, three numbers of cubes are stored for obtaining time series temperature variation data and another 12 cubes are kept for evaluating compressive strength for 1, 3, 5, and 7. As per the given conditions, 15 cubes for each mix, design is casted, and maturity indices are evaluated.

2.2 *Thermography*

Thermography is mainly conducted by employing Infrared Thermal camera. Infrared thermal imaging sensor is an active thermal device, which can capture the radiated

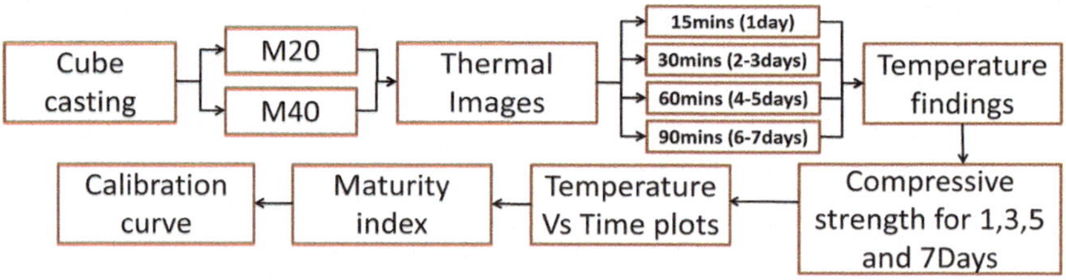

Fig. 1 Methodology flow chart for strength estimation of concrete

Table 1 No of images obtained for particular time intervals

Sl no	Grade of concrete inspected	Time interval (min)	Time span (h)	No of images obtained
1	M20 & M40	15	0–24	590
2	M20 & M40	30	25–72	570
3	M20 & M40	60	73–120	318
4	M20 & M40	90	121–168	186

thermal energy from body. The active thermal sensor gives the true surface temperature of the member. As hydration process is more active in the initial period, more priority is given during the first week of casting. To conduct this study FLUKE thermal imager Ti450 is used for producing the thermal images. The time intervals are considered according to the temperatures variations inside the concrete. For the first 24 h span, the time interval is adopted as 15 min due to the occurrence of major temperature variations in that particular span. Later on, time interval has been increased to 30 min for 25–72 h (3 days), 60 min for 73–120 h (5 days), and 90 min for 121–170 h (7 days). With the given time intervals and spans 1664 images are obtained, out of these 832 images belong to M20 and 832 images to M40.

With the help of a thermal imager, surface temperature history of the concrete cube is collected in particular time intervals as given in Table 1. As per the given time interval and time spans, three cubes per each grade is imaged. Time intervals are selected in such a way that every temperature variation is recorded. 1664 thermal images are acquired for this study. This investigation helps in finding the temperature gradient and plotting the time versus temperature graphs. During this process, compressive strength has been evaluated at 1, 3, 5, and 7 days for both the grades of concrete. The images acquired are initially processed in thermal image processing software. Thermal image processing is done in FLUKE SMART VIEW 4.0 software for extracting temperature and temperature gradients for understanding energy dissipations of the cube.

2.3 Strength Estimation

Strength estimation is a crucial part of the study. This can be calculated by the temperature investigations conducted on concrete samples. According to [11] (ASTM C1074-11), there are well-established methods for calculating the maturity index from temperature investigations. Among them is the *temperature-time factor* method which is also called as Nurse-Saul maturity function is the simplest and prominently

$$M(t) = \Sigma(T_\alpha - T_0)\Delta t \tag{1}$$

where $M(t)$= maturity index, (°F-days), T_{α}= average concrete surface temperature in (°F) during the time interval Δt, T_0 = datum temperature (usually taken to be 14 °F) and Δt = time interval (hours or days).

This equation is simple to calculate. The assumption made is that the initial strength gain varies linearly with temperature, but independent of changes in temperature. It is applicable accurately when curing temperatures does not vary much with curing conditions. Generally, the datum temperature (T_0) is the temperature. Where, hydration process cannot occur in concrete, it is assumed to be 10°C (14 °F).

3 Observations

3.1 Thermal Image with Histogram

With the help of thermal imager, 1664 images have been acquired within 7 days of concrete cube casting. These images are processed in FLUKE SMART VIEW 4.0 software for obtaining the surface temperatures. Figures 2, 3 and 4 are given as reference images obtained during the study. As shown in Fig. 2 thermal images will have spectral variation according to the surface temperature of the imaging area. This image is obtained after initial setting time, at this particular time the concrete specimen is at its lowest temperature with 66.2 °F. Histogram is also put on to the show with the corresponding image to visualize the highest number of pixels confined to a particular temperature.

Similarly, Fig. 3 and 4 show the average and high-temperature thermal images having 74.2 and 75.7 °F along with its histograms. During image processing, only the cube area from the thermal image is considered for temperature investigation. These investigations are noted and plotted for accounting the temperature changes on the concrete surface in controlled climatic conditions.

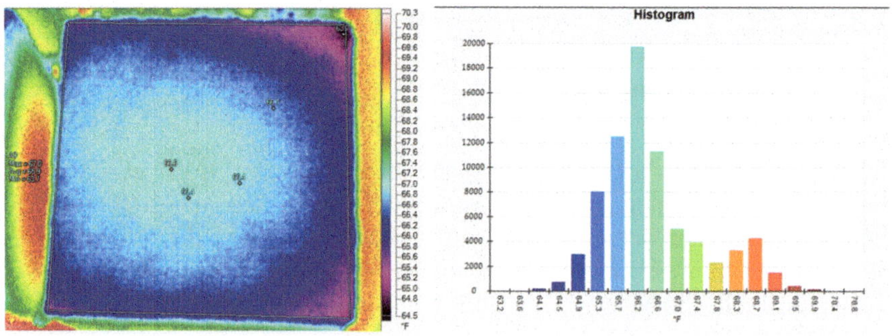

Fig. 2 Low temperature thermal image with histogram

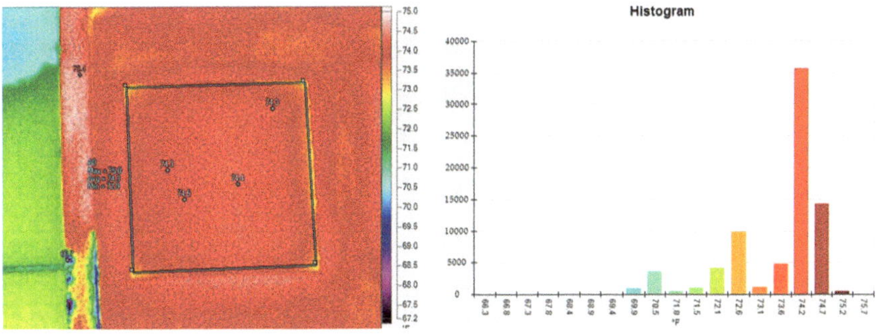

Fig. 3 Average temperature thermal image with histogram

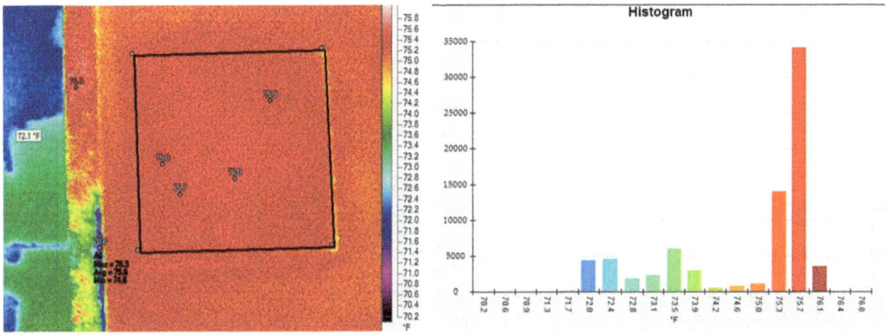

Fig. 4 High temperature thermal image with histogram

3.2 Time Versus Temperature Plots

Thermal image acquisition is carried out from the time of casting to completion of 170 h (7 days). Temperature is obtained from image processing of thermal images.

Tα is calculated from the produced graphs by taking the average of temperatures observed in 24, 25–72, 73–120, and 121–168 h. Tα is a crucial parameter in developing the calibration curve. Based on the plotted graphs of temperature with respective time for M20 grade, it states that in the initial 24 h span there is a drastic change in surface temperature. The steep downfall in Fig. 5 is due to the completion of the initial setting time. The surface temperature at that particular time on the specimen attained a maximum of 75.6 °F in M20 concrete grade. Figure 6 demonstrates the time-temperature graph of M40 grade. It clearly shows that the increase in surface temperatures is slow and it attained the maximum peak of 76.23 °F at 48 h. There is a variation in hydration process for different mix designs.

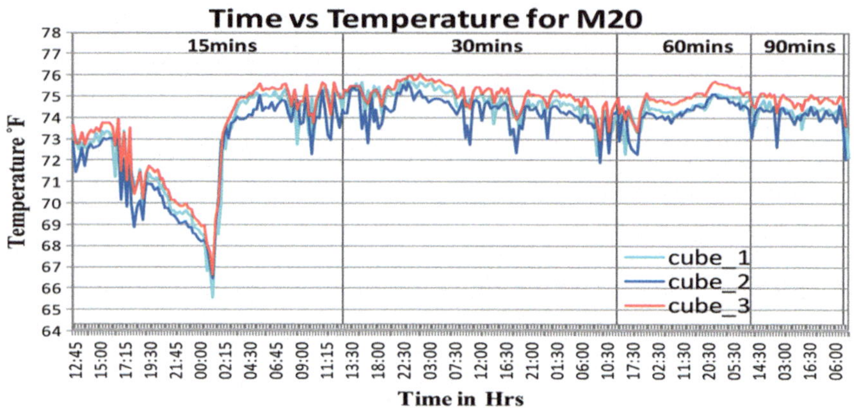

Fig. 5 Time versus temperature graph for M20

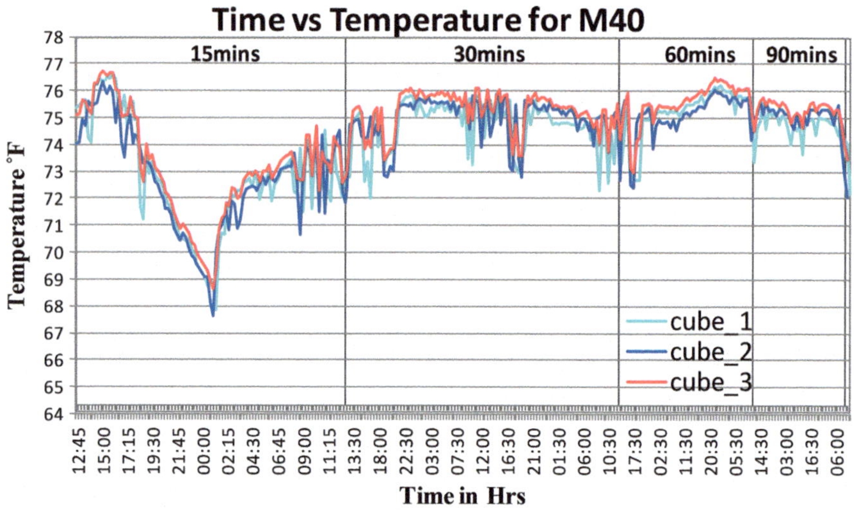

Fig. 6 Time versus temperature graph for M40

3.3 Compressive Strength

The lab cured concrete cubes of M20 is shown in Fig. 7 and M40 is shown in Fig. 8 grades are evaluated for compressive strength. Three concrete specimens are tested for each strength evaluation at 1, 3, 5, and 7 days. Compressive strength in Table 2 is accounted to know the strength gain of concrete at that particular temperature. This helps in developing the maturity calibration curve.

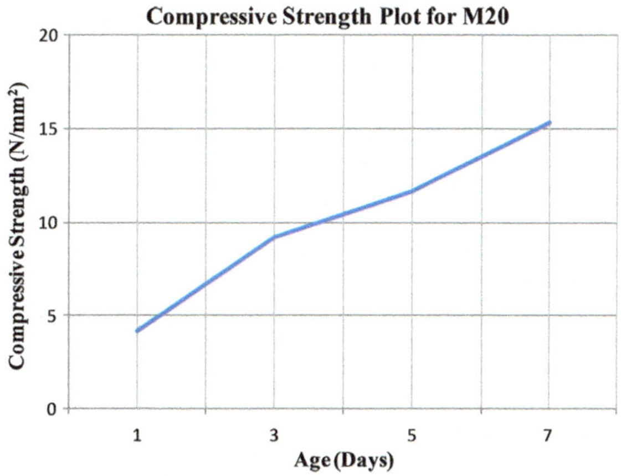

Fig. 7 M20 compressive strength graphs

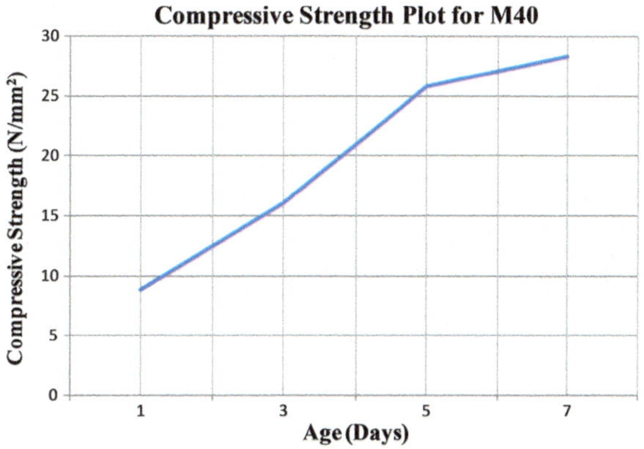

Fig. 8 M40 compressive strength graphs

Table 2 Compressive strength of M20 and M40 concrete specimens

SI no	Concrete age (Days)	Compressive strength (N/mm^2)	
		M20 grade	M40 grade
1	1	4.185	8.756
2	3	9.153	16.015
3	5	11.665	25.78
4	7	15.275	28.22

4 Results and Discussions

4.1 Nurse-Saul Maturity Calibration Curve

The maturity index is calculated from the parameters given in Table 3 using Eq. 1.

The average temperature (Tα) of M20 and M40 is obtained from the temperature investigations at 1, 3, 5, and 7 days, respectively. The maturity index is plotted in the x-axis and compressive strength on the Y-axis of the calibration graph. Figures 9 and 10 is given as corresponding M20 and M40 calibration curves.

4.2 *Validation*

The Nurse-Saul calibration curve generated by calibrating the setup can be used for early age strength estimation of concrete at a controlled environment. The calibration curve is validated by calculating the maturity index from observed temperature for estimating the strength of concrete. The temperature investigation of 80–96 h (4th Day) time span is accounted to calculate the maturity index of M20 concrete grade. At that particular time span, the average temperature (Tα) is observed as 75.404 °F. Maturity index (M_t) for observed Tα at 96 h is calculated as 5894.8 °F-Hrs and the corresponding compressive strength is 10.8 N/mm^2

Table 3 Parameters for computing the maturity index (M(t))

Time in Hrs (Δt)	24	72	120	168
M20 temp in °F (Tα)	72.345	74.515	74.305	74.11
M40 temp in °F (Tα)	72.75	74.87	75.14	74.875
Datum temperature in °F (T_0)	14	14	14	14
M20 maturity indices in °F-Hrs M(t)	1400.28	4357.08	7236.6	10098.48
M40 maturity indices in °F-Hrs M(t)	1410	4382.64	7336.8	10,227

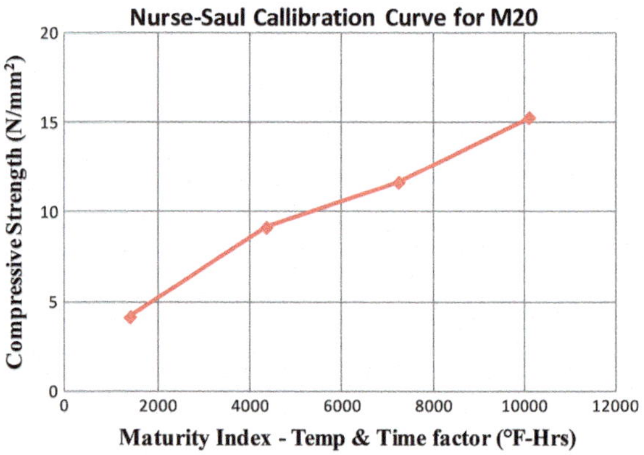

Fig. 9 M20 nurse-saul calibration curve

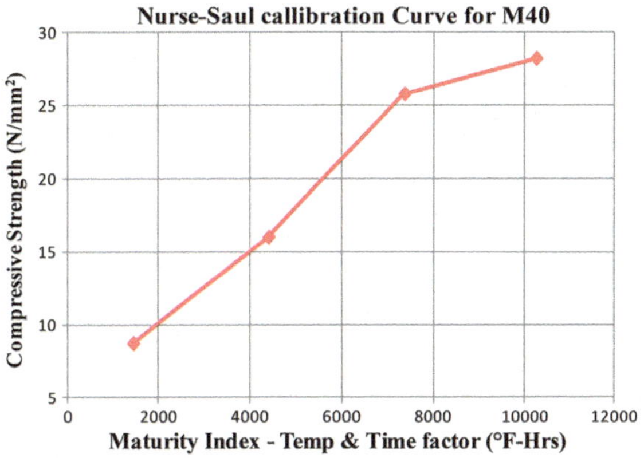

Fig. 10 M40 nurse-saul calibration curve

5 Conclusion

Thermography has shown better results in early age strength estimation of concrete. The obtained and displayed results of CMM by thermography can also be used in early age strength estimation of concrete under a controlled environment. The generated Nurse-Saul calibration curve can be used for nominal mix designs of M20 and M40 specimens, which are cured under controlled climatic conditions. The surface temperatures of concrete cubes started increasing when it is placed for a water bath in thermocol box. It is also proved that surface temperature dissipation from concrete specimens is varying according to the characteristics strength of concrete. When compared to M20, the surface temperature in M40 is gradually increasing with time. With this study, it is also understood that thermography investigations can also be extended for 14 days of strength evaluation as there is no decrease in temperature for 7 days.

References

1. Hampel U (2010) Crack detection during load tests in civil engineering material testing with digital close-range photogrammetry–algorithms and applications. Int Arch Photogr Remote Sens Spat Inform Sci 38(Part 5):268
2. Kadiri ZO, Nden T, Avre GK, Oladipo TO, Edom A, Samuel PO, Ananso GN (2014) Causes and effects of accidents on construction sites (a case study of some selected construction firms in Abuja FCT Nigeria). J Mech Civil Eng 11(5):66–72
3. Carino NJ, Malhotra VM (2003) Handbook on nondestructive testing of concrete, 2nd ed. CRC Press
4. Balaras CA, Argiriou AA (2002) Infrared thermography for building diagnostics. Energy Build 34(2):171–183
5. Peres LDP, Barbosa MP, Pinto RCA (2003) Determinação da energia de ativação para cimentos nacionais aplicando o procedimento ASTM C1074-98. In: 45th Congresso Brasileiro do Concreto, Vitória-ES

6. American Concrete Institute (2003) In-place methods to estimate concrete strength, ACI 228 1R-03
7. Baldridge AM, Hook SJ, Grove CI, Rivera G (2009) The ASTER spectral library version 2.0. Remote Sens Environ 113(4):711–715
8. ASTM International (2013) ASTM D4788-03 (2013) Standard test method for detecting delaminations in bridge decks using infrared thermography. ASTM International, West Conshohocken, PA
9. Maierhofer C, Röllig M, Schlichting J (2010) Active thermography for evaluation of reinforced concrete structures. In: Non-destructive evaluation of reinforced concrete structures, vol 2, pp 370–402
10. Bentz DP, Stutzman PE (2006) Curing, hydration, and microstructure of cement paste. ACI Mater J 103(5):348
11. ASTM International (2011) ASTM C1074-11 standard practice for estimating concrete strength by the maturity method

3D Reconstruction: An Emerging Prospect for Surveying

Shirshendu Layek, Rajat Kumar Singh, Vasanta Govind Kumar Villuri, Radhakanta Koner, Ashish Soni and Rupali Khare

Abstract Three-dimensional (3D) reconstruction has been evolved for modern surveying techniques because it provides a visual interpretation of real-world scene. 3D model can be generated from a cluster of points, which are known as cloud points. Cloud point can be compiled from various sources like Laser scanner, Microsoft Kinnect, digital images, etc. Digital images are more easily accessible technology from others for cloud point creation, and advancements of digital camera in last few years have made the camera sensors, capable of capturing in-depth details, high-resolution digital images. This study is mainly focused on computer vision's 3D reconstruction out of commodity hardware for surveying purpose with the help of camera sensor (Sony IMX298). For validation, the ground truth has been carried out with advance surveying instrument by distributing several points around Region of Interest (ROI) and evaluate the dimensions.

Keywords Surveying · 3D reconstruction · Computer vision · Photogrammetry

S. Layek (✉) · R. K. Singh · V. G. K. Villuri · R. Koner · A. Soni · R. Khare
Department of Mining Engineering, Indian Institute of Technology (Indian School of Mines) Dhanbad, Dhanbad, India
e-mail: shirso.it7@gmail.com

R. K. Singh
e-mail: rajat.singh714@gmail.com

V. G. K. Villuri
e-mail: govindvilluri@gmail.com

R. Koner
e-mail: rkoner@iitism.ac.in

A. Soni
e-mail: ashishsoni@iitism.ac.in

R. Khare
e-mail: rupali.16dr000204@iitism.ac.in

J. K. Ghosh and I. da Silva (eds.), *Applications of Geomatics in Civil Engineering*, Lecture Notes in Civil Engineering 33, https://doi.org/10.1007/978-981-13-7067-0_5

1 Introduction

If we look around us, all we will be seeing the applications of civil engineering, like roads, houses, water damp, buildings, etc. These applications are one of the most important developments in mankind history. It has raised living life quality and continues to improve. Surveying is one of the core application of civil engineering by which a place or area is observed, measured, and analyzed. This surveying has evolved throughout the time with its techniques and development of new instruments. Surveying instruments play a vital role in terms of productivity and accuracy of surveying. Surveying is mostly done to take the measurement of linear distances and angles between established stations or positions and now remote sensing plays a vital role in mapping for civil surveying purposes [1–4].

Measurement of linear distance was initially done by surveyors with metric chains, which was 20–30 m long and has the least count of 20 cm. Later metallic tapes replaced metric chains which lighter weights, compact build and least count of 1 mm. In case of angular measurement, magnetic compass was discovered more than two years, which provides magnetic north and that includes local attraction error and the least count is 30 s [5]. This issue has been resolved with fine instruments like theodolite and total stations with the least count of 1 s. In the course of the most recent 50 years, the overview business has experienced massive changes because of innovation progressions [6]. Including the presentation of electronic distance meters (EDMs), add up to stations, global navigation satellite system (GNSS), and robotic total stations [6]. Digitization is one of the core inventions in modern science and technology. Capturing and processing as well as rendering a real-world object or scene have been a blessing for mankind [7]. Recent progress in electronics, especially in the section of compact digital cameras and their implementation in smartphones, causing it more accessible with a cheaper price [8]. With the modern computer hardware and software, computer vision is used for processing images, this combination of capturing and processing images has been used in many fields, and here are a few examples [9]

- Medical science
- Defense and Intelligence
- Virtualization
- Digital Mapping
- Social networking and amusements
- Games
- Visual Effects in the film industry.

Today, point-cloud creation advances are testing those conventional mine surveying working pattern, with surveyors now taking a look at arrangements able to do rapidly delivering precise point cloud information of mine is in-progressing stage. Technology, for example, unmanned aerial vehicles (UAV) consolidated with photogrammetric forms are currently used to make point-cloud information rapidly, successfully, and securely. In numerous locales, laser scanning work processes have been ease back to establish a connection inside the mining business. Be that as it may, the previous couple of years have seen an altogether different pattern with different laser scanner like terrestrial (TLS), airborne (ALS) and mobile (MLS) frameworks ending up noticeably ever exhibit inside the business [10–12].

3D reconstruction has been studying and getting improved with each decade. Since then there are numerous algorithms that have been implemented and are getting improved with time, some of the popular algorithms are Triangulation, Sampling 3D space, Consistency Function, Snakes, Graph Cut, Silhouettes, Multi-scale Approaches and many [9]. 3D reconstruction of a scene involves several steps

- Image Acquisition
- Camera Calibration
- Feature Matching
- Reconstruction of the scene.

It is intriguing to watch the development of modern technologies and its applications, past two decades have been most remarkable [8]. The world first phone with a digital camera lens was launched in the year 2000 from a Japanese Company named Sharp [13] and since then this technology has evolved in every possible direction. Smartphones can be found very easily. Roughly 1 out of 3 people in the world possess a phone, this is estimated by KPCB [14], which showed 2.5 billion smartphones are active and will increase to 6.1 billion within year 2020. This study aims to deploy the advantages of 3D reconstruction with smartphone cameras into the field of civil surveying.

2 Related Work

3D digital copy of a scene or object has been done by many technologies like laser (ground), Lidar (aerial), structured light, photogrammetry, etc. These instruments have their advantages and disadvantages. Lidar and Laser are very precise and verified up to millimeters of accuracy but they are expensive. Photogrammetry is a technology which is getting more accessible with the smartphones and available commodity hardware [15, 16].

Snavely et al. [17] show how the world can be modeled in 3D using Internet photographs, it showed that images are used from Internet of several sites and building models of those sites with the help of 3D scene modeling and visualization. Later on, with this thought, a new study showed that the city Rome can be modeled in 3D in a day with the help of cluster computing and internet photographs hosted by various users from an image hosting website called Flickr [15, 18].

3D digital copy of a scene or object has been done by many technologies like a laser (ground), Lidar (aerial), structured light, photogrammetry, etc. These instruments have their advantages and disadvantages. Lidar and Laser are very precise and verified up to millimeters of accuracy but they are expensive. Photogrammetry is a technology which is getting more accessible with the smartphones and available commodity hardware [10–15].

3 Methodology

A 3D model is made of the heritage building of IIT (ISM). The acquired photos are taken from Sony IMX 298 sensor. This sensor is commonly available in many smartphones like Oneplus 3, Oneplus 3T, Xiaomi Mi5, Huweii Mate 8, Asus Zenfone 3, etc. The specification of the sensor is given in the following table [19]

- CMOS image sensor
- Image size: Diagonal 6. 521 mm (Type 1/2.8)
- Total number of pixels: 4720 (H) × 3600 (V) approx. 16.99 M pixels
- Number of effective pixels: 4672 (H) × 3520 (V) approx. 16.44 M pixels
- Number of active pixels: 4656 (H) × 3496 (V) approx. 16.28 M pixels
- Chip size: 6.433 mm (H) × 4.921 mm (V)
- Unit cell size: 1.12 μm (H) × 1.12 μm (V)
- Phase Detection Auto Focus (PDAF)
- Single Frame High Dynamic Range (HDR) with equivalent full pixels
- High signal to noise ratio (SNR)
- Full resolution @30 frame/s (Normal/HDR). 4K2 K @30 frame/s (Normal/HDR) 1080p @60 frame/s (Normal/HDR)
- Output video format of RAW10/8, COMP8
- Pixel binning readout and H/V sub-sampling function
- Advanced Noise Reduction (Chroma noise reduction and RAW noise reduction).

For unbiased experiment photographs are taken auto mode, on every 2–3 m gap six photos are taken covering front left side and right side at 45° angle with upside and downside by of the first floor and the second floor of the heritage building by focusing the center of the viewfinder to the center of the floors, respectively (Fig. 1).

3.1 Image Acquisition

Image acquired in the IIT Dhanbad campus of the front surface of the Heritage building the Mining department, i.e., our Interest Area.

3.2 Importing the Images

The first input for the project was importing the images to the system for further procedures.

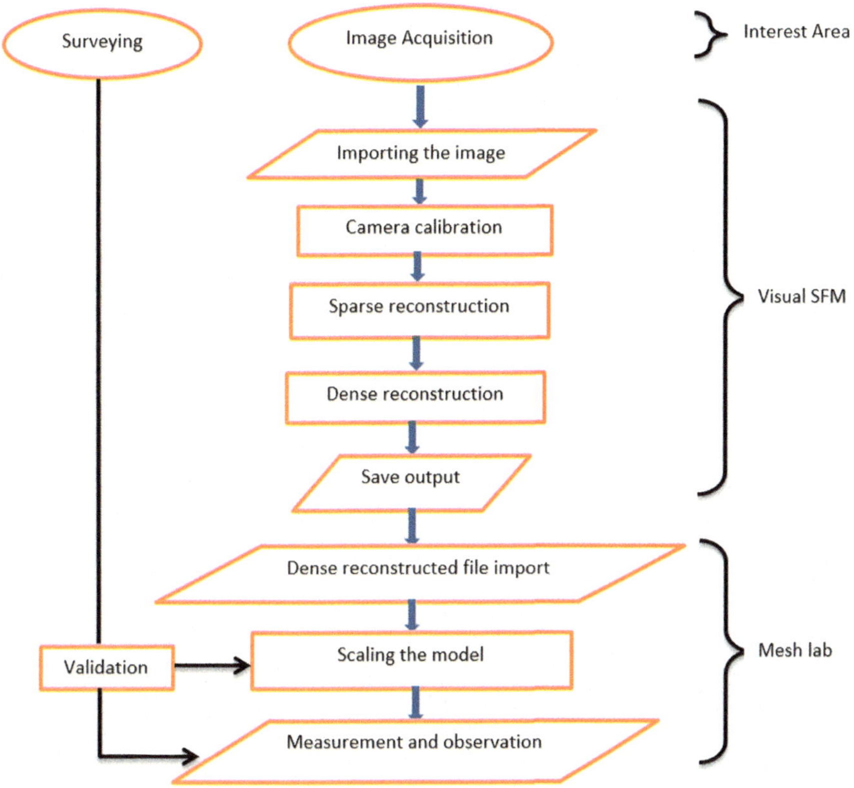

Fig. 1 Methodology of 3D reconstruction model

3.3 Camera Calibration

Calculation of the focal point and image sensor parameters leads an image to improve accuracy for lens distortion, measure the size of an object, or determine the position of the camera in the scene.

3.4 Sparse Reconstruction

The construction of 3D model of the interested area by feature matching with the help of computer vision.

3.5 Dense Reconstruction

It is the process of capturing the shape and appearance of real objects.

3.6 Save Output

Save the output generated for further procedures.

3.7 Dense Reconstructed File Import

The output generated by the visual SFM will be the input to the Meshlab so we import the file.

3.8 Surface Reconstruction

Formulation considers all the points, poisson approach permits an order of privately upheld premise capacities and in this manner, the arrangement lessens to an all-around adapted meager direct framework. This portrays a spatially versatile multi-scale algorithm whose time and space complexities are relative to the span of the recreated model. This helps to remake of surfaces with more noteworthy detail than beforehand achievable.

3.9 Scaling the Model

The models are scaled so that it can accurately represent the original model. This is done by surveying with Total Station.

3.10 Measurement and Observation

The final step to confirm the credibility of the model.

3.11 *Surveying*

The study of the structure through 3D modeling of the structure starting from acquiring images via sparse, dense, surface reconstruction to scaling, and measuring the model.

3.12 *Validation*

To verify the validity of all the data acquired through the 3D reconstruction process.

4 Experimental Results

The construction of 3D model of the interested area by feature matching with the help of computer vision (Figs. 2 and 3).

In the procedure for feature extraction and matching for 1 ton and following 2 ton and so on, this initial step takes most of the time for 3D reconstruction of a real scene. More the number of images, better the result is. For some obstruction, we were able to take 322 usable photos to form a 3D model of the heritage building of Indian Institute of Technology (ISM) (Figs. 4 and 5).

For scaling, we used total station data by surveying the ROI. It has been observed that the leaner distance calculated by Total Station and the 3D reconstruction is minimal. For better observation and visual interpretation, this technology is very much helpful instead of reconnaissance survey. And also slope stability is a factor, which can benefit from this technology (Fig. 6).

The analyses of the technology and the gadgets we used with respect to the result we get during the period of this project are as follows:

Fig. 2 Sparse reconstruction made from the imagery of Oneplus 3T

Fig. 3 Sparse reconstruction made from the imagery of Nokia Lumia 1020

Fig. 4 Dense reconstruction made from the imagery of by Oneplus 3T

- Total number of photos taken as input: 328
- Total size of the photos: 1.65 GB
- Resolution of the photos: 4640 * 3480
- Loading time of photos: 136 s
- Feature matching time: 95.117 min
- 3D reconstruction (sparse): 115 s
- 3D reconstruction (dense): 90.417 min
- Total points matched: 53,956 point.

This is not a perfect model made from digital images, however, in terms of virtualization, this technology can be very useful for surveying purpose for both interpretation and measurements as the ROI has been shown with dimension (Fig. 7).

Fig. 5 Dense reconstruction made from the imagery of by Nokia Lumia 1020

Fig. 6 Measurement is done on the 3D model created from Nokia Lumia 1020

Fig. 7 Measurement is done on the 3D model created from Oneplus 3T

5 Conclusions and Future Scope

As per the outcome of this study conclude that this technology is far from perfection but the features and advantages of this technology cannot be neglected. With time this technology is getting better and with this prospective it can be said that this can be a future of surveying, with the assets like monitoring, interacting with the 3D model of mines, measurements of required sights, hazard monitoring, slope, and mine stability, etc., many factors can be easily handled with this technology. Present study shows with a waiting time of few hours every day a model of the mining area can be built with a commodity hardware and a smartphone which are very common these days, so basically by using a smartphone and a commodity computer hardware we were able to build 3D model of the heritage building of IIT (ISM) within a few hours. With the generated model, measurement and observation are done with new prospects for surveying purpose.

References

1. Turner T (2004) Landscape planning and environmental impact design. Routledge
2. Sarris Z, Atlas S (2001) Survey of UAV applications in civil markets. In: Proceedings of the 9th mediterranean conference on control and automation
3. Petrie G, Kennie TJM (1987) Terrain modelling in surveying and civil engineering. Comput Aided Des 19(4):171–187
4. Tang X, Cong N (2011) Present situation and development of surveying and mapping satellites in China. Geomatics World 2:013
5. History of the Compass, Wikipedia. https://en.wikipedia.org/wiki/History_of_the_compass

6. Rüeger JM (1980) Recent developments in electronic distance measurement. Aust. Surv. 30(3):170–177
7. Lanham RA (2010) The electronic word: democracy, technology, and the arts. University of Chicago Press
8. Attewell J (2005) Mobile technologies and learning. Learning and Skills Development Agency, London
9. 3D Reconstruction, Wikipedia. https://en.wikipedia.org/wiki/3D_reconstruction
10. Lin Y (2012) Tree height growth measurement with single-scan airborne, static terrestrial and mobile laser scanning. Sensors 12(9):12798–12813
11. Demantke J (2011) Dimensionality based scale selection in 3D lidar point clouds. Int Arch Photogr Remote Sens Spat Inform Sci 38(Part 5):W12
12. Yang B (2015) An automated method to register airborne and terrestrial laser scanning point clouds. ISPRS J Photogr Remote Sens 109:62–76
13. Camera Phone, Wikipedia. https://en.wikipedia.org/wiki/Camera_phone
14. Internet Trends, Wikipedia. http://www.kpcb.com/internet-trends
15. Agarwal S (2011) Building rome in a day. Commun ACM 54(10):105–112
16. Suveg I, Vosselman G (2004) Reconstruction of 3D building models from aerial images and maps. ISPRS J Photogr Remote Sens 58(3–4):202–224
17. Snavely N, Seitz SM, Szeliski R (2006) Photo tourism: exploring photo collections in 3D. ACM Trans Graph (TOG) 25(3)
18. Geiger A, Ziegler J, Stiller C (2011) Stereoscan: dense 3D reconstruction in real-time. In: 2011 IEEE intelligent vehicles symposium (IV)
19. Sony IMX298 Product Brief, Sony. https://www.sony-semicon.co.jp/products_en/IS/sensor1/img/products/ProductBrief_IMX298_20160210.pdf

Monitoring of Concrete Hydration Using Resin Jacketed Embedded PZT Sensors

Rajat Chhabra, Prateek Negi, Naveet Kaur and Suresh Bhalla

Abstract Effective monitoring of concrete hydration makes construction of civil structures efficient and safe by indicating the strength gain of concrete under curing. Due to the complexity in construction practices, the condition monitoring of concrete is still a challenge for engineers. Hence, even after the curing of concrete as per the recommendations by standards codes, expected strength is not always achieved. In this paper, concrete hydration has been monitored using two types of sensors, i.e., concrete vibration sensor (CVS) and resin jacketed piezo sensors (RJP). Three concrete cubes of M35 grade are cast with both types of sensors embedded at equal distances from the center. The admittance signatures are recorded over a period of 28 days. The experiment shows variation of signatures with strength gain of the cubes over time. CVS has shown better results in monitoring the concrete hydration in comparison to RJP sensors.

Keywords Concrete vibration sensor (CVS) · Electromechanical impedance (EMI) technique · Hydration · Lead zirconate titanate (PZT) · Resin jacket piezo (RJP) sensor

R. Chhabra (✉) · P. Negi · S. Bhalla
Civil Engineering Department, Indian Institute of Technology (IIT) Delhi, Hauz Khas, New Delhi 110016, India
e-mail: rajat71093@gmail.com

P. Negi
e-mail: negidynamic@gmail.com

S. Bhalla
e-mail: sbhalla@civil.iitd.ac.in

N. Kaur
CSIR-CRRI, Mathura Road, New Delhi 110025, India
e-mail: naveet.kaur1985@gmail.com

J. K. Ghosh and I. da Silva (eds.), *Applications of Geomatics in Civil Engineering*, Lecture Notes in Civil Engineering 33, https://doi.org/10.1007/978-981-13-7067-0_6

1 Introduction

Effective hydration of concrete is important in achieving the desired strength of a structural member to function effectively. The monitoring of concrete hydration can make it efficient and safe but it is rarely being done due to the complexity involved in the construction practices. Predicting the strength and hydration related cracking pattern could help in reducing the prospective damage. Researchers have been using various classical techniques over the years for hydration monitoring like ultrasonic techniques, thermal analysis, and X-ray diffraction, and scanning electron microscopy. The electro-mechanical impedance (EMI) technique is a relatively new nondestructive evaluation (NDE) technique which uses a lead zirconate titanate (PZT) patch as an actuator and a sensor simultaneously for structural health monitoring (SHM) of civil structures over the past years. But, these have been recently extended to monitor the hydration of the concrete under curing.

The EMI technique uses a piezo sensor excited at ultrasonic vibrations (in 30–400 kHz range) to derive the impedance signature of the structure in terms of admittance signatures. The recorded admittance signatures consist of two parts, i.e., conductance (real part) and the susceptance (imaginary part). Bhalla and Soh (2014) derived the admittance signal for a 2D structure (Eq. 1)

$$\bar{Y} = G + B = 4\omega j \frac{l^2}{h}\left(\overline{\varepsilon_{33}^{T}} - \frac{2d_{31}^2 \overline{Y^E}}{(1-\nu)}\right) + \frac{8\omega d_{31}^2 \overline{Y^E} l^2}{h(1-\nu)}\left(\frac{Z_{a,eff}}{Z_{s,eff} + Z_{a,eff}}\right)\bar{T}_l \quad (1)$$

where G is the conductance, B is the susceptance, j is $\sqrt{-1}$, d_{31} is a piezoelectric strain coefficient, $\overline{Y_E} = Y_E(1 + \eta j)$ is the complex Young's modulus of the PZT patch (at constant electric field), $\overline{\varepsilon_{33}^T} = \varepsilon_{33}^T(1 - \delta j)$ is the complex electric permittivity of the patch (at constant stress field), ν is the Poisson's ratio, l is the length of PZT patch and h is the thickness of PZT patch. η and δ denote the mechanical loss factor and the dielectric loss factor of the patch, respectively. $Z_{a,eff}$ and $Z_{s,eff}$ are the effective mechanical impedance of the PZT patch and host structure, respectively, which includes the effect of mass, stiffness, damping, and boundary conditions.

Any changes in the host structure can be visualized in changes of the admittance signatures predominantly in the real part, i.e., conductance signatures. Shin et al. [1] used surface bonded PZT transducers for long term hydration monitoring but not the initial hydration as they bonded the PZT patch to the hardened concrete surface. Yang et al. [2] developed a reusable PZT patch based setup to monitor the initial hydration of concrete. It had been observed that over a period, the sensitivity of the PZT patches had reduced. Quinn et al. [3] developed an embedded wireless sensing system for monitoring initial curing and health of concrete structures. Kong et al. [4] monitored the initial concrete hydration using piezo-ceramic-based smart aggregates.

Bhalla and Gupta [5] defined CVS as a packaged and ready to use PZT-cement mortar composite sensor specially designed for concrete structures. It is commercially manufactured by Central Electronics Limited (CEL, 2018) [6]. This paper presents the resin jacketed piezo (RJP) sensor that is proposed by Negi [7] in comparison to the concrete vibration sensor (CVS) by Kaur and Bhalla [8] to monitor the initial hydration of the concrete under curing. The EMI technique has been proved to be effective in monitoring the hydration of concrete and hence, can be used in practice.

2 Experimental Work Done

The CVS and RJP sensors were fabricated as shown in Fig. 1. The sensors were embedded in three concrete cubes (Cube-I, Cube-II, Cube-III) of M35 grade at equidistant from the center as shown in Fig. 2.

The cubes with embedded sensors were wrapped inside a plastic sheet to prevent loss of water due to evaporation. And additional water was provided by covering them in the wet clothes to simulate the actual conditions. The cubes were maintained at a temperature of 20 ± 2 °C in the lab conditions.

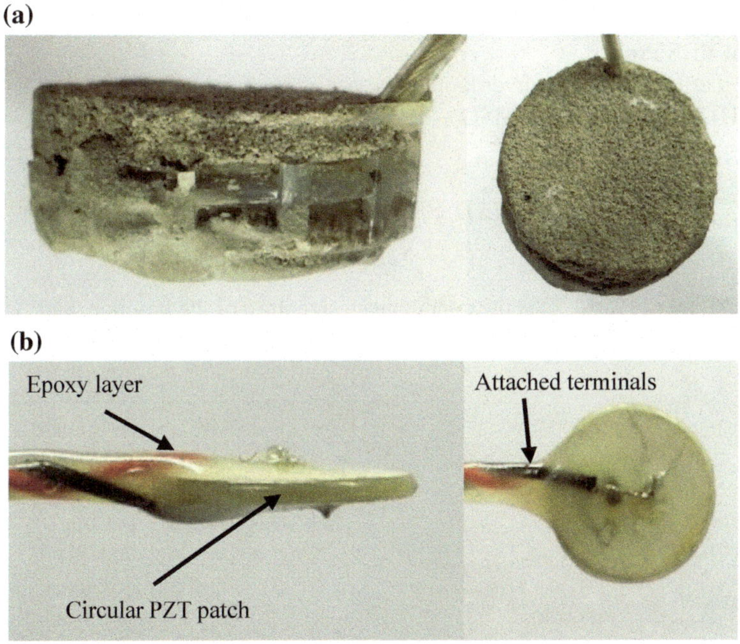

Fig. 1 **a** Concrete vibration sensor (CVS). **b** Resin jacketed piezo (RJP) sensor

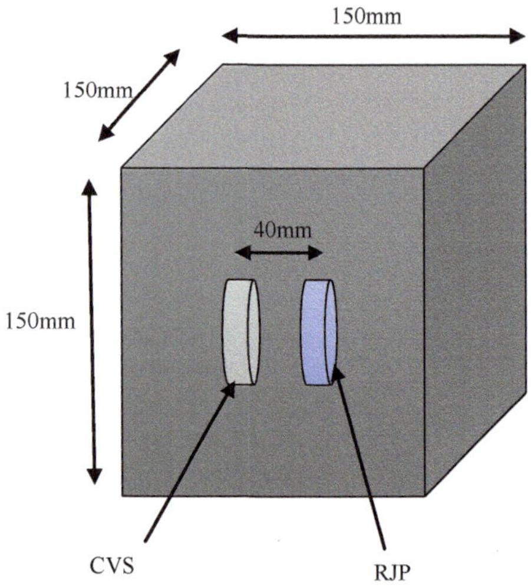

Fig. 2 CVS and RJP embedded in concrete cube

The experimental setup was done as shown in Fig. 3 where the Agilent E4980A 20 Hz–2 MHz precision LCR meter was connected to a laptop and a specimen through a USB cable. The values of conductance (G) and susceptance (B) were obtained by connecting in frequency range 1–1000 kHz with an increment of 1 kHz through the Agilent VEE Pro 9.0.

In order to quantify the hydration of the concrete, the root means square deviation (RMSD) index was used. RMSD is defined as

$$RMSD = \sqrt{\frac{\sum_{i=1}^{n}\left(G_i^1 - G_i^0\right)^2}{\sum_{i=1}^{n}\left(G_i^0\right)^2}} \tag{2}$$

where, G_i^0(i= 1, 2, 3, …) is the conductance signature of the baseline, and G_i^1 is the conductance signature after hydration.

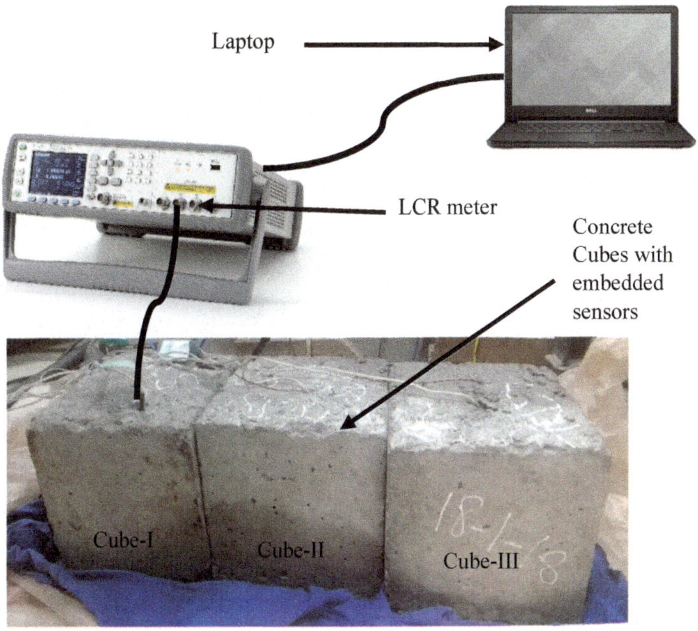

Fig. 3 Experimental setup

3 Experimental Results

The conductance signatures for each of the sensors were plotted for the 3rd, 7th, 14th, 21st, and 28th day. The first peak of the signatures is the significant peak; thus, it was considered for the signature comparison. The considered frequency range for the CVS was 170–240 kHz and for the RJP was 200–280 kHz as shown in Fig. 4 and Fig. 5, respectively.

The conductance signature of the third day of curing was considered as the baseline signature for the calculation of RMSD values of the 7th, 14th, 21st, and 28th day based on the above frequency range.

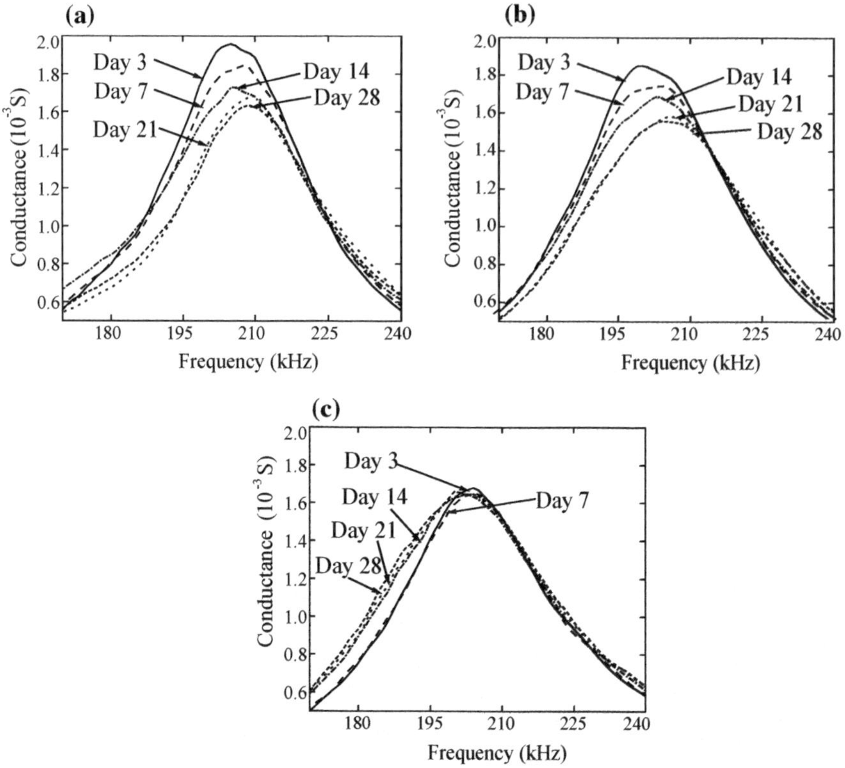

Fig. 4 Comparison of conductance signatures for CVS embedded in **a** Cube-I, **b** Cube-II, **c** Cube-III

From the plots, it was clearly visible that with increase in the curing time, the signature is shifting toward the right, thus indicating an increase in the stiffness of the concrete. Also, it was observed that shifting of the curve is more between the third and seventh day as compared to the later signatures, which indicates that the strength of the concrete increases at the higher rate in the initial period of curing as compared to the later stages. This can also be observed with the values of the RMSD in Table 1.

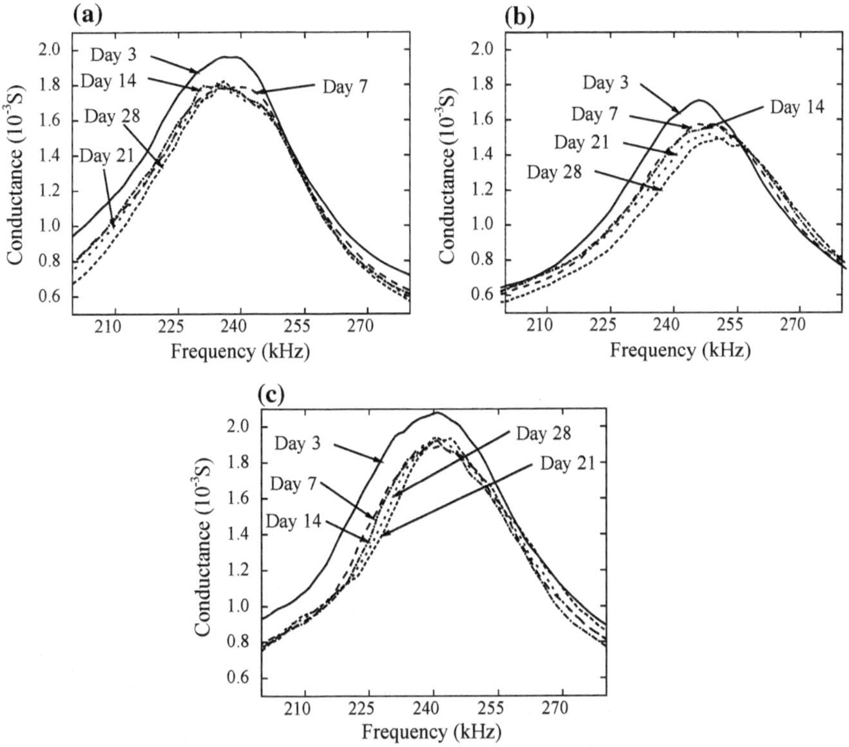

Fig. 5 Comparison of conductance signatures for RJP embedded in **a** Cube-I, **b** Cube-II, **c** Cube-III

Table 1 RMSD variation in signatures of CVS and RJP sensors

No. of days	RMSD (%)					
	Cube-I		Cube-II		Cube-III	
	CVS	RJP	CVS	RJP	CVS	RJP
Day 7	5	8.81	4.09	7.28	1.4	10.9
Day 14	9.6	8.89	7.59	8.09	7.2	12.57
Day 21	16.45	10.48	14.33	10.59	8.5	12.68
Day 28	16.75	12.75	14.36	14.77	10.79	13.5

The RMSD plots in Fig. 6 shows that RJP provides more accurate results as compared to the CVS sensors.

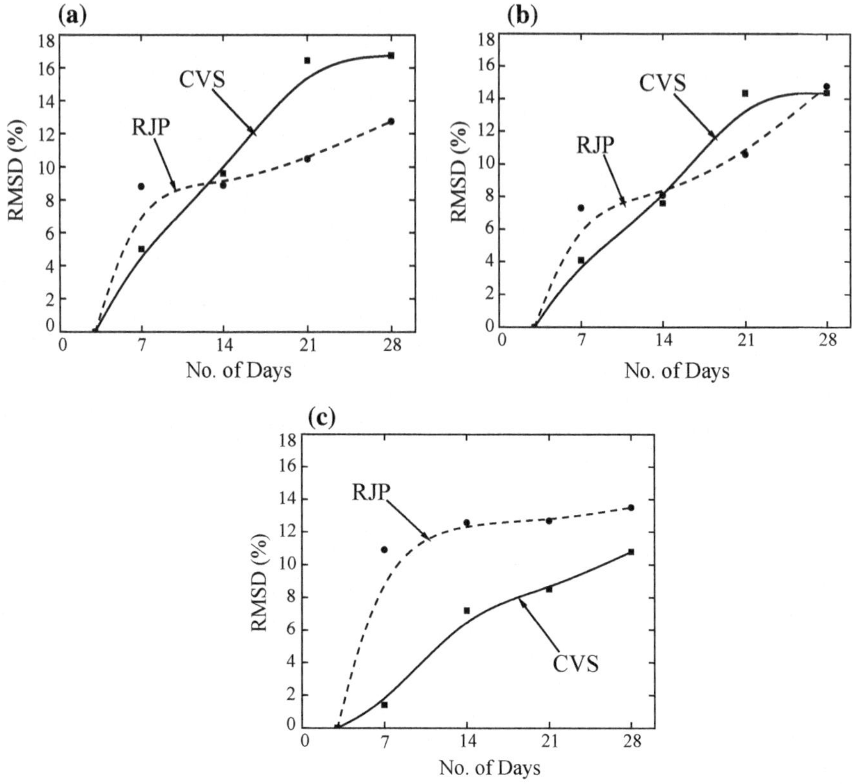

Fig. 6 Comparison of RMSD variation in signatures for **a** Cube-I, **b** Cube-II, **c** Cube-III

4 Conclusions

Based on the present study, following points are concluded:

i. The sensors can be used effectively in monitoring of the concrete hydration. Thus, providing a better sense to the hydration process in achieving the desired strength.
ii. The curve shifts toward the right, thus indicating an increase in the resonant frequency and thereby, indicating increase in the stiffness of the concrete.
iii. The RJP provides accurate results as compared to the CVS sensors.
iv. Due to the low cost of the sensors, these have a bright future in the construction practices.

References

1. Shin SW, Qureshi AR, Lee JY, Yun CB (2008) Piezoelectric sensor base nondestructive active monitoring of strength gain in concrete. Smart Mater Struct 17(5)
2. Yang Y, Divsholi B, Soh CK (2010) A reusable PZT transducer for monitoring initial hydration and structural health of concrete. Sensors 10(5):5193–5208
3. Quinn W, Kelly G, Barrett J (2012) Development of an embedded wireless sensing system for the monitoring of concrete. Struct Health Monitor 11(4):381–392
4. Kong Q, Hou S, Ji Q, Song G (2013) Very early age concrete hydration characterization monitoring using piezoceramic based smart aggregates. Smart Mater Struct 22(8)
5. Bhalla S, Gupta A (2007) A novel vibration sensor for concrete structures. Patent Application No. 1011/DEL/2011, Invention Disclosure No. FT/IPR/CE/SB/2007/0570, Foundation for Innovation and Technology Transfer (FITT), IIT Delhi, India
6. Central electronics limited. http://www.celindia.co.in/drupal7/. Accessed 24 Feb 2018
7. Negi P (2018) Application of piezo sensors in condition monitoring of underground structures. PhD thesis, Department of Civil Engineering, IIT Delhi, New Delhi
8. Kaur N, Bhalla S (2015) Combined energy harvesting and structural health monitoring potential of embedded piezo-concrete vibration sensors. J Energy Eng ASCE 141(4):D4014001 (1–18)

Part III
Geomatics in Geotechnical Engineering

Spatial Variability of Depth to Weathered Rock for Chennai Using Geostatistical Kriging

B. Divya Priya and G. R. Dodagoudar

Abstract The objective of this study is to evaluate the spatial variability of depth to weathered rock in Chennai city using field test data. The database consists of nearly 400 borehole data covering the entire city with stratigraphic information and the same is used in the study. Ordinary kriging technique, which uses the spatial correlation of the data, is applied to estimate the depth at locations where the field measurements are not available. In addition, the variance of the estimated data is also computed. The developed map indicates that the depth to weathered rock in the study area varies from 3 to 30 m from the ground level. The estimated depths to weathered rock are compared and validated with a few selected borehole data and Multichannel Analysis of Surface Wave (MASW) test results. These maps can be used in the ground response analysis, foundation analysis, and design studies.

Keywords Spatial variability · Depth to rock · Borelog · Kriging · MASW

1 Introduction

The knowledge about surficial and subsurface conditions plays a vital role in the analysis and design of geotechnical structures and facilities. Due to the complex nature and diverse distribution of the soils along the subsurface, the assessment of its geotechnical properties is necessary in establishing geotechnical design parameters for design and construction of underground structures. In addition, it also helps in planning and execution of soil exploration programme. Local site conditions such as depth to bedrock, soil layer information, location of water table, and various geologic parameters are the factors influencing the amplification of earthquake ground motions which induce earthquake-related hazards.

B. Divya Priya (✉) · G. R. Dodagoudar
Computational Geomechanics Lab, Department of Civil Engineering, Indian Institute of Technology Madras, Chennai 600036, India
e-mail: divyapriya03@gmail.com

G. R. Dodagoudar
e-mail: goudar@iitm.ac.in

J. K. Ghosh and I. da Silva (eds.), *Applications of Geomatics in Civil Engineering*, Lecture Notes in Civil Engineering 33, https://doi.org/10.1007/978-981-13-7067-0_7

Spatial variability of depth to weathered rock from the ground surface is important information for numerous applications in geotechnical engineering and urban geosciences. A depth to rock information at a particular site is very much needed in foundation design to provide the complete three-dimensional (3-D) subsurface information and seismic ground response analysis to estimate the damage potential of an earthquake. In the present study, the spatial variability of depth to weathered rock is evaluated in Chennai city using borehole data. The estimated depths to weathered rock have been cross-validated with a selected borehole data and the Multichannel Analysis of Surface Wave (MASW) test results.

2 Study Area

The study has been carried out in Chennai district covering an area of 178.2 km^2 and it is located between 12.75°–13.25°N and 80.0°–80.5°E. The city is a low-lying area, situated at an average altitude of 6.7 m above the Mean Sea Level (MSL). Chennai is India's fifth largest metropolitan city, the capital of Tamil Nadu state, situated on the southeast coast of India and in the northeast corner of Tamil Nadu, described by coastal plains of the Bay of Bengal. Chennai is the fastest growing and highly urbanized city undergoing various infrastructure development projects and urban planning activities. The general geology of the city comprises mostly of sandy clay, shale, and sandstone as shown in Fig. 1 [1].

The geology of the study area consists of shallow bedrock on the east and south, and Gondwanas below the alluvium in the north and west. Almost the entire area is covered by the pleistocene/recent alluvium, which is deposited by the Cooum and Adyar Rivers [2]. The coastal region of the city is fully covered by marine sediments of thickness varying from 4 to 12 m. The riverbanks and shorelines are embedded with sandy regions. Igneous and metamorphic rocks are found in the southern area; marine sediments containing clay–silt sands and charnockite rocks are found in the eastern and northern parts and the western parts are composed of alluvium and sedimentary rocks [2]. Clayey deposits cover most of the city area. With respect to seismicity, Tamil Nadu is generally considered to be less seismically active state compared to states in the northern and western parts of the country. Moderate earthquakes have occurred in the past in the study area. Chennai lies in the Zone III as per the Bureau of Indian Standards classification [3]. A few tremors have been felt in the city during the Pondicherry earthquake of magnitude, M_w 5.5 on September 25, 2001 centered 100 km from Chennai and a Sumatra earthquake of magnitude, M_w 9.1 on December 26, 2004 [2].

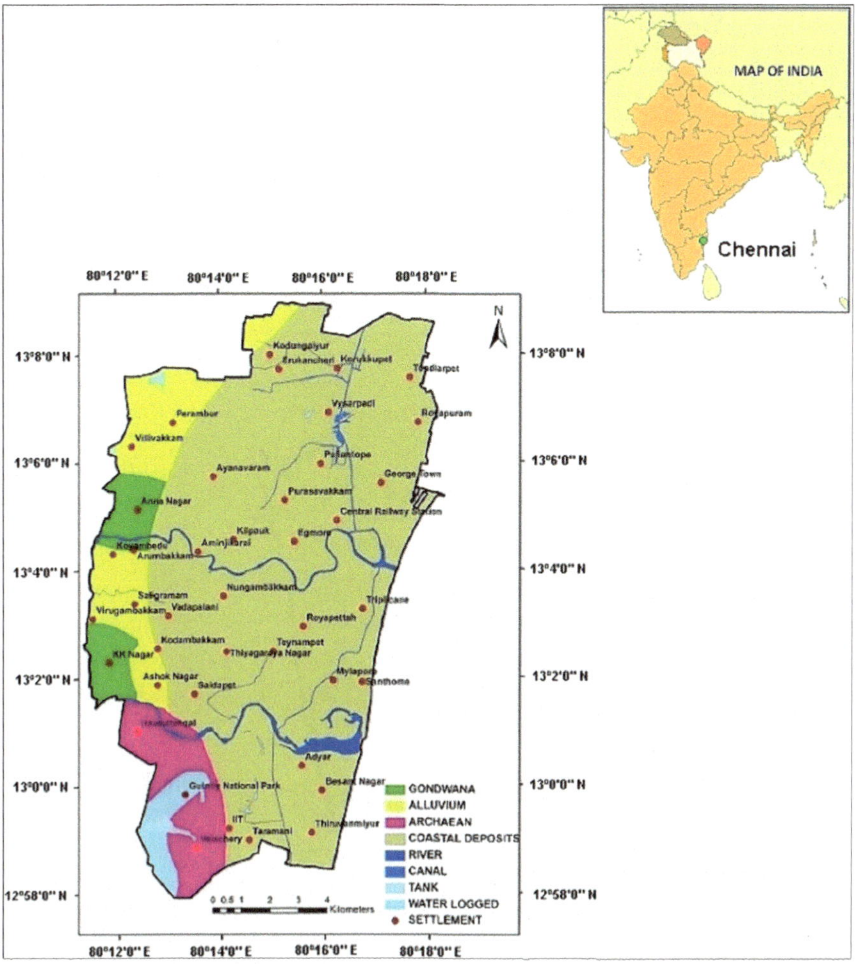

Fig. 1 Geological map of Chennai

3 Data Input and Subsurface Profiling

In this study, nearly 400 borelogs have been collected from the reputed geotechnical agencies of Chennai to establish the subsurface profile of the study area. Details of the borehole locations and their corresponding coordinates are obtained from the field. Borehole data along with the index and engineering properties of the subsoil layers are collected for different locations in the study area as shown in Fig. 2. The elevation values of the study area are estimated from the CartoDEM data with 2.5 m spatial resolution. The elevation data have a horizontal accuracy of 15 m and a relative vertical accuracy of ±5 m. A base map of the study area is prepared using ArcGIS® software, ESRI (Fig. 2). All the elements in the map layers are georeferenced with

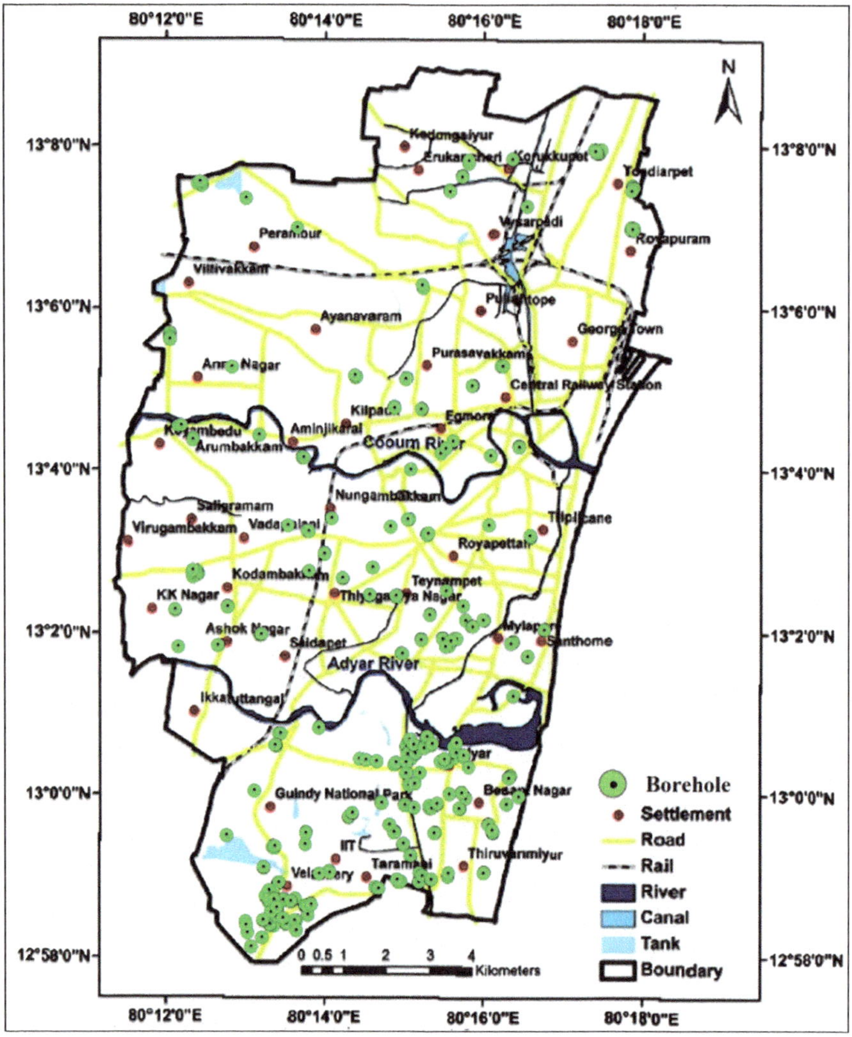

Fig. 2 Borehole locations in the study area

minimum Root Mean Square (RMS) error and projected to WGS 1984 UTM Zone 44N (World Geodetic System 1984) using the WGS 1984 spheroid.

Based on the collected borelogs, the geotechnical characterization is carried out in the study area through validation and standardization of borehole data. The borehole data are organized based on the developed borehole standard and stored in a comprehensive geotechnical database as a Relational Database Management System (RDBMS). The database is then integrated within the Geographic Information System (GIS) environment for real-time visualization and spatial analysis [4]. The process of development of GIS-based geotechnical database is shown in Fig. 3. The

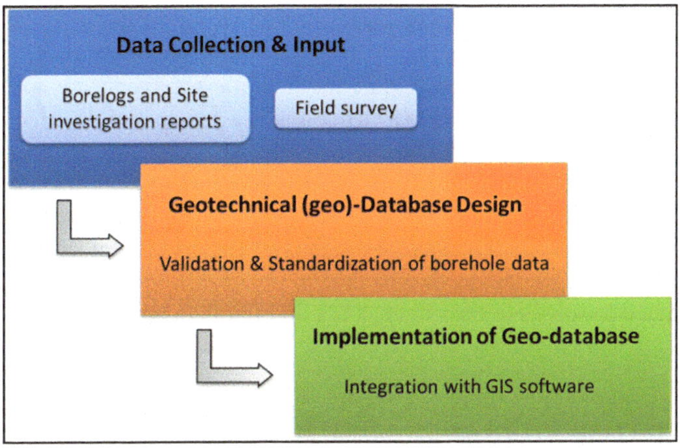

Fig. 3 Development of GIS-based geodatabase

GIS-based geodatabase ensures the availability of a single source of useful information for all developmental activities related to "urban geosciences". In addition, it helps experts to address the practical geotechnical problems through interpretations and intuitive analysis.

4 Spatial Variability of Depth to Weathered Rock

As GIS-based geotechnical database contains geotechnical feature class with location and elevation information (z-value), the boreholes of the study area are represented spatially to understand the stratigraphic variation with respect to location. Additionally, the subsurface information like thickness of the overburden soil, depth to bedrock can be mapped from the database which will be of great value in foundation analysis and design studies and location of underground utility structures. Representation of these information using thematic and contour maps provides a user-friendly tool to interpret the subsurface information for infrastructure development projects. The knowledge of depth to rock from the ground surface for the specific region is a vital parameter for geotechnical engineering activities and seismic site characterization studies.

The main soil types that exist in the Chennai region are identified and depth to weathered rock information is extracted from the developed geodatabase. The data between the sampled locations where the field information is not available are interpolated using geostatistical kriging to represent the variation of rock levels in the study area. The geostatistical kriging, an interpolation technique, is considered as the best linear unbiased estimate and optimal for geological and geotechnical predictions in space because it uses a linear combination of weighted sample values with minimum variance [5]. In this study, the knowledge of the semivariogram of the borehole data is used with ordinary kriging to estimate the depths to weathered rock for Chennai city at locations where the field measurements are not available. The ordinary kriging is chosen due to its simplicity and spatial dependency of the available data. A spherical model has been used to get a reasonable fit to the depth values as it has minimum nugget effect and minimum sill with maximum range compared to other models [6]. Figure 4 represents the spatial variation map of depth to weathered rock in the study area. The estimated depths to weathered rock values are cross-validated with the depth information obtained from the selected borehole data at selected locations in the Chennai city and found to match reasonably well.

5 Comparison with Geophysical Test Results

Shear wave velocity (V_S) of the soil profile is an important parameter for estimating the stiffness of soil layers which has a major significance in various earthquake engineering projects and seismic hazard studies. Most commonly, surface wave methods are used for shear wave velocity profiling, as it is cost-effective and environmental friendly. Among the various geophysical methods for near-surface characterization and measurement of shear wave velocity, the most widely used techniques are (a) Spectral Analysis of Surface Waves (SASW) and (b) Multichannel Analysis of Surface Waves (MASW) tests. The MASW test is found to be the most efficient seismic method for geotechnical characterization of near-surface materials [7]. It is a non-invasive geophysical method which measures V_S to characterize the dynamic properties of the underlying soil. It also aids in discrimination of stratigraphic material boundaries and spatial variation of V_S with depth.

In the present study, the MASW tests that record Rayleigh-type surface waves on a multichannel mode have been conducted at selected locations in the study area to establish the V_S profile of the subsurface. The process involves three steps: (i) acquisition of ground roll, (ii) construction of dispersion curve (phase velocity vs. frequency), and (iii) back calculation (inversion) of the V_S profile from the calculated dispersion curve [7].

5.1 *Experimental Setup*

In MASW test, the ground roll is generated by an impulsive source, coupled with collection of motion by a series of receivers. The term "Multichannel record" indicates an acquisition of seismic waves using multiple channel geode seismic recorder. The recorded multichannel surface wave data are processed using phase shift method to estimate the variation of V_S with depth. The field survey of MASW test consists of a source, receiver, and an acquisition system as shown in Fig. 5.

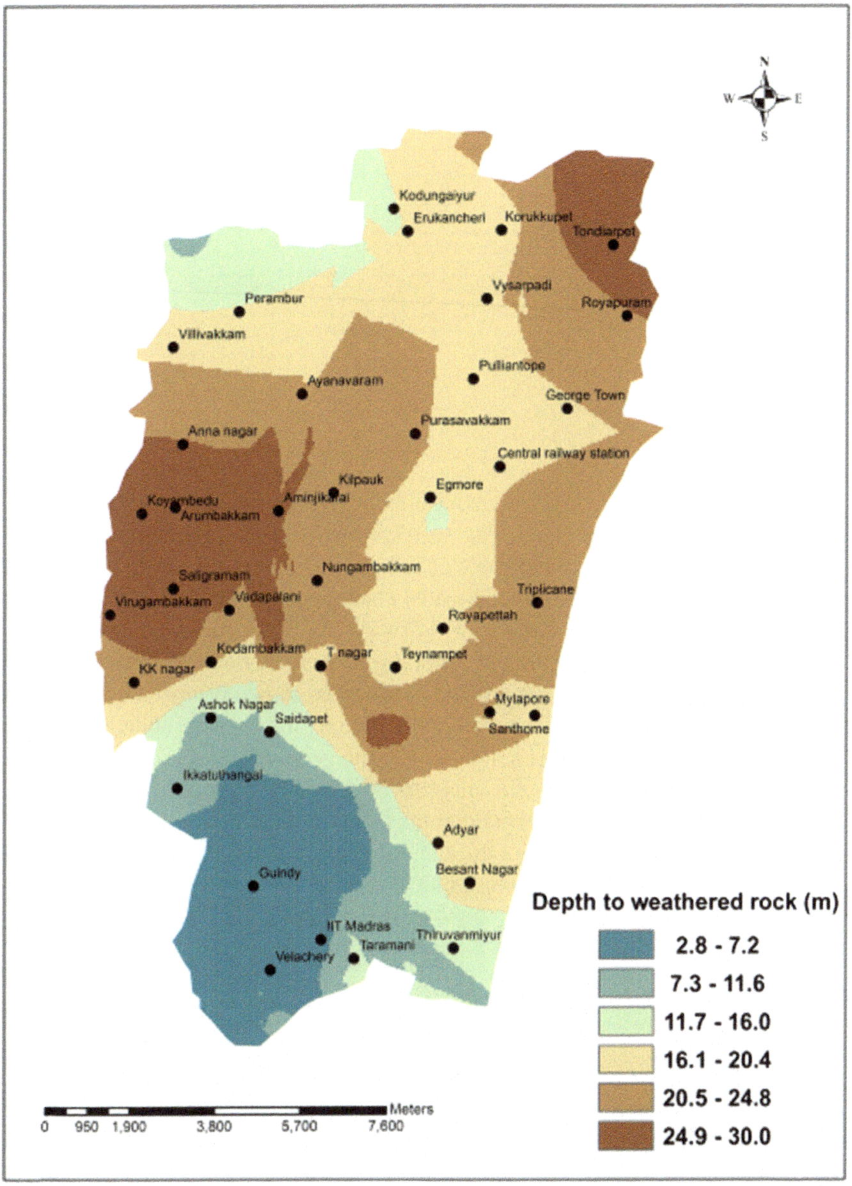

Fig. 4 Depth to weathered rock map of Chennai city

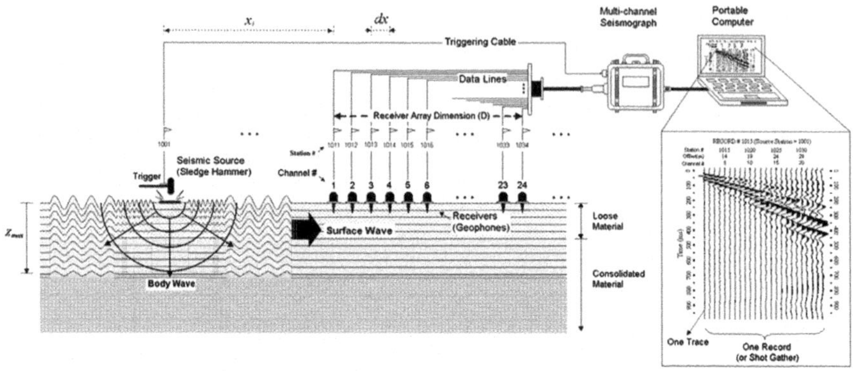

Fig. 5 Schematic of the MASW test

The motion is generated, when a controlled active source of 8 kg sledgehammer hits against the metal base plate. The corresponding signals are detected simultaneously by 4.5 Hz frequency receivers called geophones arranged in a linear array. The raw data (wiggle plot) recorded by a 24 channel geometrics make seismograph (Geode) with Single Geode Operating Software (SGOS) and stored in a portable computer. For efficient acquisition of ground roll data, the field configurations such as receiver spacing and the source offset range need to be optimized based on the near-field conditions. At locations, where preliminary information about the subsurface like thickness of the soil is not available, the geophone spacing can vary from 0.5 to 2 m with minimal source offset in order to obtain the V_S profile close to reality. During each test, the source is applied at the front, middle, and end of the receivers spread at nearest offsets to acquire the consistency of the field data. Three shots are stacked to improve the signal-to-noise ratio.

5.2 *Data Analysis and Vs-Profiling*

Raw field data (delay in travel time versus receiver distance) are further analyzed using surface wave analysis software. The shear wave velocity profile is generated by the following procedure: (i) filter and analyze the acquired data to obtain the frequency range of Rayleigh waves, (ii) development of dispersion curves by calculating phase velocities of Rayleigh waves, and (iii) inversion to estimate the one-dimensional (1D) shear wave velocity profile by comparing with the theoretical dispersion curves iteratively. A typical V_S profile corresponds to a particular site at Egmore, Chennai, which is shown in Fig. 6. It has been observed that the shear wave velocity of the subsurface profile ranges from 140 to 300 m/s.

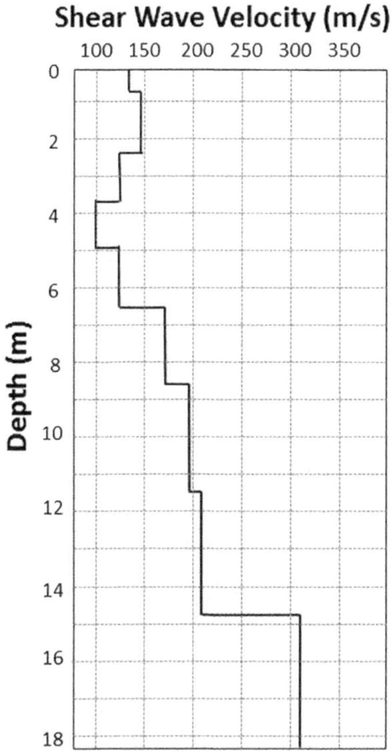

Fig. 6 Shear wave velocity profile at Egmore site

As per the National Earthquake Hazard Reduction Programme (NEHRP) site classification, the average shear wave velocity in the top 30 m [$(V_S)_{30}$] for very dense soil and soft rock ranges from 360 to 760 m/s and rock has the $(V_S)_{30}$ of 760–1500 m/s. The rock which has $(V_S)_{30}$ greater than 1500 m/s is classified as hard rock. Several studies have been carried out considering different ranges of V_S for geomaterials [8–10].

In the present study, V_S of 330 ± 30 m/s is considered as the criterion for the identification of weathered rock. The depths corresponding to this range are identified from V_S profiles and the same depths are recognized as depths to weathered rock at test locations in the study area. The depths estimated from the measured shear wave velocities from the MASW tests are compared with those mapped using the borehole data. It is found that both the results agree well with each other (Fig. 7).

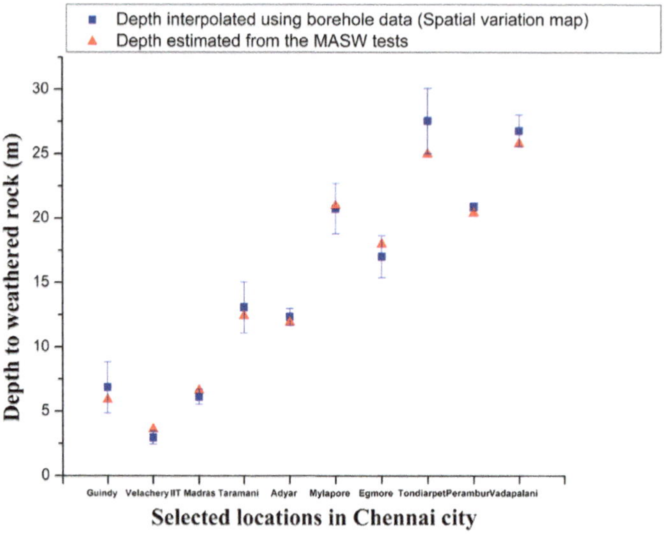

Fig. 7 Comparison of depths to weathered rock data

6 Conclusions

Spatial variability of depth to weathered rock has been evaluated using field tests data and kriging technique for the Chennai city. The study indicates that the ordinary kriging is the optimal method for mapping the spatial variability of depth to weathered rock in terms of the reduced kriging variance. The developed thematic map indicates that the depth to weathered rock varies from 3 to 30 m in the study area (Fig. 4). In general, the weathered rock has been found at shallow depths, i.e., within 7 m from the ground surface in the southern parts of the Chennai city, at deeper depths (about 20–30 m) in the western parts and at moderate depths (8–15 m) in the central parts of the city. The mapped depth to weathered rock using borehole data has been compared with the depth to weathered rock estimated from the MASW tests and the differences between them lie in the acceptable range (Fig. 7) indicating that both the results are in good agreement. Thus, the combination of geotechnical and MASW test data along with the kriging can be used effectively to map the depth to rock surface for urban geosciences studies. The spatial distribution map of depth to weathered rock for Chennai city will be of immense use in the foundation analysis and design and seismic ground response studies.

References

1. GSI (1999) Explanatory brochure on geological and mineral map of Tamilnadu and Pondicherry. Geological Survey of India, New Delhi (India)
2. Uma Maheswari R, Boominathan A, Dodagoudar GR (2010) Seismic site classification and site period mapping of Chennai city using geophysical and geotechnical data. J Appl Geophys 72(3):152–168
3. BIS IS 1893 (Part 1): Indian standard criteria for earthquake resistant design of structures (2016)
4. Divya Priya B, Dodagoudar GR (2017) Building 3D subsurface models and mapping depth to weathered rock in Chennai, South India. J Geomatics 191–200
5. Sun CG, Kim HS, Chung CK, Chi HC (2014) Spatial zonations for regional assessment of seismic site effects in the seoul metropolitan area. Soil Dyn Earthq Eng 56:44–56
6. Abdideh M, Bargahi D (2012) Designing a 3D model for the prediction of the top of formation in oil fields using geostatistical methods. Geocarto Int 27(7):569–579
7. Park CB, Miller RD, Xia J (1999) Multi-channel analysis of surface waves. Geophysics 64(3):800–808
8. Anbazhagan P, Sitharam TG (2009) Spatial variability of the depth of weathered and engineering bedrock using multichannel analysis of surface wave method. Pure Appl Geophys 166:409–428
9. Trupti S, Srinivas KNSSS, Pavan Kishore P, Seshunarayana T (2012) Site characterization studies along coastal Andhra Pradesh-India using multichannel analysis of surface waves. J Appl Geophys 79:82–89
10. Pegah E, Liu H (2016) Application of near-surface seismic refraction tomography and multichannel analysis of surface waves for geotechnical site characterizations: a case study. Eng Geology 208:100–113

Surface Soil Moisture Retrieval Using C-Band Synthetic Aperture Radar (SAR) over Yanco Study Site, Australia—A Preliminary Study

G. Punithraj, Umesh Pruthviraj and Amba Shetty

Abstract The motto of this work is to evaluate retrieval of surface soil moisture (5 cm) using Sentinel-1a C-band data Synthetic Aperture Radar (SAR). Data for this study is collected from Yanco Study site, Australia. Yanco study site consists of 37 soil moisture measuring stations at every 20 min interval for various soil depths and it also provides other Hydro-Meteorological information. SAR backscattered energy is a function of soil roughness and soil moisture. Surface roughness is eliminated using change detection approach. The R^2 performance statistics revealed that between Backscattered energy and NDVI there is no relation. Volumetric soil moisture and backscattered energy showed a positive correlation with $R^2 = 0.57$ and 0.43 for VH and VV polarization. Dielectric constant also showed a positive correlation with backscattered energy having $R^2 = 0.62$ and 0.38 for VH and VV polarization respectively. By taking into account of all these affecting parameters, a regional Semi-empirical model is developed to retrieve surface Soil moisture over the study area.

Keywords SAR · Sentinel-1 · Soil moisture · Semi-empirical model

1 Introduction

Soil moisture plays an important role in various hydrological processes and acts as a medium between Earth and atmosphere for heat and water exchange. The retrieval of surface soil moisture is important because of its various applications in the field of hydrology, forest fire prediction, agriculture, drought, and meteorology. It is the main parameter for separating initial conditions of infiltration and runoff rates during

G. Punithraj (✉) · U. Pruthviraj · A. Shetty
NITK, Surathkal, Mangalore 575025, India
e-mail: puniththeraj@gmail.com

U. Pruthviraj
e-mail: pruthviu@gmail.com

A. Shetty
e-mail: amba_shetty@yahoo.com.in

J. K. Ghosh and I. da Silva (eds.), *Applications of Geomatics in Civil Engineering*, Lecture Notes in Civil Engineering 33, https://doi.org/10.1007/978-981-13-7067-0_8

surface flow modeling [1]. However, availability of long-period and large area in situ soil moisture data are restricted due to practical constraints. Along with that, regional-scale differences in soil properties, topography, and land cover impacts on surface soil moisture, which makes it inconvenient to fully assess regional surface soil moisture conditions based on in situ point measurements.

Technological developments in remote sensing have given different methods for measuring surface soil moisture temporally and spatially across large areas [2]. Various remote sensing platforms supporting soil moisture retrieval are ground based, airborne based, or space based. Spaceborne satellites are given best provision to cover larger areas with a short period of repetition. Synthetic aperture radar (SAR) can be used to retrieve surface soil moisture. The basis for radar-based soil moisture investigations is the linear relationship between backscattered energy (σ°) and volumetric soil moisture content (M_v) in the top 5 cm of soil profile [3, 4].

SAR is helpful to monitor surface soil moisture because of its penetration power and suitability to all weather conditions [3, 5–7]. The backscattered energy is a function of geometric properties and dielectric properties of soil for barren/grassland/agriculture lands [8]. Spaceborne and airborne SAR has already shown its capability in retrieving soil moisture and roughness and, to a smaller extent, to the soil's textural composition [1–4, 9–15]. The motto of this work is to develop a regional semi-empirical model to retrieve the top surface soil moisture (5 cm) using Dual Polarized C-band data.

2 Study Area and Data Products

2.1 Yanco Study Site, Australia

Yanco study site lies in Murrumbidgee catchment, New South Wales of Australia. It is one of the test site of Oz Net hydrological monitoring network of Australia. Out of so many flux data, Oz net data is selected because it is the only one which has measured soil moisture data at every 20 min interval from 2009 to till date and is available at free of cost. The areal extent of the study area is 60 km × 60 km^2 with gentle slope lies to the south-west of the Yanco Research Station. There are totally 37 soil moisture measuring sites are distributed across the study area (Fig. 1) [16]. The study area is mainly composed of Agricultural and barren/Grasslands. The annual average rainfall of this area is 400 mm and temperature varies from 11 to 24 °C. In spite of the availability of extremely good field measured data not much scientific studies have been reported at this site for regional scale.

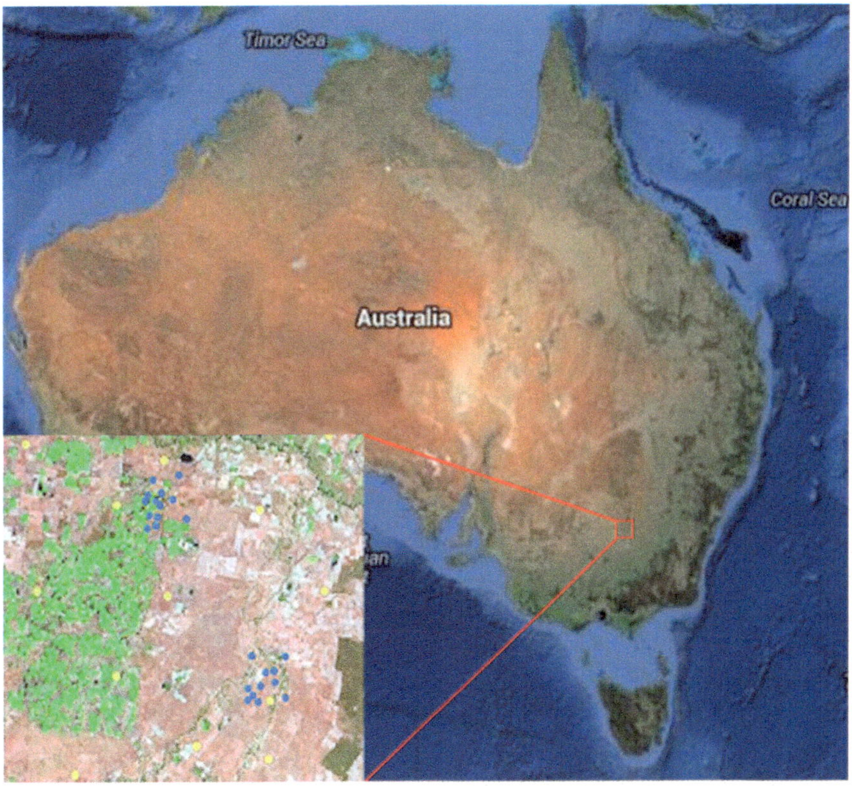

Fig. 1 Yanco study site, Australia (*Source* oznet.org.au)

2.2 *Satellite and Field Data*

Sentinel-1a: Sentinel-1a is a SAR in C-band (frequency: 5.405 GHz), that provides continuous imagery (day, night and all weather).It provides dual polarized data with short revisit times and precise measurements of spacecraft altitude and position are also available. The Data characteristics of Sentinel-1a is explained in Table 1.

Sentinel-1a image has been procured from Alaska satellite facility data portal (https://vertex.daac.asf.alaska.edu/). The acquisition dates of the scene are 20/12/2014 and 13/10/2014 with look direction Right. The collected image is of interferometric wide swath mode ground range detected multi-looked image from which Yanco study area is clipped. The data is preprocessed and Geocoded using well-established tools in Sentinel Application Platform (SNAP) environment.

Landsat-8: Land use/land cover map had developed using Landsat-8 which is downloaded from USGS Earth Explorer website (https://earthexplorer.usgs.gov) having Ground Resolution of 30 m and the image is acquired on 04/10/2014. The data

Table 1 Data characteristics of Sentinel-1a

Satellite/Sensor	Sentinel-1a
Frequency (Ghz)	5.405
Frequency Band	C-band
Polarization	VV and VH (Dual-pol)
Orbit direction	Ascending
Incidence angle (°)	23
Resolution (m)	15 * 20
Swath (km)	100–150
Period	2014–present
Repetivity	12 days

is layer stacked and preprocessed using ERDAS imagine software and projected to GDA 1994 MGA Zone 54.

Field Measured Data: At, Yanco study site measured soil temperature and soil moisture data are available overall 37 Locations at 20 min interval and is showed in Fig. 2 along with their location names. Rainfall data and soil texture data is available only at 13 and 5 locations, respectively. Soil moisture is measured at various depths whereas in this study top 5 cm measured soil moisture data is used. Out of 37 measuring locations, 2 locations (Y2 and Y3) are out of the image tile and 1 location (Y4) data is not available for the whole period. In some locations due to operational conditions data are not available. So totally 24 locations are used for this study.

3 Results

3.1 Rainfall

Rainfall data is collected at 13 locations of the study area using tipping bucket rain gauge arrangement. From the data, it is verified that there was no effect of rainfall on soil moisture in the study area prior to the days of sampling during satellite overpass.

3.2 Land Use and Land Cover

Land use/land cover map for the study area was developed using Landsat-8 images. The procured data is layer stacked, preprocessed and classified using Maximum likelihood algorithm in ERDAS imagine environment. Level-1 classification was carried out for the study area. Land use/land cover map of the study area is presented in Fig. 3. Major covers are agricultural and barren land.

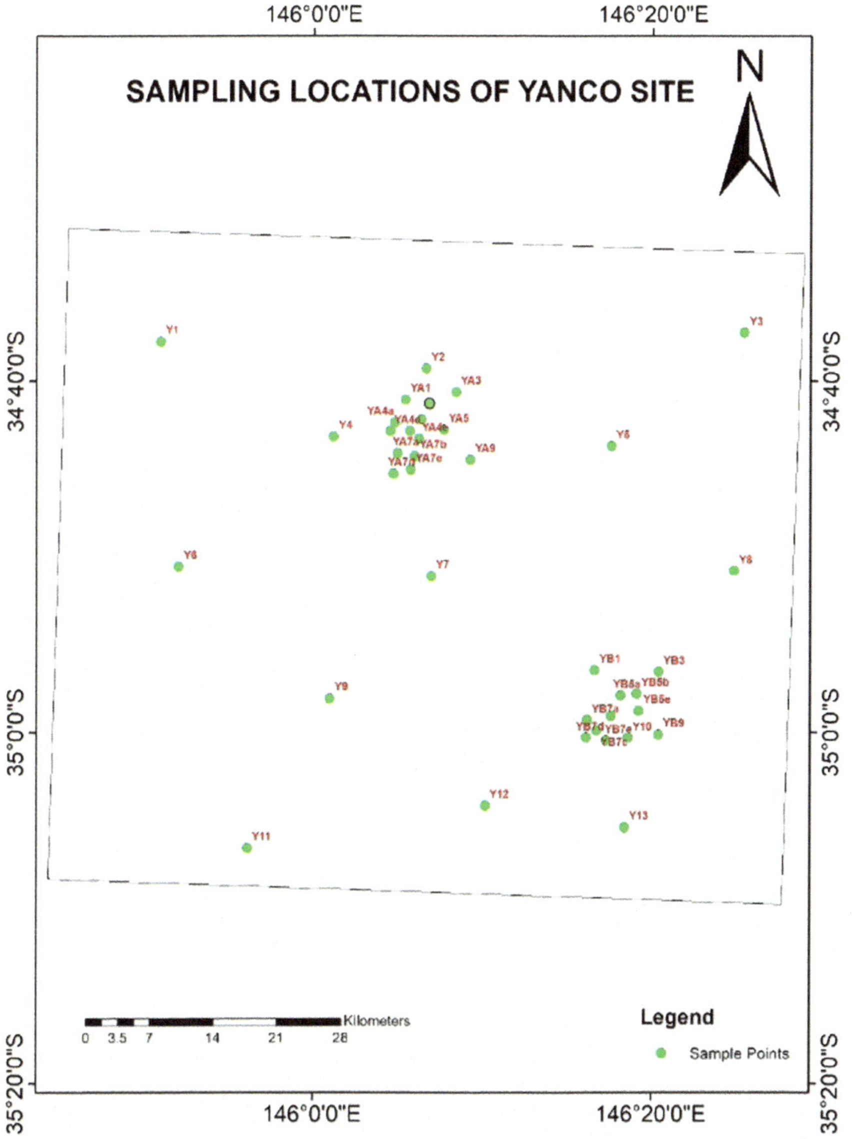

Fig. 2 Representation of sample location

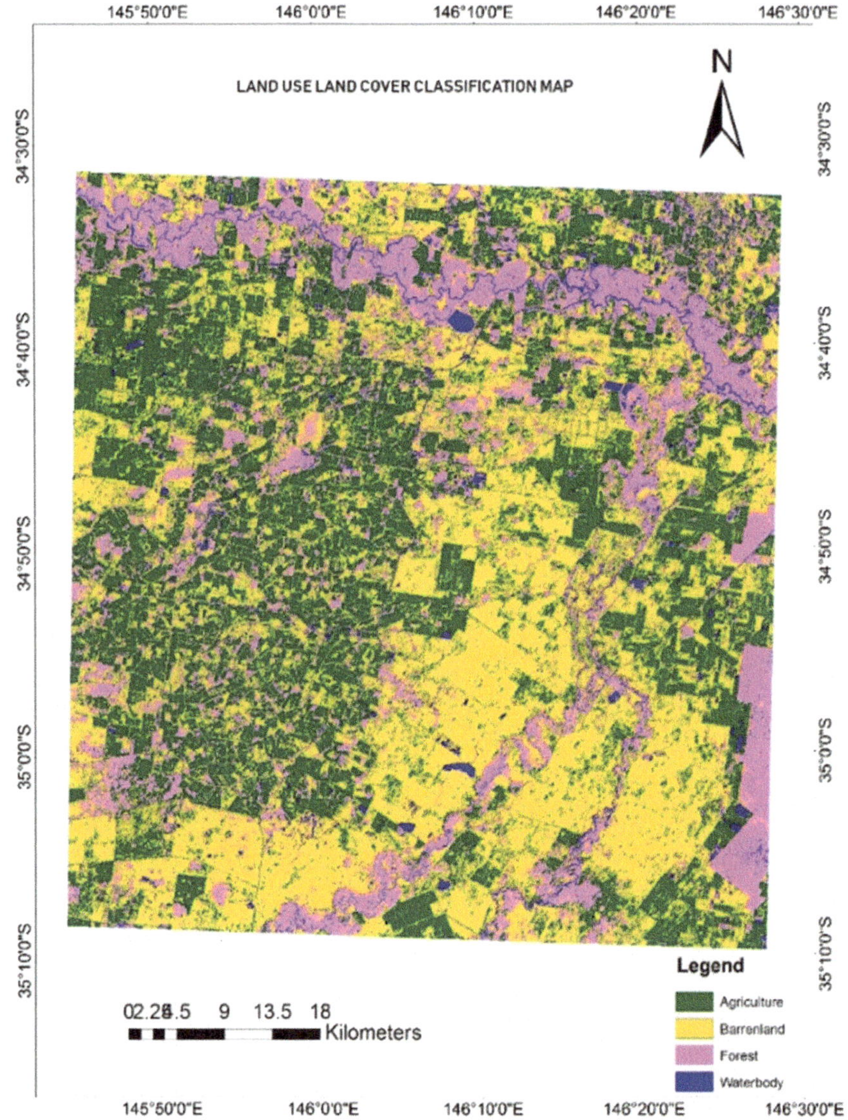

Fig. 3 Land use/land cover map of Yanco study site

3.3 Normalized Difference Vegetation Index (NDVI)

NDVI map for the study area was prepared using Landsat-8 image and NDVI values for the sample locations extracted. NDVI map is developed to check the interference of the vegetation in radar signal. The NDVI value of sample location varies from −0.0268 to 0.0222 (Table 2). The values of NDVI are within the limit where backscattered energy is not much affected [5] accordingly vegetation correction are not applied. The performance of R^2 Statistics is found to be 0.0319 and 0.0344 for VH and VV polarization. The field wise distribution of NDVI is shown in Fig. 4. The correlation between NDVI and $\sigma°$ VH and $\sigma°$ VV is shown in Fig. 5. The NDVI map of Yanco area is presented in Fig. 6.

Table 2 Descriptive statistics of observed variables

Variable	Minimum	Maximum	Mean	Standard deviation
$M_v(\mathrm{m}^3/\mathrm{m}^3)$	5.41	27.8	12.03	6.67
σ° VH	−24.37	−15.11	−20.83	2.36
σ° VV	−18.81	−9.33	−14.99	2.89
ε	4.35	11.91	7.9	3.08
NDVI	−0.027	0.022	0.0031	0.013

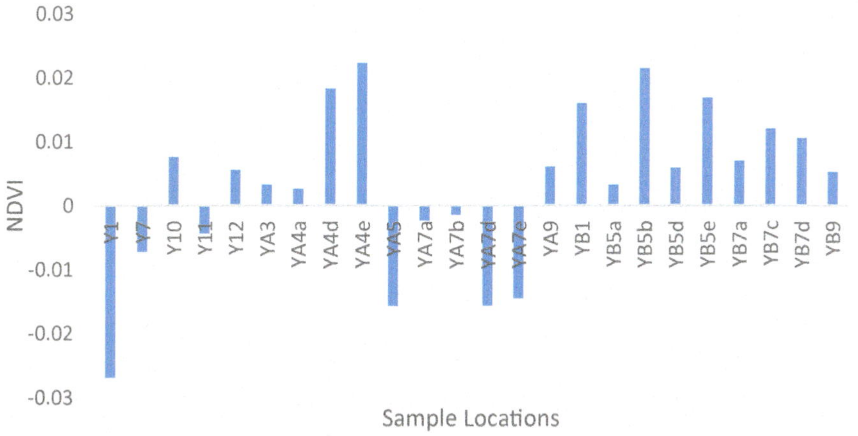

Fig. 4 Field wise NDVI value distribution

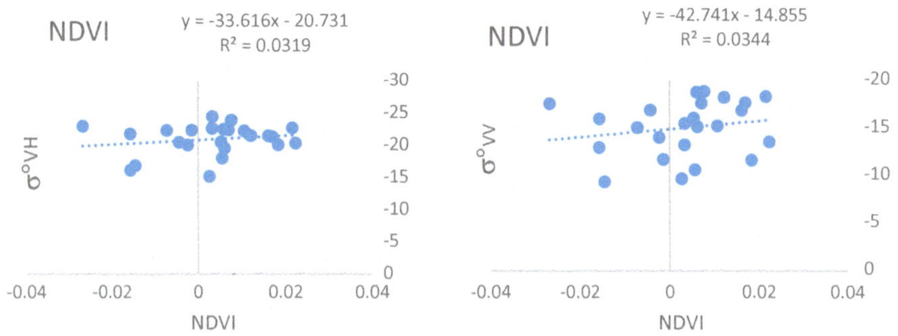

Fig. 5 Relationship between NDVI and backscattered energy of VH and VV

3.4 *Parameters Required for Soil Moisture Retrieval*

3.4.1 Backscattered Coefficient (σ°)

Two SAR data procured on 20/12/2014 and 13/10/2014 are preprocessed and geocoded using SNAP software. The backscattered energy of the study area is calculated using Eq. 1 and described in Table 2. The backscattered energy varies from −15.11 to −24.98 (dB) for VH polarization and −9.33 to −18.82 (dB) for VV polarization. The backscattered energy of Yanco site for both VV and VH polarization are shown in Figs. 7 and 8.

$$\sigma^\circ(\mathrm{dB}) = 10 * Log_{10}(DN_{p)} \tag{1}$$

3.4.2 Soil Moisture

Soil moisture is directly provided in terms of volumetric by the Oz Net hydrological monitoring network of Australia and the same is used in this study. Soil moisture value varies from 5.41 to 27.8 presented in Table 2.

3.4.3 Dielectric Constant

Dielectric constants of each sample were quantified using Eq. 2 [6]. Dielectric constants vary from 4.35 to 11.91 as details shown in Table 2.

$$\varepsilon = (a_0 + a_1 s + a_2 c) + (b_0 + b_1 s + b_2 c)M_v + (c_0 + c_1 s + c_2 c)M_v^2 \tag{2}$$

where

S = Percentage of sand by weight

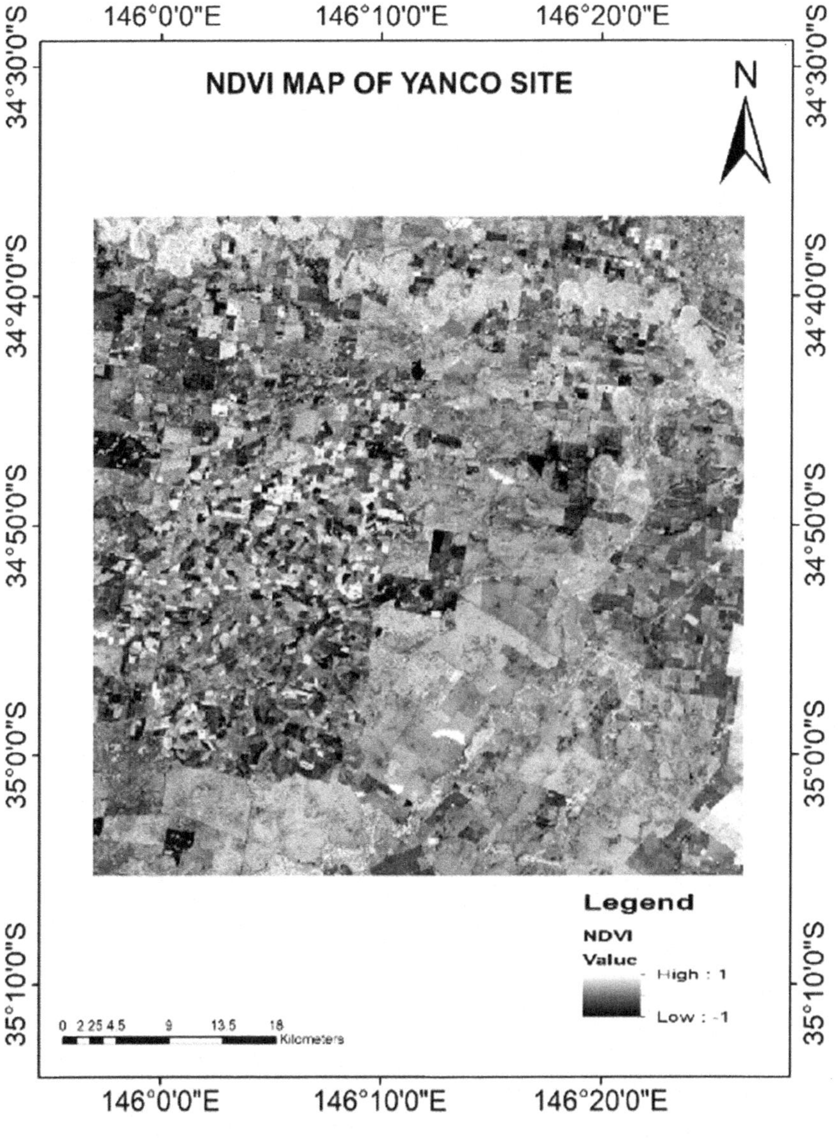

Fig. 6 NDVI map of Yanco study site

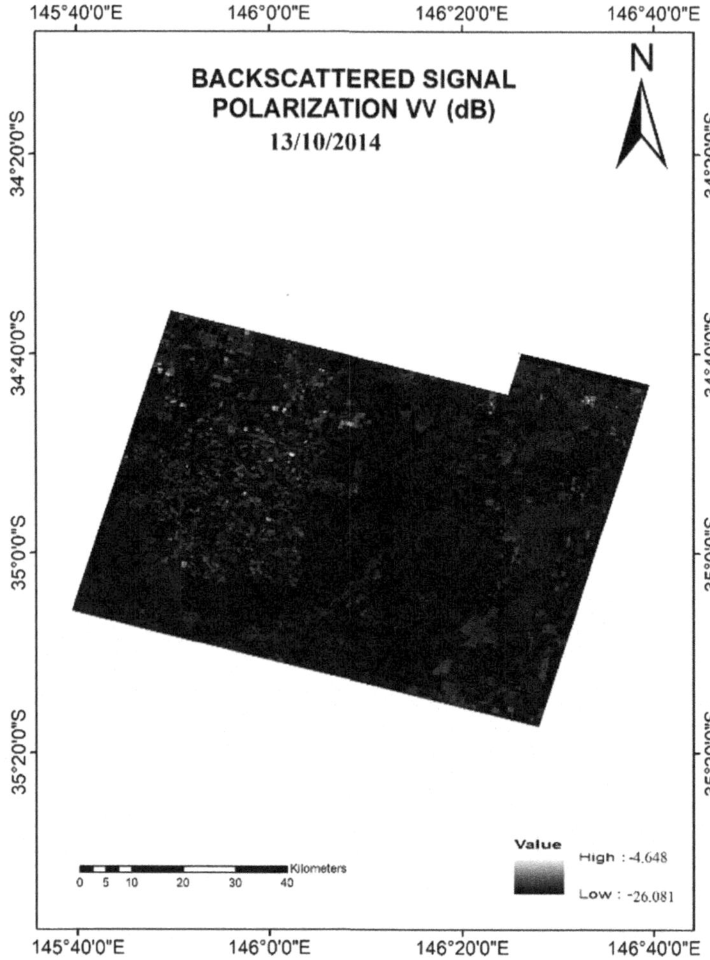

Fig. 7 Backscattered energy image of VV polarization (dB)

C = Percentage of Clay by weight
a_i, b_i and c_i = Frequency dependent coefficients adopted from [6].

3.4.4 Surface Roughness

Surface roughness of study area is eliminated using change detection approach. The difference in backscattered energy of two SAR images of the same area can eliminate the effect of roughness on backscattered energy [17].

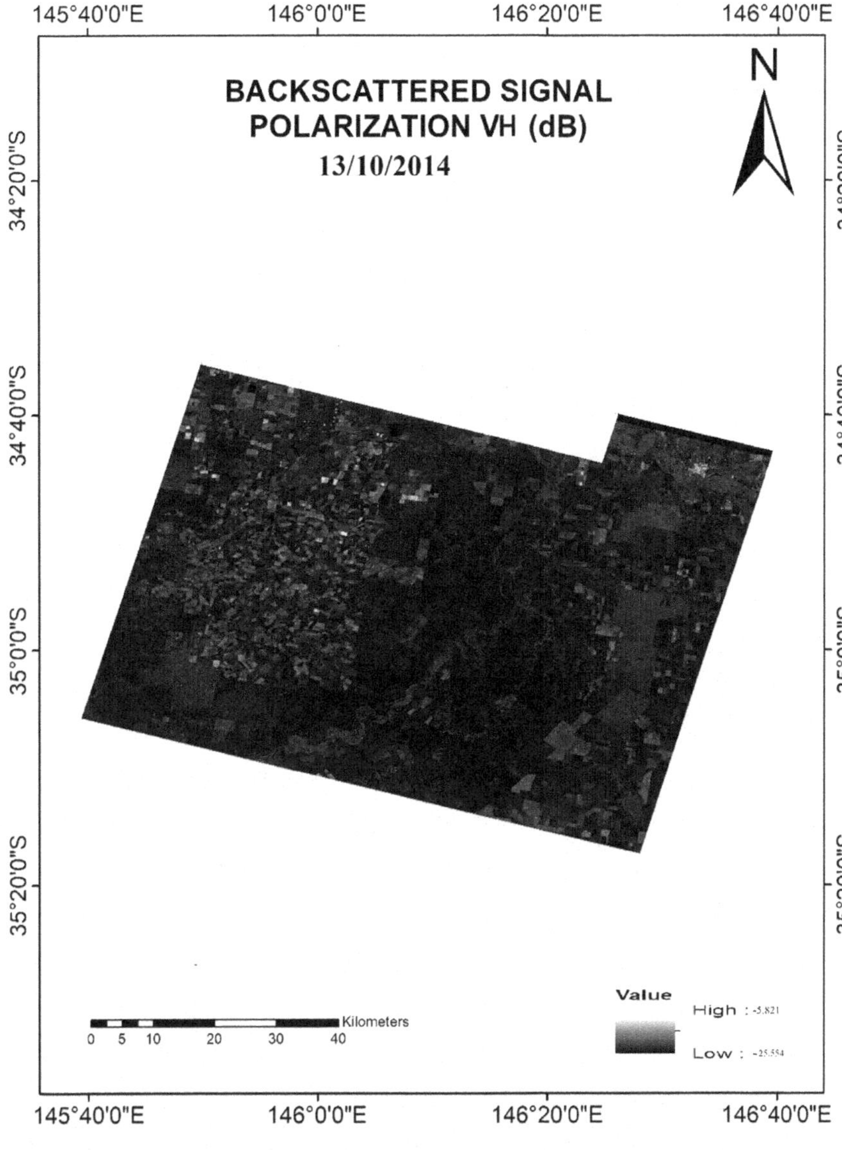

Fig. 8 Backscattered energy image of VH polarization (dB)

4 Analysis

4.1 Relationship Between Backscattered Energy (σ°) and Volumetric Soil Moisture (M_v)

The relationship between σ° and M_v were examined by scatter plot for both VH and VV polarization. The positive correlation was found between σ° and M_v with $R^2 = 0.578$ and $R^2 = 0.432$ for VV and VH polarization, respectively. The SAR backscatter is directly correlated to soil moisture of the study area. The relationship σ° and M_v for both VV and VH are shown in Fig. 9.

4.2 Relationship Between σ° and ε

Dielectric constant (ε) of dry soil is around 3 and for water is 80 [18]. So the mixture of water and soil varies between these ranges. The distribution of dielectric constant of soil depends upon the texture of soil and its water holding capacity [6, 11]. The relationship between σ° and ε were examined by scatter plot for both VH and VV polarization. The positive correlation was found with $R^2 = 0.62$ and $R^2 = 0.38$ for VV and VH polarization, respectively. The relationship between σ° and ε for both VV and VH are shown in Fig. 10.

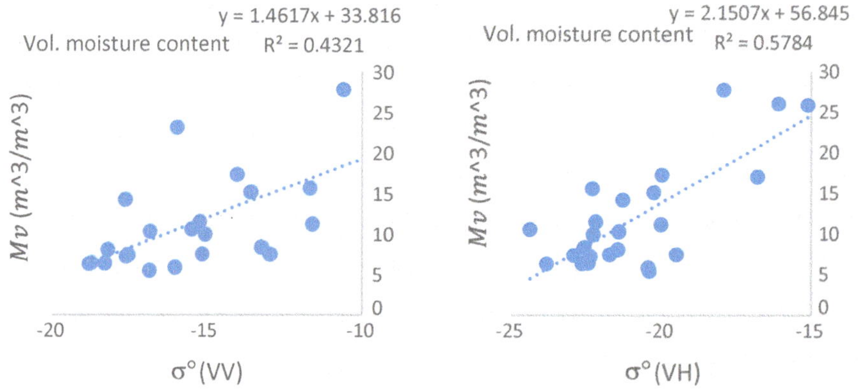

Fig. 9 relationship σ° and M_v for both VV and VH

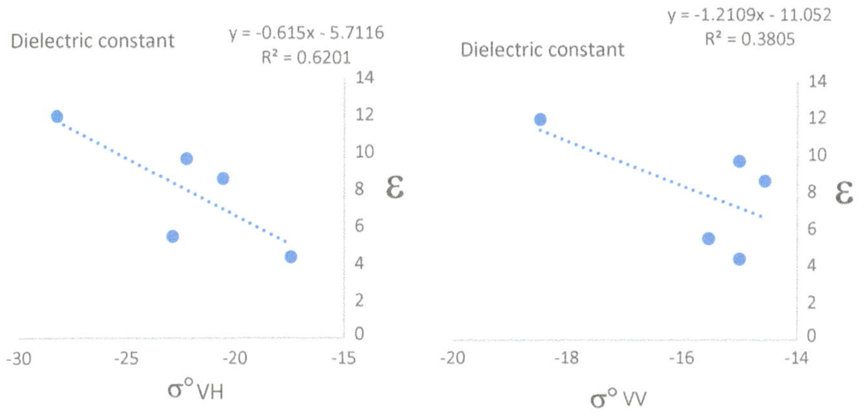

Fig. 10 Relationship between σ° and ε

4.3 Topp Model

Dielectric constant is derived using Eq. 2. This model is used to validate the ground truth soil moisture collected from the field. Topp model equation is described in Eq. 3. Since only 5 location texture values are available so those soil moisture values are validated. The relationship between estimated and measured soil moisture is presented in Fig. 11.

$$M_v = -5.3 * 10^{-2} + 2.92 * 10^{-2}\varepsilon - 5.5 ** 10^{-4}\varepsilon^2 + 4.3 * 10^{-6}\varepsilon^3 \quad (3)$$

where

ε = Dielectric constant
M_v = Volumetric moisture content.

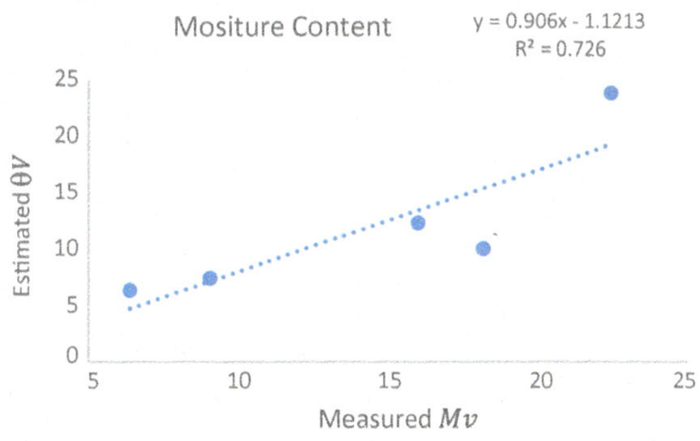

Fig. 11 Relationship between estimated and measured soil moisture

5 Semi-Empirical Model

To derive semi-empirical model to retrieve soil moisture major factors influencing backscattered energy are considered. To derive a model multilinear regression analysis is carried out where M_v is considered as dependent and σ° (VV) and σ° (VH) as independent variables. Since ε is of lesser samples it is not considered in the multiple linear regression analysis (MLR). This MLR model has adjusted $R^2 = 0.6$ and RMSE of 0.010 at 95% confidence level. Derived semi-empirical model is described in Eq. 4.

$$M_v = -40.44 + 1.81\sigma^{\circ}_{VH} + 0.47\sigma^{\circ}_{VV} \tag{4}$$

6 Conclusions

Even though so many inversion models to retrieve surface soil moisture from active or passive microwave remote sensing have been developed regardless of point measurements. SAR has shown its capability to retrieve surface soil moisture. However, backscattered energy is governed by land surface features so it is an ill-posed problem to extract surface soil moisture from single imagery. The developed method here is quite simple and practical for the evaluation of soil moisture. To develop this model major factors considered are *VH* and *VV*. It was found as follows:

1. Backscattering coefficient (σ°) has a positive relationship with volumetric moisture content (Mv) both for σ°_{VH} and σ°_{VV} with $R^2 = 0.58$ and $R^2 = 0.43$ respectively.
2. Dielectric constant (ε) has a positive relationship with σ°_{VH} having $R^2 = 0.62$ and σ°_{VV} having $R^2 = 0.38$.
3. Topp model derived Soil moisture (θ_V) was validated with the ground truth soil moisture value. It has been found that θ_V versus M_v having $R^2 = 0.72$.
4. Developed semi-empirical model is dependent on only backscattered energy hence model performance R^2 is 0.6 by using the values of dielectric constant this can be improved.

References

1. Baghdadi N, Zribi M (2006) Evaluation of radar backscatter models IEM, OH and Dubois using experimental observations Int. J Remote Sens 27(18):3831–3852
2. Engman ET (1992) Soil moisture needs in earth sciences. In: Proceedings of international geoscience and remote sensing symposium (IGARSS), pp 477–447
3. Craig DM, Ulaby FT (1986) Active microwave soil moisture research. IEEE Trans Geosci Remote Sens GE-24:23–36

4. Engman ET, Chauhan N (1995) Status of microwave soil moisture measurements with remote sensing. Remote Sens Environ 51:189–198
5. Dobson MC, Ulaby FT (1986) Preliminary evaluation of the SIR-B response to soil moisture, surface roughness, and crop canopy cover. IEEE Trans Geosci Remote Sens GE-24(4):517–526
6. Hallikainen MT, Ulaby FT, Dobson MC, El-Rayes MA, Wu LK (1985) Microwave dielectric behavior of wet soil, I, empirical models and experimental observations. IEEE Trans Geosci Remote Sens 23(1):25–34
7. Kornelsen KC, Coulibaly P (2013) Advances in soil moisture retrieval from synthetic aperture radar and hydrological applications. J Hydrol 476:460–489
8. Ulaby FT, Percy PB (1976) Optimum radar parameters for mapping soil moisture. IEEE Trans Geosci Elect Ge-14(2):81–93
9. Baghdadi N, Aubert M, Cerdan O, Franchistéguy L, Viel C, Eric M, Zribi M, Desprats JF (2007) Operational mapping of soil moisture using synthetic aperture radar data: application to the touch basin (France). Sensors 7:2458–2483
10. Srivastava HS, Patel P, Manchanda ML, Adiga S (2003) Use of multi incidence angle RADARSAT-1 sar data to incorporate the effect of surface roughness in soil moisture estimation. IEEE Trans Geos Remote Sens 41(7):1638–1640
11. Srivastava HS, Patel P, Navalgund RR (2006) Incorporating soil texture in soil moisture estimation from extended low-1 beam mode RADARSAT-1 SAR data. Int J Remote Sens 27(12):2587–2598
12. Srivastava HS, Patel P, Sharma Y, Navalgund RR (2009) Large-area soil moisture estimation using multi-incidence-angle RADARSAT-1 SAR data. IEEE Trans Geosci Remote Sens 47(8):2528–2535
13. Zribi M, Dechambre M (2002) A new empirical model to inverse soil moisture and roughness using two radar configurations. In: IEEE international geoscience and remote sensing symposium, vol.4. Toronto, Ontario, Canada, pp 2223–2225
14. Oh Y (2004) Quantitative retrieval of soil moisture content and surface roughness from multipolarized radar observations of bare soil surfaces. IEEE Trans Geosci Remote Sens 42:596–601
15. Shi J, Wang JR, O'neill P, Engman ET (1997) Estimation of bare surface soil moisture and surface roughness parameters using L-band SAR image data. IEEE Trans Geosci Remote Sens 35(5):1–13
16. Smith AB, Walker JP, Western AW, Young RI, Ellett KM, Pipunic RC, Grayson RB, Siriwidena L, Chiew FHS, Richter H (2012) The Murrumbidgee soil moisture monitoring network data set. Water Resour Res 48 W07701, 6. https://doi.org/10.1029/2012wr011976
17. Zribi M, Saux-Picart S, André C, Descroix L, Ottlé C, Kallel A (2007) Soil moisture mapping based on ASAR/ENVISAT radar data over a Sahelian region. Int J Remote Sens 28(16):3547–3565. https://doi.org/10.1080/01431160601009680
18. Dubois PC, van Zyl J, Engman T (1995) Measuring soil moisture with imaging radars. IEEE Trans Geosci Remote Sens 33(4):915–926

Landslide Hazard Mapping Using Geo-Environmental Parameters—A Case Study on Shimla Tehsil, Himachal Pradesh

C. Prakasam, R. Aravinth, Varinder S. Kanwar and B. Nagarajan

Abstract A landslide is defined as the movement of a mass of rock, debris, or earth down a slope. Landslides are a type of "mass wasting," which denotes any down-slope movement of soil and rock under the direct influence of gravity. The present research paper is an attempt to assess the vulnerability of Shimla Tehsil to landslides. Five causative factors such as land use and land cover, slope morphometry, relative relief, lithology, and hydrogeology are used to calculate the landslide vulnerability. Survey of India, Geological Survey of India Toposheets, ASTER GDEM, and LANDSAT 8 OLI/TIRS sensor were used as data sources. The causative factors were analyzed and processed in GIS environment. The weightages were assigned based on ratings derived from Bureau of Indian Standards, pp 1–19, 1998, [1] for macroscale landslide mapping. Finally, the factors were integrated using weighted overlay method to produce final vulnerability map. The final output was categorized into four types based on the Total Hazard Estimated Index (THED) ranging from very low hazard to high hazard. The output reveals that the entire Shimla Tehsil falls under four categories of vulnerability ranging from very low vulnerability to high vulnerability. 62.02% of the Shimla Tehsil is prone to low hazard followed by very low at 26.10%, moderate at 11.55%, and high covering only 0.32%. Most of the major settlements are located along moderately vulnerable area.

Keywords Landslide hazard evaluation factor · Total hazard estimate index · Land use and land cover · Weighted overlay · Slope morphometry · Relative relief

---

C. Prakasam (✉) · R. Aravinth · V. S. Kanwar
Department of Civil Engineering, Chitkara University, Baddi, Himachal Pradesh, India
e-mail: cprakasam@gmail.com

B. Nagarajan
Department of Civil Engineering, Indian Institute of Technology Kanpur, Kanpur, Uttar Pradesh, India

J. K. Ghosh and I. da Silva (eds.), *Applications of Geomatics in Civil Engineering*,
Lecture Notes in Civil Engineering 33, https://doi.org/10.1007/978-981-13-7067-0_9

1 Introduction

Hill ecosystem is one of the most fragile ecosystems in the world, leading to various geohazard and environmental problems due to various natural and anthropogenic causes [2]. Landslides are among one of the important catastrophic disasters occurring in mountain region leading to the change of surface geomorphology [3]. In India, about 0.42 million km^2 of the land area excluding snow-covered areas are prone to landslide hazards. Out of this 0.14 million km^2 of the area falls under north-east Himalayas (Darjeeling and Sikkim) (GSI website). The Himalayan region is more prone to slope instability due to its rugged terrain [4]. Each year the state is affected by one or more major landslides causing social and economic losses. Loss of life, damages to houses, roads, communication lines, and agricultural lands are some of the examples. The fragile nature of rocks forming mountain along with climatic and various anthropogenic causes has made state vulnerable to landslides. Deforestation, unscientific road construction, terracing and water-intensive agricultural practices, and encroachment have led to increased intensity and frequency of landslides State Disaster Management Plan (SDMP), Himachal Pradesh. Recent incidents that occurred on Mandi–Pathankot NH 154 with 46 life loss and damage to the road, and landslides that occurred near Dhalli Tunnel along the Kalka–Shimla NH 5A leading to damaged buildings, parked vehicles, and roads were some of the examples for the intensity of landslides.

Prakash [5] compiled a list of historical socioeconomically significant landslides around India, where 371 events of reported landslides event occurred for the year 1800–2011. Based on his findings he concluded that west and northwest of Himalayas are highly prone to landslides accounting for about 49% of the people killed and 51% of the landslide occurrences. The data on various occurrences and fatalities is given in Table 1.

Many researchers and scientists have previously analyzed landslide vulnerability for multiple areas. Anbalagan and Parida [6] studied the geo-environmental problems of Harmony Landslide in Uttarakhand. Anbalagan and Singh [7] studied landslide vulnerability mapping for Kumaun Himalayas using various causative factors such

Table 1 Data on various events and fatalities between (1800–2011)

Sl. no.	Year	No. of socioeconomically significant events	Persons killed	No. of fatal event
1	2011	26	74	19
2	2010	85	368	53
3	2009	47	270	46
4	2008	36	220	30
5	2007	54	409	39
6	1800–2007	123	2630	61

Source Prakash [5]

as geomorphology, geology, slope, soil, rainfall, etc. Kanungo et al. [8], Martha et al. [2], Naithani [9], Rawat et al. [10], Saha et al. [11], Panikkar, and Subramaniyan [12] were some other researchers who made notable contribution in the field of landslide vulnerability mapping.

1.1 Objectives

1. To prepare various thematic layers such as land use and land cover, relative relief, slope morphometry, etc.
2. To categorize the various layers into vulnerable hazards using Landslide Hazard Evaluation Factor (LHEF) scheme.
3. To delineate landslide vulnerable zones using weighted overlay method through remote sensing and GIS techniques.

2 Study Area

Shimla Tehsil is located between 30°59′3″ and 31°14′10″ North latitude and 76°58′19″ to 77°19′21″ East longitude. The total geographical extent of Shimla Tehsil is 36,830 ha (in Fig. 1). According to 2011 census Shimla Tehsil consists of 576 villages. As of 2011 census data, the total population of Shimla Tehsil is 1,71,640 people, among which 1,69,578 of them reside in Shimla municipal corporation and 2,062 of them belong to Shimla Rural monitored by Jutogh cantonment board [13]. The literacy rate stands at 93.63% which is greater than the literacy rate of Himachal Pradesh 82.8%. Sutlej, Pabbar, and Giri are the three main rivers that drain through the Shimla Tehsil. Timber and charcoal are the important forest produce. Besides them resins, grass, medicinal herbs, and bamboos are also produced at minor quantity. Agriculture and horticulture are the important economic activities in the Shimla Tehsil. Most of the Tehsil's agriculture is dependent on rainfall. The climate is suitable for growing cereals, off-season vegetables, temperate, and stone fruits. The soil is sandy loam at the valleys and skeletal in the mountainous areas. Geologically, the rock formations in Shimla district ranges from Pre-Cambrian to Quaternary period. The climate is subtropical in the valleys and temperate in the hilltops. The average annual rainfall of the district is 999.4 mm out of which 75% occurs during the monsoon period June to September. The temperature can go as low as 0 °C during winter times and as much as 40 °C during summer times.

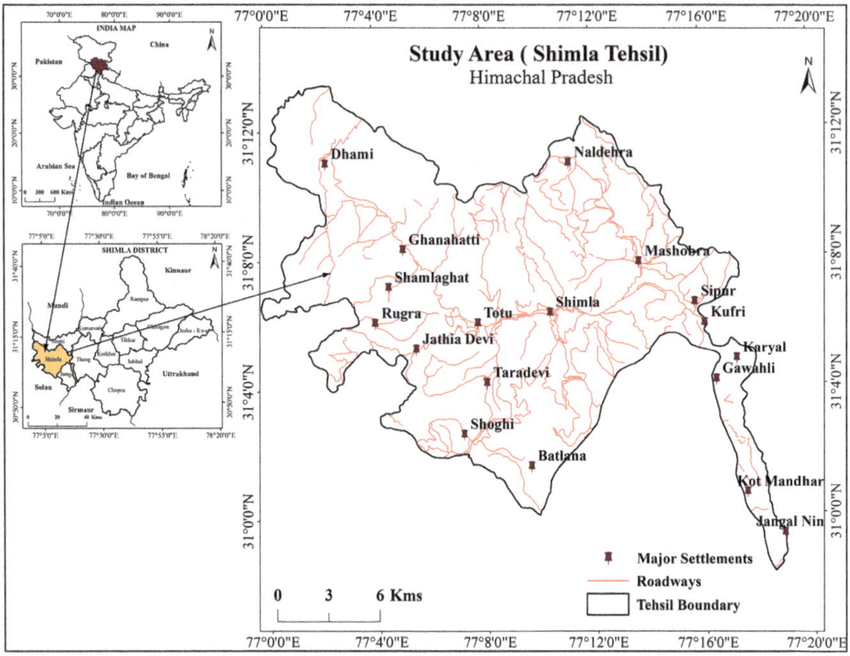

Fig. 1 Study area

3 Data Used

The base map was created using Survey of India Toposheets 53E/04, 53E/08, 53F/05 at 1:50,000 scale. The land use and land cover were prepared using Landsat 8 OLI/TIRS sensor obtained from United States Geological Survey (USGS) website. The lithology map was prepared using geology data obtained from Geological Survey of India. The relative relief and slope morphometry were prepared from ASTER GDEM data obtained from USGS website. The hydrogeological conditions were obtained from Central Ground Water Board (CGWB) brochure Shimla. Table 2 depicts the data used for the study.

4 Methodology

The methodology adopted in the research is weighted overlay analysis. It is a simple bivariate analysis model where weights are assigned to each causative factor and integrated to produce a final vulnerability map based on the rating assigned to the factors. The factors adopted in this study are referred from report published by Bureau of Indian Standards IS (14496) Part 2: 1998. The report is provided with the

Table 2 Data used

Sl. no	Data	Source	Year	Resolution
1	Toposheets	Survey of India	1974	1:50,000
2	Landsat 8 OLI	Earth explorer (USGS)	17/10/2017	30 m
3	Lithology	Geological Survey of India	1975	1:50,000
4	Hydrogeology	Central Ground Water Board	March, 2013	1:50,000
5	Slope morphometry	ASTER GDEM	–	30 m
6	Relative relief	ASTER GDEM	–	30 m

Table 3 Maximum LHEF rating for causative factors

Sl. no	Causative factors	Maximum LHEF rating
1	Lithology	2
2	Slope morphometry	2
3	Relative relief	1
4	Land use and land cover	2
5	Hydrogeology	1

Source BIS (14496) Part 2:1998

guidelines to map Landslide Hazard Zonation on macroscale. The Landslide Hazard Evaluating Factor Scheme (LHEF) uses six major causative factors such as lithology, slope morphometry, relative relief, land use and land cover, and hydrogeology to estimate the instability of the area. The factors and rating assigned to each factor are provided in Table 3.

The data were analyzed and processed in GIS environment to find the vulnerability extent of the landslide in Shimla Tehsil. The vulnerable zones were categorized into five zones ranging from very low hazard zone to very high hazard zone. The overall flow of research methodology was shown in Fig. 2 (Fig. 3).

5 Data Analysis

5.1 Land use and Land cover

Land use and land cover are an important causative factor in landslide vulnerability. Extensive deforestation and construction of buildings may reduce the cohesion between soils and can cause slope instability leading to mass movements and landslides. Land use and land cover were prepared from LANDSAT 8 OLI/TIRS sensor for the year 2017. The data was layer stacked and processed through supervised classification using maximum likelihood classifier method. Shimla Tehsil was classified

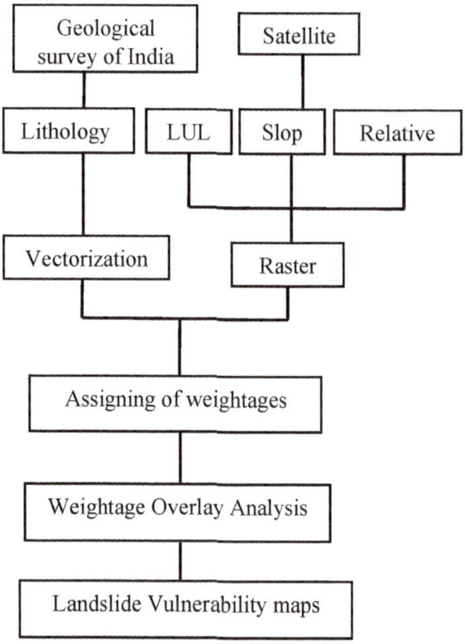

Fig. 2 Methodology

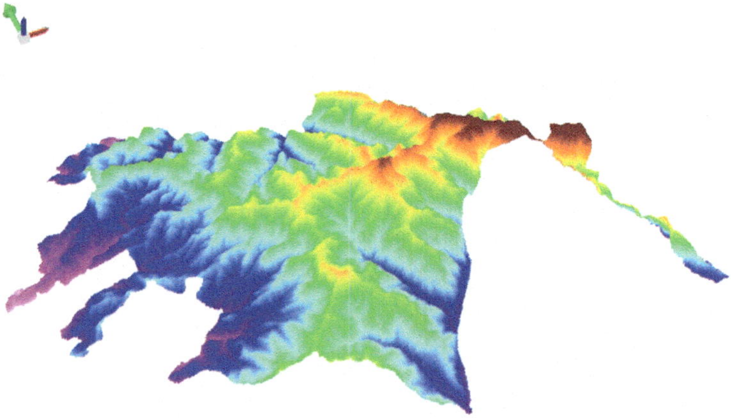

Fig. 3 3D digital elevation model of Shimla Tehsil

into four classes, namely, forest, agriculture, barren land, and built-up area. Land use and land cover were classified using NRSC Level 1 classification system. 63% of the area falls under forest cover, 14% of the area falls under agriculture, slope, and settlement covers 12 and 9%, respectively. Land use and land cover of Shimla Tehsil are described in Fig. 4 (Table 4).

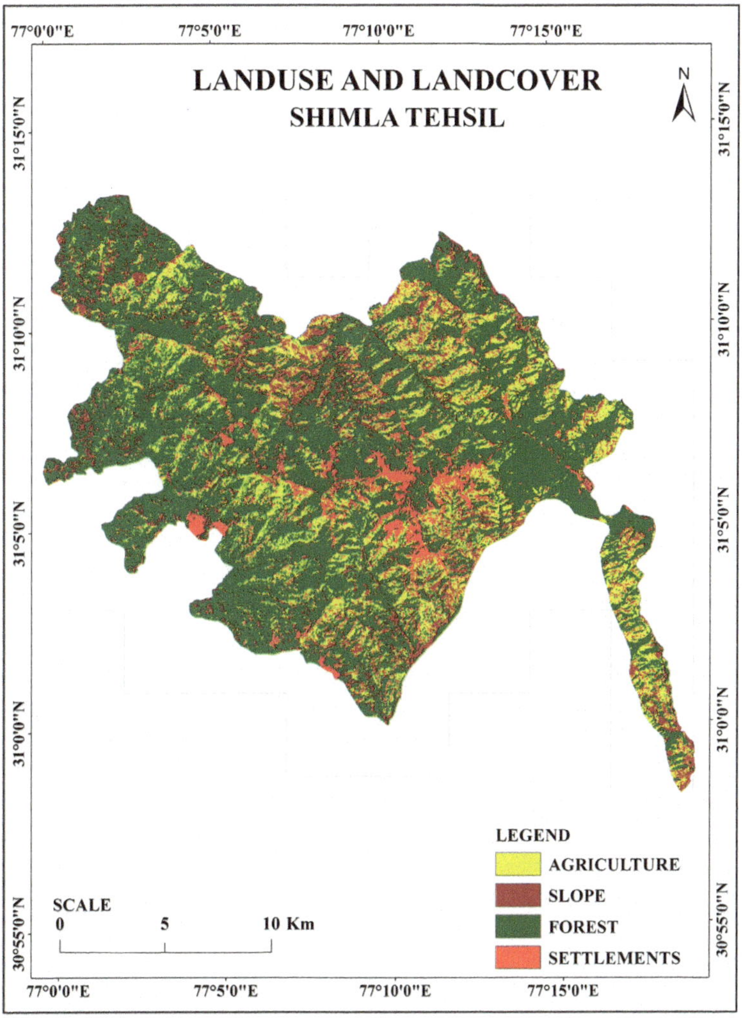

Fig. 4 Land use and land cover

Table 4 Area coverage LULC

Sl. no	Class	Area (ha)	Percent (%)
1	Agriculture	5,225.31	14.19
2	Forest	23,354.53	63.42
3	Settlements	3,605.37	9.79
4	Slope	4,639.24	12.60
	Total	36,824	100.00

(*Source* Authors calculation)

5.2 *Lithology*

Lithology is study of physical characteristics of rocks such as size, texture, core samples, etc. Lithology was digitized from Geological Survey of India maps at 1:50,000 scales, etc. Shimla Tehsil is composed of 11 major rock-forming groups such as Bhotli, Basanti, Chhaso, Sanjauli, Taradevi, Subathu formations, etc. Major rock types found in the Tehsil were quartzite, phyllite, slate, limestone, shale, granite, dolomite, etc. The ratings were assigned based on LHEF scheme. Rocks such as quartzite, granite, gabbro, and limestone were classified as Type 1 rocks and were assigned lower values. Type 2 was classified as well-cemented and poorly cemented sedimentary rocks. Slate, phyllite, schist, and shale were categorized as Type 3 rocks and were given higher ratings. Lithology of Shimla Tehsil is shown in Fig. 5 (Table 5).

5.3 *Slope Morphometry*

Slope angle was considered as a parameter as it has considerable influence in the landslide hazards. The slope map was derived from ASTER GDEM data with the resolution of 30 m. Shimla Tehsil was highly dominated by gentle and steep slopes. The slope map was processed in GIS environment and categorized into four types, namely, very gentle, gentle, moderate, steep slope, etc. Figure describes the slope variations along the Shimla Tehsil. Slope morphometry of Shimla Tehsil is described in Fig. 6 (Table 6).

5.4 *Relative Relief*

Relative relief is defined by the range between minimum and maximum elevations in a given area. In other words, it is the difference in height between highest and

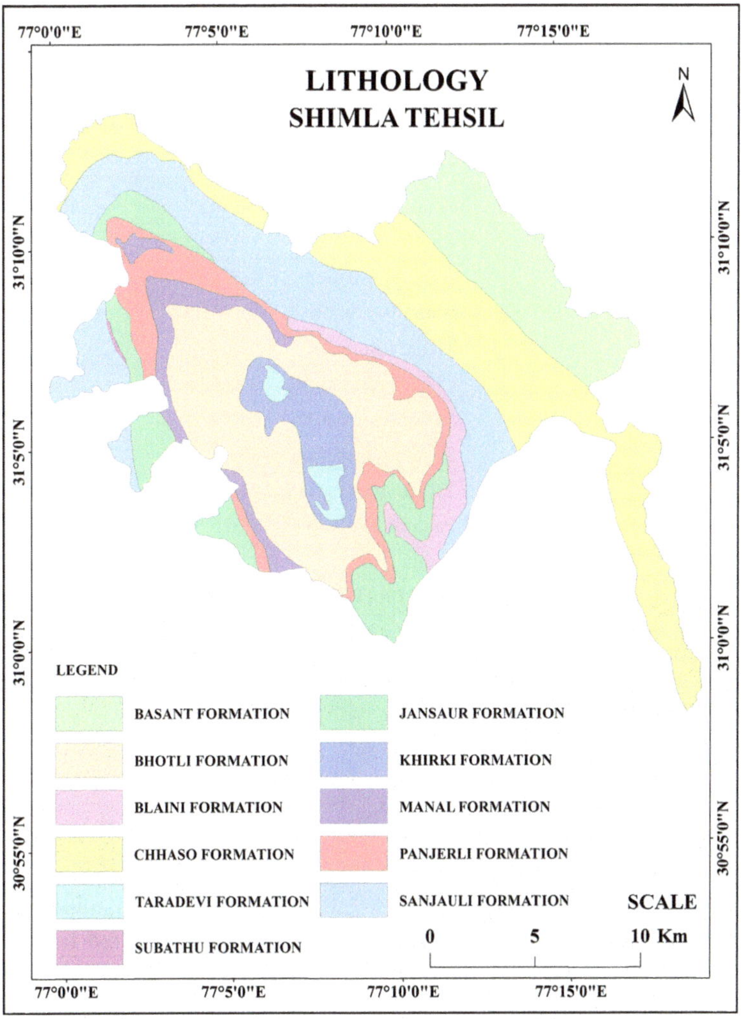

Fig. 5 Lithology

lowest points in a determined surface grid. A 10,000 m^2 grid was created for Shimla Tehsil in the GIS environment and the relative relief was calculated for every grid. The relative relief ranges from 0 to 188 in the study area. The relative is classified as low and moderate based on the LHEF rating. Relative relief of Shimla Tehsil is described in Fig. 7 (Table 7).

Table 5 Area coverage lithology

Sl. no	Rock type	Rock formation	Area (ha)	Percent (%)
1	Type 1 rock	Jansaur formation	3,272.96	30.25
2		Khirki formation	1,677.58	15.50
3		Manal formation	459.99	4.25
		Total	5410.53	50.00
1	Type 2 rock	Bhotli formation	7,335.11	11.67
2		Basant formation	4,197.47	6.68
3		Blaini formation	1,183.73	1.88
4		Manal formation	936.93	1.49
5		Chhaso formation	7,810.64	12.43
6		Sanjauli formation	6,824.21	10.86
7		Subathu formation	30.01	0.05
8		Taradevi formation	419.50	0.67
9		Panjerli formation	2,676.51	4.26
		Total	31,414.09	50.00
		Type 1 rock	5,410.53	14.69
		Type 2 rock	31,414.09	85.31
		Total	36,824.63	100.00

(*Source* Authors calculation)

5.5 Hydrogeological Conditions

Hydrogeology deals with the movement of groundwater between soils and rocks in the earth crust. Geologically, the rock formations in Shimla ranges from Pre-Cambrian to Quaternary Period Hydrogeological data was acquired from Central Ground Water Board (CGWB), Shimla. Shimla district is covered by soft and hard rock aquifers. Most part of the district is covered by hard rock aquifers of the Proterozoic age and soft rock aquifers are found along the Sunni Tehsil in a smaller amount. Based on the landslide reports from Geological Survey of India and Central Ground Water Board, the rating assigned for the hydrogeological condition is 0.8.

6 Result and Discussion

In the present study, Landslide Hazard Evaluation Factors scheme was used to calculate the landslide vulnerability. Factors such as lithology, LULC, slope, relative relief, and hydrogeological conditions were taken into account to calculate the landslide vulnerability (Fig. 8). The factors were converted to raster datasets and the

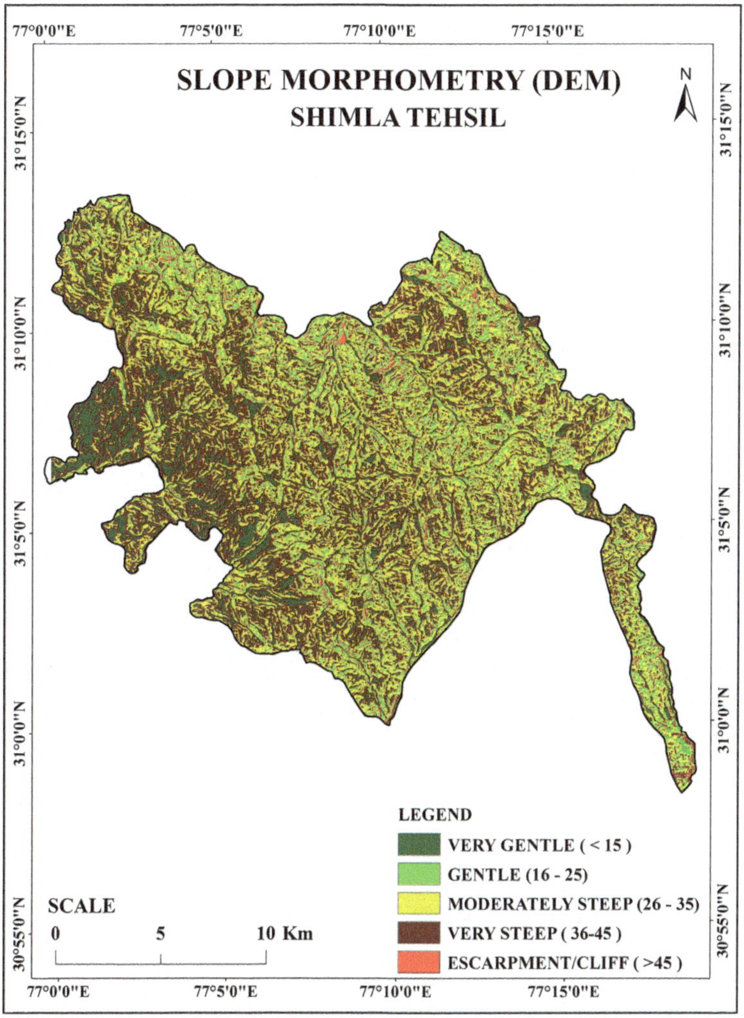

Fig. 6 Slope morphometry

Table 6 Area coverage slope

Sl. no	Class	Degrees	Area (ha)	Percent (%)
1	Very gentle	<15°	4,998.71	13.57
2	Gentle	16°–25°	12,362.31	33.57
3	Moderately steep	26°–35°	13,269.57	36.03
4	Steep	36°–45°	5,198.73	14.12
5	Escarpment	>45°	995.51	2.70
		Total	36,824	100.00

(*Source* Authors calculation)

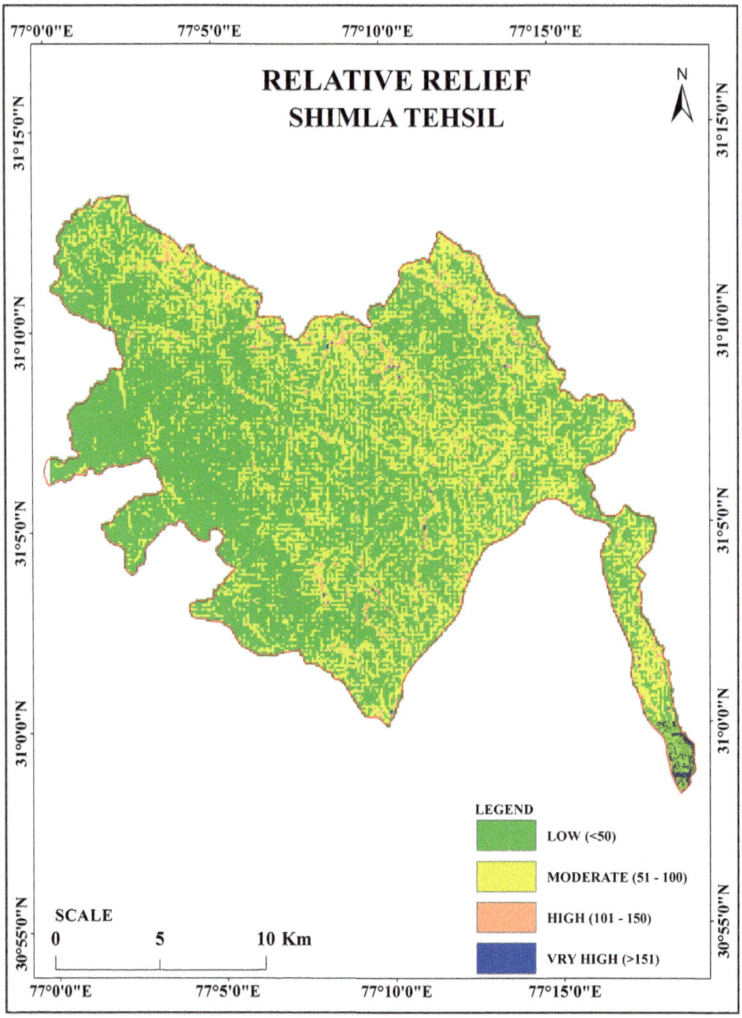

Fig. 7 Relative relief

Table 7 Area coverage relative relief

Sl. no	Class	Range	Area (ha)	Percent (%)
1	Low	<50 m	21,965.25	59.65
2	Moderate	51–150 m	14,140.32	38.40
3	High	101–150 m	681.82	1.85
4	Very high	>150 m	37.26	0.10
		Total	36,824	100.00

(*Source* Authors calculation)

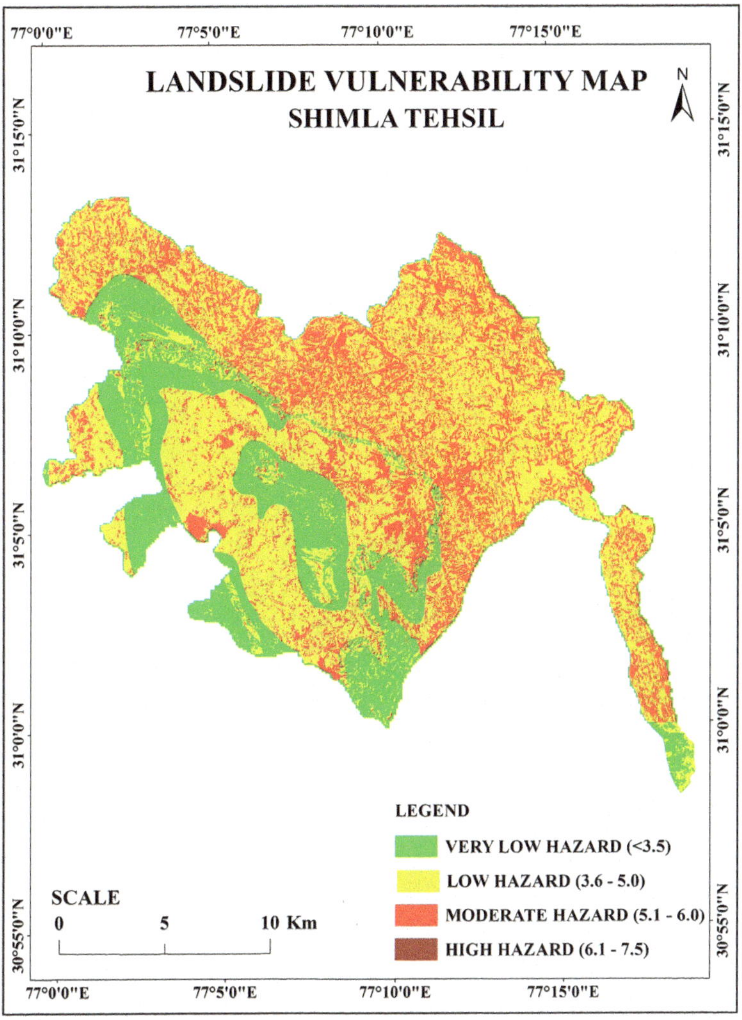

Fig. 8 Landslide vulnerability map

weightages were processed using weighted overlay method. Given in Table 8 is the description of vulnerability zones and amount of are fall within them (Table 9).

Table 10 represents the hazard vulnerability and total area contributed by each subfactors. Most of the area in LULC (23,254.53 ha), slope morphometry (13,269.57 ha), and relative relief (14,140.32 ha) falls under moderate hazard, whereas lithology (27,380.17 ha) and hydrogeology (36,284 ha) fall under very high and high category. The Total Hazard Estimation Index (THED) is calculated by overlaying all factor maps using weighted overlay methods. The final LHEF rating varies from 0.3 to 6.6. The output reveals that the entire Shimla Tehsil falls under four categories of

Table 8 Area coverage landslide hazards

Sl. no	Zone	Zone description	THED value	Area (ha)	Percent (%)
1	I	Very low	<3.5	9,613.20	26.10
2	II	Low	3.6–5.0	22,840.10	62.02
3	III	Moderate	5.1–6.0	4,254.41	11.55
4	IV	High	6.1–7.5	118.00	0.32
			Total	36,825.70	100.00

Table 9 It depicts the relative class for various factors and their related subfactors present in their ground surface and the weightage assigned to them

Land use class	Subclass	THED value	Hazard categorization
LULC	Agricultural land	0.6	Very low hazard
Slope morphometry	Very gentle	0.5	
Relative relief	Nil	0	
Lithology	Jansaur, Khirki, and Manal formation	0.2	
Hydrometeorology	Dry	0	
LULC	Built-up land	0.6	Low hazard
Slope morphometry	Gentle	0.8	
Relative relief	Low	0.3	
Lithology	NIL	0	
Hydrometeorology	Damp	0.2	
LULC	Forest area	0.8	Moderate hazard
Slope morphometry	Moderate	1.2	
Relative relief	Medium	0.6	
Lithology	Nil	0	
Hydrometeorology	Wet	0.5	
LULC	Slope/barren land	2	High hazard
Slope morphometry	Steep slope	1.7	
Relative relief	High	1	
Lithology	Taradevi and Panjerli formation	1.2	
Hydrometeorology	Dripping	0.8	
LULC	Nil	0	Very high hazard
Slope morphometry	Escarpment	2	
Relative relief	Nil	0	

(continued)

Table 9 (continued)

Land use class	Subclass	THED value	Hazard categorization
Lithology	Bhotli, Basant, Blaini, Chhaso, Sanjauli, and Subathu formation	2	
Hydrometeorology	Flowing	1	

Table 10 Relative vulnerability classes and the factors contributing to each vulnerability class along with their area

Hazard vulnerability	LULC (ha)	Slope morphometry (ha)	Relative relief (ha)	Lithology (ha)	Hydrogeology (ha)
Very low	5,225.31	4,998.71	0	6347.46	0
Low	3,605.37	12,362.31	21,965.25	0	0
Moderate	23,354.53	13,269.57	14,140.32	0	0
High	4,639.24	5,197.73	680.82	3,096.01	36,824
Very high	0.00	995.51	37.26	27380.17	0
Total	36,824	36,824	36,824	36,824	36,824

vulnerability ranging from very low vulnerability to high vulnerability. 62.02% of the Shimla Tehsil is prone to low hazard followed by Very low at 26.10%, moderate at 11.55%, and high covering only 0.32%. Most of the major settlements are located along moderately vulnerable area.

Figure 9a represents the recently occurred landslide along Dhalli tunnel in Shimla town. It is a rainfall-induced debris slide with a running distance of about 100 m. Figure 9b, c shows damaged buildings along the downside of the landslide. Some buildings and nearby parked vehicles are heavily damaged. Figure 9d, e shows some of the landslides noted along the Shimla to Jhakri National Highway 22. The rock structure present along these NH is highly weathered and loose soil. These roads connect some of the remote corners of villages in Shimla Tehsil. Landslide occurrences along these area not only possess physical problems but also great risks in times of disaster and medical emergencies.

7 Conclusion

The present study demonstrates the application of weighted overlay model for landslide hazard mapping using remote sensing and GIS for Shimla Tehsil. Integration of remote sensing and GIS datasets provides a reliable and updated database on land and water resources, and can be used in the process of decision-making. High-resolution satellite imageries and real-time climatic data have increased the level of accuracy of LHZ maps in time. Increase in human population which may lead to haphazard

Fig. 9 **a** Rock slide occurred along the NH 5A in Shimla town. **b** and **c** settlements damaged due to the landslide. **d** and **e** landslide occurred along a place called Luhri located in Shimla—Rampur National Highway

settlement and uncontrolled expansion can induce the occurrence of landslides. The landslide vulnerability can be used as a reference when constructing suitable stabilization measure and can also provide an insight in determining the safest location for the constructional activities. Even though the high hazard areas are only few percent compared to moderate and low hazard, they can easily become prone to landslide through unplanned anthropogenic activities. This approach of landslide hazard zonation using remote sensing and GIS is advantageous for rapid assessment when the area is not accessible. Use of high-resolution images and DEM data can increase the accuracy of the landslide hazard assessments.

Acknowledgements The research work presented in this conference is a part of NRDMS-DST funded research project entitled "Landslide Prone area mapping—Geospatial Approach: A Case study of Shimla district, Himachal Pradesh". We would like to express our sincerest gratitude to NRDMS-DST, GOI, New Delhi, India for funding the research project.

References

1. Preparation of landslide hazard zonation maps in mountainous terrain—guidelines (Part2-Macrozonation). In: Bureau of Indian Standards, vol 14496, 2nd edn. BIS, New Delhi, pp 1–19 (1998)
2. Martha TR, Kerle N, van Westen CJ, Jetten V, Vinod Kumar K (2012) Object-oriented analysis of multi-temporal panchromatic images for creation of historical landslide inventories. ISPRS J Photogramm Remote Sens 67(1):105–119. https://doi.org/10.1016/j.isprsjprs.2011.11.004
3. Gupta RP, Joshi BC (1990) Landslide hazard zoning using the GIS approach—a case study from the Ramganga catchment, Himalayas. Eng Geol 28(1–2):119–131. https://doi.org/10.1016/0013-7952(90)90037-2
4. Panikkar SV, Subramanyan V (1996) A geomorphic evaluation of the landslides around Dehradun and Mussoorie, Uttar Pradesh, India. Geomorphology 15(2):169–181. https://doi.org/10.1016/0169-555X(95)00121-K
5. Prakash S (2016) Historical records of socio-economically significant landslides in India. J South Asian Stud 4(2):177–204. https://doi.org/10.1080/01431160010014260
6. Anbalagan R, Parida S (2013) Geoenvironmental problems due to harmony landslide in Garhwal Himalaya, Uttarkhand, India. Int J Emerg Technol Adv Eng 3(3):553–559
7. Anbalagan R, Singh B (1996) Landslide hazard and risk assessment mapping of mountainous terrains-a case study from Kumaun Himalaya, India. Eng Geol 43:237–246. https://doi.org/10.1016/S0013-7952(96)00033-6
8. Kanungo D, Arrora M, Sarkar S, Gupta R (2009) Landslide susceptibility zonation (LSZ) mapping-a review. J South Asia Disaster Stud 2:81–105
9. Naithani A (2007) Macro landslide hazard zonation mapping using univariate statistical analysis in parts of Garhwal Himalaya. J Geol Soc India 70:353–368
10. Rawat MS, Uniyal DP, Dobhal R, Joshi V, Rawat BS, Bartwal A, Aswal A et al (2015) Study of landslide hazard zonation in Mandakini Valley, Rudraprayag district, Uttarakhand using remote sensing and GIS. Current Sci 109(1):158–170
11. Saha AK, Gupta RP, Arora MK (2002) GIS-based landslide hazard zonation in the Bhagirathi (Ganga) Valley, Himalayas. Int J Remote Sens 23(2):357–369
12. Panikkar S, Subramaniyan V (1997) Landslide hazard analysis of the area around Dehra Dun and Mussoorie, Uttar Pradesh. Curr Sci 73:1117–1123
13. Census 2011 list for Himachal Pradesh. http://www.census2011.co.in/census/city/4-shimla.html
14. Central ground water board brochure. http://www.cgwb.gov.in/District_Profile/HP/Shimla.pdf
15. District state disaster management plan. http://www.hpsdma.nic.in/DisasterManagement/SDMP.html

Part IV
Geomatics in Hydraulics Engineering

Estimation of Groundwater Recharge Rate Using SWAT MODFLOW Model

K. N. Loukika, K. Venkata Reddy, K. H. V. Durga Rao and Amanpreet Singh

Abstract Groundwater, which is in aquifers below the surface of the Earth, is one of the most important natural resources. Due to the increased population, drought, improper irrigation practices, and pollution, depletion of groundwater takes place more rapidly. With the advent of remote sensing data and Geographical Information Systems (GIS) and integrating SWAT with MODFLOW model, estimation of groundwater recharge became much easier. The HRU's in the SWAT model are exchanged with MODFLOW cells by using SWAT-MODFLOW interface to simulate the groundwater head distribution and groundwater recharge rates. The study is conducted in Chintalapudi village, which is located in West Godavari district, Andhra Pradesh. To estimate spatiotemporal distribution of groundwater recharge, it requires DEM, LULC map, Soil map, rainfall data, aquifer parameters such as aquifer thickness, hydraulic conductivity, and specific yield. Model results are shown by month and year to determine seasonal and decadal trends in groundwater recharge rates. The SWAT-MODFLOW was calibrated and validated using groundwater levels and streamflow data.

Keywords Grounwater recharge · SWAT-MODFLOW · GIS

K. N. Loukika (✉)
NIT Warangal, Warangal 506004, India
e-mail: loukika.iiit@gmail.com

K. Venkata Reddy
Department of Civil Engineering, NIT Warangal, Warangal 506004, India
e-mail: kvreddy229@gmail.com

K. H. V. Durga Rao · A. Singh
NRSC, Hyderabad, India
e-mail: durgarao_khv@nrsc.gov.in

A. Singh
e-mail: amanpreetsingh@nrsc.gov.in

J. K. Ghosh and I. da Silva (eds.), *Applications of Geomatics in Civil Engineering*,
Lecture Notes in Civil Engineering 33, https://doi.org/10.1007/978-981-13-7067-0_10

1 Introduction

Water is precious and the most commonly used resource. Groundwater is the quantity of water added to the aquifer. Water is an important natural resource which supports life. The major source for groundwater is due to the rainfall. Other sources for groundwater recharge include recharge from rivers, streams, irrigation water, etc. There has been an increase in groundwater development and utilization, particularly in developing countries, for agriculture, industry, and rural water supply. There is an increase in population, urbanization, and it directly implies the demand for groundwater. To maintain balanced groundwater levels, it is required to do groundwater modeling, which includes simulation and estimation of groundwater recharge.

To optimize the utilization of groundwater and surface water, there is a need to estimate the groundwater recharge for sustainable water management. To know the annual, decadal changes in the groundwater recharge several techniques were implemented to access the groundwater recharge for smaller scales like Hydrus 1D [5]. For spatiotemporal scales, numerical modeling has been developed. In this paper, groundwater recharge is estimated by integrating SWAT with the MODFLOW model. The HRU's are disaggregated, i.e., converting from multipart to single part polygons [3]. There is a need to spatially locate the HRU's which are generated from the SWAT model [2]. The linking of SWAT and MODFLOW can be carried out through SWAT MODFLOW interface if there is one soil layer. If the number of layers is more, should have to separately run both the models and linking should be done. For finding the recharge, the unsaturated flow package and evaporation flow package in MODFLOW should be turned off. The outputs from SWAT are exchanged with spatially discretized MODFLOW grid cells [7]. After exchanging, the model results are spatially visualized in GIS.

2 Location and Extent

Chintalapudi is located in West Godavari district of Andhra Pradesh in India is taken as a study area. The total area of the study area is 668 km^2. It lies between 17°30′N to 16°15′N latitude and 81°30′E to 80°50′E longitude. The highest elevation range is 772 m and the lowest elevation range is 20 m. It consists of clay and loamy soils. Most of the area is covered with plantations and forests. High-capacity sets are operated more comparatively to tube wells. Ground water circulation is limited to 100 m depth. Ground water levels are depleted to 180 m. The entire study area is shown in Fig. 1.

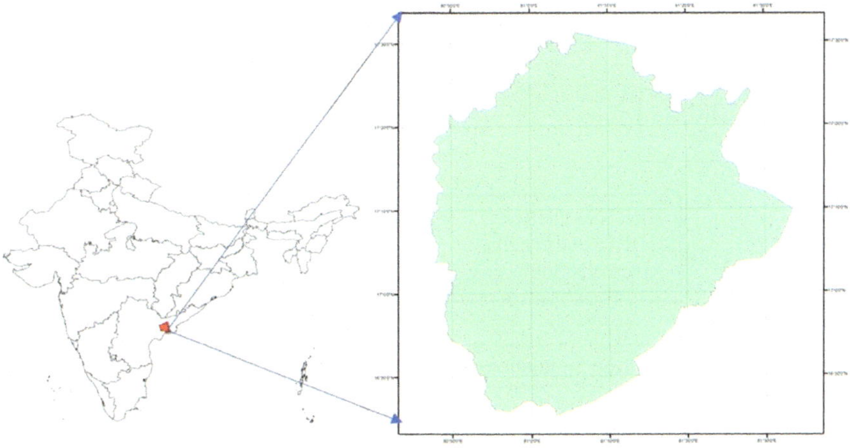

Fig. 1 Geographical area of Chintalapudi in West Godavari district

3 Methodology

The methodology for the study area is the coupling of SWAT model with MODFLOW model. Watershed delineation is done through SWAT using Digital Elevation Model (DEM). In SWAT the entire area is divided into subbasins and HRU's using Land Use and Land Cover (LULC) map and Soil map. The total study area is divided into 430 subbasins and subbasins are then divided into HRU's. The threshold for creating HRU's should be zero and 1792 HRU's were generated. The weather data, precipitation data, and temperature data are given to run the SWAT model to get recharge values, i.e., percolation to shallow aquifer and deep aquifer values. The methodology is shown in Fig. 2.

MODFLOW is the finite difference groundwater model used for subsurface flows [6]. The spatial discretization of the two models is different. Integration of SWAT and MODFLOW for estimating groundwater recharge based on water balance components. The HRU's and subbasins which are generated from SWAT model are linked with the finite difference grid of MODFLOW model. There is a need to give the spatial location or geolocate the HRU's which are derived from SWAT model and to link with the grid of MODFLOW which is spatially discretized. After linking both the models we will get a unique spatial framework.

The interface consists of four steps (1) Spatial setup: Loading SWAT model, specifying grid cell size, performing linking operations. (2) MODFLOW: Loading required data to construct MODFLOW model. (3) RT3D: Loading required data to

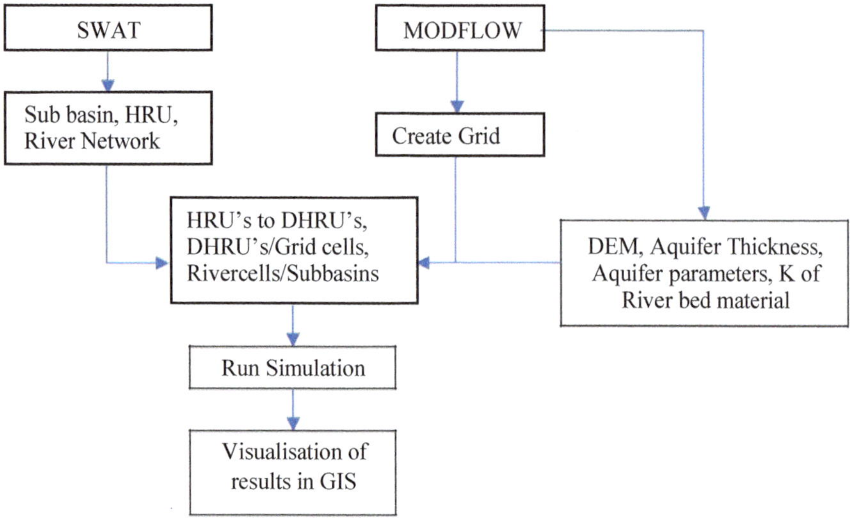

Fig. 2 Methodology for the coupled SWAT-MODFLOW model

construct RT3D (optional). (4) Simulation: Specifying timing information, output options, and running the simulation. There is a need to give the spatial location or geolocate the HRU's which are derived from SWAT model and to link with the grid of MODFLOW which is spatially discretized. The exchange of recharge values which we got through simulation from SWAT can be done with the MODFLOW grid. Then it is linked with DHRU's and it is spatially visualized in GIS environment.

3.1 Input Data

DIGITAL ELEVATION MODEL (DEM): The Carto DEM which has a resolution of 2.5 m is taken for the given study area. The maximum and minimum elevations of study area are 714 and 24 m, respectively. The soil map for the study area is clipped from SWAT Indian datasets and the entire area has 4 types of soils, i.e., loam, clay, sandy loam, and sandy clay. The land use and land cover map are given by NRSC, ISRO Hyderabad. The necessary inputs to generate a SWAT model are considered as land use and land cover map, soil map, DEM, and weather data which include the precipitation and temperature data for the past 15 years. DEM, LULC map, and Soil map are shown in Figs. 3, 4, and 5.

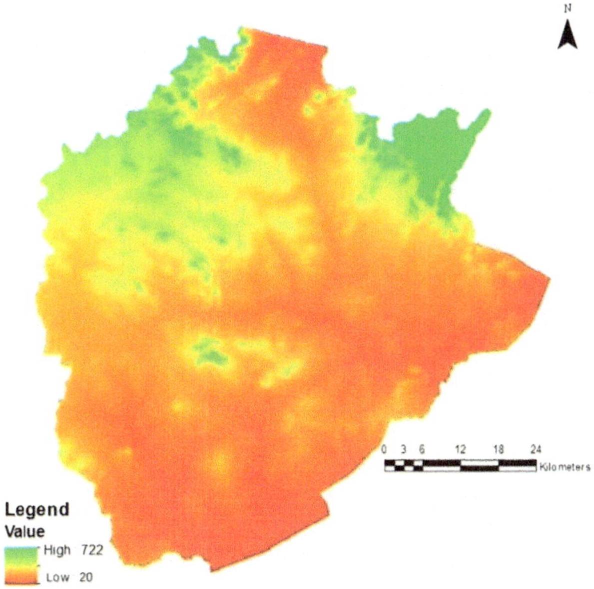

Fig. 3 DEM for the study area

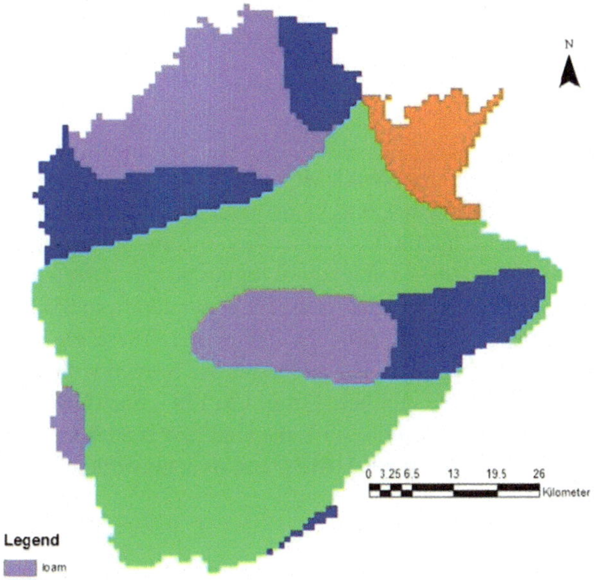

Fig. 4 Soil map

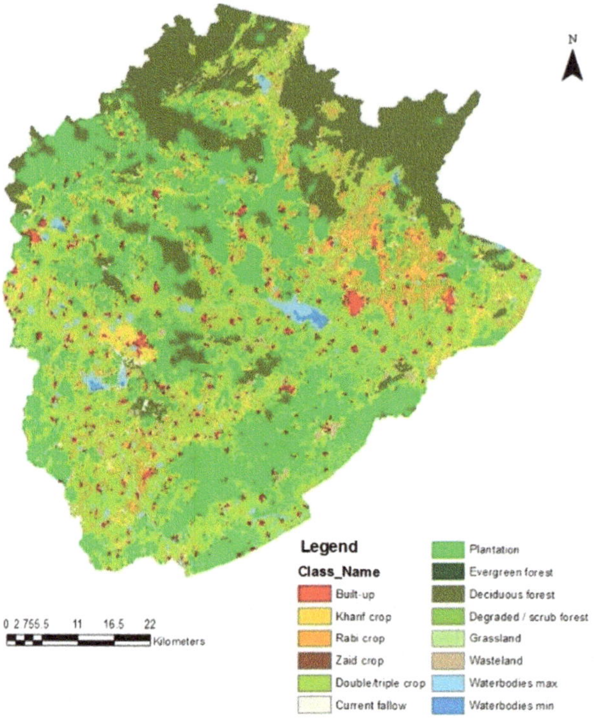

Fig. 5 Land Use and Land Cover map of the study area

The necessary input data for SWAT-MODFLOW model are the outputs obtained from SWAT Model are Subbasins, HRU's and river network data, aquifer parameters such as aquifer thickness, hydraulic conductivity, specific storage, specific yield are taken from the literature review and textbook values based on the aquifer conditions.

The subbasins, HRU's, and river streams are generated from SWAT model which are used for integrating with the MODFLOW model. These are generated by the setup of SWAT Model. A total of 431 subbasins are formed with a dominant HRU count of 1793. The HRU's are numbered from 1 to 1793. Each HRU has its own land cover, soil, and its slope. For a subbasin 1, the land covers are forest and agricultural land with soil loam. The Subbasins, HRU's, and river network are shown in Figs. 6, 7, and 8 (Table 1).

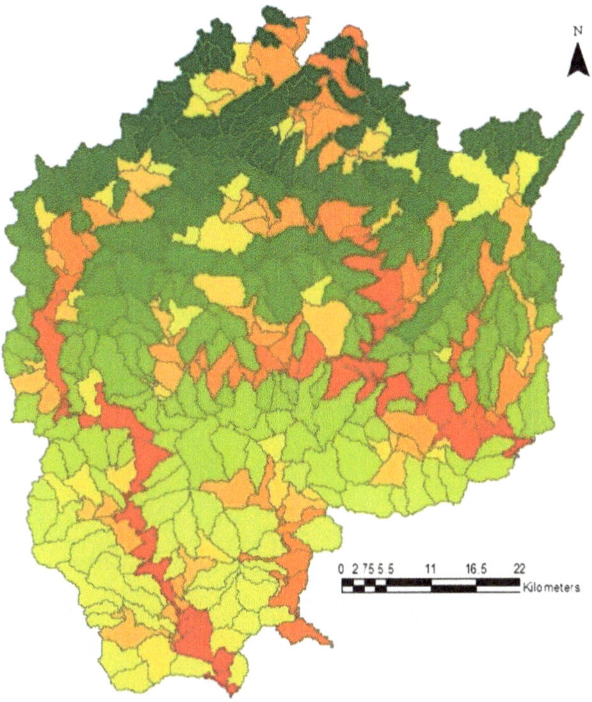

Fig. 6 Subbasins map obtained from SWAT model

From the above table, a clear understanding can be made about the land cover distribution in the study area. The Agricultural land is occupying nearly 43.74% of entire watershed whereas Evergreen Forests are very less occupying 0.01% of the watershed. The study area is agriculture dominant and plantations are considered as next dominant with 24.88%.

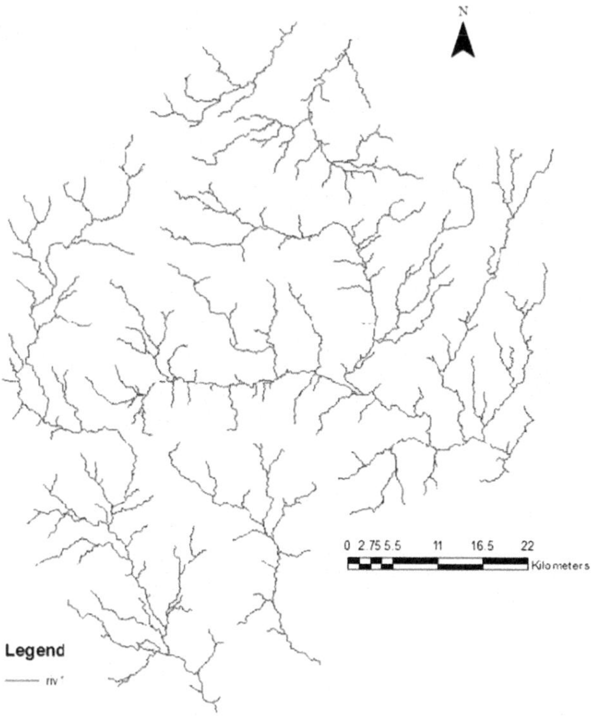

Fig. 7 River network map obtained from SWAT model

4 Results and Discussions

The study area mostly consists of agriculture area. By applying the coupled SWAT-MODFLOW model, groundwater recharge is estimated. The parameters considered to estimate groundwater recharge are precipitation, temperature and aquifer parameters like hydraulic conductivity, specific storage, specific yield, aquifer thickness. The groundwater recharge is simulated for the period of 1990–2005. The estimated groundwater recharge in the month of January was 16.5272 mm and in the month of February was 17.5596 mm. The simulated groundwater recharge is shown in Figs. 9 and 10.

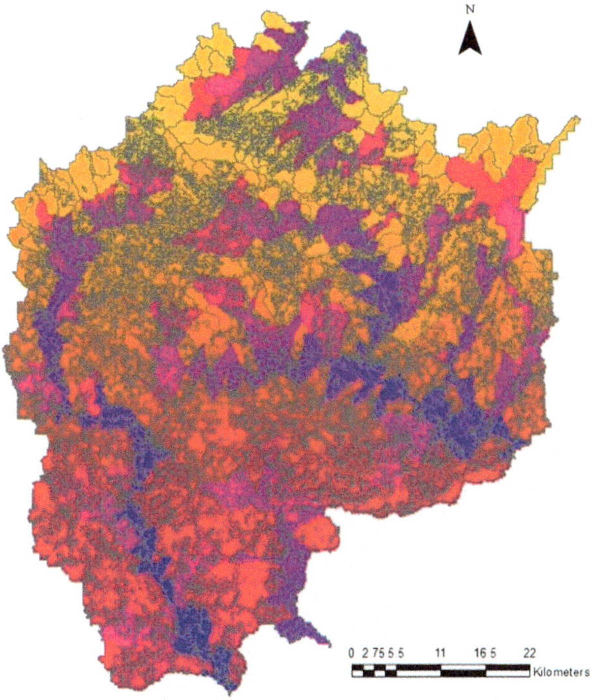

Fig. 8 HRU's obtained from SWAT model

Table 1 The area of land cover type and percent of area occupied in the entire watershed of study area

Land use	Area	% of watershed
URLD	9900.63	2.56
AGRL	169145.01	43.74
BARR	16362.00	4.23
AGRR	96218.28	24.88
FRST	77505.66	20.04
RNGE	1301.67	0.34
WATR	13995.18	3.62
FRSE	26.73	0.01

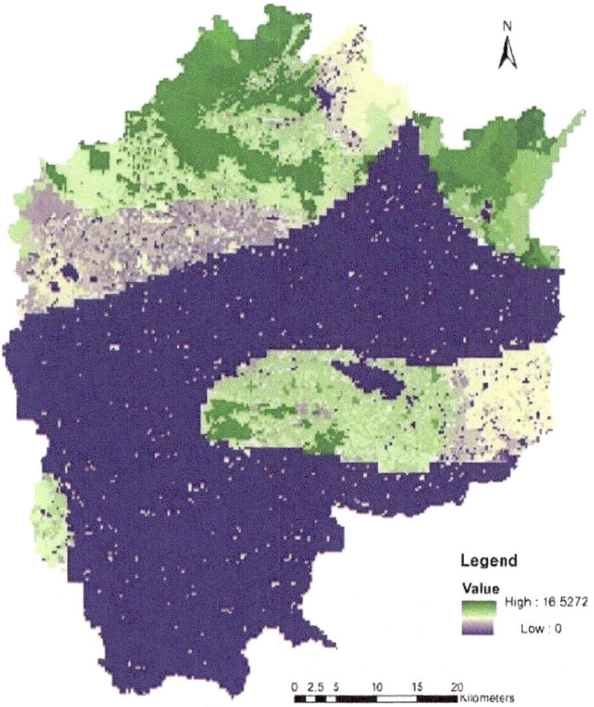

Fig. 9 Ground water recharge layer for the month of January

5 Conclusions

Groundwater recharge is estimated using SWAT-MODFLOW model [1]. The spatial discretization of two models is different and we will get a unique spatial framework after linking both the models. The sustainable management of groundwater is required for future generations. Remote sensing and GIS makes work easier for the estimation of groundwater recharge rates. The aquifer parameters such as hydraulic conductivity, specific yield, and specific storage show the impact on groundwater recharge [4]. The results obtained from SWAT-MODFLOW model are visualized monthly.

Acknowledgements The work presented in this paper is done for Chintalapudi in West Godavari district as a part of the project given by NRSC, Hyderabad. We would like to thank K H V Durga Rao scientist at NRSC and Aman Preet Singh for providing data used in this paper.

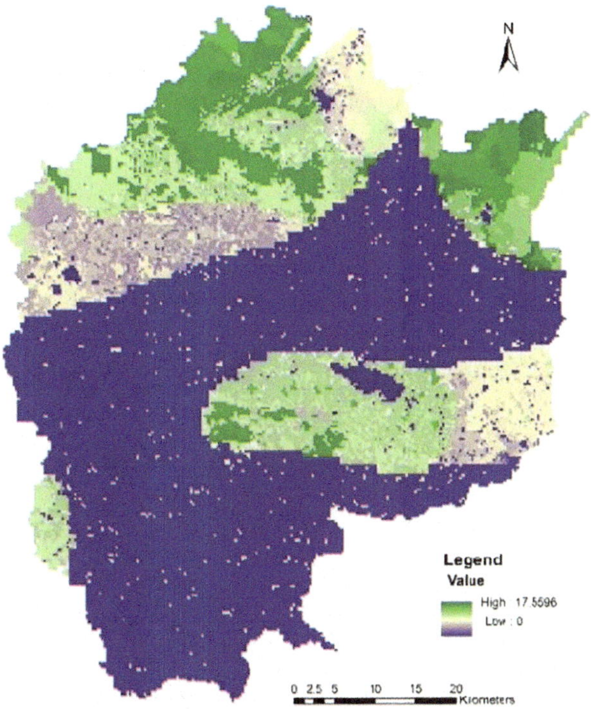

Fig. 10 Ground water recharge layer for the month of February

References

1. Bailey R, Rathjens H, Bieger K, Chaubey I, Arnold J (2017) SWATMOD-Prep: graphical user interface for preparing coupled SWAT-MODFLOW simulations. J Am Water Resour Assoc 53:400–410
2. Bailey RT, Wible TC, Arabi M, Records RM, Ditty J (2016) Assessing regional-scale spatio temporal patterns of groundwater–surface water interactions using a coupled SWAT-MODFLOW model. Hydrol Process 30:4420–4433
3. Chung IM, Kim NW, Lee J, Sophocleous M (2010) Assessing distributed groundwater recharge rate using integrated surface water-groundwater modelling: application to Mihocheon watershed, South Korea. Hydrogeol J 18:1253–1264
4. Ehtiat M, Mousavi SJ, Srinivasan R (2018) Groundwater modeling under variable operating conditions using SWAT, MODFLOW and MT3DMS: a catchment scale approach to water resources management. Water Resour Manag 1–9
5. Guzman JA, Moriasi DN, Gowda PH, Steiner JL, Starks PJ, Arnold JG, Srinivasan R (2015) A model integration framework for linking SWAT and MODFLOW. Environ Model Softw 73:103–116

6. Harbaugh AW, Banta ER, Hill MC, McDonald MG (2000) MODFLOW-2000, the U.S. geological survey modular ground-water model-Users guide to modularization concepts and the ground-water flow process. Open-File Report 00–92. Reston, Virginia: U.S. Geological Survey, 121
7. Kim NW, Chung IM, Won YS, Arnold JG (2008) Development and application of the integrated SWAT-MODFLOW model. J Hydrol 356:1–16

Flood Hazard Risk Assessment and Mapping of a Sub-watershed of Imphal River Basin, Manipur, India: A Multi-resolution Approach

Maisnam Bipinchandra, Ngangbam Romeji and Chandramani Loukrakpam

Abstract Flood is one of the major disasters that prevail in the northeastern India. Mapping of spatial inundation patterns during flood events is regarded for environmental management and disaster monitoring. This study emphasized on the assessment and flood inundation mapping of sub-watershed of Imphal River Basin of Manipur. Multispectral datasets from Landsat-8 OLI and Sentinel-2 and Digital Elevation Models (DEM) from ASTER and Cartosat with 30 and 10 m resolutions, respectively, are used for producing flood inundation maps. Hydrologic Engineering Center's River analysis system (HEC-RAS) and RAS mapper are used in this study to generate flood inundation map. The study involved the assessment of flood map derived from multi-resolution two-dimensional hydrodynamic model approach. The result from the study provides the required assessment of impacts of flood events in the study area and will help the policy-makers to take an effective decision-making for mitigation and preventive measure of the region.

Keywords Flood · Inundation mapping · HEC-RAS

1 Introduction

Flood is a natural disaster. It is inevitable. It is a part of the hydrological cycle. It causes havoc to life and properties.

It also endangers the economic development of a country. *"Flooding is a general temporary condition of partial or complete inundation of normally dry areas from overflow of inland or tidal waters or from unusual and rapid accumulation or runoff"* [1]. Floods cannot be prevented totally but preventive measures and mitigation can

M. Bipinchandra · N. Romeji · C. Loukrakpam (✉)
Department of Civil Engineering, NIT-Manipur, Imphal, India
e-mail: chandramani106@gmail.com

M. Bipinchandra
e-mail: bipinmaisnam@gmail.com

N. Romeji
e-mail: ngromezi@gmail.com

J. K. Ghosh and I. da Silva (eds.), *Applications of Geomatics in Civil Engineering*, Lecture Notes in Civil Engineering 33, https://doi.org/10.1007/978-981-13-7067-0_11

be done. As per the Floods Directive [2], flood risk is the definition of risk as the product of "hazard" and the "vulnerability". It is necessary to focus on the prevention and risk management. There should be emphasis on risk analysis, management, and assessment so as to come up with measures to decrease the risk in the areas of interest. The comprehensive analysis and assessment of flood risk is an essential part of the whole risk management concept [3]. Flood risk may increase due to human activity and may decrease by appropriate flood management and planning [4]. Flood hazard assessment results can be adopted by land use and development planners as part of an integrative approach to improve flood preparedness that can improve future land developments and raise community awareness.

The Imphal River Basin, Manipur, India is subjected to frequent floods in the past decades causing damage to human, livestock, and agricultural crops. This in turn affects the small economy of the state of Manipur. Thus, it has become necessary to study the different factors contributing to the frequent floods so as to assess the flood risk in the river basin. For this purpose, a small part of the Imphal River Basin has been considered to analyze the extent of flood in this area so as to visualize the probable area of flood-affected areas in this basin. Apart from the causative factors of the floods, this study mainly focuses on the effectiveness or accuracy of spatial resolution of the satellite imagery in analyzing the inundation during floods. The result of the analysis depends on the spatial resolution of the satellite images which in turn affects the procedure of assessment of the flood risk in the areas likely to be inundated during floods.

2 Study Area

The Manipur is part of the eastern extended range of Himalayas, namely, Purvanchal and shares the international boundary with Myanmar in the eastern front. The state constitutes valley region in the center which is surrounded by the hill ranges in all direction and has the Total Geographic Area (TGA) of 22,327 km^2. The states extend from 92.6019450°E to 95.1753321°E longitudes and 23.4759838°N to 25.5746684°N latitude. The valley region comprises almost 10% of the TGA of the state and the remaining area is covered with hill ranges. The average elevation of the valley is about 790 m above the sea level and that of the hills ranges between 1500 m and 1800 m above mean sea level. Barak River Basin, Manipur River Basin, and Yu and Lanye River Basin are the main river basins of the state. The southwest monsoon climate is prevailing in the region and the region receives an average rainfall of 1500 mm per year and temperature ranges from subzero to 36 °C [5]. The population of the state as per 2011 census is 2,855,794. The density of population is 130/km^2. The study area is a part of the Imphal River Basin in the state of Manipur, India. The economy of the state is primarily agriculture, forestry, cottage, and trade.

The Imphal River is a major river which originates in Senapati district at the hills of *Karong*. In this study, total length of 69.28 km stretch of the river up to *Minuthong Bridge* of Imphal East is considered. The Gaging station under the

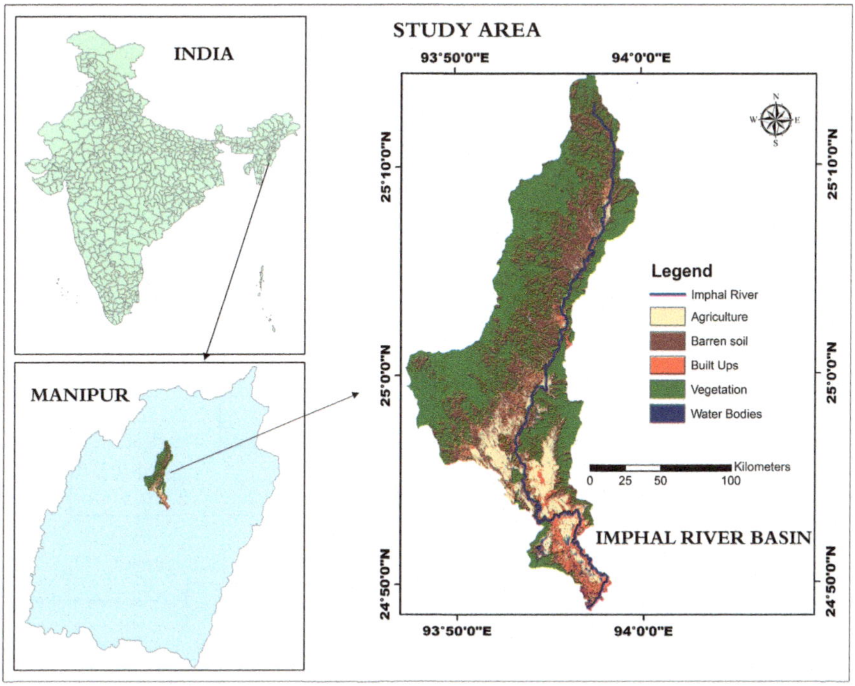

Fig. 1 Study area

Minuthong Bridge is considered as the outlet point and it flows toward the southern part of Manipur. The area of the basin under study is 374.27 $km^{2.}$ It lies between the latitudinal extent of 24°48′41.838″ N and 25°14′34.039″ N and longitudinal extent from 93°47′22.159″ E to 93°59′41.302″ E (as shown in Fig. 1).

3 Dataset and Methodology

In this study, datasets of 30 and 10 m spatial resolutions are employed for the generation of flood inundation map. Landsat-8 OLI and Sentinel-2 multispectral imageries are used for generating the LULC map. Digital elevation models of ASTER and Cartosat with 30 and 10 m spatial resolutions are used for extracting the terrain layers. As the study area is located in the Intertropical Convergence Zone (ITCZ), cloud contaminations of the satellite imagery prevail. To generate cloud-free image, Spatiotemporal Image Fusion Model (STI-FM) is employed [6] as shown in Table 1.

The DEM is preprocessed to remove the redundant values and hydrological corrected DEM is used. The Landsat-8 and Sentinel-2 are processed to convert the DN values to surface reflectance [7, 8]. The datasets after preprocessing are the input in the Geo-RAS and the input layers are generated. After exporting the processed

Table 1 Dataset

Dataset	Path/Row	Date
MOD09A1 (Base image)	H26/V06	03/12/2017
MOD09A1 (Auxiliary image)	H26/V06	01/11/2017
MOD09A1 (Base image)	H26/V06	27/12/2017
MOD09A1 (Auxiliary image)	H26/V06	09/11/2017
Landsat-8 (Base image)	135/43	05/12/2017
Landsat-8 (Auxiliary image)	135/43	03/11/2017
Sentinel-2 (Base image)	135/43	30/12/2017
Sentinel-2 (Auxiliary image)	135/43	10/11/2017

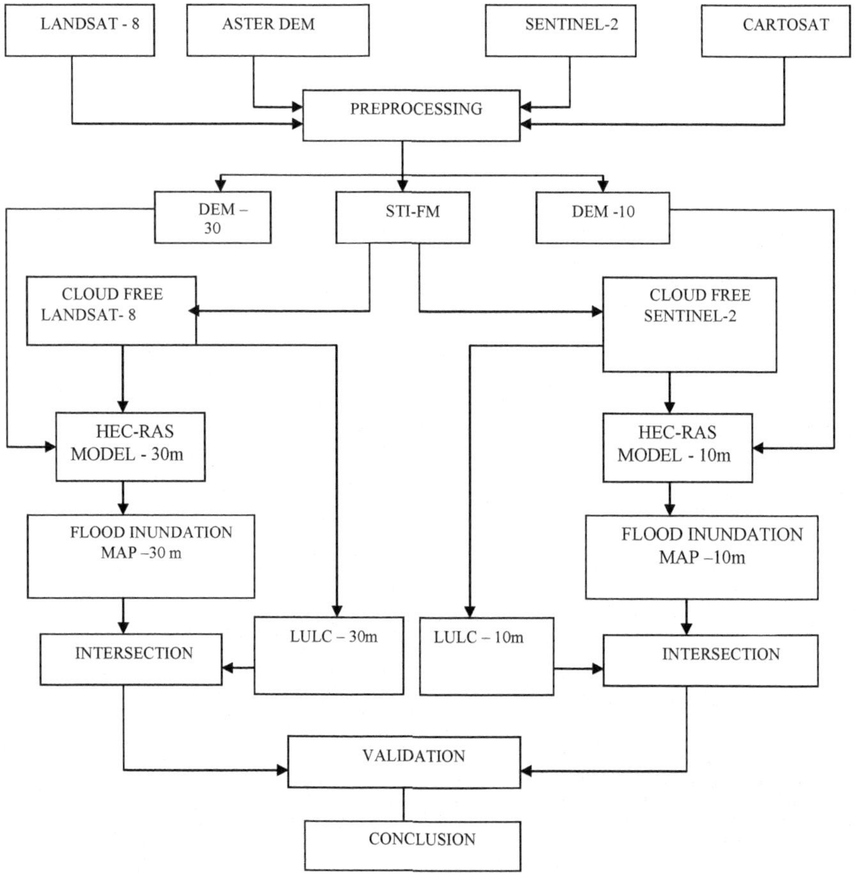

Fig. 2 Methodological flowchart for generating the flood inundation map of the Imphal River Basin, Manipur

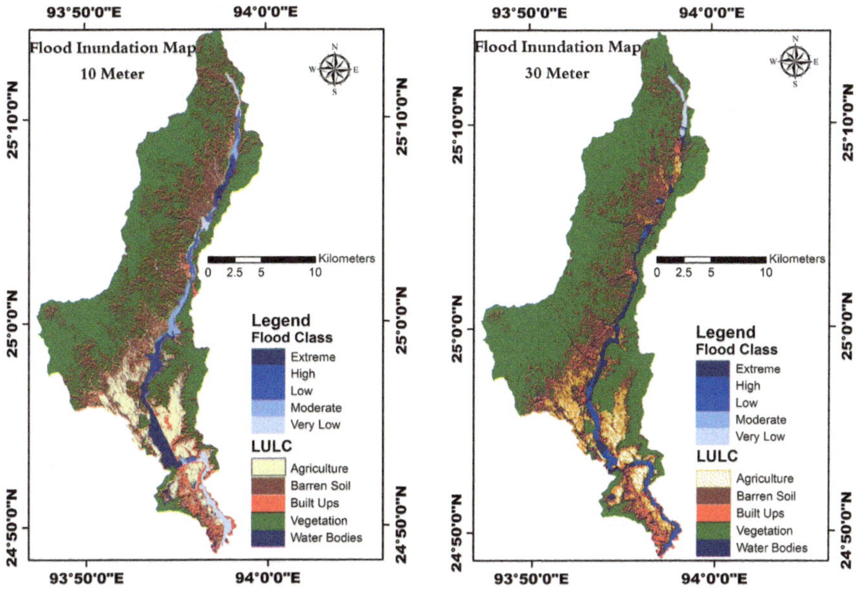

Fig. 3 Flood inundation map (10 and 30 m)

terrain layers, the two-dimensional hydrodynamic model is executed in HEC-RAS [9]. The output result from HEC-RAS is then employed as the input to the RAS mapper and thus the flood-inundated area of the study area is generated as shown in Fig. 2 (Fig. 3).

4 Results

After the generation of flood inundation map, it then overlays with the LULC class and the area under the influence of various flood class is calculated. The intersection table for the flood inundation from 10 to 30 m spatial resolution is given in Tables 2 and 3.

Table 2 Intersection table of flood inundated from 10 m resolution (Area in km^2)

	Barren	Vegetation	Built-ups	Agriculture	Water bodies
Very low	1.755699992	1.114869952	2.437259912	1.193150043	0.524001002
Low	1.81467998	1.11401999	0.513608992	0.563069999	0.393115997
Moderate	1.773489952	0.774062991	0.425680012	1.106549978	0.29824999
High	2.580840111	0.579051018	0.316897988	1.660609961	0.625528991
Extreme	2.247839928	0.414826006	1.329699993	2.28008008	0.878027976

Table 3 Intersection table of flood inundated from 30 m resolution (Area in km^2)

	Barren soil	Vegetation	Water bodies	Built-ups	Agriculture
Very low	0.79816002	0.654501021	0.376581997	0.0576073	0.01575
Low	0.0198037	0.019346301	0.019054901	0.00982073	0.00037439
Moderate	0.0276049	0.0230939	0.030146301	**0.0234037**	**0.00105122**
High	1.83958006	0.595206022	2.390350103	**3.04906011**	**0.505913973**
Extreme	3.04116988	1.299329996	2.034389973	**1.15612996**	**0.453613997**

This study mainly focuses on the flood inundation and its intensity on the land use class of agriculture and built-ups. Tables 2 and 3 show the variation in the spatial extent of the flood-affected region of the study area for the above mentioned two land use classes.

The comparison is made in between the two results obtained from the two different spatial resolutions. As the result of the comparison, the area under "extreme flood" is observed to be **2.28008008** and **0.453613997 km²** for 10 and 30 m, respectively. For "high flood" and "moderate flood", the area is found to be **1.660609961** and **1.106549978 km²** from the 10 m resolution. Similarly, in case of 30 m resolution, the extent is observed to be **0.505913973** and **0.00105122 km²**, respectively.

As in case of "Agriculture class", similar analysis is considered for the "Built-ups class" and result is found in similar pattern. The extreme flood class is observed to expand over more spatial extent in case of 10 m than the 30 m. In case of "High class", the result from 30 m has more spatial extent than of 10 m and the pattern is vice versa in case of the "Moderate flood class".

5 Validation

In this study, the flood inundation of a region is considered to be based on the topography of the study area. Slope in degree and elevation in meters are classified into five classes based on the National Remote Sensing Centre classification. The region with low slope and low elevation is considered to have maximum risk of flood and area with high slope and elevation is considered to have least risk of flood.

The classified slope and elevation (Figs. 4 and 5) layers of both 10 and 30 m overlay, respectively, after setting equal influence to both the parameters. The resultant overlay flood hazard risk map is given in Fig. 6.

The flood inundation map generated from the RAS mapper is then intersected with the flood hazard map generated as mentioned earlier. The accuracy assessment is performed in the region of intersection and thus the more accurate flood inundation map is inferred.

The intersected imageries are validated using error matrix and accuracy assessment is performed. The accuracy assessments of the intersected images are done using the flood hazard map acquired by integrating slope and elevation of the study

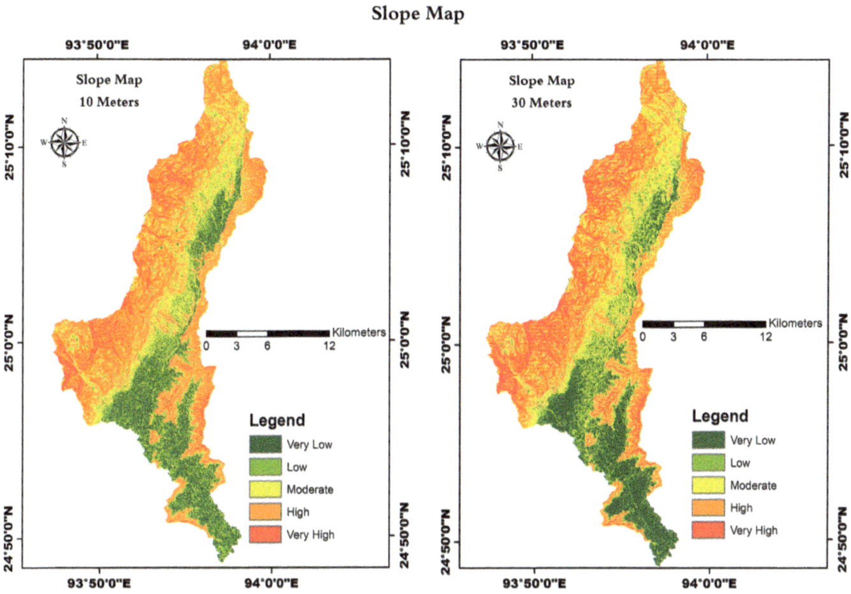

Fig. 4 Slope map

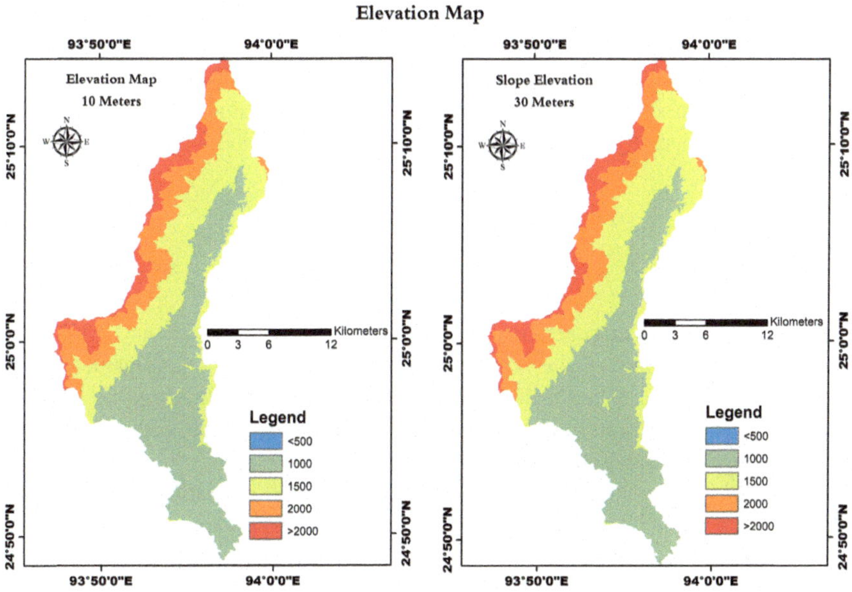

Fig. 5 Elevation map

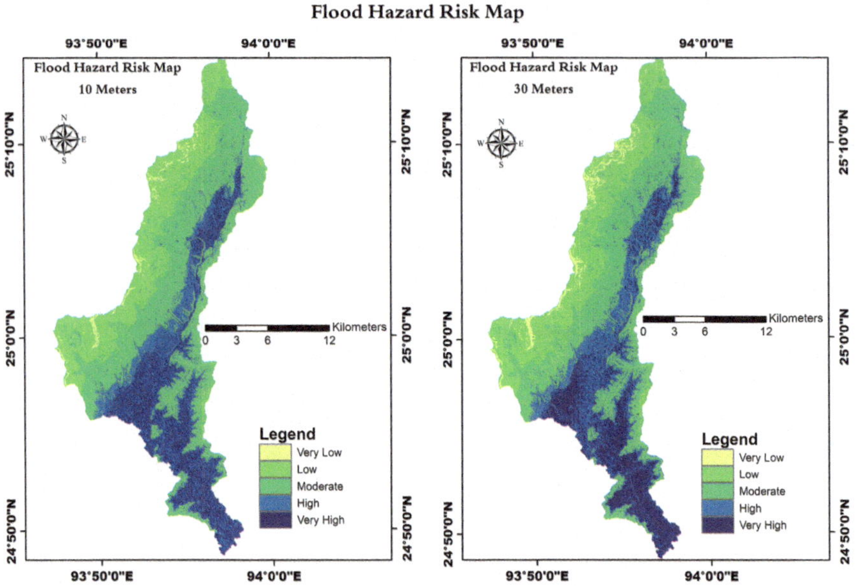

Fig. 6 Flood hazard risk map

area. The overall accuracy and kappa coefficient of that flood inundation map generated using 10 m spatial resolution are obtained as 84.45% and 0.80, respectively. In case of flood inundation map generated using 30 m spatial resolution, the overall accuracy and kappa coefficient are observed to be 70.13% and 0.67, respectively.

6 Conclusion

The flood inundation map generated using 10 m spatial resolution is found to be more accurate than that of the one with 30 m, spatial resolution. Considering the same flood event, the value of the affected inundated area on the basis of land use classes is found to be more accurate in case of the model result with finer spatial resolution as compared to the coarser resolution.

From this study, considering the model result with 10 m spatial resolution irrespective of the flood inundation intensity class, 16.02% of the total agricultural area and 26.13% of the total built-up area of the study area are affected by flood. It is expected that the result of this study would help policy- and decision-makers to take up necessary preventive and mitigation measures.

Acknowledgements The author gratefully acknowledges the United States Geological Survey (USGS) and European Space Agency (ESA) for providing the Landsat, MODIS dataset, and Aster DEM and Sentinel dataset, respectively. The gratitude is also shown to the Northeast Space Application Centre, Shillong, Meghalaya, India for providing the fine spatial resolution DEM.

References

1. Jeb DN, Aggarwal SP (2008) Flood inundation hazard modeling of the River Kaduna using remote sensing and geographic information systems. J Appl Sci Res 4(12):1822–1833
2. European Commission (EC): Directive 2007/60/EC of the European Parliament and of the Council of 23 October, 2007 on the assessment and management of flood risks. Off J Euro Union (2007)
3. Meyer V, Scheuer S, Haase D (2009) A multicriteria approach for flood risk mapping exemplified at the Mulde river, Germany. Nat Hazards 48:17–39
4. Simonovic, SP (2009) A new method for spatial and temporal analysis of risk in water resources management. J Hydro Inform **11**(3,4):320–329
5. Desertification and Land Degradation Atlas of India (Based on IRS AWiFS data of 2011–2013 and 2003–2005). Ahmedabad: Space Applications Centre, ISRO, Ahmedabad, India (2016)
6. Hasan, KH, Quazi K (2015) Spatiotemporal image-fusion model for enhancing the temporal resolution of Landsat-8 surface reflectance image using MODIS image. J Appl Remote Sens 9
7. Department of the Interior, USGS (2017). https://landsat.usgs.gov/sites/default/files/documents/ledaps_product_guide.pdf
8. ESA standard document, Sentinel-2 User Handbook (2015). https://sentinel.esa.int/documents/247904/685211/Sentinel-2_User_Handbook
9. Brunner GW (2016) HEC-RAS, river analysis system hydraulic reference manual, Us army corps of engineers, hydraulic engineering centre (HEC). http://www.hec.usace.army.mil/software/hec-ras/documentation/HEC-RAS%205.0%20Reference%20Manual.pdf

Hydraulic Modeling of River Discharge Subjected to Change in Riverbed Morphology

Pramodkumar Kappadi and M. K. Nagaraj

Abstract In natural channels such as rivers, the flow behavior is highly complicated phenomenon due to unsteady and nonuniform flow. Hydraulic modeling is essential for the study of characteristics of unsteady flow in rivers. Changes in riverbed morphology influence the increase in depth of flow in rivers. In the present study, the roughness coefficient is varied to match the natural condition. The objective of the paper is to study the effect of the change in riverbed morphology on the river discharge using a hydraulic model. The study is carried out using the Hydrologic Engineering Center—River Analysis System (HEC-RAS). A river length of 12 km of the Nethravathi river regime, Karnataka from Uppinangadi to Bantwal is considered for the study. Daily river stage and discharge data are collected from Central Water Commission (CWC) gaged at Bantwal station. GIS interface of HEC-GeoRAS is also used to extract the cross section, bed slope, and length of the river channel from Digital Elevation Model (DEM) data of resolution 30 m. The cross sections are represented for each kilometer length of the river. Since accurate data is unavailable in the study area, cross sections are simplified. HEC-RAS model was used for the simulation of surface water levels and discharge values. Manning's roughness coefficient and river cross sections were defined for the calibration of observed river stage and discharge data. The predicted discharge was in good agreement with that of the observed discharge value. Study results illustrated that accuracy of predicted maximum water depth, and surface water level depends upon the precise representation of Manning's roughness. The study is useful for the prediction of the flood dynamics in the river regime with the change in riverbed morphology.

Keywords Hydraulic modeling · Riverbed morphology · Manning's roughness coefficient

P. Kappadi (✉) · M. K. Nagaraj
National Institute of Technology Karnataka, Surathkal, Mangalore 575025, India
e-mail: pramodkappadi@gmail.com

M. K. Nagaraj
e-mail: rajumkn@yahoo.com

J. K. Ghosh and I. da Silva (eds.), *Applications of Geomatics in Civil Engineering*, Lecture Notes in Civil Engineering 33, https://doi.org/10.1007/978-981-13-7067-0_12

1 Introduction

Water is most precious resource in the world due to its reduction in availability. Rivers are the veins of nature which are one of the world's most valuable natural resources and are important to the livelihood in many ways. In natural channels such as rivers, the flow behavior is highly complicated phenomenon due to unsteady and nonuniform flow [1–3]. The incessant process of unsteady flow in the rivers can result in flooding. Unsteady flow in a river due to dynamics of rainfall discharge, modifications of channel geometry, and reservoirs and sluices operation are significant effects on river discharge [4]. The comprehension and accurate prediction of river flow by hydraulic modeling are essential. Changes in riverbed morphology influence the increase in depth of flow in rivers. 1D Saint-Venant equations are commonly used in hydraulic modeling such as discharge routing [5], flood prediction [6], and surface and subsurface runoff [7]. The Saint-Venant equations consist of mass and momentum conservation, and it describes the gradually varied and transient flow, in a channel with irregular cross sections.

Developed hydraulic model for floodplain mapping and flood forecasting, calibrated channel roughness for different rivers [8–12]. Using optimization method, single channel roughness value was estimated by taking the boundary condition as constraints [13]. In this regard, calibrated roughness of the channel was evaluated for semiarid river of Western Australia [14].

The objective of the paper is to study the effect of the change in riverbed morphology on the river discharge using a hydraulic model in Nethravathi River Regime.

2 Materials and Methodology

2.1 Study Area

The study area under investigation is a humid, tropical river basin of the southern Indian peninsula [15]. Among the number of rivers, the Nethravathi River is one of the important west flowing rivers of Western Ghats of Karnataka state in India. In the study area, the range of altitude varies approximately 0–1,200 m from mean sea level. The area of basin is 3,657 km^2 (CWC 2006). The length of the river is 103 km. The geographical location of the Nethravathi River lies from 12°29′11″ to 13°11′11″ N latitudes and 74°49′08″ to 75°47′53″ E longitudes as shown in Fig. 1.

The annual rainfall over the area varies between 1500 and 8000 mm. Rainfall in the Western Ghats occurs during three separate seasons, northeast monsoon (October to December), southwest monsoon (June to September), and pre-monsoon (March to May), and these three seasons, respectively, contribute about 6, 90 and 4%, of the total annual rainfall. The area has an adequate rainfall and numerous streams flow through the Western Ghats, in small meandering channels within wide valley floors. During rain streams occupy the whole valley floors and expand their channel.

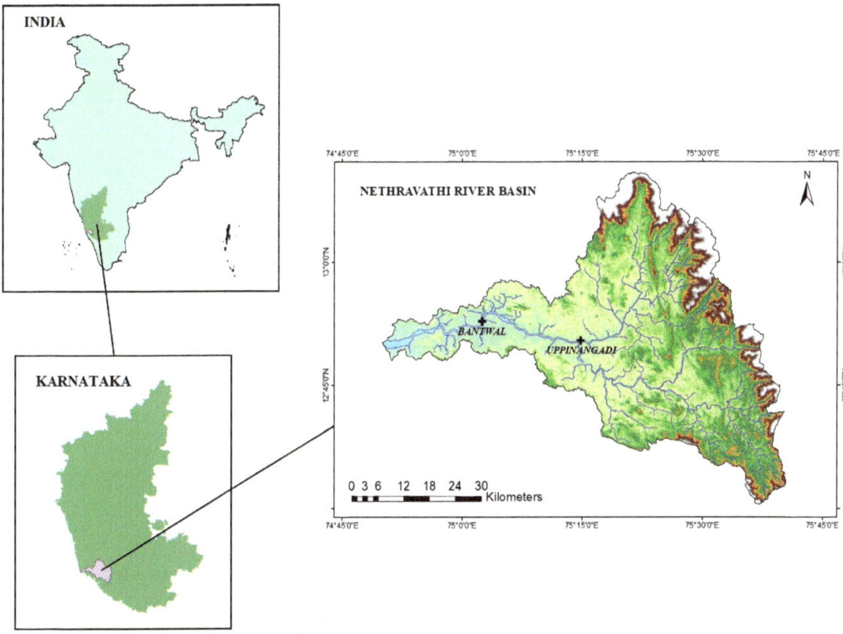

Fig. 1 Geographic location of the Nethravathi basin

2.2 *Dataset*

Terrain data is downloaded from USGS National Elevation Dataset (NED) to extract a river cross-section geometry from DEM using resolution of 30 m (1 arc), and hydrological data (stage and discharge) are collected from Central Water Commission (CWC) India. Map and toposheet are collected from Department of Mines and Geology Karnataka and Survey of India to delineate the river.

2.3 *HEC-GeoRAS*

HEC-GeoRAS has a set of procedures, tools, and utilities for processing geospatial data in ArcGIS. The GeoRAS software helps in the preparation of input data which can be imported to HEC-RAS. In order to create an import file, DEM of the river system is used. GeoRAS assists in creating a series of points, lines, and polygonal layers which are required for HEC-RAS model. The important layers which are created are flow path centerlines, stream centerline, main channel banks, and cross-section cut lines as RAS layers. One can also provide the details regarding bridge, levee, culverts, and storage area if any are present in the river course. After preparing the input geometric file with the help of HEC-GeoRAS, data are imported to HEC-RAS. Imported geometric data file will be saved in the format of .sdf.

2.4 HEC-RAS

HEC-RAS is a hydraulic model developed in 1964 by Hydrologic Engineering Centre (HEC) of the U.S. Army Corps of Engineers, which is dependent on implicit finite difference solutions of the Saint-Venant equations (Eqs. 1 and 2).

Continuity equation:

$$\frac{\partial y}{\partial t} + D_h \frac{\partial v}{dx} + V \frac{dy}{dx} = 0 \tag{1}$$

Momentum equation:

$$\frac{\partial y}{\partial t} + V \frac{\partial v}{dx} + g\left(\frac{dy}{dx} - S_b + S_f\right) = 0 \tag{2}$$

where V is the flow velocity, $D_h = \frac{A}{B}$ is hydraulic depth, y is the flow depth, y is the flow depth, A is flow area, S_f is the slope of energy grade line, S_o is the channel bottom slope, x is the distance along the channel length, g is the acceleration due to gravity, and t is time.

HEC-RAS is a widely accepted tool for the floodplain determination due to its user-friendly credibility and it is a software which requires minimum data to yield the water spread area and works well for the case of steady flows. HEC-RAS is software which is freely available and thereby makes it a different one compared to others. It can also be used to perform unsteady calculations, sediment transport as well as water quality analysis. Also, HEC-RAS can be effectively coupled with other software to generate the maps of flood extent, depthwise, velocitywise, etc. To conduct flow simulation through HEC-RAS, the following input data are given and they are channel resistance, channel geometry, and upstream and downstream boundary conditions. Extracted the cross-section data at 1 Km intervals from Uppinangadi to Bantwal over a length of 12 km. The discharge data is used for simulation and the friction slope is considered as boundary condition for the downstream reach. The observed flow hydrograph at Bantwal has been used for validation. For adequate simulation of flow, the different values of Manning's coefficient (n) (from 0.04 to 0.020) are used.

3 Flow Characteristic Calculation

In this study, the unsteady stateflow condition has been used. Flow hydrograph and normal depth are used for upstream and downstream boundary conditions, and 50 m^3/s for initial condition. Upstream gaging station is at Uppinangadi downstream gaging station at Bantwal. Flow hydrograph has been fixed at upstream corresponding to Uppinangady. For a given discharge, flow characteristics include top width, energy grade line, water surface profiles, flow velocity, flow area, wetted perimeter, water surface elevation, Froude number, shear total, specific force, slope, and power total, which have been computed.

4 Result and Discussion

4.1 Water Surface Profiles

Water surface profiles are shown in Fig. 2 along the river, for the stream flows at the upstream (RS 12,000). Depth of water from RS 12,000 to RS 11,000 is more due to bed slope which is nearly flat, and it is significantly varying between RS 11,000

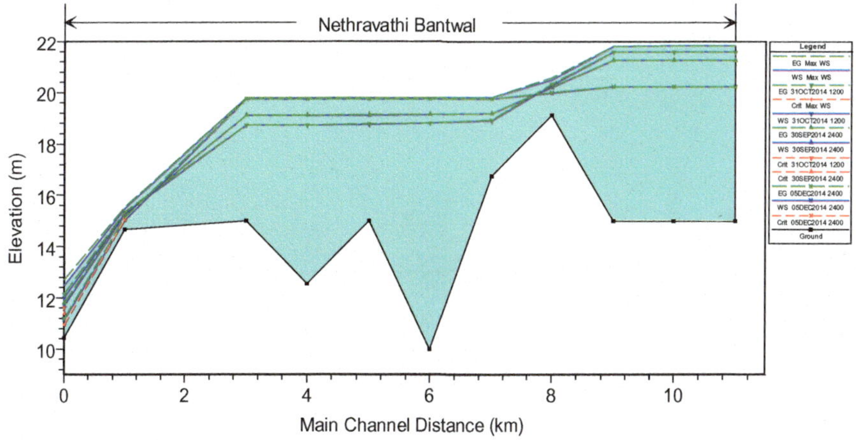

Fig. 2 Water surface profiles along the river

and RS 1000 as bed gradient varies, and these results show that the Bantwal area is sensitive to deformation of bed slope of the river and water level fluctuations.

4.2 Wetted Cross Section

The wetted cross section is shown in Figs. 3, 4, 5, respectively, at Uppinangadi (Section RS 11,000), Intermediate station (Section RS 4000), and Bantwal (Section RS 1000). Result shows that the wetted cross section is wider at RS 11,000, and larger at cross sections RS 4000 and narrow at RS 1000 and the flow is decreasing upstream to downstream. These results show that the increase of roughness coefficient water level profile is nonuniform and shows as a gradual variable. Thus, the water level decreases from upstream cross sections to downstream, and this consequently leads to the decrease of flow capacity in the floodplain.

4.3 Volume of Water

Figure 6 shows that the water of volume in upstream is more than down streams due to the width of flow, flow velocity, and is more in upstream.

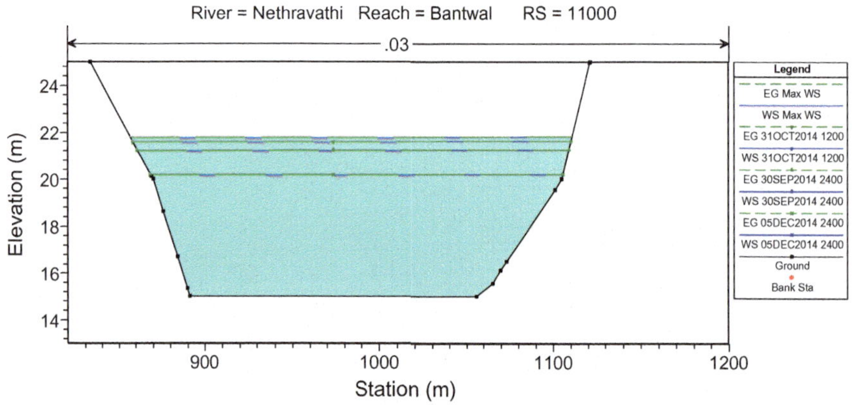

Fig. 3 Wetted cross section at RS 11,000 (upstream section)

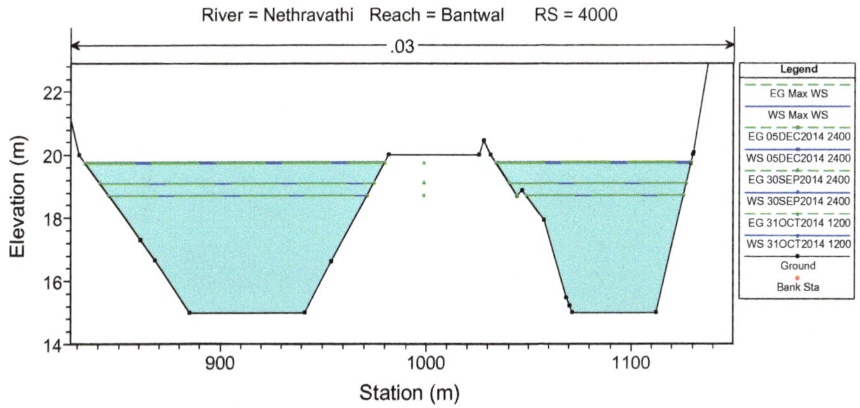

Fig. 4 Wetted cross section at RS 4000 (intermediate section)

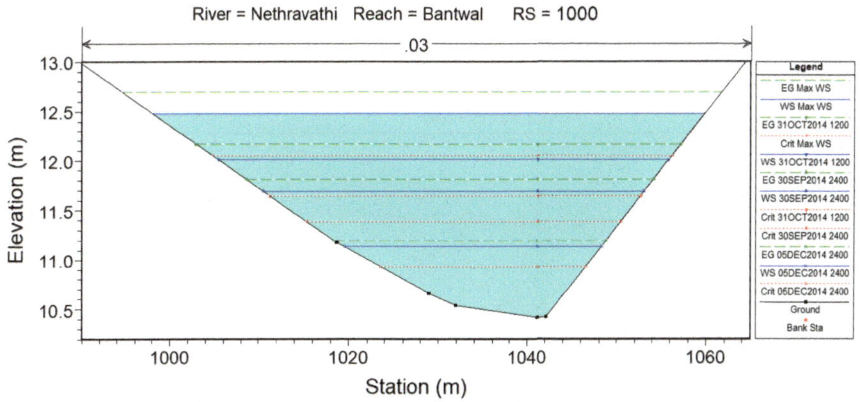

Fig. 5 Wetted cross section at RS 1000 (downstream section)

4.4 Flow Velocity

Figure 7 shows the flow velocity along the river. Velocity of flow in river is less in section RS 12,000 to RS 11,000. In section RS 1000, highest velocity was observed. These results show that the variations in the flow velocity are due to the significant changes in the bed morphology of Nethravathi River and the roughness coefficient effect.

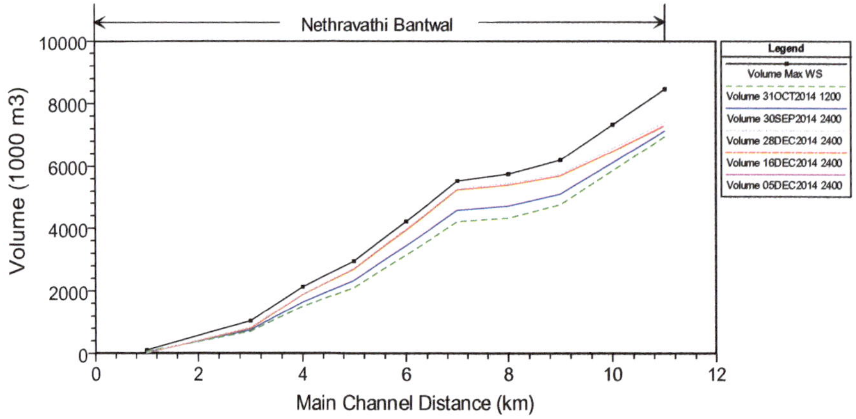

Fig. 6 Volume of water along the river

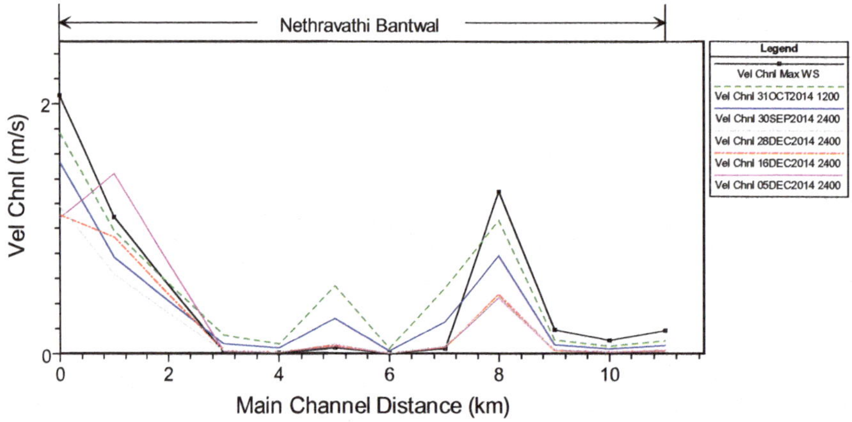

Fig. 7 Flow velocity along the river

4.5 *Rating Curve*

Figure 8 shows the rating curve for different Manning's coefficient. From the figure, it is observed that the accuracy of the result depends on accurate representation of channel geometry and Manning's coefficient.

Figure 9 shows comparison of predicted discharge versus observed discharge. From the figure, it is observed that they are in good agreement.

Figure 10 shows stage discharge relationship at RS 1000.

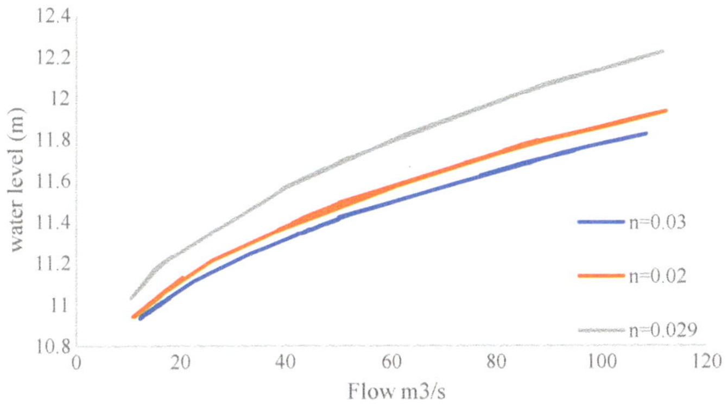

Fig. 8 Rating curve for different Manning's coefficient

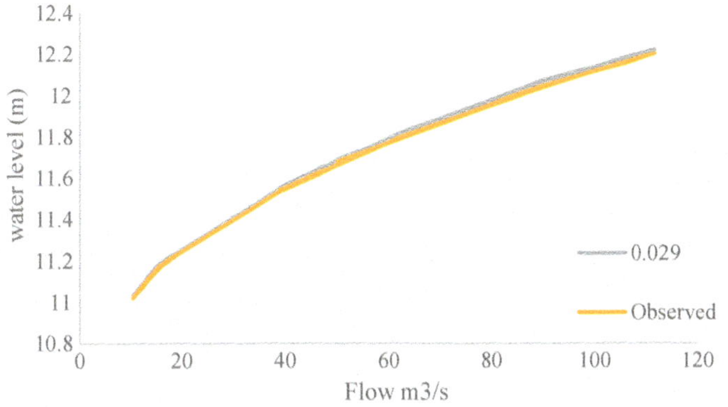

Fig. 9 Downstream rating curve of observed and estimated stage and discharge data at section RS 1000

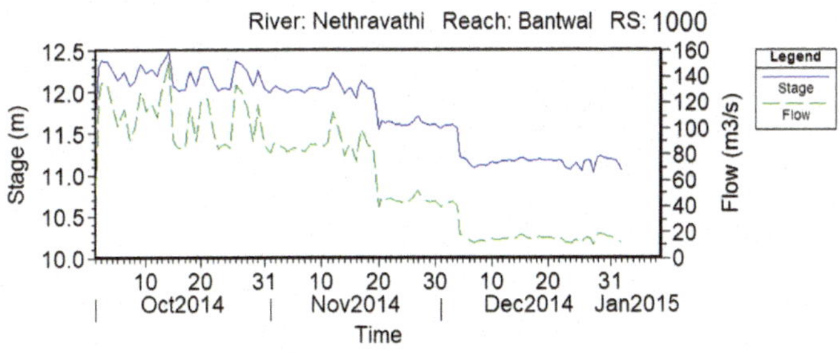

Fig. 10 Stage and flow versus time at RS 1000 estimated from HEC-RAS

5 Conclusions

The simulation shows that the computed rating curve matches well with the observed rating curve. Bantwal area is sensitive to deformation of bed slope of the river and water level fluctuations. The variations in the flow velocity are due to the significant changes in the bed morphology of Nethravathi River. The accuracy of the result is depending on accurate representation of channel geometry and Manning's coefficient. Changes in riverbed morphology influence the increase in depth of flow in rivers

References

1. Wohl Ellen E (1998) Uncertainty in flood estimates associated with roughness coefficient. J Hydraul Eng 124:219–223
2. Hin LS, Bessaih N, Ling LP, Ghani AA, Zakaria NA, Seng MY (2008) Discharge estimation for equatorial natural rivers with overbank flow. Int J River Basin Manag 6(1):13–21
3. Ballesteros J, Bodoque J, Díez-Herrero A, Sanchez-Silva M, Stoffel M (2011) Calibration of floodplain roughness and estimation of flood discharge based on tree-ring evidence and hydraulic modelling. J Hydrol 45:103–115
4. Wang G, Yang H, Wang L, Xu Z, Xue B (2014) Using the SWAT model to assess impacts of land use changes on runoff generation in headwaters. Hydrol Process 28:1032–1042
5. Serrano SE (2016) Propagation of nonlinear flood waves in rivers. J Hydrol Eng 21
6. Alekseevskii NI, Krylenko IN, Belikov VV, Kochetkov VV, Norin SV (2014) Numerial hydrodynamic modeling of inundation in Krymsk on 6–7 July 2012. Power Tech Eng 48:179–186
7. Hughes JD, Langevin CD, White JT (2015) MODFLOW- based coupled surface water routing and groundwater-flow simulation. Ground Water 53:452–463
8. Vijay R, Sargoankar A, Gupta A (2007) Hydrodynamic simulation of river Yamuna for riverbed assessment: a case study of Delhi region. Environ Monit Assess 130:381–387
9. Patro S, Chatterjee C, Mohanty S, Singh R, Raghuwanshi NS (2009) Flood inundation modeling using mike flood and remote sensing data. J Indian Soc Remote Sens 37:107–118
10. Wasantha Lal AM (1995) Calibration of riverbed roughness. J Hydraul Eng 121:664–671
11. Parhi PK, Sankhua RN, Roy GP (2012) Calibration of channel roughness of Mahanadi river (India) using HEC-RAS model. J Water Resour Prot 4:847–850
12. Usul N, Burak T (2006) Flood forecasting and analysis within the Ulus basin, Turkey, using geographic information systems. Nat Hazards 39:213–229
13. Ramesh R, Datta B, Bhallamudi S, Narayana A (1997) Optimal estimation of roughness in open-channel flows. J Hydraul Eng 126:299–303
14. Doherty R (2010) Calibration of HEC-RAS models for rating curve development in semi-arid regions of western Australia. In: AHA 2010 conference
15. Shetkar RV, Mahesha A (2010) Tropical, seasonal river basin development: hydrogeological analysis. J Hydraul Eng 16:280–291

Effect of Temporal-Based Land Use–Land Cover Change Pattern on Rainfall Runoff

B. Aneesha Satya, M. Shashi and Deva Pratap

Abstract High-resolution satellite imagery was often used to study the spatial patterns and temporal changes by analyzing Land Use–Land Cover (LULC) changes using rapid urbanization, thereby increasing the impervious areas has an increasing effect on runoff in urban areas. The objective was to estimate the effects of temporal change based LULC on rainfall runoff. A LULC classified map of the year 1995 derived from the Landsat TM satellite image was used as input for a Runoff model with the precipitation data. A similar procedure is carried out with a LULC classified map of the year 2005 as an input into the model, along with similar precipitation data and calibration parameters. The comparison of the outcomes shows an increase in runoff volume and peak discharge between the times because of LULC.

Keywords LULC · Satellite imagery · Rainfall · Runoff

1 Introduction

Urban change is the result and consequence of the urbanization process. Urbanization is a population migration from rural to urban areas and happens because of the high natural urban population increase. In developing countries population increase as well as immigration from the rural area toward urban area [1]. The urban change brought severe losses for cropland, vegetation cover, and water bodies and responsible for decreasing of air quality, increasing of local temperature, and deterioration of water quality of ground and surface water. The world is becoming rapidly urbanized and this process needs to be monitored so that decision makers have up-to-date information concerning land use to make sound decisions on future development. Geographic

B. Aneesha Satya · M. Shashi (✉) · D. Pratap
Department of Civil Engineering, NIT Warangal, Warangal 506004, India
e-mail: mshashi@nitw.ac.in

B. Aneesha Satya
e-mail: aneeshasatya@gmail.com

D. Pratap
e-mail: prataprec@yahoo.com

J. K. Ghosh and I. da Silva (eds.), *Applications of Geomatics in Civil Engineering*, Lecture Notes in Civil Engineering 33, https://doi.org/10.1007/978-981-13-7067-0_13

Information System (GIS) combined with Remote Sensing (RS), Global Positioning System (GPS) have been widely applied and found as powerful tools for detecting and analyzing urban change at all scales. It can be used in inventorying the environment, observing, and assessing the changes as well as forecasting the changes based on the existing situation and it is a standard tool for management of natural resources [4]. RS and GIS for urban change analysis require the acquisition of data set, classification of images using efficient algorithm and analysis of the change. Remotely sensed data is a useful and effective tool for urban change detection analysis because it is easily accessible and some images are available free of cost at USGS website [5]. Change Detection (CD) is the process of identifying differences in the state of an object by observing it at different times and it can be applied in many areas. In general objective, CD in RS is considered to identify the geographic location and detect the changes, identify the nature of change, measure the area of change, and assess the accuracy of change [6]. The objective was to estimate the effects of temporal change based LULC on rainfall runoff. A LULC classified map of the year 1995 derived from the Landsat TM satellite image was used as input for a Runoff model with the precipitation data. A similar procedure is carried out with a LULC classified map of the year 2005 as an input into the model, along with similar precipitation data and calibration parameters. The comparison of the outcomes shows an increase in runoff volume and peak discharge between the times because of LULC. The paper highlights the key inputs required for simulation and assessment of the behavior of temporal changes in LULC.

2 Materials and Method

2.1 Data Source

The Landsat data with 30 m resolution is downloaded for the study area for the years 1995, 2005 from the website USGS. Supervised classification is carried out to prepare the LULC map of the respective years in ERDAS IMAGINE. Soil data for the area was collected from the irrigation department, GWMC, Warangal, Telangana and the types of soil distribution were mapped.

2.2 Study Area

Warangal Municipal Corporation (WMC) shown in the Fig. 1, with an area of about 80 km^2, lies between 17°46′ and 18°13′, is a tri-city comprising of the areas of Kazipet, Hanmakonda, and Warangal. While Kazipet is the railway center with the location of Locomotive sheds, Hanmakonda is the central part of the city with predominantly educational institutes and the new residential and commercial areas devel-

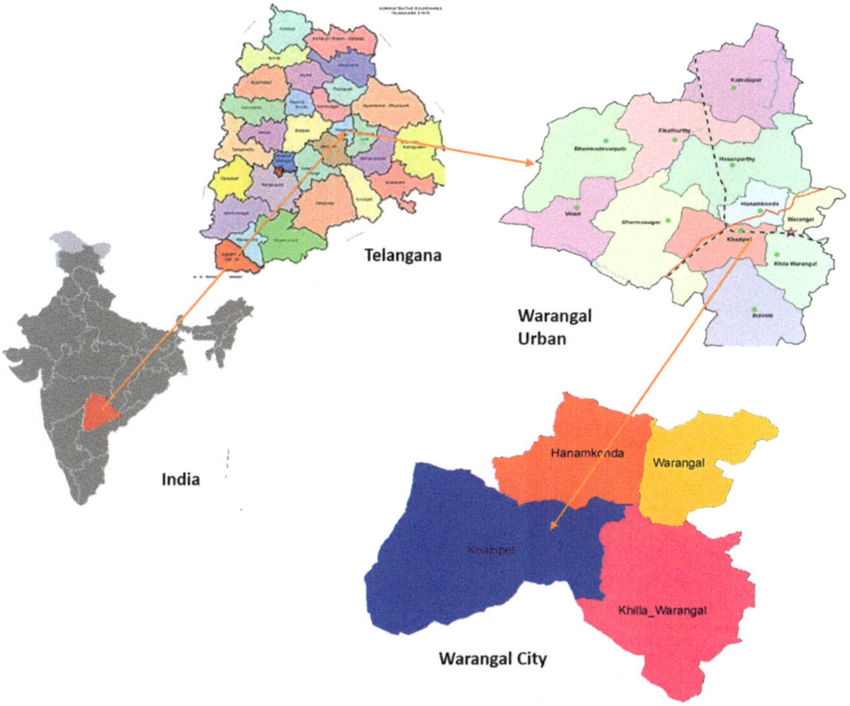

Fig. 1 Study area location map

oping around. The total population of Warangal city was 5.3 lakhs in 2001 which increased to 6.3 lakhs in 2011, registering a decadal growth of about 19%. The population of the urban spread was 7,33,278 as per 2001 census while it was 8,75,242 as per 2011 census with a decadal growth of 19.36%. Average annual precipitation is 970 mm, mainly occurs during the rainy season of the year.

3 Methodology

For the study area, the SRTM Digital Elevation Model (DEM) was downloaded from USGS and Fig. 2 presents the elevation map of the study area. The soil map of the area is prepared using ArcGIS. In the study area as shown in Fig. 3, three types of soils (clay loamy, sandy clay loam, sandy loam) were present. LULC layers for the years 1995 and 2005 were prepared. Using the SCS Curve Number method, the respective layers as input runoff are calculated for the area. Finally, the spatial and temporal changes were compared and produced for analysis.

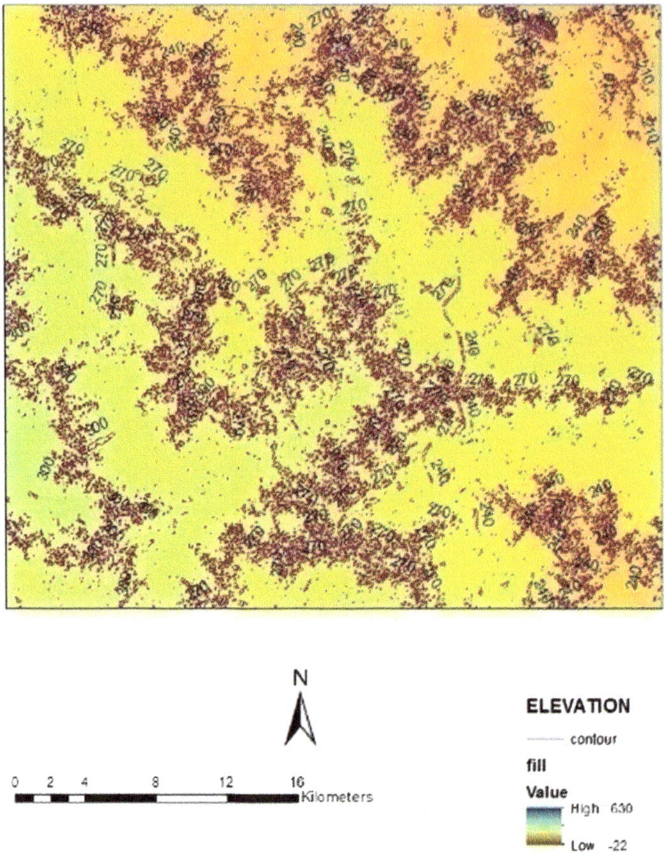

Fig. 2 DEM considered for the study area

3.1 Generation of CN Map

In order to generate the CN map of WMC, the extension of CN in ArcMap is used. The soil raster and LULC map generated were given as inputs to intersection tool and a hydrological soil group was noted, where we get a polygon shapefile as output which is a merged soil-land map. Then the CN value for every polygon of the soil-land map was allotted appropriately based on.

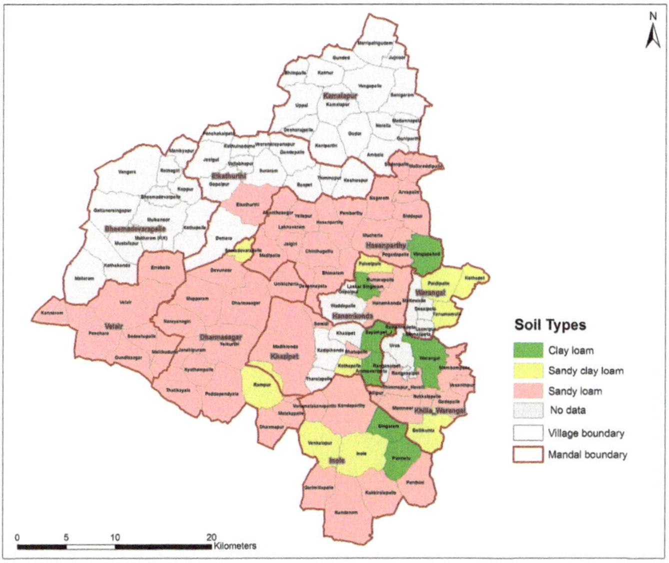

Fig. 3 Soil map prepared for Warangal district

$$\mathrm{CN} = \left(\sum (\mathrm{CN_i} \times \mathrm{A_i})\right)/\mathrm{A}$$

where

CN	weighted curve number.
CN_i	curve number from 1 to any no. "n".
A_i	area with curve number.
CN_i A	the total area of the watershed.

3.2 Usage of SCS Model for the Estimation of Runoff

The Soil Conservation Service Model otherwise called the Hydrologic Soil Cover Complex Model is an adaptable and generally utilized system for runoff estimation. The model uses runoff-creating ability communicated by a numerical value (Curve Number) changing between 0 and 100. Soil Conservation Service (SCS) Model is used to assess runoff potential. LULC, soil, precipitation information were utilized as the contribution to create yearly runoff potential and the diagrams were plotted. The CN unadjusted value is higher in contrast with CN balanced with a slant. In this way, the created bend numbers might be utilized for the forecast of runoff from an ungauged watershed.

Table 1 Percentage of the LULC classes areas in the years 1995 and 2005

Classes	Percentage (%) of the area in the year 1995	Percentage (%) of the area in the year 2005
Built-up	11.172	16.79046
Agricultural land	82.33112	56.41422
Wastelands	4.161711	24.16947
Water bodies	2.334348	2.625852

4 Result and Discussion

Giving DEM as an input to the hydrological tool in ArcMap, WMC was delineated into five sub-watersheds. Supervised classification is carried out on the satellite data for the study area and is presented in Table 1. Based on supervised classification, the classes namely water bodies (2.334%), wasteland (4.161%), agriculture land (82.331%) and settlements (11.172%) were identified in the year 1995 as shown in Fig. 4 while in the year 2005 it is observed as water bodies (2.6258%), wasteland (24.169%), agriculture land (56.414%) and settlements (16.790%) as shown in the Fig. 5. Further LULC and Soil map merged data was used for the generation of CN. Curve Number: The USDA curve number [9] modified as per Indian conditions was used to assign the curve number to individual polygons created based on the hydrological soil groups and land use classes of respective areas. When the 1995 LULC was given as input for runoff modeling the surface runoff was 154 mm in the year 1995 and 223 in the year 2005, the variations in the surface runoff during the time periods considered from 1995 to 2005 is shown in the Fig. 6. When the 2005 LULC was given as input for runoff modeling the surface runoff was 208 mm in the year 1995 and 294 in the year 2005, the variations in the surface runoff during the time periods considered from 1995 to 2005 is shown in the Fig. 7 showing an increase in the peak as Table 1 shows that there is an increase in built-up area, increasing impervious surface.

5 Conclusions

Remote detecting and GIS technique is an incredibly strong choice or time tested traditional steady system to our standard strategy for examination, arranging, observing, demonstrating, information putting away and basic leadership process. Satellite picture is basic for recognizable proof of the wide physical features, for stream arrange, arrive use/arrive cover, soils surface, water bodies and so forth. LULC is a basic parameter commitment of SCS model that could be settled accurately with the help of this framework. With the help of remote Sensing, GIS and SCS demonstrate, it is possible to make organization game plans for use and headway of the region.

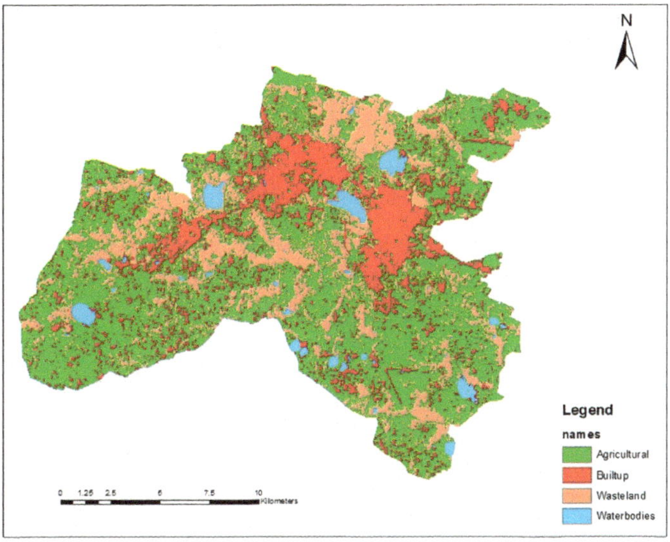

Fig. 4 LULC of the year 1995

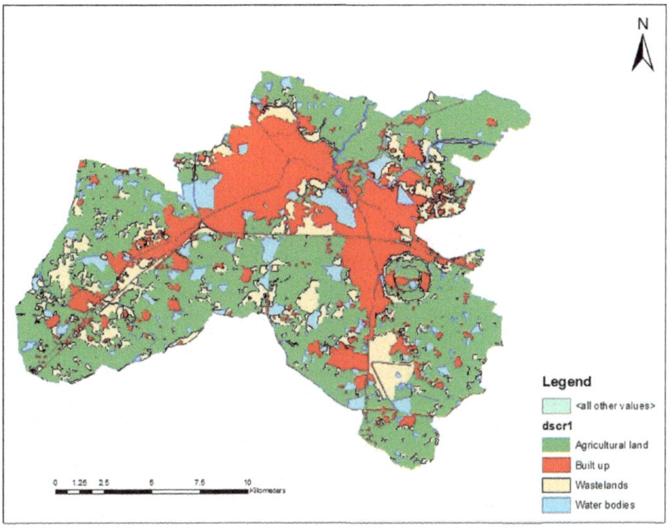

Fig. 5 LULC of the year 2005

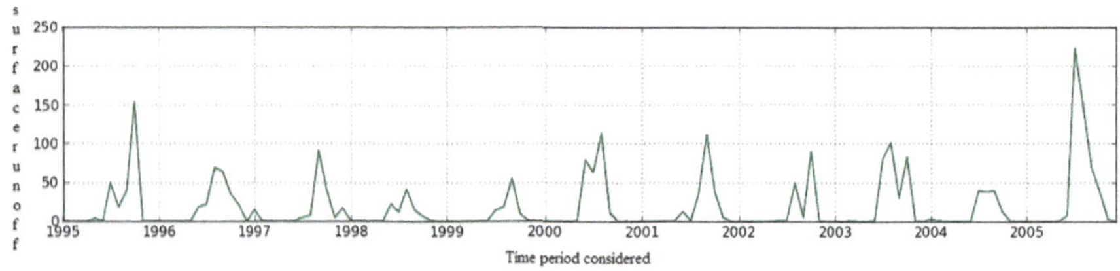

Fig. 6 Surface runoff during the period when LULC of the year 1995 is given as input in runoff modeling

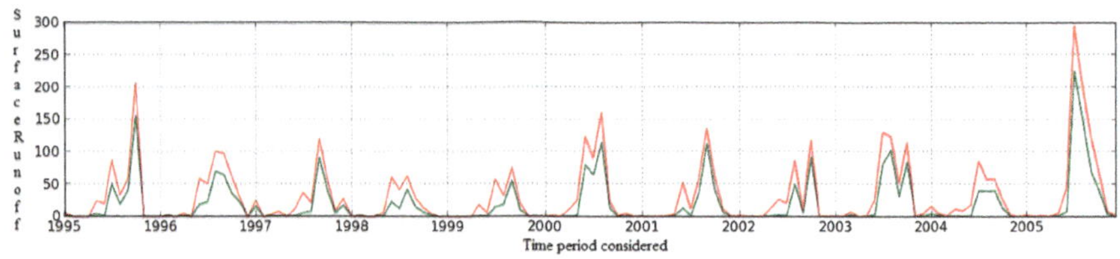

Fig. 7 Surface runoff during the time period when LULC of the year 1995 is given as input in runoff modeling

References

1. Amutha R, Porchelvan P (2009) Estimation of surface runoff in Malattar sub watershed using SCS-CN method. J Indian Soc Remote Sens 37:291–304
2. Ebrahimian M, See LF, Ismail MH, Malik IA (2009) Application of NRCS curve number method for runoff estimation with GIS in the Kardeh watershed, Iran. Eur J Sci Res 34(4):575–590
3. Gajbhiye S (2014) Estimation of rainfall generated runoff using RS and GIS. Lambert Academic Publishing, Germany. ISBN 978-3-659-61084-4
4. Gajbhiye S, Sharma SK (2012) Land use and land cover change detection of Indra river watershed through remote sensing using multi-temporal satellite data. Int J Geomat Geosci 3(1):89–96
5. Gajbhiye S, Mishra SK, Pandey A (2013) Effects of seasonal/monthly variation on runoff curve number for selected watersheds of Narmada basin. Int J Environ Sci 3(6):2035–2046
6. Gajbhiye S, Mishra SK, Pandey A (2013) Prioritizing erosion prone area through morphometric analysis: an RS and GIS perspective. Appl Water Sci (2013). https://doi.org/10.1007/s13201-013-0129-7
7. Gandini ML, Usuoff EJ (2004) SCS-CN estimation using RS NDVI in a GIS environment. J Environ Hydrol 12 (2004)
8. Mishra SK, Gajbhiye S, Pandey A (2013) Estimations of design runoff curve numbers for Narmada watershed (India). J Appl Water Eng Res 1(1):67–79
9. Tripathi MP (1999) Hydrological modeling of small watershed. PhD thesis, Department of Agricultural and Food Engineering, Indian Institute of Technology, Kharagpur

Effect of Land Use–Land Cover Change on Runoff Characteristics in Mumbai City

Sahoo Biswa Manaschintan and Sahoo Sanat Nalini

Abstract Change in Land use and Land cover (LU/LC) impacts the runoff attributes of a drainage basin to a huge degree, which thus, influences the surface and groundwater accessibility of the zone, and hence prompts furthermore change in LU/LC. Consequently, it becomes essential to survey the impact of variation in LU/LC on the runoff characteristics of an area. Such an examination can adequately be accomplished by utilizing watershed simulation models with coordinated GIS framework. In this study, the change in land use–land cover of Mumbai city is detected and analyzed. Hydrological modeling is performed by SWAT to simulate runoff from the basin. The results of SWAT modeling were calibrated and validated using SWAT-CUP by SUFI-2 method. It was observed that the exponential increase in urbanization resulted in an increase in simulated runoff in Mumbai for the past two decades (from 1995 to 2016).

Keywords Geoinformatics · RS · GIS · LU/LC · SWAT · Runoff

1 Introduction

To suit the amplified urban populace and in addition their exercises, individuals have prompted changes in land use–land cover (LU/LC) quickly in temporal and spatial scales. LU//LC change assumes an essential part in the runoff generation due to adjustment in other hydrological processes. This has brought about the decrease in peak time for the flood hydrograph. The peak discharges reach the outlet in a short interim of time causing a surge in intensity and frequency of flooding. Hence, to accomplish a superior planning, administration and practical improvement of the watershed, proper understanding, learning and evaluation of the effects of LU/LC change on the watershed hydrological process are of extraordinary significance to foresee the flood potential, flood hazard, and flood risk. To evaluate changes in the overflow generation at different spatial and temporal scales, LU/LC change exam-

S. B. Manaschintan · S. S. Nalini (✉)
Department of Civil Engineering, NIT Rourkela, Rourkela, Odisha, India
e-mail: sahoosanat@nitrkl.ac.in

J. K. Ghosh and I. da Silva (eds.), *Applications of Geomatics in Civil Engineering*, Lecture Notes in Civil Engineering 33, https://doi.org/10.1007/978-981-13-7067-0_14

ination is particularly required. Land use and land cover changes may have four noteworthy direct effects on the hydrological cycle and water quality: they can cause floods, droughts, changes in stream and groundwater systems, and they can influence water quality [10]. The investigation of effects of land use change has fascinated researchers [5, 9] who, through their studies, tried to comprehend land use change, its circumstances, and end results. Ott and Uhlenbrook [8] opined that LU/LC may have both immediate and long-lasting impacts on terrestrial hydrology, altering the balance between precipitation, evapotranspiration (ET) and the resultant runoff. Dams [3] stated that since the era of industrialization and rapid population growth, land use change phenomena have strongly accelerated in many regions. Coutu and Vega [2] conducted a study on the East Branch of the Brandywine Creek Watershed and observed that surface runoff has risen about 12.15–20.8% in the year 1992–2000 due to the significant increase in urban area.

The current study examines the land use and land cover change and its effect on the hydrological regime of Mumbai, India. The fundamental goal of the present study is to evaluate the process of urbanization with reference to the spatial and temporal variation of the LU/LC change for the area of Mumbai in India. The satellite image of the year 1995 through the satellite image for the year 2016 is used to find the spatiotemporal LU/LC changes. This LU/LC change is used to generate the hydrological model in ArcSWAT. The impact of LU/LC change on flood peak discharges and runoff volume is assessed. The trend of land use change between 1995 and 2016 was used because during that time the effects of land use change in Mumbai Region had been severed. Hence the objectives of the current study are (i) To analyze the change in land use and land cover from the year 1995 to 2016. (ii) To estimate the daily and monthly surface runoff in Mumbai City and (iii) To analyze the impact of LU/LC change on runoff characteristics during this period (1995–2016).

2 Study Area

Mumbai city in India as shown in Fig. 1, lies at 18°55′ North and 72°54′ East, consists of four rivers such as Mithi, Poisar, Dahisar, and Oshiwara. These rivers originate from the hilly region and flow toward west to join the Arabian Sea through creeks. The city is surrounded by the Arabian Sea and is intercepted by creeks namely Malad, Mahim, Mahul, and Thane creeks. The city consists of two distinct regions: Mumbai City district and Mumbai Suburban district, which form two separate revenue districts of Maharashtra. The city district region is also commonly referred to as the Island City or South Mumbai. The total area of Mumbai is 603.4 km^2 of this, the island city spans 69 km^2, while the suburban district covers 368.1 km^2, together accounting for 437.71 km^2 under the administration of Municipal Corporation of Greater Mumbai (MCGM).

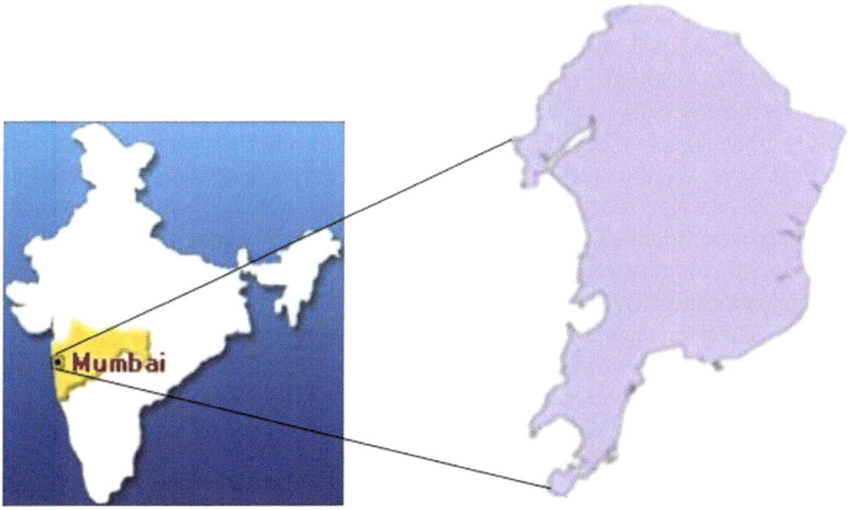

Fig. 1 Location map of the study area

3 Data Collection

The following dataset is collected for the study purpose.

- Landsat imageries of spatial resolution 30 m were collected from USGS (United States Geological Survey) for the years 1995, 1998, 2000, 2002, 2008, 2011, 2013, and 2016.
- The SRTM (Shuttle Radar Topography Mission) 90 m Digital Elevation Model (DEM) of the study area was obtained from Consortium for Spatial Information (CGIAR-CSI).
- Soil data was collected from the Food and Agricultural Organisation of the United Nation (FAO).
- Meteorological data like daily precipitation, minimum and maximum temperature were collected from the National Climate Data Centre (NCDC).
- The daily discharge data at gauge station was collected from India Water Resources Information System (WRIS).

4 Methodology

The collected imageries of Mumbai city for the years 1995, 1998, 2000, 2002, 2008, 2011, 2013 and 2016 were processed in ArcGIS platform through the processes like preprocessing, classification and post-processing. Preprocessing includes geo-referencing, mosaicking, image clipping and generating the shapefile for the study area. False Color Composite (FCC) was produced for the collected satellite images for different years and further analysis was carried out. Maximum likelihood classification (MXL) technique was employed to classify the study area into four different types of land covers like vegetation, water body, barren land, and urban land. High-density urban area, low-density urban area, roads, and airport were categorized into one land cover, i.e., urban land. In water body class, rivers and lakes were taken into account. Dense vegetation, light vegetation, and forest were included under vegetation land cover. Mountain, unused fields, barren lands were categorized in barren land type. As a part of post-processing, the classification result obtained above through the application of MXL is assessed for its accuracy. Accuracy assessment is advantageous for checking the validity for the classification approach for evaluating errors. Two sourced elements of information are then compared: the classified map and the ground information (reference data) from Google earth. The partnership amongst the classified map and the reference data is summarized in a mistake matrix or confusion matrix or a contingency table [4, 6]. Thus, LU/LC maps for the study area were obtained for different years by using geospatial technologies and the change in LU/LC was also detected. Runoff is simulated by employing these LU/LC changes through Soil and Water Assessment Tool (SWAT).

SWAT is a consistent, long term, physically based conceptual model. This model operates at basin scale on daily time step [1]. It is a hydrologic model with ArcView GIS interface which has been developed by the USDA-ARS and the Blackland Research and Extension Centre. Within SWAT model, the catchment is primarily split into sub-basins or sub-watersheds predicted on topographic criteria accompanied by further division into a series of Hydrological Response Units (HRU) on the basis of soil type, slope, and land use combinations. Generally, water enters by means of precipitation into the SWAT watershed system. In the model, flow routing and water quality analysis is carried out based on HRU to each sub-basin and eventually into the outlet of the watershed. In the current study, SWAT 2012 model is integrated with Arc GIS software for runoff simulation in the study area.

The model was set up with the aid of ArcSWAT that runs under the ArcGIS environment. The setup consisted of preparation of input data like delineation of the watershed using the digital elevation model (DEM), HRU definition using soil, slope, land use, and agricultural practice data and preparing LU/LC maps for different years. Meteorological data like precipitation, minimum and maximum temperatures collected from NCDC and solar radiation, humidity and wind speed obtained from weather generator tool in ArcSWAT were fed to the model. After the input of the data, calibration and validation are performed. The detailed methodology is explained in terms of a flowchart shown in Fig. 2.

For SWAT simulation, a warm-up period of 2 years (1992–1994) was taken. The model was calibrated for 7 years (1995–2002) of data and validated for 4 years (2004–2007). The calibration and validation of the model were done using SWAT-CUP tool using SUFI-2 (Sequential Uncertainty Fitting version 2) algorithm. Prior to the application of SUFI-2 for calibration, the most sensitive parameters were selected by running the sensitivity analysis. Out of all parameters considered for

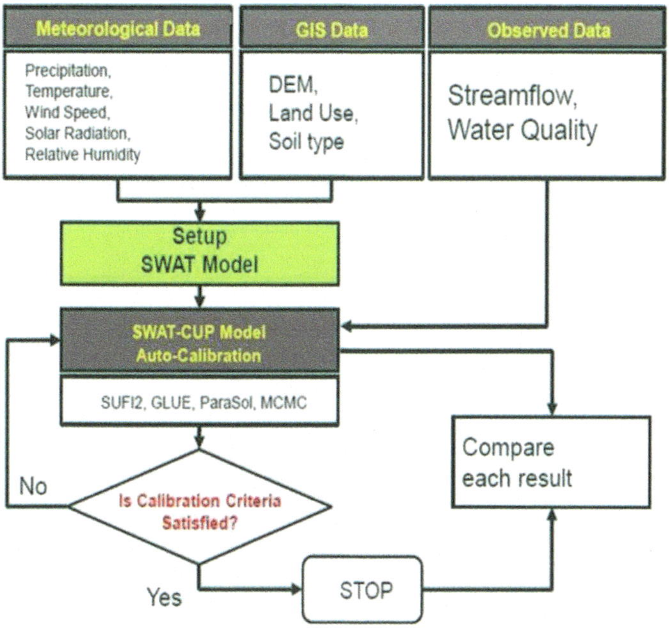

Fig. 2 SWAT model flow diagram

sensitivity analysis, only 24 were found effective for monthly flow simulation. 24 most sensitive parameters for the stream flow assessment were considered for the model parameterization and calibration process.

5 Result and Discussion

Using the methods described above, LU/LC maps were prepared for the study area over two decades. The remotely sensed data reveal that substantial changes took place over the study area during these years i.e. from 1995 to 2016. Figure 3 and Table 1 indicate the change in LU/LC during these years. An increase of 16.4 km^2 of urban land is observed from 1995 to 2016. The average increase rate was 0.713% annually. During this period, Mumbai City has undergone a massive expansion. The forest area and barren land decreased from 22.16 km^2 and 23.07 km^2 in 1995 to 18.37 km^2 and 10.88 km^2, respectively, in 2016 (Table 1). From Fig. 4 it can be observed that the Mumbai city shows an exponential growth trend in urbanization from the year 1995 to 2016.

It is evident from Table 1 that, there has been a significant land use and land cover change in the area where the forest and vegetation land covered 32.11% in 1995 decreased to 28.46% in 2002 and further decreased to 26.62% in 2016. However,

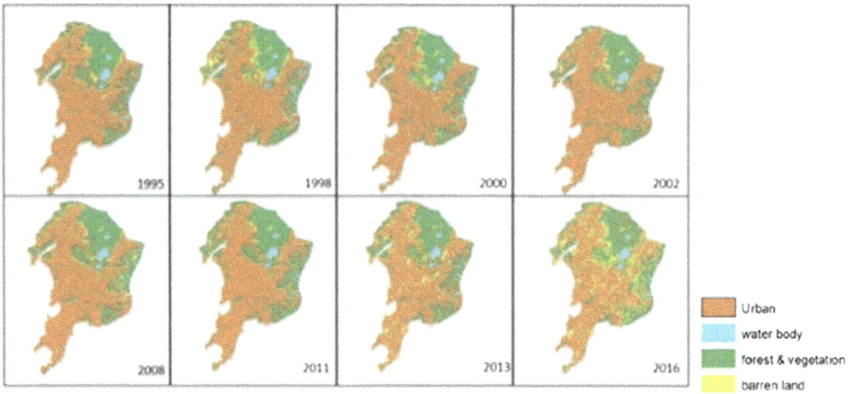

Fig. 3 LU/LC maps of Mumbai city

Table 1 LU/LC of Mumbai city from 1995 to 2016

LU/LC (%)	Year							
	1995	1998	2000	2002	2008	2011	2013	2016
Urban	31.98	34.01	36.58	38.25	45.24	47.46	51.11	55.74
Water body	2.48	2.33	2.21	2.14	1.94	1.96	1.79	1.87
Forest and vegetation	32.11	29.76	28.67	28.46	27.22	28.17	27.34	26.62
Barren land	33.43	33.91	32.54	31.16	25.6	22.41	19.76	15.77

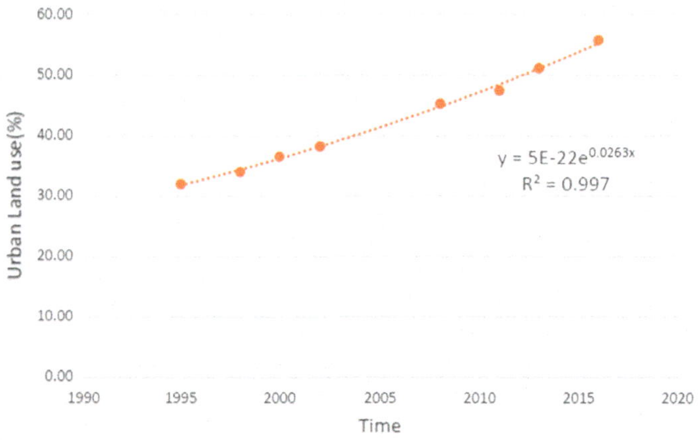

Fig. 4 Urbanization growth trend of Mumbai

the area occupied by urban land cover increased from 31.98% in 1995 to 36.58% in 2000, 45.24% in 2008 and 55.74% in 2016. The area occupied by barren land showed a slight increase from 33.43% in 1995 to 33.91% in 1998, then it decreased to 25.6% in 2008 and 15.77% in 2016. The variation showed by water body land cover is very little, it decreased from 2.48% in 1995 to 1.87% in 2016. These changes have been attributed to the expansion and intensification of urban areas to satisfy the demands of the increasing population. For the period covered by the study, the classified Landsat images across the study area revealed the decrease by 5.5% of forest and vegetation land, the built-up area increased by 23.77% while barren land decreased by 17.66%. During this period the land cover occupied by water body decreased by 0.61%.

Figure 5 shows a comparison between calibrated and observed discharges from 1995 to 2002. From SWAT-CUP analysis, Nash–Sutcliffe efficiency (NSE) was found to be satisfactory for both the calibration period and validation period. NSE was found to be 0.88 and the correlation coefficient R^2 was found to be 0.95 (Fig. 6) for the calibrated data for discharge. The same was found as 0.59 and 0.77 (Fig. 7) respectively for the validated discharge. Thus, the result is quite satisfactory [7] and can be accepted.

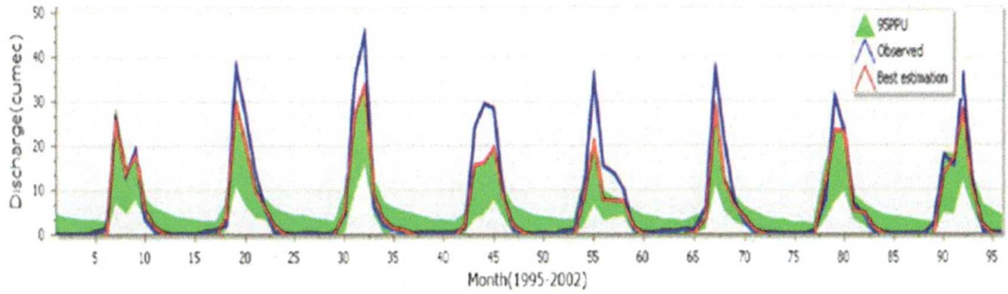

Fig. 5 Calibrated versus observed data for monthly time step

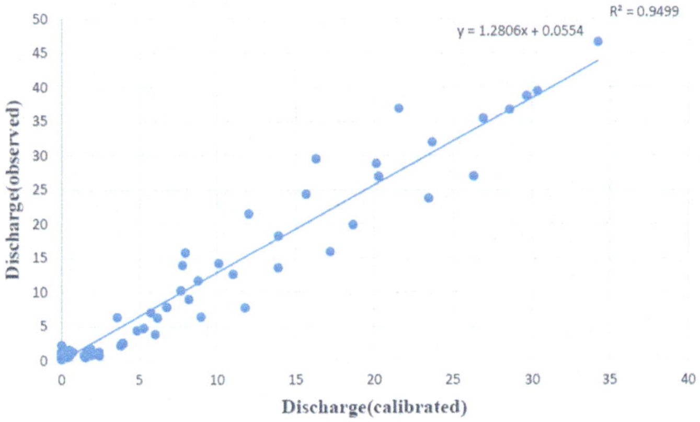

Fig. 6 Correlation between calibrated and observed discharge

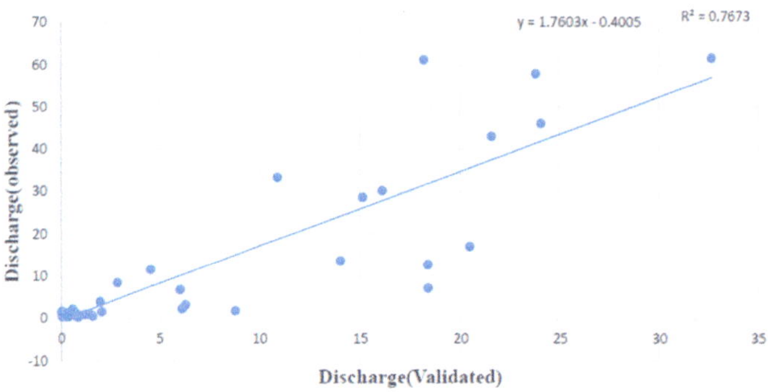

Fig. 7 Correlation between observed and validated discharge

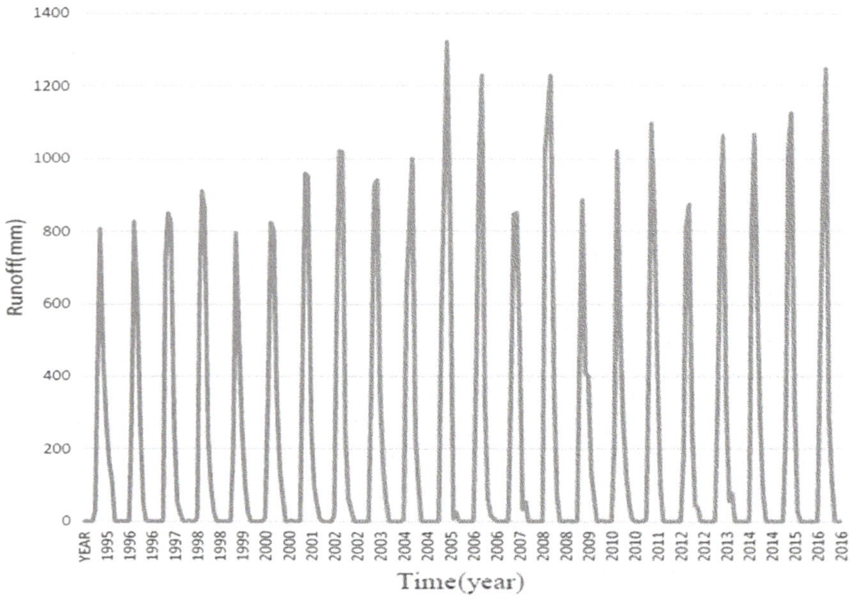

Fig. 8 Simulated runoff for the study area from 1995 to 2016

From Figs. 8 and 9, it was observed that the surface runoff of the study area has significantly increased from 1995 to 2016 although the change in precipitation intensity in these years was very minimal. Due to abnormally high rainfall in 2005, the study area witnessed a maximum surface runoff during the study period. The maximum monthly rainfall and surface runoff in 1995 was 1107.88 mm and 807.65 mm, and for 2016 it was 1097.7 mm and 1248.7 mm, respectively. The increase in runoff of 441 mm while rainfall quantity is almost same can be related to the variation in land use–land cover of Mumbai city or the rise in urban land use which promotes impervious land that leads to less infiltration.

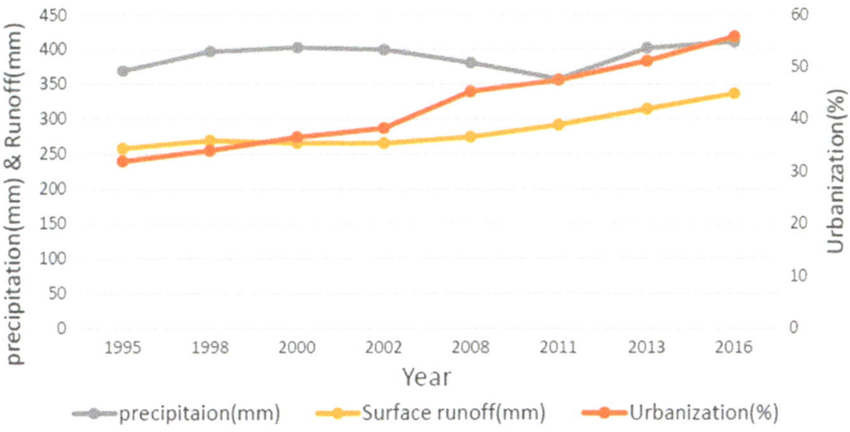

Fig. 9 Comparison between precipitation and runoff with urbanization from 1995 to 2016

6 Conclusions

In this study, the land use and land cover classification of Mumbai city were performed by employing capabilities of RS and GIS. Then it was analyzed for variation in land use and land cover. A hydrologic model was made in ArcSWAT for study area and simulated for the time period spanning from 1995 to 2016. The hydrologic model was calibrated and validated in SWAT-CUP and the effect on runoff due to land use and land cover change was examined. The following conclusions are drawn from the current research work:

- The urban land cover increased by 23.77%, forest or vegetation and barren land decreased by 5.50% and 17.66%, respectively, from 1995 to 2016.
- The surface runoff of Mumbai city was increased significantly from 1995 to 2016, although precipitation remained almost the same. It was caused by an increase in urbanization, which leads to a surge in impervious land.

Acknowledgements The authors would like to acknowledge the Science and Engineering Research Board (SERB), India for providing the financial support for this research program vide project no. ECR/2016/000057.

References

1. Arnold JG, Srinivasan R, Muttiah RS, Williams JR (1998) Large area hydrologic modelling and assessment; part I: model development. J Am Water Resour Assoc 34(1):73–89
2. Coutu GW, Vega C (2007) Impacts of land use changes on runoff generation in the east branch of the brandywine creek watershed using a GIS-based hydrologic model. Middle States Geogr 40:142–149

3. Dams J (2007) Predicting land-use change and its impact on the groundwater system in the Grote-Nete catchment, Belgium. Hydrogeol J 15:891–901
4. Jensen JR (2005) Introductory digital image processing: a remote sensing perspective. Prentice Hall, Upper Saddle River, New Jersey
5. Lambin EF, Turner BL, Geist HJ, Samuel BA, Angelsen A, Bruce JW, Coomes OT, Dirzo R, Fischer G, Folke C, George PS, Homewood K, Imbernon J, Leemans R, Li X, Moran EF, Mortimore M, Ramakrishnan PS, Richards JF, Skanes H, Steffen W, Stone GD, Svedin U, Veldcamp TA, Vogel C, Xu J (2001) The causes of land-use and land cover change: moving beyond the myths. Glob Environ Change 11(4):261–269
6. Lilleland T, Kiefer R, Chipman JR (2004) Remote sensing and image interpretation, 5th edn. Wiley
7. Me W, Abell JM, Hamilton DP (2015) Effects of hydrologic conditions on SWAT model performance and parameter sensitivity for a small, mixed land use catchment in New Zealand. Hydrol Earth Syst Sci 19, 4127–4147
8. Ott B, Uhlenbrook S (2004) Quantifying the impact of land-use changes at the event and seasonal time scale: a process-oriented catchment model. Hydrol Earth Syst Sci 8:62–78
9. Verburg PH, Overmars KP, Huigen MGA, Groot WT, Veldkamp A (2006) Analysis of the effects of land-use change on the protected areas in the Philippines. Appl Geogr 26(2):153–173
10. Weng Q (2001) Modeling urban growth effects on surface runoff with the integration of remote sensing and GIS. Environ Manag 28(6):737–748

Snowmelt Runoff Estimation Using Energy-Balance Approach

Tripti Khanduri and Praveen K. Thakur

Abstract The energy balance of a snowpack derives the production of snowmelt water. Proper budgeting of this energy driven processes is required for the efficient management of water resources. The study area is a part of North-Western Himalayas, i.e. Manali sub-watershed (area = 350 km^2), situated in the Himachal Pradesh state, India. The region experiences snowfall from December to February with January being the coldest month. Meteorological data and remotely sensed data from Landsat ETM+, IRS P-6 LISS-III and MODIS 8-day snow cover data product of the watershed has been used as input. For simulation of energy-exchange process and estimation of the amount of snowmelt generated, a physically based Utah Energy Balance model was used. The model performances were found good and an agreement between the observed and simulated values was obtained with few over- and under-predictions and the coefficient of determination being 0.718 at the Manali outlet.

Keywords Snowmelt runoff · Energy balance · UEB model · NRCS-CN · VIC routing model

1 Introduction

Computation of snowmelt from a watershed is made using either an energy balance approach or some index approaches. For energy balance approach information regarding radiation energy, sensible and latent heat of the snowpack is required. Apart from this, the energy transferred through rainfall over the snow and heat conduction from the ground to the snowpack is also an important pre-requisite. On the other hand, an index approach estimates snow cover energy exchange through an empirical formula using one or more variables. These variables include air temperature, net radiation, vapour pressure, wind speed and solar radiation. One of the most popular

T. Khanduri (✉)
Civil Engineering Department, Indraprastha Engineering College, Ghaziabad, Uttar Pradesh, India
e-mail: tripti.khanduri@ipec.org.in

P. K. Thakur
Indian Institute of Remote Sensing, Dehradun, Uttarakhand, India

J. K. Ghosh and I. da Silva (eds.), *Applications of Geomatics in Civil Engineering*,
Lecture Notes in Civil Engineering 33, https://doi.org/10.1007/978-981-13-7067-0_15

methods for index approach is degree-day method. This method uses air temperature for representation of energy flux. Air temperature is an easy parameter to measure, extrapolate and forecast. It is found from the researches that vapour pressure, net radiation and wind speed can significantly improve the snowmelt prediction [7]. A recent advancement had been the use of the hybrid approach, which uses inputs of both the energy balance and index methods [6].

A comparison of temperature index and energy budget method reveals that the first approach has the advantage that air temperature is commonly observed, and its areal variability can be easily estimated. However, being an index approach there is a significant diversion between the index variable and snow cover heat exchange. The advantage of this theoretical heat transfer approach is that the observed diversion can be reduced. The disadvantage of this approach is that it relies on only one variable. Most of the times the variables, such as solar radiation, long wave radiation and albedo, are not observed but are estimated based on literature available and induce errors [7]. Temperature index method simplifies the inputs required considering the limitations of the basin in the availability of basic data. Though the temperature index method is the most practical method for daily streamflow forecasting, the thermal budget approach is needed for design flood determinations.

Energy-balance approach considers the actual energy exchanges taking place in the process of snowmelt. Snow predictions can be improved using energy-balance approach as it considers precipitation, wind, temperature, latent heats, net radiation, albedo, energy content, water equivalent, cloudiness, vegetal cover, etc. instead of temperature alone [7]. SEBAL is a similar energy balance approach given for evapotranspiration calculation, but it is applicable to agricultural fields. Variable Infiltration Capacity (VIC) model (snowmodel) also uses an energy balance approach to represent snow accumulation and ablation on the ground. It assumes that the low-lying vegetation is completely covered by snowfall and does not affect the ground snowpack energy balance. Flerchinger [1] developed the SHAW model and used it to simulate the energy and mass balance of the soil and snow cover as a single system [7]. In this model, snowmelt functions accurately represent the energy balance processes involved but are too complicated to explicitly distribute over a grid. Flerchinger [1] used the SHAW model in point mode, to do multiple simulations over small drainage. This implied that if a physically-based model is distributed it can provide detailed information on the impacts of spatially and temporally varying snowmelt basin hydrology and soil moisture. SNTHERM model [2] can simulate snow cover energy and mass balance accurately but requires extensive forcing and snow cover structure data. SNTHERM also requires the input of snow structure and conditions, during deposition events, so it has difficulty simulating the development of the snow cover unless detailed deposition data are available.

Tarboton and Luce [8] showed that both snow deposition and ablation could be simulated. In general, for wide applicability over mountain drainages parameterized models are needed [4]. Utah Energy Balance (UEB) Model is used for understanding snow processes and for runoff, erosion and water balance forecasting and modelling. Among the various energy-balance models available, Utah Energy Balance Model (UEB) is chosen for it being open source software, simple to operate, can be modified, calibrated and validated as per the requirement of the area. The model proves to be simple in application to the area for the first time since this energy balance approach has not been applied to the study area. Remote sensing approach, which has not been implemented using the UEB model, is used in this study.

2 Study Area

The study area is the Manali sub-watershed (350 km^2) situated at the north of Kullu district between 32.23°–32.42°N and 77.05°–77.28°E, Himachal Pradesh (HP) in India. The sub-watershed is a part of the Beas River, one of the major tributaries of the Indus river system. The Beas River originates from Beas Kund glacier (elevation of 4038 m) on the eastern slope of Rohtang pass in the Western Himalayas. It flows in north–south direction and takes a turn near Larji (elevation of 957 m) towards the west and then it maintains its flow again north–south and west direction. The area is surrounded by districts of Kangra and Mandi in the east, Lahul in the north and Spiti in the northeastern side, Kinnaur and Shimla in the southeastern part [3].

The area has a varied climate owing to its elevation difference (1800–5932 m). The area generally experiences a normal low monthly maximum temperature due to its altitude. The mean daily minimum temperature ranges between −15 and 0 °C in January to mean daily maximums temperature between 20 and 30 °C in June. June being the hottest month with a temperature above 20 °C and January being the coldest month with temperatures below 2 °C. After the month of June, which is the hottest month, the temperature continues to fall, and the lowest temperature is observed in the month of January, which is the coldest month. The mean temperature falls and goes in negative in January. The snowfall occurs in the month of December and January [5] (Fig. 1).

The relative humidity is high in the months of May–June, which is the per-monsoon period and July–September, which is the monsoon period and lower in the months of October–January, which is the winter season. 70% of the annual rainfall occurs during monsoon season with frequent cloud burst events. Average annual rainfall observed in the sub-watershed is 100 cm. Snowfall generally occurs in the months of December and January at elevations above 2500 m. Heavy snowfall causes high-altitude regions to cut off from the lower region as most of the mountain passes are closed. August is observed the wettest month throughout the sub-watershed. The maximum rainfall occurs during the months of July and August and minimum rainfall occurs during the months of October–January. The maximum amount of snowfall occurs in the months of December and January. In general, the climate is cold in the entire watershed. The intensity of cold varies and is due to the differences in elevation and aspects. The effect of global warming is also observed from the breakout of increasing diseases, insects and other hydro-meteorological events. Sudden cloud burst events, heavy intensity rainfall in the monsoon season cause devastating floods in this fan-shaped basin [3].

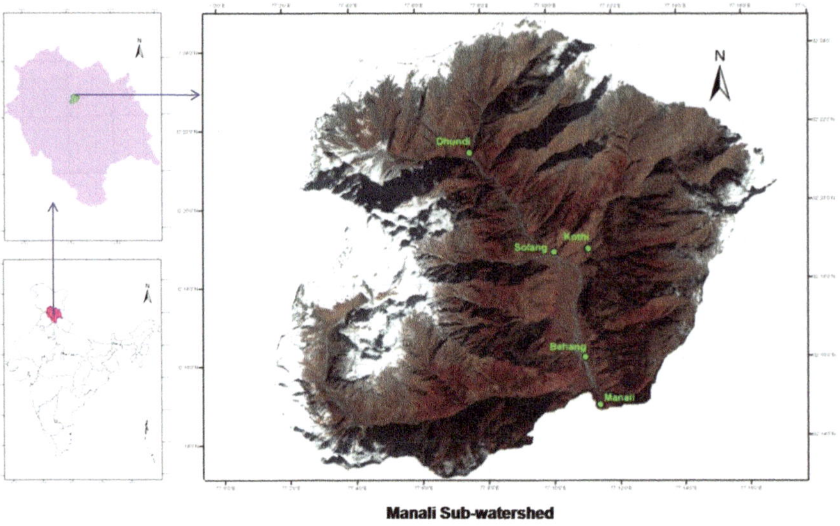

Fig. 1 Study area

The soils of this area are young and variable in depth, which increases from inclined hill slopes to the valley. The categories of soil type found in the area are rock outcrops, grasslands, forest soils and fluvial valley soil. The prevailing soil types are categorized into three textural groups, coarse-loamy, loamy-skeletal and rock outcrop. All these are formed according to the nature of parent rock material. This categorization is also affected by the elevation, slope and aspect and prevailing rainfall, temperature, vegetation covers etc. This soil classification is not based on elevation alone but other mentioned aspects also. However, the river channel and valley soil is mostly affected by location and elevation. Soil thickness in the valley areas is more influenced by human activities [3] (Fig. 2).

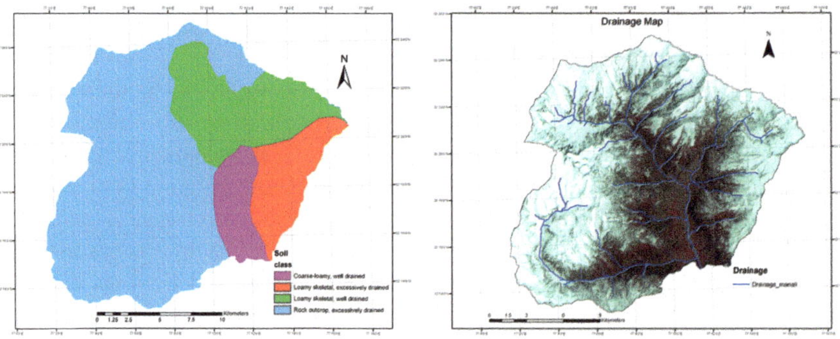

Fig. 2 Soil texture map and drainage map

3 Materials and Methodology

3.1 Materials Used in the Study

The remote sensing data and data products used in the study are enlisted in Table 1. These products are available at no cost to the users worldwide. These are referenced to the WGS84/EGM96 geoid.

Hydro-meteorological data required for the study was obtained from Automatic Weather Stations (AWS) established by Snow and Avalanche Establishment (SASE), DRDO in the sub-watershed. The data included information related to air temperature, precipitation, wind speed and relative humidity at three sites, namely, Dhundi, Solang and Bahang. Table 2 enlists the software used in the study.

The overview of the methodology followed is given by Fig. 3.

4 Result and Discussions

See Tables 3 and 4, Figs. 4 and 5.

Table 1 Remote sensing data used for the study

Satellite/Sensor	Date	Spatial resolution (m)
ASTER GDEM	2009	30
Landsat ETM+	2012	30
MODIS L-3 product (8-Day Snow Cover)	2012	500
IRS P-6 LISS-III	29 October 2011	23.5
ALOS PRISM	26 September 2011	2.5

Table 2 Softwares used

Softwares
Erdas Imagine
ArcMap
Envi
Utah energy balance (UEB) model
Variable infiltration capacity (VIC) routing model

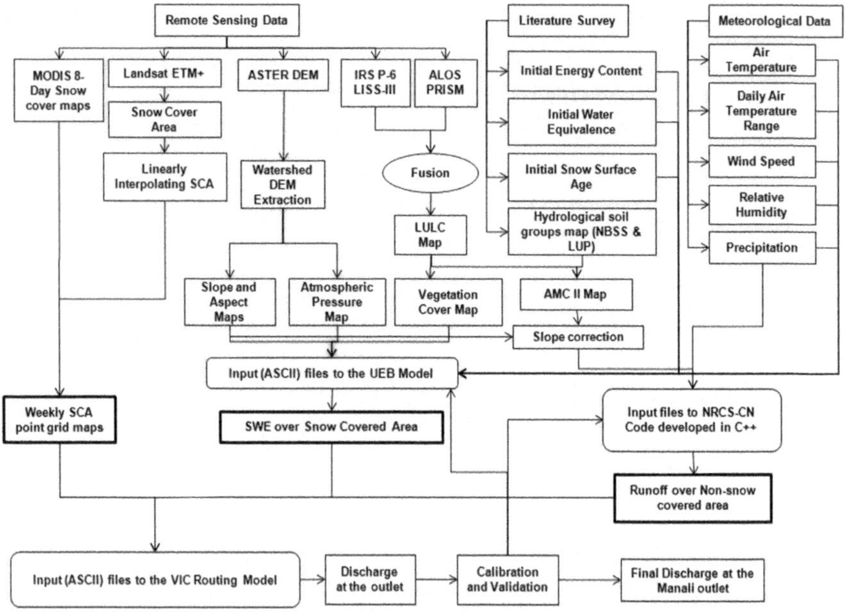

Fig. 3 Methodology followed

Table 3 Efficiency table before and after calibration

Efficiencies	Pre-calibration values	Post-calibration values (after removing extreme rainfall events)
R^2	0.5468	0.7130
wR^2	0.5157	0.6163
E	0.5099	0.5032
D	0.8719	0.8931
lnE	0.3676	0.3477
E_1	0.4923	0.5189
D_1	0.7646	0.7822
E_{rel}	0.3955	0.2865
D_{rel}	0.8420	0.8465

Table 4 Efficiency table for validation period (2013)

Efficiencies	Validation values
R^2	0.7889
wR^2	0.6891
E	0.7023
D	0.9320
lnE	0.6218
E_1	0.5482
D_1	0.7926
E_{rel}	0.6188
D_{rel}	0.9129

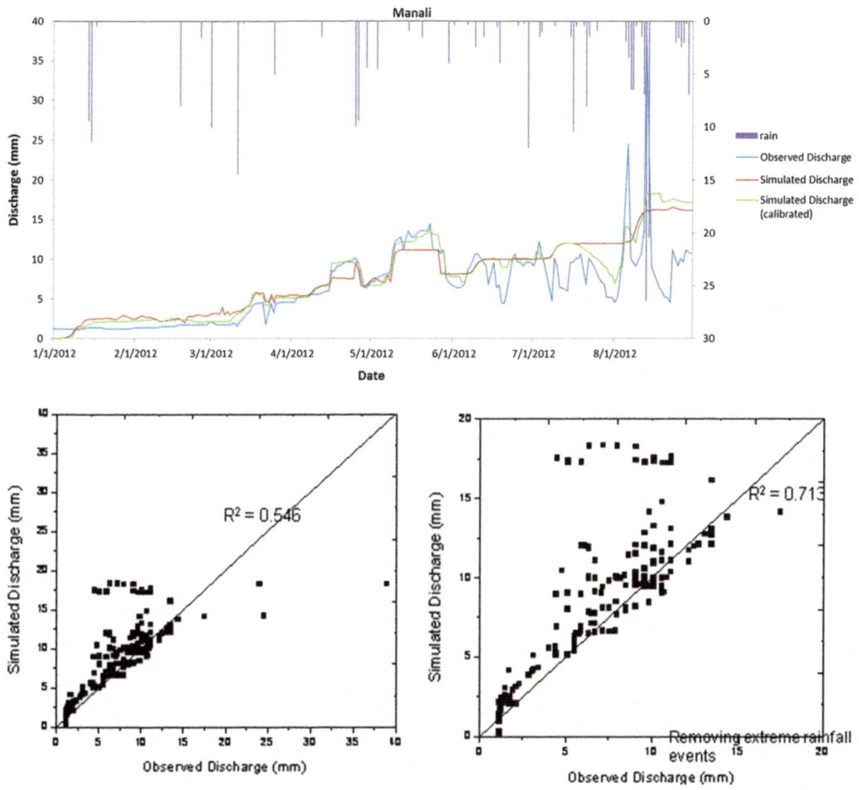

Fig. 4 Pre-calibration and post-calibration results

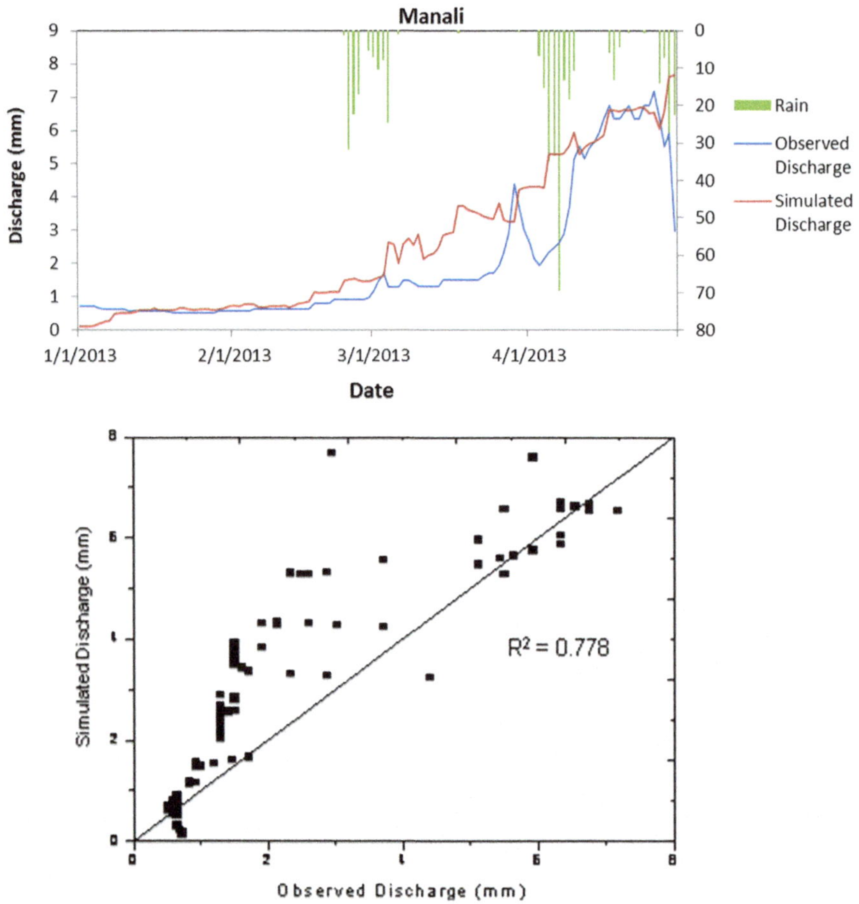

Fig. 5 Validation results

5 Conclusions

In the first part of the paper, we applied an image fusion technique to improve the LULC classification of the study area. The accuracy was found to improve by 15% and the Kappa coefficient increased from 0.5918 to 0.8039. This was verified on field and found to be appropriate with overall accuracy being further improved by 5% and the kappa coefficient increased to 0.9036.

The paper also provides a technique to derive snow cover area (SCA) by incorporating vegetation cover (NDVI) within the NDSI approach. This has improved the accuracy of extracted snow cover area, especially in the forested area of Manali sub-basin. This was interpolated to compare with data products of MODIS on a weekly basis. However, there were missing data in MODIS data product.

In the final part of the paper, energy-balance of the snowpack was simulated using the UEB Model. The model was improved by incorporating a variation of snow density due to freshly fallen snow and variation of thermal conductivity of fresh snow. Diurnal variation was obtained in the results with afternoon value being greater than the forenoon value. These results follow the flow trends prevailing in the sub-watershed. Further, this model needs improvement in the field of snow vegetation interaction (detailed) and coupling with the flow routing model.

References

1. Flerchinger GN (2009) The simultaneous heat and water (SHAW) model: technical documentation. Northwest Watershed Research Centre USDA Agricultural Research Service Boise, Idaho, Technical Report NWRC 2000–09
2. Jordan R (1991) A one-dimensional temperature model for a snow cover: technical documentation for SNTHERM.89. US Army Corps of Engineers Cold Regions Research and Engineering Laboratory: hanover, New Hampshire, 49, Special Report 91–16
3. Maity DK (2009) Hydrological and 1 D hydrodynamic modelling in Manali sub-basin of Beas river, Himachal Pradesh, India. M.Sc thesis, Indian Institute of Remote Sensing, Jan 2009
4. Marks D, Domingo J, Susong D, Link T, Garen D (1999) A spatially distributed energy balance snowmelt model for application in mountain basins. Hydrol. Process. 13:1935–1959
5. Negi HS, Kulkarni AV, Semwal BS (2009) Estimation of snow cover distribution in Beas basin, Indian Himalaya using satellite data and ground measurements. J Earth Syst Sci 118(5):525–538
6. Şensoy A (2005) Physically based point snowmelt modelling and its distribution in upper Euphrates basin. PhD thesis
7. Singh P, Singh VP (2001) Snow and glacier hydrology. Kluwer Academic Publisher, Netherlands
8. Tarboton DG, Luce CH (1996) Utah energy balance snow accumulation and melt model (UEB)

Assessment of Reservoir Sedimentation and Identification of Critical Soil Erosion Zone in Kodar Reservoir Watershed of Chhattisgarh State, India

Champat Lal Dewangan and Ishtiyaq Ahmad

Abstract Intense rainfall in a shorter span of time causes a huge amount of erosion resulting in rapid deposition of sediment, consequently reducing the capacity of reservoir. Thus, it is necessary to estimate reservoir sedimentation and identify critical zones susceptible to erosion. With the help of RS and GIS technique, reservoir sedimentation and soil erosion have been assessed in Kodar river watershed. LANDSAT-8 imagery is used to calculate the capacity at various elevations of the reservoir. 14.39% of the reservoir capacity has been lost due to sedimentation in the last 36 years when compared to the designed storage. RUSLE model is being applied to estimate erosion. Thematic maps were prepared for various factors and overlaid in GIS resulting in computation of erosion. The estimated average soil erosion is 5.50 tons/hectare/year. To identify the critical zones, watershed is isolated into 11 sub-watersheds out of which 5 are under critical zone which needs best management practices.

Keywords RS and GIS · Sedimentation · Erosion · RUSLE

1 Introduction

In the current scenario, there are water crises particularly in the agricultural sector as the rainfall intensity and rainy days are not uniform, resulting in maximum runoff within a short period of time. It also causes the erosion of topsoil when the water enters a reservoir, its velocity starts decreasing as the cross-sectional area of the river increases toward the dam. Sediments from muddy water and bed load from the river settle down in the bed of the reservoir, consequently reducing the storage capacity of the reservoir and nowadays, there is a constraint for the construction of new major reservoirs as feasible sites are not easily available. In such a condition, we need to

C. L. Dewangan (✉) · I. Ahmad
Department of Civil Engineering, NIT Raipur, Raipur, India
e-mail: champatdewangan@gmail.com

I. Ahmad
e-mail: iahmad.ce@nitrr.ac.in

J. K. Ghosh and I. da Silva (eds.), *Applications of Geomatics in Civil Engineering*,
Lecture Notes in Civil Engineering 33, https://doi.org/10.1007/978-981-13-7067-0_16

enhance resiliency, reliability, and efficiency of the existing reservoir for the supply of water at right time to the users. In order to assess the reservoir sedimentation, RS and GIS-based method has been applied by many researchers [1–4]. Jain et al. [1] determined the sedimentation in the Bhakhara reservoir with the help of multi-date remote sensing data IRS-1B, LISS II and found that the average rate of sedimentation is 25.23 Mm^3/year, which is closer to the result obtained from Hydrographic survey. Tiwari et al. [4] estimated the sedimentation rate of the Tehri reservoir with the help of LANDSAT-5 TM imagery and SRTM DEM and found that the average sedimentation rate is 12.52 Mm^3/year. Pandey et al. [3] with the help of IRSP6/LISS III imagery and NDWI method has used to identify open water feature and to enhance the water pixels in the satellite imagery and found that 11.76% capacity has been lost by Patratu reservoir in the life span of 44 years. Similarly, Narasayya [2] has carried out reservoir sedimentation of Srisailam reservoir with the help of multispectral data from IRS 1C, 1D, and P6 for LISS III sensor. The sediment index computed including total sediment deposition of 1960.842 Mm^3 in the last 28 years. Soil erosion is a process of detachment, transportation, and deposition of soil particle from one place to another [5], and sedimentation is the result of erosion of soil in the watershed area of the reservoir. If erosion is less, sedimentation is less else sedimentation will be more if erosion is more. Soil erosion from the watershed is determined by Ahmad et al. [6] using RUSLE model and factor maps are prepared from remote sensing data for the calculation of soil erosion in Tandula watershed. Pandey et al. [7, 8] used MUSLE model and remote sensing data for the calculation of soil erosion in Karso watershed. In this paper, the eroded soil of Kodar watershed has been evaluated and estimated by using similar techniques and is found to be very effective.

2 Materials and Methods

2.1 Study Area

Kodar reservoir watershed covers the area of Mahasamund district, of Chhattisgarh state, which is located in between latitude of 21° 01′ 00″ N–21° 14′ 30″ N and longitude of 82° 10′ 30″ E–82° 23′ 30″ E. Kodar watershed have catchment area of 326.45 km^2 and the average annual rainfall is 1066.8 mm. The main river that flows across the Kodar watershed is Kodar River, which is a tributary of Mahanadi River. The Kodar reservoir was commissioned in the year 1980–81. The gross, net, and dead storage capacities are 160.35 Mm^3, 149.02 Mm^3, and 11.33 Mm^3, respectively, and dead, minimum drawdown, and full reservoir levels are 286.04 m, 288.49 m, and 295.336 m, respectively. Figure 1 shows the location map of the study area.

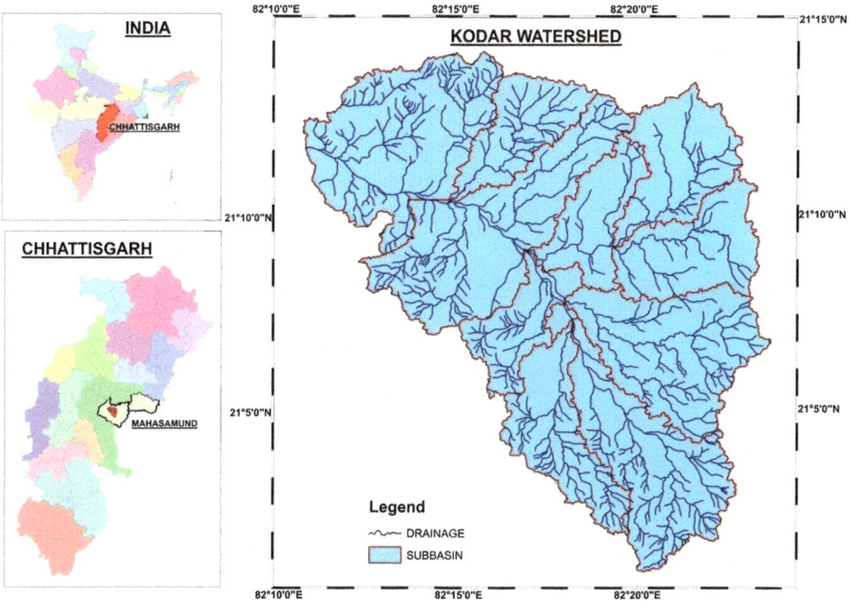

Fig. 1 Location map of the study area

2.2 *Data Used*

For the computation of sediment volume deposited in the reservoir, the satellite imagery LANDSAT 8 OLI and LANDSAT 7 ETM are used which are downloaded from USGS earth explorer from the year 2013 to 2017. The imagery was used to extract the water spread area at different elevations of the reservoir. Data for the water level, original area-elevation-capacity table of reservoir, rainfall data from the year 1994 to 2016 and soil, and C and P-factor data is taken from the State Water Resources Department, Chhattisgarh. DEM is downloaded from Alaska satellite facilities of NASA and ESA.

2.3 *Methodology*

2.3.1 Estimation of Sediment and Revised Reservoir Volume

For estimation of sediment volume and revised reservoir volume, Normalized Differential Water Index (NDWI) method is used which identifies and enhances water pixel in the Remote Sensing Imagery. The equation given by Mcfeeters is used for NDWI calculation is

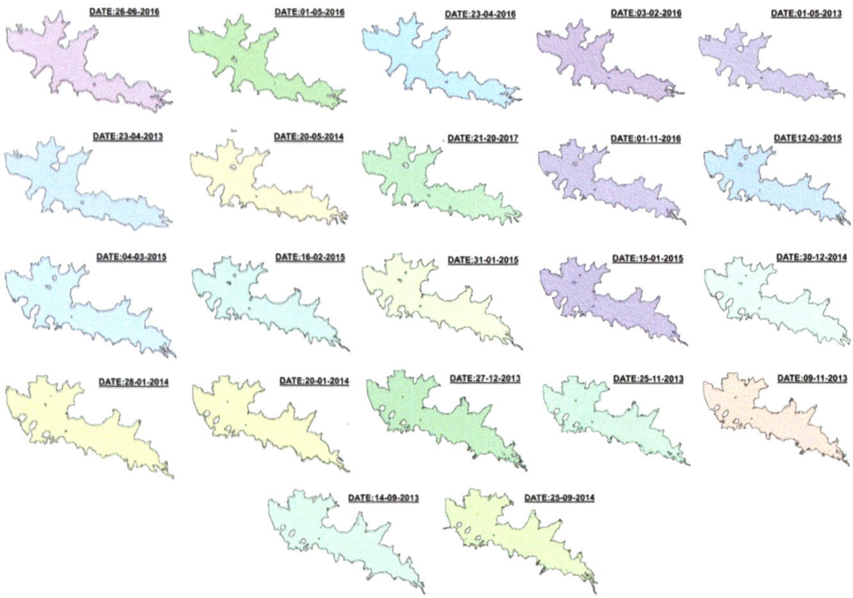

Fig. 2 Extracted water spread area at different dates

$$NDWI = \frac{Green - NIR}{Green + NIR} \tag{1}$$

where NIR is digital number in near-infrared band and green is digital number in green band. This wavelength maximizes the reflectance properties of the water. The index value lies between -1 and 1 having a threshold value of zero [9]. If NDWI is greater than zero, then cover in the imagery is water and if equals to or less than zero, the type of cover is other than water. McFeeters [9] states that "if digital number value in near infra-red spectral region then digital number value of the water pixel is comparatively less than the digital number value of the band blue and red, then it should be classified as water otherwise not". Then, water pixel boundary has been digitized and water spread/submergence area at various levels of the reservoir and storage is computed correspondingly with the help of trapezoidal formula. The computation of water spread/submergence area and revised volume from imagery is shown in Fig. 2 and Table 1.

2.3.2 Identification of Critical Soil Erosion Zone

For the identification of critical soil erosion zone, Revised Universal Soil Loss Equation (RUSLE) model, established by the Agriculture Research Service scientists Wischmeier and Smith is used in which annual soil loss (A) of the area in tons per

Table 1 Computation of water spread/submergence area and revised volume from imagery

S. no	Date	R. L. in meters	Submergence area in km^2	Designed volume in MCM	Revised volume in MCM	Volume lost in percentage
1	LSL	286.04	4.15	0	0	0
2	26/6/2016	287.27	5.67	7.42	6.01	0.94
3	1/5/2016	288.18	7.92	15.41	12.17	2.17
4	23/4/2016	288.52	7.98	18.79	14.87	2.63
5	3/2/2016	288.85	9.04	22.41	17.68	3.18
6	1/5/2013	289.56	10.9	30.76	24.74	4.04
7	23/4/2013	290.07	11.89	37.58	30.55	4.72
8	20/5/2014	290.59	13.24	44.94	37.08	5.27
9	21/2/2017	291.05	13.28	51.86	43.18	5.82
10	1/11/2016	291.54	15.58	59.86	50.25	6.45
11	12/3/2015	292.05	17.14	69.06	58.59	7.02
12	4/3/2015	292.27	17.67	73.06	62.42	7.15
13	16/2/2015	292.72	18.81	82.14	70.62	7.73
14	31/1/2015	293.09	19.04	89.78	77.62	8.16
15	15/1/2015	293.43	19.46	97.04	84.17	8.64
16	30/12/2014	293.73	19.5	103.85	90.01	9.29
17	28/1/2014	294.07	20.47	111.63	96.81	9.95
18	20/1/2014	294.31	22.76	117.58	101.99	10.46
19	27/12/2013	294.61	23.85	125.67	108.99	11.19
20	25/11/2013	294.71	24.08	130.99	111.38	13.16
21	9/11/2013	294.89	25.82	133.78	115.87	12.01
22	14/9/2013	295.19	26.32	143.7	123.7	13.42
23	FRL	295.336	26.83	149.02	127.58	**14.39**
24	25/9/2014	295.47	27.29	–	–	–

hectare can be computed by the multiplication of the six factors, which affect the soil erosion in GIS platform methodology as shown in Fig. 3. The formula used is

$$\mathbf{A} = \mathbf{R} \times \mathbf{K} \times \mathbf{L} \times \mathbf{S} \times \mathbf{C} \times \mathbf{P} \tag{2}$$

Rainfall Erosivity Factor (R)

It refers to the Rainfall Erosivity index, which expresses the capability of rainfall to erode the soil surface open in the atmosphere. From the various field studies, it is found that Rainfall Erosivity index is a numerical value, which is proportional to the product of kinetic energy of rainfall event and 30 minute maximum intensity

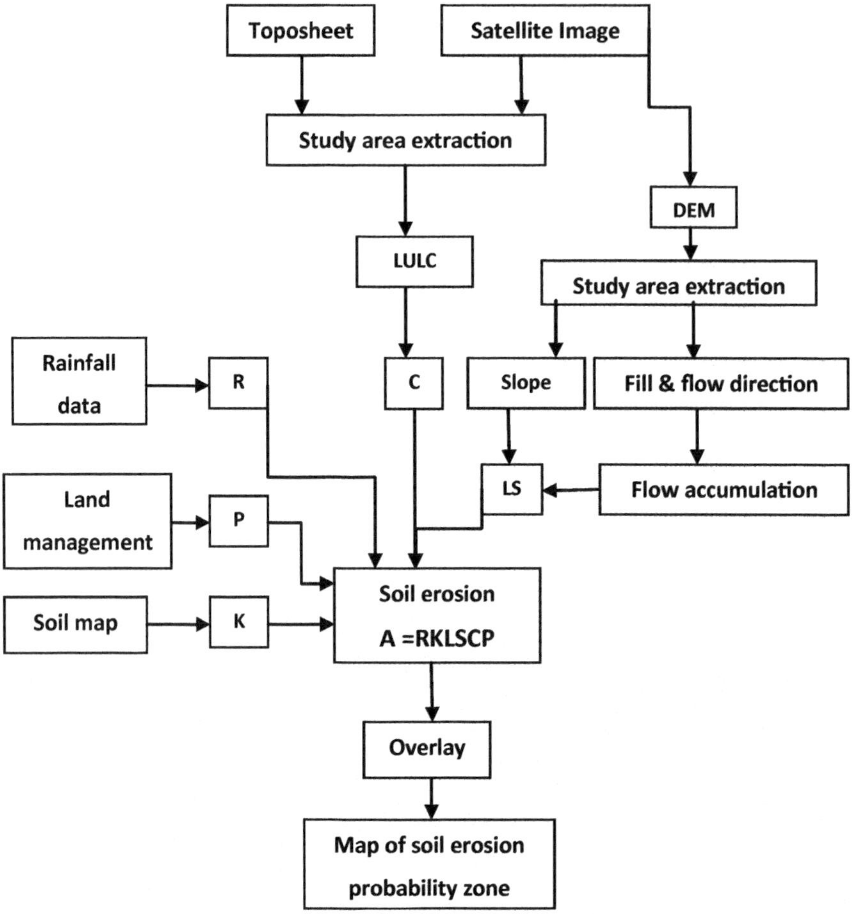

Fig. 3 Methodology of RUSLE model

[5]. For Indian condition in the wake of analyzing the information gathered from 45 stations conveyed in various zones all through India, a simple connection among the Rainfall Erosivity index (R) and annual/seasonal rainfall (X) has been established by Singh et al. [10] is

$$R_a = 79 + 0.363 \times X_a \quad (3)$$

$$R_a = 50 + 0.363 \times X_a \quad (4)$$

Accordingly, the computation of R-factor for Kodar watershed has been done in Table 2.

Table 2 Computation of *R*-factor for Kodar watershed

Station	Rain gauge name	Longitude	Latitude	Mean annual rainfall (mm)	Annual R-factor (MJ mm ha^{-1} h^{-1})
1	2	3	4	5	6
1	Kodar	82.18	21.2	1283.37	544.86
2	Bagbahara	82.43	21.05	985.26	436.65
3	Pithora	82.52	21.26	1022.49	450.16

R factor map of the investigation region is shown in Fig. 4

Soil Erodibility Factor (K)

It refers to the various soil properties on which a soil becomes susceptible to get eroded by water/wind. It is essentially the rate of a vulnerability of soil particles to the erosion per unit of Rainfall Erosivity index (R) for a predetermined soil in a unit plot having 9% uniform slope and length of 22 m. The value of K depends upon the particle size parameter (M) of soil, which can be calculated by summation of percentage of silt and percentage of fine sand multiplied by 100 minus percentage of clay, percentage of organic matter (a), soil structure code (b), and permeability class (c) [11]. The formula used for computation of K is

$$K = 2.8 \times 10^{-7} \times M^{1.14}(12 - a) + 4.3 \times 10^{-3}(b - 2) + 3.3 \times 10^{-3}(c - 3) \quad (5)$$

Accordingly, the computation of K-factor for Kodar watershed has been done in Table 3.

Topographic Factor (LS)

The erosive potential of a particular soil with slope length and slope steepness is represented by Topographic Factor (LS). Transportation of detached soil particle due to rainfall and surface runoff is affected by this factor. On steep ground, erosion is more as compared to flat region due to flow velocity is more in steep ground and less in flat ground. Due to the accumulation of runoff from larger area in longer slope, erosion is more as compared to shorter slope. The degree of slope (S) factor is the connection of soil loss in real slope to the form having 9% slope under the indistinguishable circumstance [6]. With the help of Digital Elevation Model, LS factor has been prepared in GIS. The formula used for the computation of Topographic Factor LS factor is

$$L \times S = \left(\frac{\lambda}{22.13}\right)^{0.5} \times \left(0.065 + 0.045 \times S + 0.0065 \times S^2\right) \quad (6)$$

Table 3 Computation of *K*-factor for Kodar watershed

Location no	Soil code	% of fine sand	% of silt	% of clay	*M*	*a*	*b*	*c*	*K*-factor
1	**2**	**4**	**5**	**6**	**7**	**8**	**9**	**10**	**11**
1	657	11.03	11.32	1.80	2668.59	1.62	3	1	0.15
2	670	11.03	11.32	1.80	2668.59	1.62	3	1	0.15
3	670	11.03	11.32	1.80	2668.59	1.62	3	1	0.15
4	746	3.20	26.87	3.22	2910.32	1.97	3	2	0.20
5	746	3.20	26.87	3.22	2910.32	1.97	3	2	0.20
6	746	3.20	26.87	3.22	2910.32	1.97	3	2	0.20
7	689	8.60	23.87	12.22	2850.38	2.03	3	1	0.20
8	689	8.60	23.87	12.22	2850.38	2.03	3	1	0.20
9	733	4.47	14.12	2.14	1819.22	1.21	3	3	0.15
10	747	10.03	19.83	0.00	3086.00	0.86	3	2	0.24
11	747	10.03	19.83	0.00	3086.00	0.86	3	2	0.24

K factor map of the investigation region is shown in Fig. 4

where λ is flow accumulation in each raster cell, and LS factor map of the investigation region is shown in Fig. 4

Crop Management Factor (C)

The proportion of soil loss from ground under particular crop to the soil loss from a continuous uncultivated land, provided that the rainfall conditions, soil type, and slope are identical is known as Crop Management Factor. The cropping pattern influences the soil erosion in many ways by their various features like the type of crop, cover quality, root growth, and water used by rising plants [5]. C-factor modifies the volume of runoff, drop size distribution, and kinetic energy of rainfall also reduce the velocity of the runoff. C-factor value is shown in Table 4 and the map of the investigation region is shown in Fig. 4.

Conservation Practice Factor (P)

The proportion of soil loss under a given conservation practice to the soil loss from up and down the slope. The conservation practice mainly consists of bunding, strip cropping, contouring, and terracing in which the most effective practice in medium slope range of 2–7% is contouring. Generally when land slope decrease from medium to zero, the effectiveness of contour tillage to reduce the soil loss get reduced as compared to the non-contoured tillage form. Similarly when land slopes increases

Table 4 C and P-factor for Kodar watershed corresponding to land use

S. no	Land use	Area (km^2)	Percentage area	*C*-factor	*P*-factor
1	2	3	4	5	6
1	Agricultural Land	235.23	72.06	0.28	1
2	Forest	6.23	1.91	0.011	0.8
3	Scrubland	3.57	1.09	0.21	1
4	Water bodies	27.72	8.49	0.009	0
5	Build up	0.11	0.03	0.024	1
6	Deg. forest	47.00	14.40	0.011	0.8
7	Other wasteland	6.57	2.01	0.28	1

from medium to steep, the contour row diminishes its capacity to reduce the soil erosion as it is having a very few capacities to retain the water on the soil surface [5]. P-factor value is shown in Table 4 and map of the investigation region is shown in Fig. 4.

The raster layer of the above parameters is prepared in GIS platform and by overlay operation, the average annual soil erosion is computed sub-watershed wise. Sediment rate of the Kodar watershed is 164,164 tons/year. Sub-watershed which is

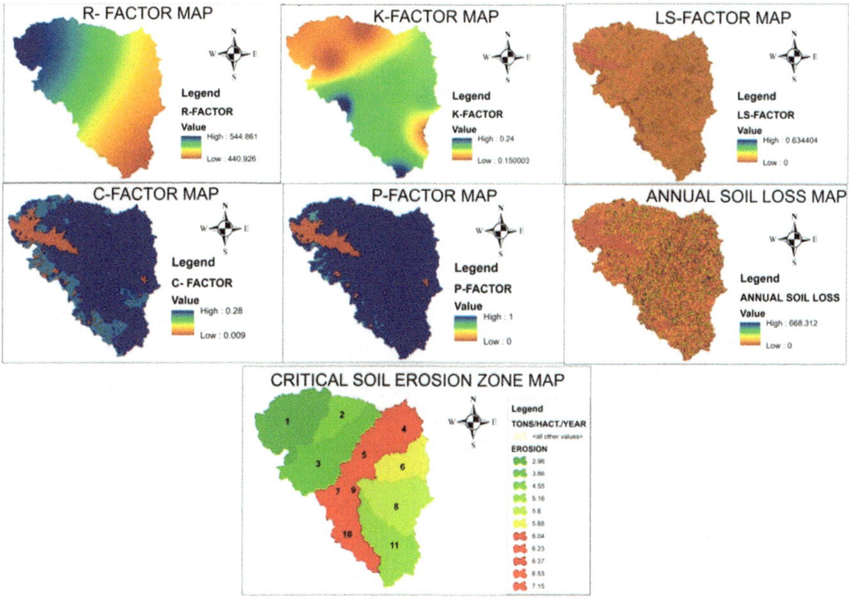

Fig. 4 Various factor map of RUSLE model

having maximum erosion is taken as critical of soil erosion where catchment area treatment is required at the earliest. Annual soil loss and critical soil erosion zone map of the investigation region is shown in Fig. 4.

2.3.3 Validation of the Result

Joglekar analyzed the data from various reservoirs in India and abroad and observed that the sediment yield rates are as follows:

$$q_{sv} = \frac{0.00597}{A^{0.24}} \text{Mm}^3/\text{Km}^2/\text{Year} \tag{7}$$

where A is Watershed area in km^2 [12].

By applying the Joglekar's formula, the annual sediment yield rate of Kodar watershed was found to be 178,583 tons/year.

Along these lines, the assessed soil loss from the Kodar watershed utilizing USLE contrasted and sediment yield using Joglekar's formula is 0.92 times assuming sediment density 1.10 tons/m^3 which is a good result.

3 Result and Discussion

In remote sensing and GIS-based method, the revised capacity of reservoir and sediment volume has been computed with the help of LANDSAT imagery. It is available free in the recycle period of every 16 days. In this study, the imageries are taken at approximately each 5% capacity of reservoir so that the capacity can be computed precisely. Total sediment deposited in the reservoir is 21.44 MCM at various elevations of the reservoir, which is 14.39% of the total capacity in the last 36 years (1981–2016) when compared with the original designed capacity. The comparative graph is shown in Fig. 5. With the help of RUSLE model, annual soil loss of the watershed has been calculated in which 80% area is having soil loss of 0–5 Tons/hectare/year, 8% area is having soil loss of 5–15 Tons/hectare/year, 10% area is having soil loss of 15–50 Tons/hectare/year, and 2% area is having more than 50 Tons/hectare/year soil loss. The average annual soil loss in the study area is 5.5 Tons/hectare/year. The model is validated using empirical relation and result is 0.92 time of the calculated value. For identifying the critical soil erosion zone, the whole watershed is divided into 11 sub-watersheds in which five, namely 4, 5, 7, 9, and 10 are coming under critical zone as they are having average annual soil loss more than 6.0 Tons/hectare/year. Thus, requiring catchment area treatment on priority basis.

The method used is widely accepted, effective, time, resource, and cost saving for the reservoir sediment assessment and prioritization of the erosion prone area for catchment treatment so that we can manage the reservoir system within the limited time and resources. Water Resources Department can perform this type of study time

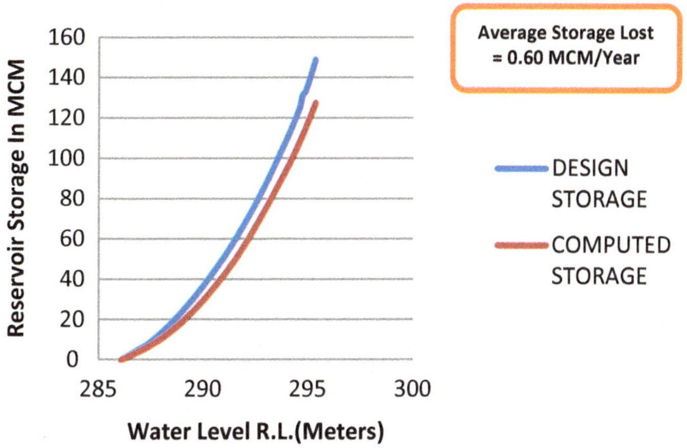

Fig. 5 Difference between designed and computed storage in the reservoir

to time in the office itself and can monitor and manage the sediment and water storage requirement for the various demands in their juridiction area. Also, the farmers of the catchment area will have awareness about the erosion of their agricultural land, how the fertility of land can be protected so that we will have control over the reservoir sedimentation and volume of water stored in the reservoir for effective and efficient use of the soil and water resources of the nation. In the current scenario of climate change, the rainfall pattern is totally affected, and in such a situation, we will have to secure our water storage in the reservoir to fulfill the demand of various users, i.e., farmers for irrigation, drinking water, industrial water, daily purpose, etc.

Acknowledgements The authors acknowledges the support provided by the State Data Centre, Water Resources Department, Chhattisgarh, Raipur, India, for providing the necessary data to complete this work and also to the Department of Civil Engineering, National Institute of Technology, Raipur for providing guidance and necessary institutional support.

References

1. Jain SK, Singh P, Seth SM (2002) Assessment of sedimentation in Bhakhra Reservoir in the Western Himalayan region using remotely sensed data. Hydrol Sci J 47(2):203–212
2. Narasayya K (2013) Assessment of reservoir sedimentation using remote sensing satellite imageries. Asian J Geoinform 12(4)
3. Pandey A, Chaube UC, Mishra SK, Kumar D (2016) Assessment of reservoir sedimentation using remote sensing and recommendations for desilting Patratu Reservoir, India. Hydrol Sci J 61(4):711–718
4. Tiwari S, Verma S Ghosh S (2011) Estimation of sedimentation rate of a Reservoir using remote sensing data: a case study of Tehri Reservoir. Int J Latest Trends Eng Technol 7(3):245–253
5. Suresh R (2016) Soil and water conservation engineering. Standard Publishers Distributers, Delhi

6. Ahmad I, Verma MK (2013) Application of USLE model & GIS in estimation of soil erosion for Tandula Reservoir. Int J Emerg Technol Adv Eng 3(4):570–576
7. Pandey A, Chowdhary VM, Mal BC (2009) Sediment yield modelling of an agricultural watershed using MUSLE, remote sensing and GIS. Paddy Water Environ 7(2):105–113
8. Pandey A, Mathur A, Mishra SK, Mal BC (2009) Soil erosion modeling of a Himalayan watershed using RS and GIS. Environ Earth Sci 59(2):399–410
9. McFeeters SK (2013) Geographic information system to detect swimming pools for mosquito abatement: a practical approach
10. Singh G, Chandra S, Babu R (1981) Soil loss and prediction research in India. Central Soil and Water Conservation Research Training Institute, Bulletin No. T-12/D9
11. Jaiswal RK (2013) Technical report on study of Reservoir sedimentation, impact assessment and development of catchment area treatment plan for Kodar Reservoir in Chhattisgarh State. Research gate. https://doi.org/10.13140/rg.2.2.32208.51203
12. Subramanya K (2013) Engineering hydrology. New Delhi, McGraw-Hill Education (India) Private Limited

Application of Geographic Information System and HEC-RAS in Flood Risk Mapping of a Catchment

Swarnadeepa Chakraborty and Sujata Biswas

Abstract Flood is one of the catastrophic outbreaks, which ravages the normal functioning of a society whose aftermath is experienced by more or less throughout the world. In this context, the present study has been focussed to analyse the flood risk based on flood hazard and vulnerability using HEC-RAS model in GIS environment for a portion of Ajay River catchment lying in the states of Bihar, Jharkhand and West Bengal. Flood hazard map has been obtained using a multi-criteria evaluation technique. Steady flow analysis has been carried out using one-dimensional hydraulic model HEC-RAS in conjunction with GIS to generate inundation area and classify the hazard map based on the depth of inundation for 10 years, 50 years and 100 years return period. Land use characteristics and population density of the study area have been taken into account for vulnerability analysis. The risk analysis classifies the area into different categories of flood hazard zones for various return periods. The study will help in preparing the strategies for flood preparedness and in turn flood mitigation of the area.

Keywords Flood risk · Vulnerability · HEC-RAS · GIS

1 Introduction

Floods are the most recurring, unpredictable devastating natural hazards in the world, unfortunately, experienced almost every year by some parts of the world of varying magnitude. As a result, the floods are responsible for claiming more lives and causing more property damage than any other natural phenomena. Overtopping of banks, topographical and hydrological frailty, failure of the existing dams, excessive rainfall in short duration are not the only factors affecting the flood. Considering the rapid urbanization assisted by deforestation, encroachment of floodplain and several

S. Chakraborty · S. Biswas (✉)
Department of Civil Engineering, IIEST Shibpur, Howrah, West Bengal, India
e-mail: sujata@civil.iiests.ac.in

S. Chakraborty
e-mail: s.chakrabortyce@gmail.com

J. K. Ghosh and I. da Silva (eds.), *Applications of Geomatics in Civil Engineering*, Lecture Notes in Civil Engineering 33, https://doi.org/10.1007/978-981-13-7067-0_17

human activities leading to the environmental conflict has stood liable for various natural hazards.

Although most of the parts of India are affected by monsoons, it varies from heavy to scanty on different parts, therefore, there exists a great regional and temporal variation in the rainfall distribution. India is one of the worst-affected flood country in the world where millions of lives are lost due to flood. About 40 million hectares or nearly one-eighth of India's geographical area is flood-prone [1]. The worst state is West Bengal where 55.43% of the total geographical area is flood-prone [2].

River Ajay appears as a wide, sand-filled channel in dry season but eventually grows voluptuous in monsoon due to excessive rainfall. Two remarkable floods in the year 1978 and 2000 marked the historic presence of the river in the downstream areas. As the river passes through the rural areas and agricultural land, flooding causes havoc losses both in terms of crops and lives.

Flood risk zoning aims at demarcating areas are likely to be affected by floods of different magnitudes, probability levels and risks associated with living and economic activities. The preparation of a flood hazard map for any region would be one of the most crucial steps for implementing non-structural remedial measures to diminish the risk of flood. The 'Flood Hazard Map' is a map that graphically provides information on inundation (predicted inundation areas, inundation depth, etc.), in an easy-to-understand format. This map delineates the flood hazard areas and also creates easily read, rapidly accessible charts and maps, which facilitates the administrators and planners to identify areas of risk and prioritize their mitigation or response efforts [3]. The goal is to quickly evacuate local residents in a safe and proper manner in the event of floods.

Remote sensing technology along with Geographic Information System (GIS) has emerged as the key tool for flood monitoring in recent years for the benefit of mankind. Satellite images can provide some information on flood inundated areas for different magnitudes of floods for the last 10 years or so. The development in this field has evolved from optical to radar remote sensing, which has provided all-weather capability compared to the optical sensors for the purpose of flood mapping. The main advantage of using GIS for flood management is that it not only generates a visualization of flooding but also creates the potential to estimate probable damage due to flood. Based on the flood extent, flood hazard maps for different return periods have been prepared using remote sensing data and *HEC-RAS*, with an objective of flood management programme that consists of flood forecasting, flood hazard mapping and vulnerability analysis [4]. Utilizing the available database and software tools, thematic maps has been ranked on the basis of the estimated significance in causing flooding, thereby integrating the data layers in GIS environment by Weighted Overlay Analysis using Raster calculator to prepare flood hazard risk zone map [5]. In recent years, risk-based approaches have drawn attention as a means of flood management. The extent of flood hazard areas in Sindh Province and procuring flood shelters have been modelled and mapped with the aid of object-based image analysis (OBIA) along with remote sensing and GIS database [6]. The spatial extent of a flood event in the Awash River basin has been portrayed in the inundation map for 5% exceedance highest flows for various return periods through GIS and *HEC-*

GeoRAS [7]. The hazard map thus prepared showed the total areas subjected to the hazards as very low, low, moderate, high and very high-risk zones.

In the present study, an attempt has been made to analyse the inundation area in part of the Ajay River catchment along with a flood hazard risk map, which has been generated employing geomorphic, topographic and hydrological data using GIS and HEC-RAS. This study specially aims at performing flood hazard and vulnerability analysis for different return periods. Subsequently, combining both of them flood risk analysis has been carried out. This analysis helps to classify different flood risk zones which will help to take appropriate mitigation measures.

2 Study Area

A portion of the catchment of River Ajay was adopted as the study area (Fig. 1). It originates from the eastern fringe of Chota Nagpur Plateau in the Munger district of Bihar, flowing through the states of Bihar, Jharkhand and West Bengal. The upper reach of the river passes through the hilly terrain with lateritic soil. It is only from Ausgram in Bardhaman district, the Ajay River flows through alluvial plains. The catchment area lies in the path of tropical depressions or cyclonic storms that originate in the Bay of Bengal, which ends up with heavy rain spells during monsoon. The average annual rainfall in the river catchment varies from 1280 to 1380 mm, with 75–80% of the rainfall occurs during the four monsoon months of June–September. The Ajay valley was densely forested with sal, piyasal and palash trees till recent times, which is gradually reducing in numbers due to the developmental activities. The extent of the study area is 5504 km^2. The study area latitudinally extends from 23° 30′ N to 24° 45′ N and longitudinally between 86° 18′ E and 87° 57′ E.

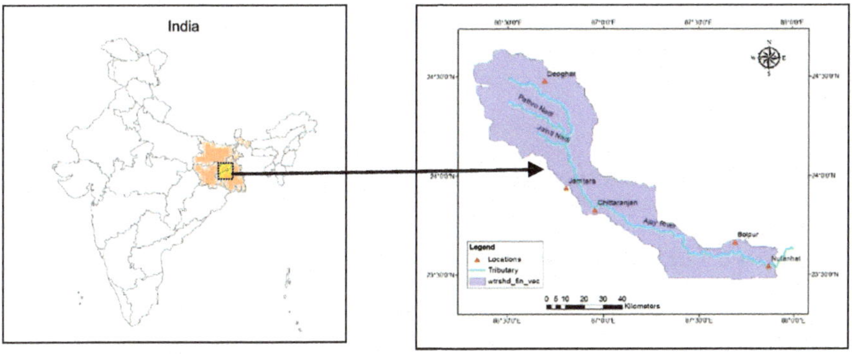

Fig. 1 Map showing the study area

3 Materials and Methods

Satellite images of IRS-P6 LISS-III (Date of Pass: 21 January, 2006) were procured from NRSC, Hyderabad, which was utilized to prepare the land use map in this present study with the aid of ERDAS IMAGINE 9.1 software. The Digital Elevation Model (DEM) is prepared from contour data, which is digitized from the Survey of India (SOI) topographical maps of 1:50,000 and 1:2,50000 scale. The population density map was prepared by utilizing the census data of 2011 obtained from the Directorate of Census Operations, West Bengal. The soil map was prepared from District Planning Map Series collected from the National Bureau of Soil Survey and Land use Planning, Kolkata. The rainfall data which is 0.5° gridded is obtained from the India Meteorological Department, Pune from 1995 to 2005 was utilized to develop the rainfall distribution map.

HEC-RAS is a Hydrologic Modelling System, which is mostly used to efficiently carry out inundation analysis using energy equations to generate flood extent and depth. HEC-GeoRAS is an interface between ArcGIS and HEC-RAS that enables to create geometric data for importing it into HEC-RAS and again processes the simulation results exported from HEC-RAS.

The Digital Elevation Model (DEM) was processed to create the Triangulated Irregular Network (TIN). TIN is an essential prerequisite for preparing all the required data sets as input for HEC-RAS simulation. Stream centerline, main channel banks (left and right), flow paths and cross section were extracted from TIN in HEC-GeoRAS interface prepared as an import file for HEC-RAS as its main input. The annual maximum discharge data of River Ajay has been utilized to carry out flood frequency analysis by Gumbel Extreme Value Type-I distribution to estimate the peak discharge of River Ajay for different return periods. Manning's n values, flood discharge for various return periods and the boundary conditions in steady flow analysis serve as major input for successful computation of water surface profiles. After post-processing of HEC-RAS results in HEC-GeoRAS, the flood hazard and vulnerability analysis has been carried out. Flood risk has been analysed subsequently combining the both.

4 Results and Discussion

4.1 Flood Hazard Map

Multi-criterion evaluation technique AHP has been applied to assign the normalized weights to the seven thematic layers (rainfall distribution, size of sub-watershed, slope, drainage density, soil, land use classes and population density) by solving pair-wise comparison matrix after determining their importance in case of flood hazard mapping, and Arithmetic Overlay Analysis has been carried out in GIS environment to classify the flood hazard risk zones. The flood hazard map shown in Fig. 2 has

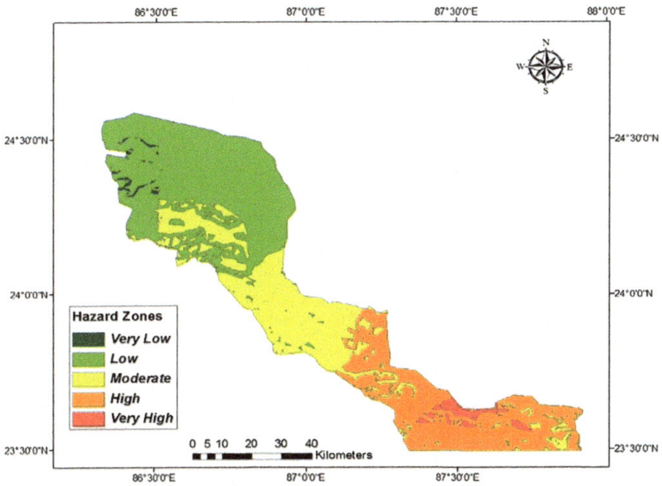

Fig. 2 Flood hazard zones

been categorized into five hazard zones, namely very low, low, moderate, high and very high enclosing an area of 38.06 km^2, 2254.95 km^2, 1418.45 km^2, 1335.66 km^2 and 103.73 km^2, respectively.

4.2 Water Surface Generation and Floodplain Delineation

Water surface profiles have been generated for individual cross sections by solving the energy equations for subcritical flow regime with an iterative process. The water surface profiles have been automatically generated within HEC-GeoRAS, which contains stream network, cross section data, bank station data and bounding polygon. Bounding polygon defines the extent of inundation mapping by connecting the endpoints of XS Cut Lines (Fig. 3). Based on water surface elevations of cross-sectional cut lines and bounding polygon layer, water surface TIN is generated, delineating a zone that connect the outer points of the bounding polygon. For each individual water surface profile, floodplain extent (depth grid) and floodplain polygon feature class was created that lead to the delineation of the floodplain. Figure 4 represents the delineated floodplain over the water surface TIN.

4.3 Flood Hazard Analysis

Both hydraulic and hydrologic parameters are responsible for flood hazard. The flood hazard analysis of the study area has been carried out for 10 years, 50 years and 100 years return period by overlaying flood depth grid with the TIN which are

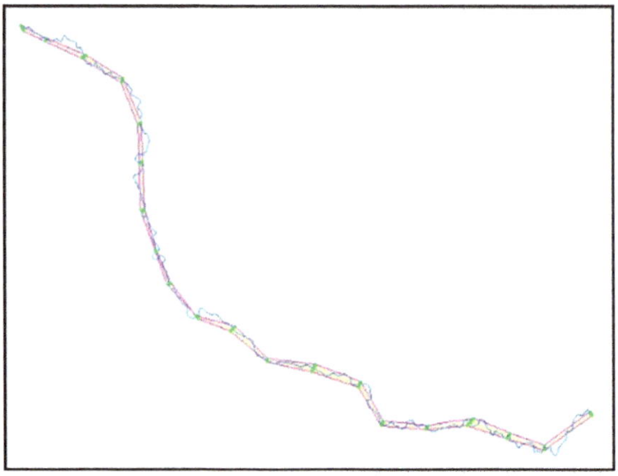

Fig. 3 Streamline, cross sections and the bounding polygon

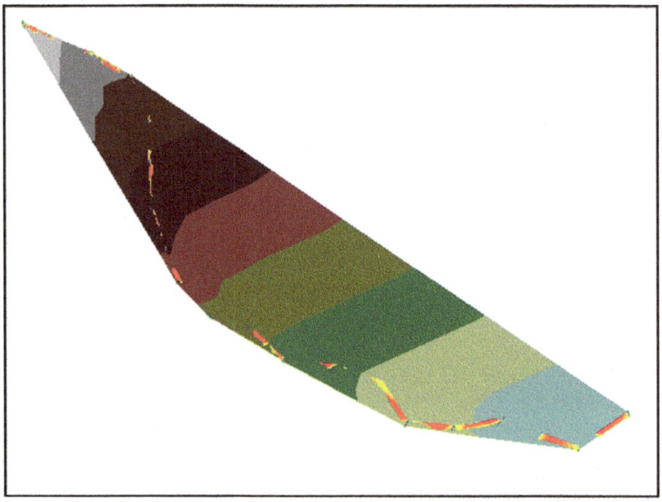

Fig. 4 Water surface TIN and floodplain delineation

shown in Fig. 5a–c, respectively. The subsequent maps show that the percentage of inundated area in the downstream is much more compared to upstream and the middle portion of the river. The percentage of total inundated area in these various return periods are graphically represented Table 1.

The above Table 1 illustrates that the percentage of total inundated area under the risk category of high and very high has increased considerably for 100 years return period with respect to 10 years and 50 years. This shows that the increased probability of flood hazard is augmented with the increase in the return period.

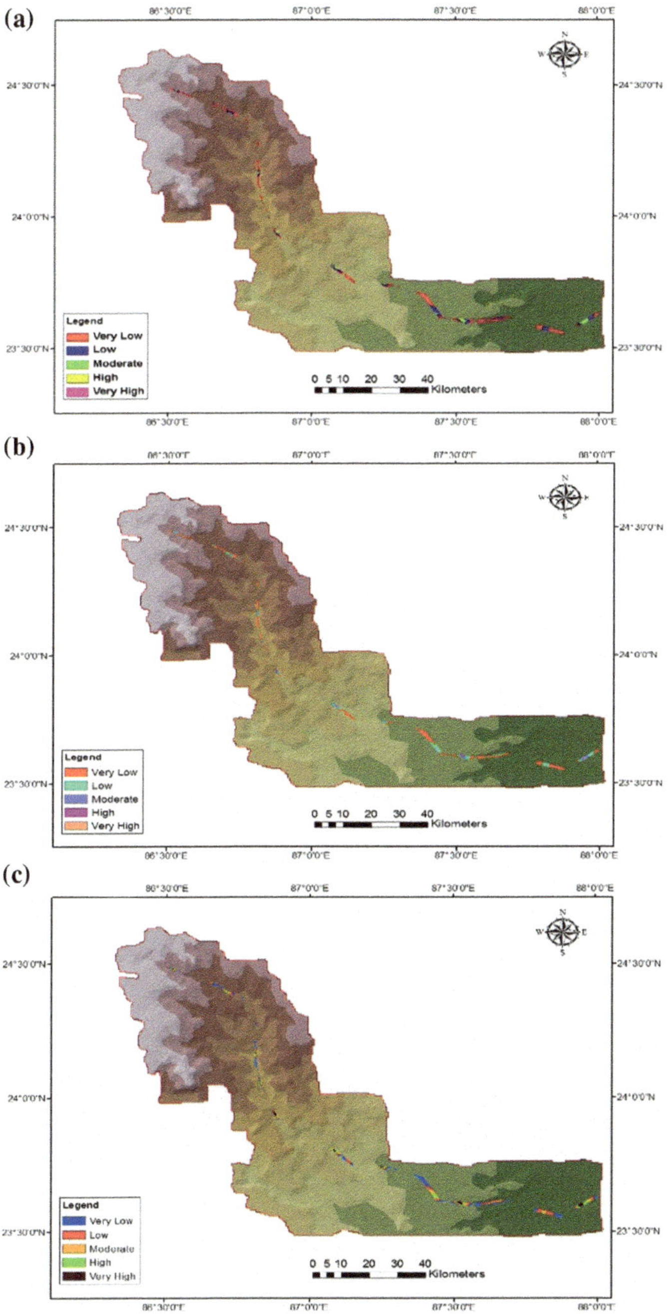

Fig. 5 Flood hazard analysis map for **a** 1 in 10 years, **b** 1 in 50 years, **c** 1 in 100 years

Table 1 Percentage of inundated area for 10 years, 50 years and 100 years

Risk category	Percentage of inundated area		
	10 years	50 years	100 years
Very low	61.88	61.73	37.75
Low	26.21	26.30	26.71
Moderate	10.29	10.33	17.82
High	1.38	1.39	12.10
Very high	0.24	0.25	5.62

Table 2 Vulnerable areas for 10 years, 50 years and 100 years return period

Land use classes	Area (ha)		
	10 years	50 years	100 years
Sand	1302.71	1336.57	1350.04
Fallow land	3348.05	3487.96	3545.45
Open forest	55.16	58.49	59.71
Dense forest	60.82	63.63	64.55
Agriculture	955.21	975.7	983.22
Settlement	2547.16	2609.06	2632.77

4.4 Flood Vulnerability Analysis

Vulnerability analysis acts as a forecast tool to enable the residents in the floodplain for evacuating them from the floodplain and relocating to a safer place. The estimation of vulnerability due to flood is mainly carried out in terms of affected area and population catering that area. The present study was focused to prepare two sets of vulnerability maps by overlaying flood area polygon with land use map and population density map of the study area. The statistics in Table 2 shows that fallow land is affected the most followed by the settlement for various return periods in comparison to agricultural land and dense forest. The graphical representation in Fig. 6 shows much of the area of Illambazar block in Bardhaman district is inundated. Some portions of both Ausgram II block in Bardhaman district and Deoghar district also shows considerable inundation as compared to the others. These results communicate about the increasing intensity of flooding with return periods.

4.5 Flood Risk Analysis

The flood risk analysis is the integration of both the vulnerability assessment and hazard assessment. It is identified by an interrelationship between the vulnerability classes and flood depth hazard classes for a particular area. Flood risk analysis has been carried out by reclassifying the flood depth grid, and then intersecting with the

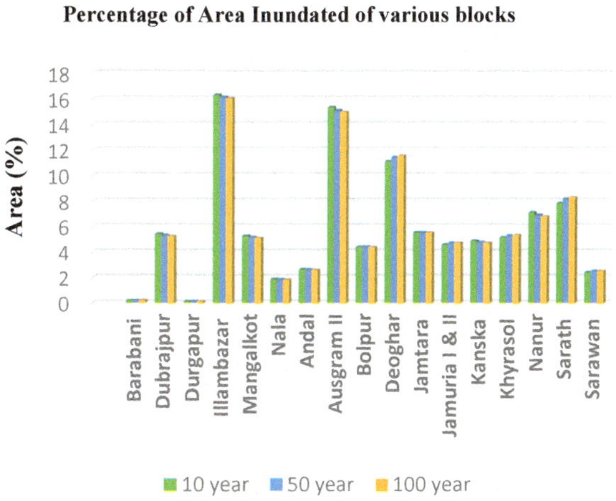

Fig. 6 Various blocks with their percentage of inundated area for defined return periods

Table 3 Area inundated with respect to flood risk classes for defined return periods

Flood hazard classes	Area (ha)		
	10 years	50 years	100 years
Very low	6406.61	6574.68	4063.43
Low	2713.56	2801.77	2874.54
Moderate	1065.32	1100.35	1917.7
High	143.14	148.39	1302.38
Very high	25.15	26.09	605.51

floodplain polygons for different return periods. The resulting attribute table reveals the portion of area inundated under the various flood hazard classes for different return periods. Table 3 shows the inundated area for flood risk classes in different return periods. Figure 7 shows the various flood risk classes with the respective inundated area along the defined return periods.

5 Conclusion

Remote sensing and GIS along with AHP has successfully created a flood hazard map for a portion of Ajay River catchment from the available database. The resulting flood hazard map of the study area obtained from the analysis has been categorized into five hazard zones, namely very low, low, moderate, high and very high enclosing an area of 38.06 km^2, 2254.95 km^2, 1418.45 km^2, 1335.66 km^2 and 103.73 km^2, respectively. With the aid of stream flow data and the inputs prepared in HEC-GeoRAS, water surface generation and floodplain delineation were successful from the simulation of HEC-RAS model for 10 years, 50 years and 100 years return period. The results were interpreted to conduct flood hazard and vulnerability analysis to quantify the extent of damage caused to the land use and population of the study area.

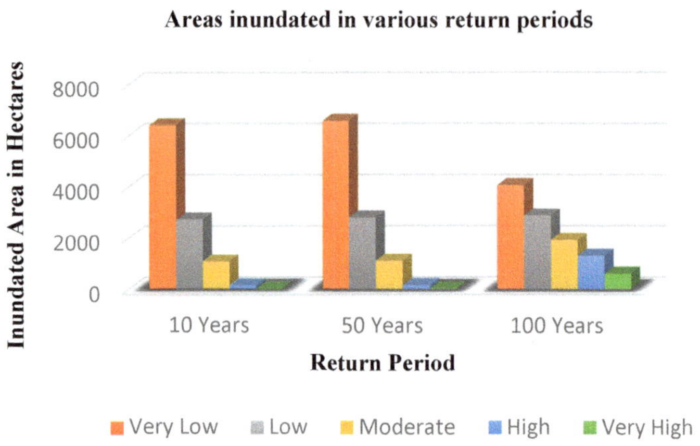

Fig. 7 Flood risk classes and its respective inundated area for various return periods

Subsequently, flood risk analysis was performed which gives a probable estimate of inundated area. The study will be helpful for taking appropriate mitigation measures in the areas susceptible to high risk, thereby reducing damages due to devastating effect of flood.

Acknowledgements The authors are grateful to India Meteorological Department, Pune, NRSC, Hyderabad, NBSS & LUP, Kolkata, Directorate of Census Operations, West Bengal and Survey of India for providing data in conducting the study.

References

1. Bapalu GV, Sinha R (2006) GIS in flood hazard mapping: a case study of Kosi River basin, India. http://home.iitk.ac.in/~rsinha/PDF's/2006_FloodGISdevelopment.pdf
2. Mukhopadhyay S (2010) A geo-environmental assessment of flood dynamics in lower Ajoy River inducing sand splay problem in eastern India. Ethiop J Environ Sci Manage 3:96–110
3. Duan M, Zhang J, Liu Z, Aekakkararungroj A (2005) Use of remote sensing and GIS for flood hazard mapping in Chiang Mai Province, Northern Thailand. www.isprs.org/proceedings/xxxviii/7-c4/203_gsem2009.pdf
4. Samarasinghe SMJS, Nandalalb HK, Weliwitiyac DP, Fowzed JSM, Hazarikad MK, Samarakoond L (2010) Application of remote sensing and GIS for flood risk analysis: a case study at Kalu-Ganga River, Sri Lanka. Int Arch Photogram Remote Sens Spat Inf Sci. XXXVIII(8):110–115 (2010)
5. Ajin RS, Krishnamurthy RR, Jayaprakash M, Vinod PG (2013) Flood hazard assessment of Vamanapuram River basin, Kerala, India: an approach using remote sensing & GIS techniques. Pelagia Res Libr Adv Appl Sci Res 4(3):263–274
6. Uddin K, Gurung DR, Giriraj A, Shrestha B (2013) Application of remote sensing and GIS for flood hazard management: a case study from Sindh Province, Pakistan. Am J Geogr Inf Syst 2(1):1–5
7. Getahun YS, Gebre SL (2015) Flood hazard assessment and mapping of flood inundation area of the Awash River basin in Ethiopia using GIS and HEC-GeoRAS/HEC-RAS model. J Civ Environ Eng 5

Drought Vulnerability Assessment Using GIS: Case of Sangli District, Maharashtra

Dhanashree Raut, Nandita Mukherjee, Zoheb Sheikh and Alolika Basu

Abstract Dry season is a crawling circumstance of shortage without energizing of assets. 66% is under dry season inclined land in Sangli. GIS is a decent apparatus for enhancing basic leadership in a fiasco administration. The point of this activity is to plan dry season memoranda, dry spell affirmation, and appraisal of the extent and help required. Number of factors is responsible in land degradation which leads to rainfall study, population and water demand and supply chain. To examine existing agrarian land utilize and its circulation in the area. We have considered the climatic state of Sangli area and its impact on the farming creation. To analyse existing agrarian land and its circulation in the area. We have also studied the climatic and topographical state of Sangli area and its impact on farming.

Keywords Mitigation · Decision-making · Drought assessment · Vulnerability analysis · Disaster management scheme

1 Introduction

Sangli is a city and the district headquarters of Sangli sistrict in the state of Maharashtra, in Western India. It is known as the Turmeric city of Maharashtra due to its production and trade of the spice.

Sangli is situated on the banks of river Krishna and Warna. The physical settings of Sangli show the contrast of immense dimensions of climate and vegetation.

D. Raut (✉) · N. Mukherjee · Z. Sheikh · A. Basu
Atur Centre, Symbiosis Institute of Geoinformatics, SIG Pune, 5th and 6th Floor, Gokhale Cross Road, Model Colony, Pune 411016, Maharashtra, India
e-mail: draut26@gmail.com

N. Mukherjee
e-mail: nandita.noni@gmail.com

Z. Sheikh
e-mail: zohebarca@yahoo.com

A. Basu
e-mail: alolikabasu123@gmail.com

J. K. Ghosh and I. da Silva (eds.), *Applications of Geomatics in Civil Engineering*, Lecture Notes in Civil Engineering 33, https://doi.org/10.1007/978-981-13-7067-0_18

The main crops cultivated in Sangli district are rice, jowar, bajra, groundnut, turmeric, soybean, sugarcane, wheat, grapes, and pomegranate. Drought is a creeping situation of scarcity without recharging of resources. 66% is under drought-prone area in Sangli. GIS helps in interpreting the results and improving decision making in disaster management and the contribution of agricultural sector to the overall economy of Sangli district. This paper also presents percentage of population engaged in the agriculture sector, cropping pattern, and production of crops. So that decision makers can undertake the studies to procure disaster management with the help of surveys, to suggest disaster management schemes, frame policies of the government and NGOs for effective implementation.

1.1 Location, Tradition and Cultural Highlights

In Sangli-Miraj region, one can see an epitome of whole Maharashtra except for seashore. This region has a strong martial tradition, which has formed an integral part of our national army. The cooperative sugar factory at Sangli has the honor of being the largest such ventures in Asia. Sangli is, therefore, known as "Sakhar Pandhari" (Table 1).

1.2 Topography Study

The area of the district is 8572 km^2. It is situated in the river basin Warna and Krishna River. According to soil, climate, and rainfall, the district could be divided into (i) Western Mountain/Hilly area with maximum rainfall, (ii) Plain land in the basins of rivers with medium-range rainfall, and (iii) Eastern plateau with inferior quality soil and minimum rainfall (drought-prone area).

Table 1 Basic information of Sangli district [1]

Area in km^2	8572
Population	28,22,149
Literacy rate	76.9
Number of municipal corporations	1, Sangli-Miraj-Kupwad Corporation
Number of municipal councils	4, Vita, Astha, Islampur, Tasgaon
Number of statutory towns	5
Number of census towns	2
Number of villages	728

1.3 Climate and Rainfall

The rainfall ranges from the rainiest in the Chandoli (Shirala) region, which has an average annual of over 4000 mm to the driest in Atpadi and Jat taluka where the average annual rainfall is about 500 ml. Sangli has a semiarid climate with three seasons, a hot, dry summer from the middle of February to the middle of June, a monsoon from the middle of June to late October, and a mild cool season from early November to early February (Map 1) [2].

Soil formation in Sangli district has been predominantly influenced by the climate. The western zone, which receives very heavy rainfall, has lateritic soils on up-ghats and reddish brown soil on hill slopes, the latter being developed on parent material of trap rock. The third is the eastern drier zone, which consists largely of regular black soils and poor shallow soils. The major type of soil found in Sangli is medium deep black soil. And, deep black soil is found near the river banks of Krishna, Varna, and Yerala. As per agro-climatic regional plan report, Sangli district comes under zone 9, which is eastern plateau and hill region. The region requires planning to maximize the use of rainwater and increase in groundwater potential. The rainfall in this region is nearly 1,300 mm but because of sloppy topography, rainwater is lost due to runoff (Table 2) [4].

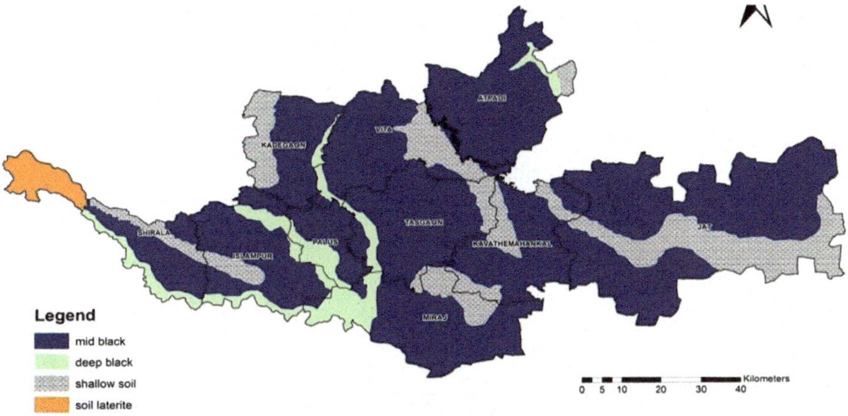

Map 1 Shows soil typology of Sangli region [3]

Table 2 The type of soil available in Sangli district

Major soil	Area (ha)
Medium deep black soil	389.4
Shallow black soil	142.7
Deep black soil	63.4

2 Methodology

1. Literature Review of Drought Vulnerability Study.
2. Secondary Data Collection
   (Agriculture, Water, Soil Conservation, Irrigation, Forestry, Rural Development, Survey of India, Google Earth, Satellite Imagery for thematic mapping, urban local bodies/village panchayats, type of Droughts—Metrological, Hydrological and Agricultural. Enhance Capacity Building, Integrated Planning to achieve drought management).
3. Analysis of datasets.

2.1 General Distribution of Rainfall

Rainfall is a very important factor, which influences the agricultural activities as well as the daily activities in the Sangli district. The region lies in the rain shadow region of Sahyadri. The important aspects of rainfall which are to be studied are general distribution of rainfall and seasonal distribution of rainfall.

The general distribution of rainfall in the Sangli district is very uneven. Shirala has the highest average yearly rainfall, whereas Palus has the lowest yearly rainfall as presented in the below map.

Seasonality is an important characteristic of the rainfall of the region other than uneven distribution and uncertainty. The principal rainy season is from June to September, and quite a large percentage above 52–54% of the mean annual rainfall of the region is received during southwest monsoon as shown in Map 2. July and August are the rainiest months throughout the region. Rainfall decreases from west

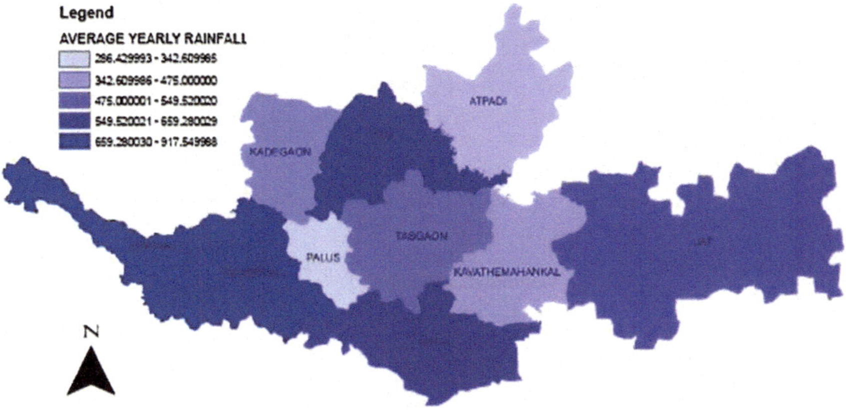

Map 2 Average yearly rainfall

to east. From the second half of September, southwest monsoon loses its strength and northeastern monsoon takes its place. Rainfall during this period is maximum (30%) in the Agrani valleys and low (26%) in Yerala valley. At the close of November, northeastern monsoon decreases its influence and cool season starts. From December to February, the rainfall is very meager through the region. The rainfall below 2% is received in eastern region [5].

2.2 *Groundwater Level*

Groundwater is the water present beneath the Earth's surface in soil pore spaces and in the fractures of rock formations. The depth at which soil pore spaces or fractures and voids in rock become completely saturated with water is called the water table. Groundwater is also often withdrawn for agricultural, municipal, and industrial use by constructing and operating extraction wells. Sangli district geology authority has created zones based on the water testing centers locations in the district. The naming of these zones are from KR-22 to KR-53, BM-103 to BM-105, and BM-112 to BM123, which comes to a total of 45 watershed zones In the following maps, the groundwater level in different zones during the start of different agricultural seasons are depicted by the presenter. The seasons selected are Kharif, Rabi, and Zaid season, which can also be named as monsoon, post-monsoon, and pre-monsoon, respectively.

July is the start of Kharif season when crops like millets, paddy, moong, cotton, and sugarcane are sown. Sangli plays a big role in sugarcane production. This is also called as monsoon season of India.

During which the season's highest rainfall is received. This rainfall replenishes the groundwater tables but it can be seen in Map 3 that even during this season Jat taluka has the lowest water table, which is due to its porous soil structure and low rainfall achieved due to its plain surface as compared vice versa to western part of the region, which receives high rainfall and is at a higher elevation than other parts, therefore, Kharif crops are more favored in eastern part of Sangli than other parts.

Rabi or winter crops are the one which includes wheat, barley, mustard, and sesame. These crops require less water as compared to Kharif crops as it can be seen that the eastern and central region still has a low water table level, therefore, Rabi crops are more attractive to farmers of these region (Map 4).

Zaid or pre-monsoon is commonly known as summer in the Indian context. This season consists of watermelon, muskmelons, and vegetables of cucurbitaceae family. Crops grown during this season are not very profitable as compared to other seasons, and also farming in this season requires a high level of irrigation which is now present in Sangli. The highest amount of rainfall is received in the Shirala area, i.e., 670 mm as compared to other talukas. The lowest amount of rainfall received is in Atpadi, i.e., 414 mm. The average rainfall received in other areas is approximately on an average of 550 mm (Map 5).

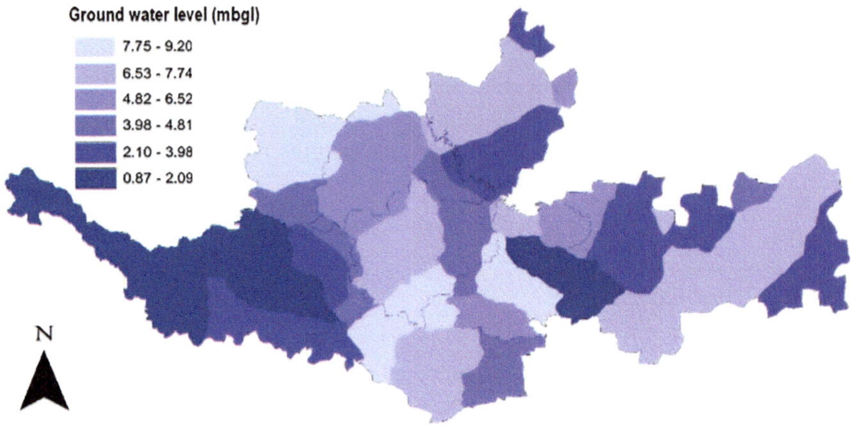

Map 3 Groundwater level in July month, Sangli 1991–2015

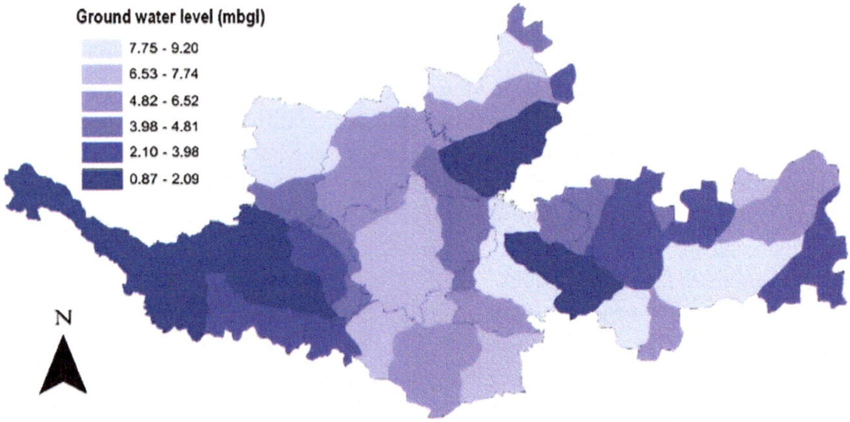

Map 4 Groundwater level in October month, Sangli 1991–2015

2.3 Sangli Crops

Kharif: The autumn crop is sown at the beginning of the summer rains. Rabi: The grain crop is sown in September. The size of the Jat taluka is more so the fallow land is also more as compared to other talukas followed by Tasgaon and Islampur (Map 6).

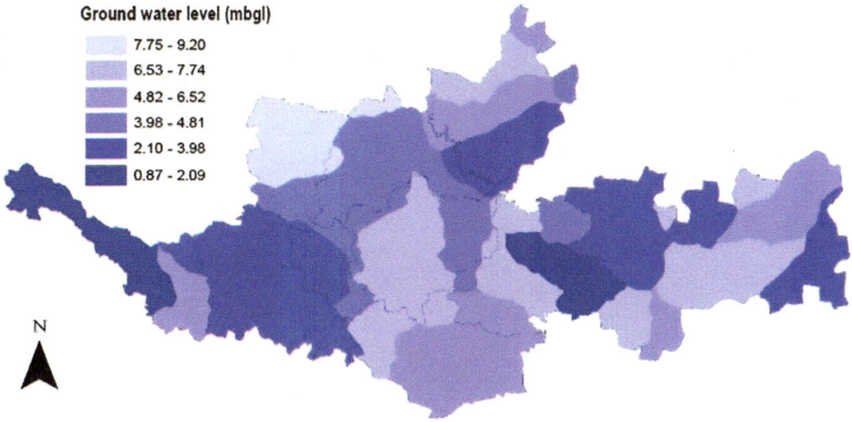

Map 5 Groundwater level in March month, Sangli 1991–2015

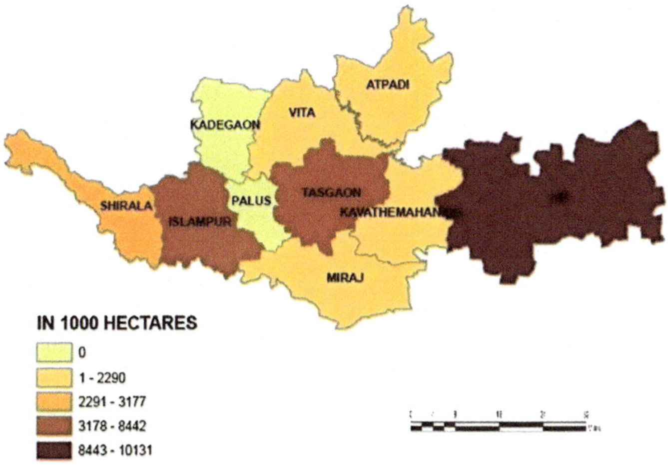

Map 6 Fallow land in Sangli district

2.4 Irrigation

Sangli has a total area of 857,200 ha of which total area under cultivation is 645,376 ha which is 74.06% of total area. Krishna River is the main river which flows through the western part of Sangli district. Eight watershed areas identified by water and agriculture development department are shown in Table 3 [5].

These watershed areas result in River Man in Atpadi taluka, Bore in Jat taluka, Agrani in Tasgaon and Kavthemahakal, Yerala, and Krishna in Khanapur. Even after having five rivers flowing from inside of region and one (Warna) from western boundary, there many drought-prone areas in Sangli is due to the presence of high

Table 3 Watershed areas

Sr. no	Watershed areas (types)	Types
1	Ghanand	Atpadi
2	Nelkaranji	Atpadi
3	Sordi	Jat
4	Belunkhi	Jat
5	Kuchi	Kavthemahakal
6	Khanapur	Khanapur
7	Manjarde	Tasgaon

Table 4 Drought-prone areas in Sangli district

Sr. no	Drought-prone area	Taluka total area (km^2)	% of drought-prone area
1	Atapadi full	872	10.12
2	Jat full	2247	26.09
3	Kavthemahakal full	707	8.2
4	Khanapur part	1326	8.83
5	Tasgaon part	1101	6.56
6	Miraj part	926	6.19
Total		7128	65.99

Table 5 Large ongoing irrigation projects in Sangli district

Sr. no	Ongoing irrigation projects	Projects coverage area
1	Krishna-Koyna lift irrigation	117320
2	Tembhu lift irrigation	83222
3	Warna project	83932
4	Krishna project	16758
Total		301232

amount of shallow black soil. Probable drought-prone area under Sangli District is shown in Table 4.

Sangli being pro-agro region many irrigation projects were implemented in these regions. These irrigation projects were divided on the basis of coverage area. The large stone being built on rivers is present in Sangli by dams or canal system. All the large irrigation projects in Sangli district are completed to their final stage as listed in Table 5 [5].

Through Table 5, it can be easily depicted that Krishna-Koyna project is the largest serving irrigation in Sangli in terms of coverage area as major part of Krishna River flows directly through the western part of Sangli region. Medium-sized irrigated projected area the one which obtained their water source from groundwater or rainwater harvesting. Jat taluka has the highest area coverage under medium irrigation projects because of the reason that Jat taluka is largely depended on these types of projects as there are no major rivers flowing through this region.

3 Proposals and Key Findings

"Water shed Management in Jat Taluka"
As seen through the analysis, Jat taluka has the lowest groundwater table level in Sangli of up to 10 m and in summer season, it was decided to identify potential areas where watershed management can be practiced in Jat taluka. The first step toward watershed management starts from analyzing topography of the selected area as seen from Map 7 Jat taluka has a very high altitude of 759 m in the central region.

There is a slope of up to 459 m in the northeastern part of Taluka. Elevation results in waterflow outside the area. Therefore, there is a need for halting this runoff in the taluka itself so that it can be absorbed by soil and help in groundwater recharge. 11 watershed areas had been identified by the Geological Department in Jat taluka that are shown in Map 8 based on the direction of flow of runoff.

It is inferred through map that watershed zone BM-123 has the lowest groundwater level during summer season. Therefore, it is going to be our target zone for watershed management. Jat taluka has no major water irrigation projects implemented by any

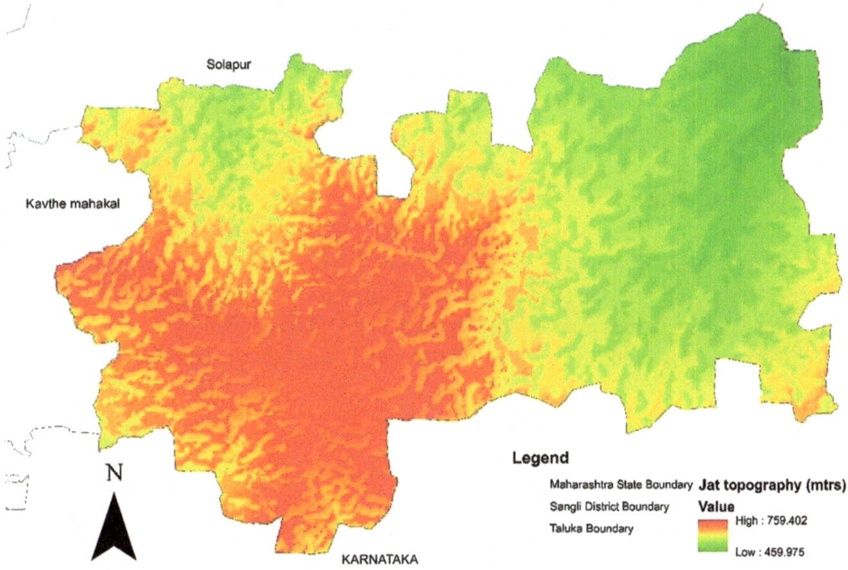

Map 7 Topography of Jat taluka

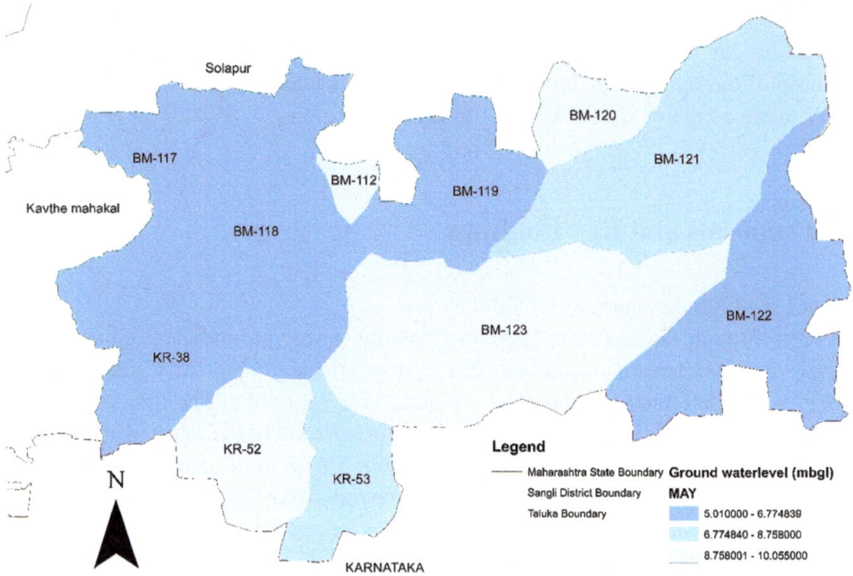

Map 8 Watershed zones and groundwater level in May of Jat taluka

agency, therefore, major usage is through groundwater level, which leads to over-exploitation of groundwater even in summer season and government has to provide water through tankers.

But when we look at the soil typology of the area in Map 9, we can infer that the area having shallow black soil have low groundwater level as compared to area having deep black soil because of the reason that shallow black soil has low percolation rate of 7 ml per minute as compared to deep black soil, which is 13 ml per minute. Due to this property, shallow black soil requires more time to allow water to seep through it, which is not happening because of the absence of any watershed management techniques. In the last step to find the areas suitable for watershed management in Jat taluka, we overlaid all the layers of analysis with land cover on the top and it was found out from the Map 10 that areas where techniques like check dams and farm ponds were in the watershed region BM-123 along the existing river and extending up to BM-121 watershed zone. Another area identified watershed management by contour bunding was the central area with highest elevation, which was one of the reasons water was

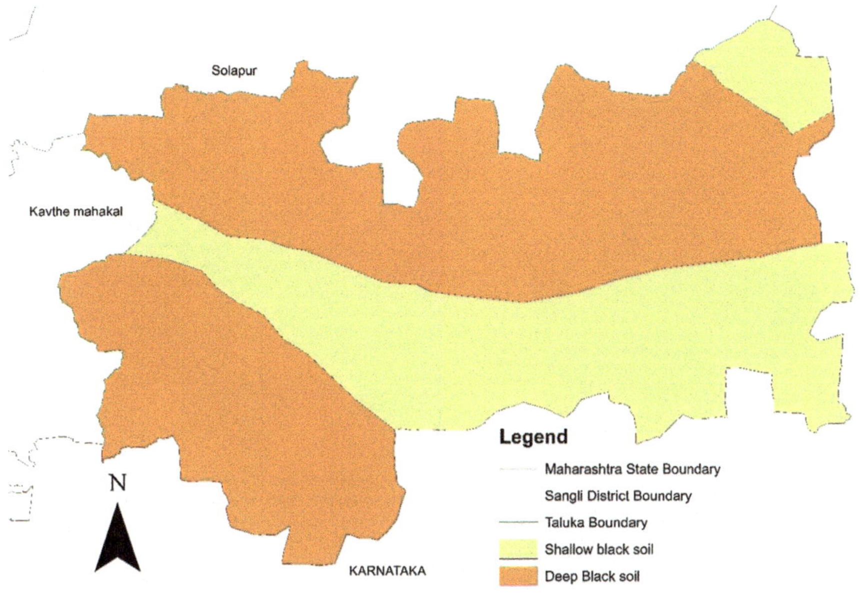

Map 9 Soil typology of Jat taluka *Source* Sangli geological department [6]

not staying in the region long enough to allow significant recharge of groundwater. By implementing suggested schemes in the identified areas, groundwater table of the Jat taluka can be brought up and helps it to grow economically and environmentally.

3.1 Schemes

Tembhu lift irrigation scheme—it has been proposed to irrigate drought-prone area above the command area of Takari lift irrigation scheme. The scheme envisages lifting of 22000 million cubic feet (TMC) water from Krishna River to irrigate 80,472 ha of lands from Karad, Kadegaon, Tasgaon, Khanapur, Atpadi, Kavthemahakal, Sangola Talukas of Satara, Sangli, and Solapur districts. Jat being one of the largest drought-prone areas occupying 26.09% of total Sangli drought-prone area, preservation, and revitalization of rivers to be done under the DPAP (Drought-prone area program 1973) [7]. **YASHADA**, State Apex Organization looks after the Integrated Watershed Development Program at district level under Zilla Parishad and at block level under Panchayat Samitii. The program will ensure the effective functioning of government-aided schemes as a major source of funds in watershed development activities.

Acknowledgements I am thankful to the experts from Sangli Irrigation Department Circle, coauthors, and also public who coordinated well in data collection as well as the must found profoundest

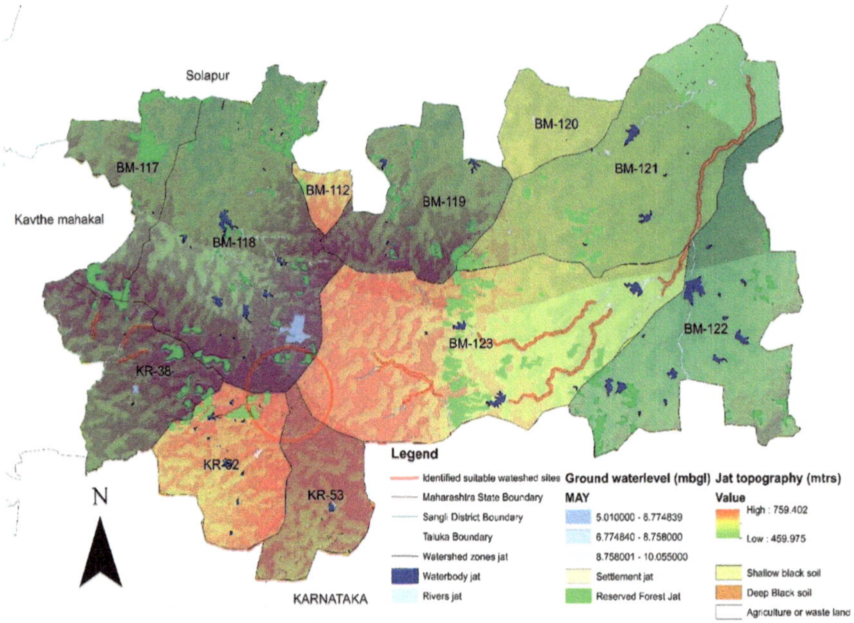

Map 10 Identified suitable watershed sites in Jat taluka

gratitude to Dr. G. S Natarajan from LIT, Nagpur who offered me suggestions and unceasing encouragement.

References

1. Census of India (2011)
2. Agro climatic regional plan report (2014)
3. National Bureau of Soil Survey and Land Utilization Plan Nagpur
4. Agriculture Statistical Information, Maharashtra State (Part II–2011)
5. Sangli Irrigation Department Data (2014)
6. Sangli Irrigation Circle and Water Department Data (2013–2014)
7. Guidelines of National Disaster Management Authority (Management of Drought)

Part V
Geomatics in Environmental Engineering

Estimation of Urban Area Surface Temperature with Landsat 8 Thermal Band Using GIS: A Case Study of Jaipur City

Lakhwinder Singh and Deepak Khare

Abstract Surface temperature is depending upon type of surface if surface is hard or made from concrete or harder matter than the temperature will be more, so it is important to study surface temperature of an urban area, with increasing urban area temperature in the central part of the city cause more air temperature, and if humidity is present, then it creates very uncomfort situation to live without any cooling system. So, the study of urban heat is mandatory to know about these types of situations in city and identify area are being in this situation [1]. This heat zone in a city center which will present in both seasons summer and winter. This situation is only uncomfort to humans in the area of earth below 32° of latitude and not in other part of world. Landsat 8 provides this facility to investigate the surface temperature using its thermal band [2]. Data collected from the thermal band is easily possible to convert into surface temperature in degree [3]. So, in this study, only year 2017 images are investigated by developing the surface temperature model in GIS. The results will show that the temperature is distributed over the surface. So, the hot area is very hot in summer. So, changing land use also have a direct connection with surface temperature if the more urban area will increase the more the surface temperature rise will increase in city [3]. If there is too much urbanization distance to increase more surface temperature. So, this study becomes important in this aspect [4].

Keywords Landsat 8 · Thermal band · Land surface temperature · Urban heat land

L. Singh (✉) · D. Khare
Department of Water Resource Development and Management, Indian Institute of Technology, Roorkee, Roorkee, India
e-mail: lakhwinder108@gmail.com

J. K. Ghosh and I. da Silva (eds.), *Applications of Geomatics in Civil Engineering*, Lecture Notes in Civil Engineering 33, https://doi.org/10.1007/978-981-13-7067-0_19

1 Materials and Methods

1.1 Study Area

Jaipur is city of Rajasthan situated in the state of India, and is located in the limits of 26° 45′ 8.7′ N–27° 1′ 55.42″ N and 75° 41″ 40.86″ E–75° 51″ 14.80″ E as shown in Fig. 1. The city area as per city boundary is considered as 316 km^2.

1.2 Data Sources

This study uses landsat satellite data for surface temperature estimation. The satellite data used is shown in Table 1 which is obtained using USGS Earth Explorer of NASA. December and January images of landsat have been used for the estimation patterns of temperature. Landsat 8 image of January as shown in Fig. 2 are considered as urban area.

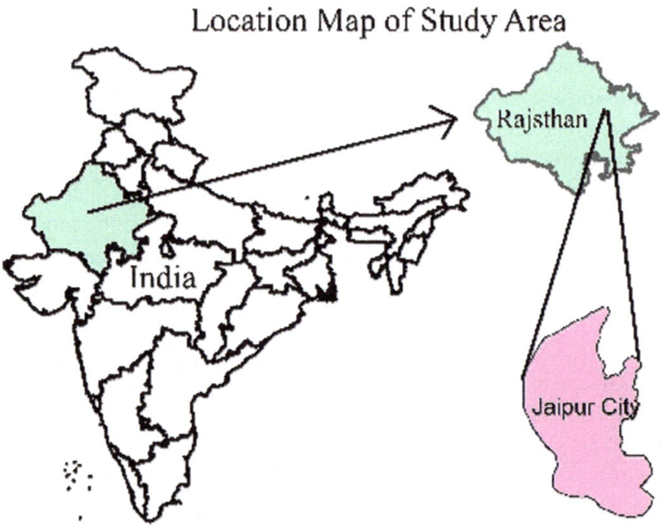

Fig. 1 Study area

Table 1 Details of satellite data used

Satellite	Sensor	Date taken	Resolution (m)
L 8	OLI and TIRS	May 15, 2017	30
L 8	OLI and TIRS	January 10, 2018	30

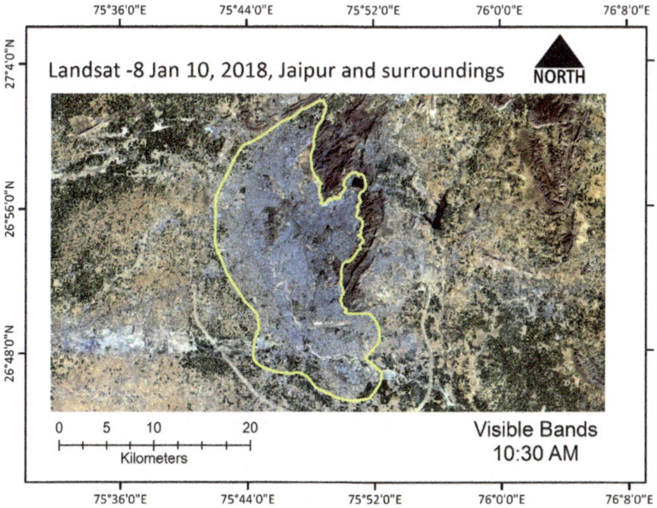

Fig. 2 Landsat 8 January 2018 visible band image with urban extent

1.3 Methodology

Standard Landsat 8 digital data provided by the USGS EROS Centre that contain quantized and calibrated scaled Digital Numbers (DN) representing multispectral image data are acquired by both the Operational Land Imager (OLI) and Thermal Infrared Sensor (TIRS) [4]. The products are delivered in 16-bit unsigned integer format, and can be rescaled to the Top of Atmosphere (TOA) reflectance and/or radiance using radiometric rescaling coefficients provided in the product metadata file (MTL file) [5]. The MTL file also contains the thermal constants needed to convert TIRS data to the top of atmosphere brightness temperature. Band 10 is more recommended then band 11 to use for surface temperature analysis [6] (Source: Landsat Site).

1.3.1 Conversion to TOA Radiance

OLI and TIRS band data can be converted to TOA spectral radiance using the radiance rescaling factors provided in the metadata file.

$$L_\lambda = M_L Q_{cal} + A_L \tag{1}$$

where

L_λ *TOA spectral radiance* (Watts/(m^2 * srad * μm))
M_L *Band-specific multiplicative rescaling factor from the metadata*

A_L *Band-specific additive rescaling factor from the metadata*
Q_{cal} *Quantized and calibrated standard product pixel values*

1.3.2 Conversion to Top of Atmosphere Brightness Temperature

IRS band data can be converted from spectral radiance to top of atmosphere brightness temperature using the thermal constants provided in the metadata file.

$$T = \frac{K_2}{\text{In}(\frac{K_1}{L_\lambda} + 1)} \quad (2)$$

where

T *Top of atmosphere brightness temperature (K)*
L_λ *TOA spectral radiance* (Watts/(m^2 * srad * μm))
K_1 *Band-specific thermal conversion constant from the metadata*
K_2 *Band-specific thermal conversion constant from the metadata*

1.3.3 Model Development

Based on Eqs. (1) and (2), an ArcGIS-based model has been developed and integrated into the toolbox of ArcGIS. This model gets input from the data, as Band 10 and all other parameters are acquired from the equations to convert to surface temperature and generate output image with value in degrees. Actual Landsat thermal equation is giving output in kelvin. But we converted kelvin to degree for better understanding purpose and integrated into the model. This independent model can be run on any PC having ArcGIS. Figure 3 shows graphical user interface of the model. All temperature classifications used in study used Natural brake interval. It considered more effective then equal interval. Because it uses distribution on spatial trend, pattern, and histogram. It means more variation has more class less variation has large class. So, this represents data better than tradition equal interval method (Fig. 3).

2 Results and Discussion

Our results show that a large difference between temperatures for both seasons has different patterns also. The results show that hard surface and cluster of settlement of industries has more heat up during summer. It also impacts the nearby area also. Landsat 8 mostly cross form India in daytime during 10:00–10:30 a.m. for better angle of viewing and reflectance. Surface temperature also make above air hot and cluster of settlement become urban heat island. In this heat island, during summer, living condition is worst if there is no cooling system. On other side, if there is cooling

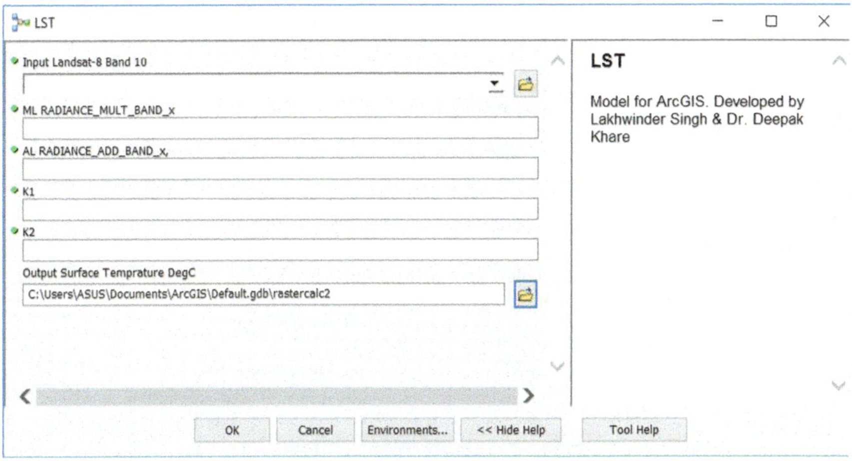

Fig. 3 GUI of LST model

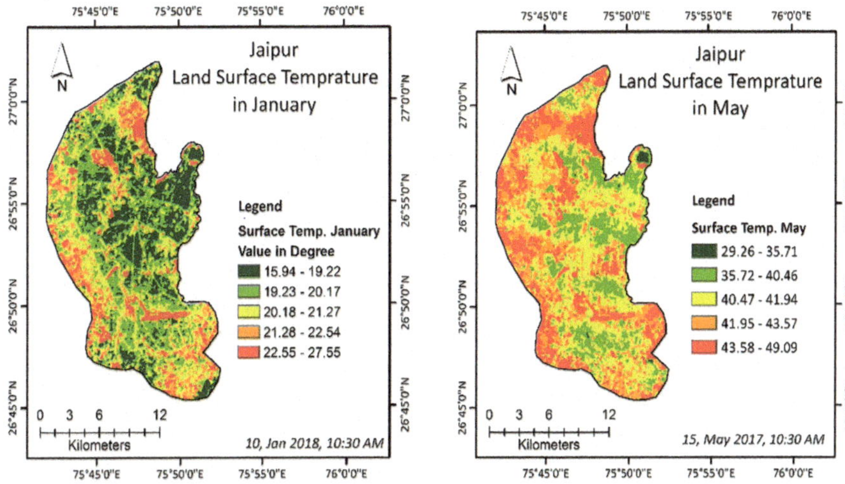

(a) January LST (b) May LST

Fig. 4 Showing land surface temperature of summer and winter

system, it also generates heat in the air. The condition becomes very worst for poor people living in this heat land who does not have cooling systems. Figure 4a, b show the surface temperature pattern during summer and winter. Considered January and May months.

From Fig. 4, it is clear that temperature is following almost the same spatial trend. But difference is between the values. If we investigate in zoomed map, we can see the impact of surface temperature more clearly. As shown in Fig. 5, Jaipur airport is hot up to 49 °C. Runway and road toward airport is clearly showing more temperature

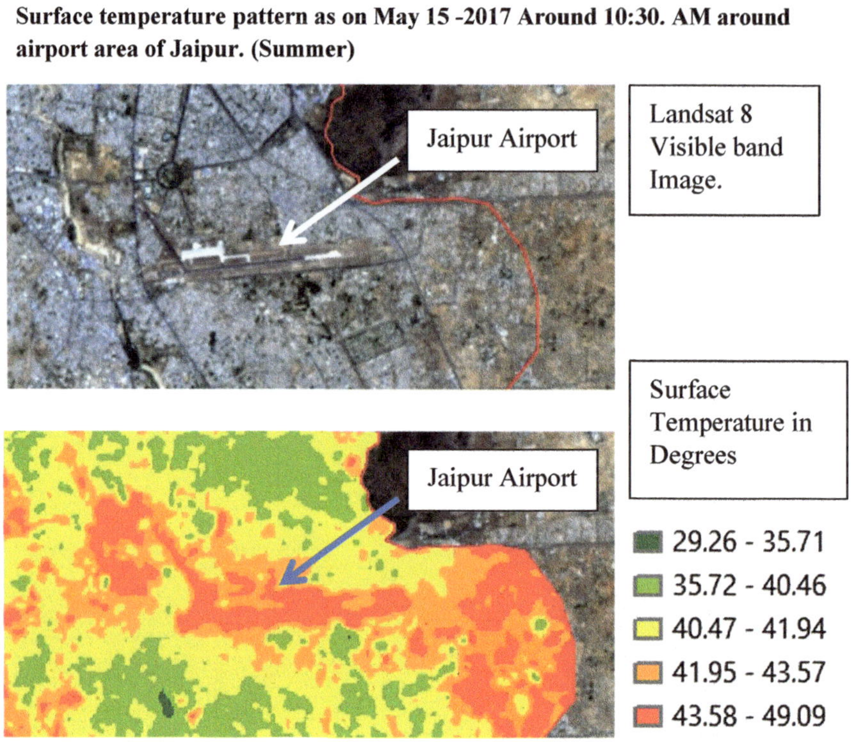

Fig. 5 Zoomed to airport

than by the surrounding area. This area is hard surface which absorb more heat also east part of city with barren/mixed land is hotter. Its more dangers condition in urban heat land for human who surviving without cooling system in these areas.

In Fig. 6, we can see temperature variation is between 29° and 49° at morning. This figure also showing both images visible and thermal. It is clearly visible how temperature creates an urban heat island. This urban heat land tends us to make good plan for these areas to prevent from being hot. Figure 6 shows the heat of industrial area and how it impacts the nearby hospital that is present on the edge of heat island. Temperature around the hospital is approaching to 43° due to heat generated by the industrial area. So, it must be away from its heat region. Its also shows a hard surface empty land near to industrial are at east of city getting hotter than surroundings. Both hot region almost merged to each other and creating one hot region. But the temperature of water body is around 29 °C.

As shown in both Images 5 and 6, heat is more effective in summer. But as shown in Fig. 8 heat is less effective in winter as image taken on January 2018. Air circulation pattern around heat island is shown in Fig. 7 Air circulation also impact on nearby area. But it follows the same spatial pattern at different temperatures.

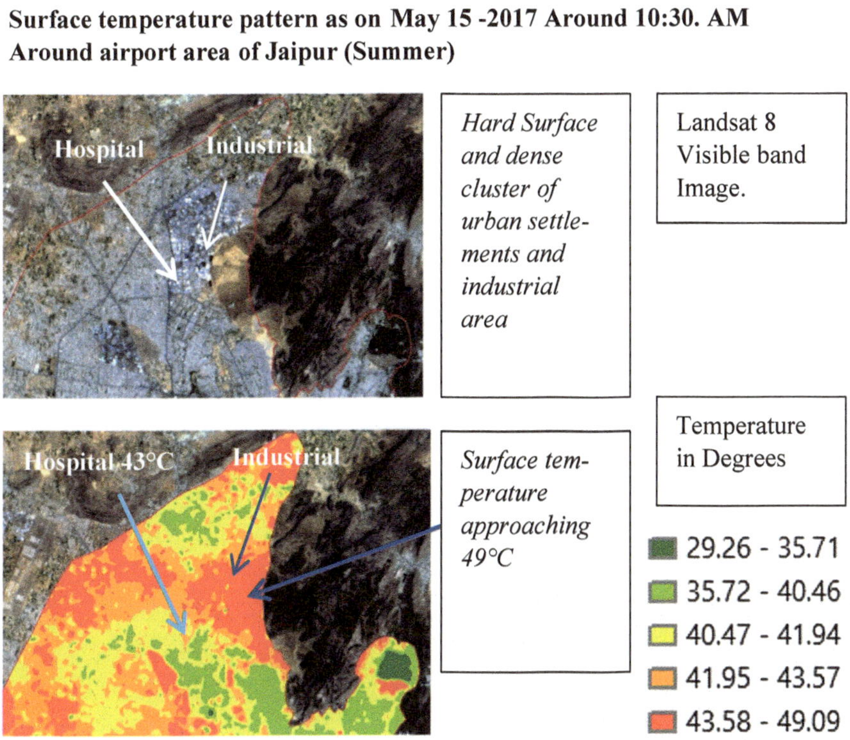

Fig. 6 Zoomed industrial area

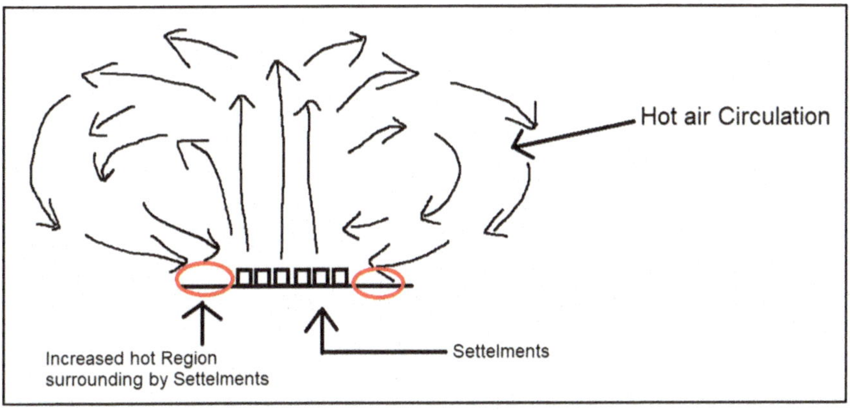

Fig. 7 Hot air circulation around urban heat land

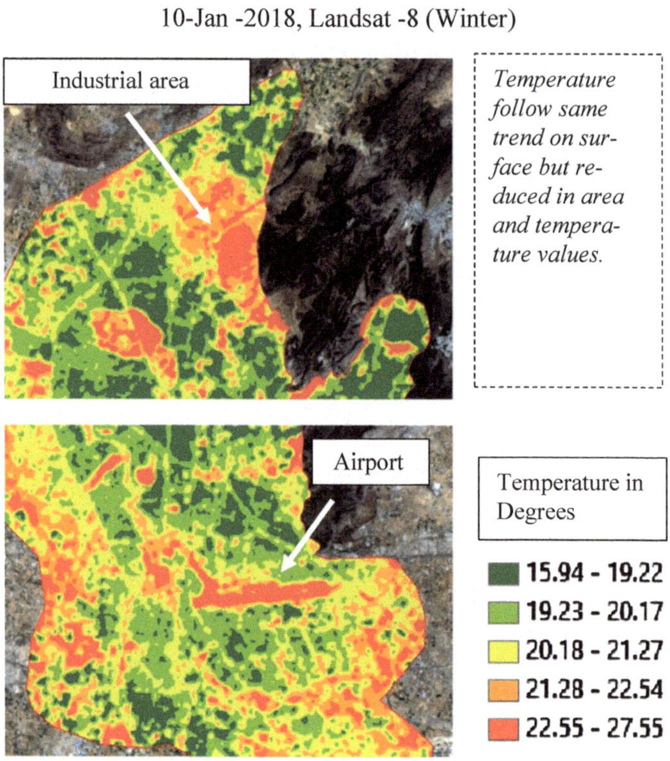

Fig. 8 Land surface temperature in January

As shown in Fig. 8, just the area of headland is reduced, and rest of the pattern is the same. Temperature ranges from 15° to 27° which is favorable for most of human being. As from results if more heat is generated in center of heat island, then it has more possibility to spread over the surroundings, less heat has less possibility. If the surface is much hot, then air just above the surface become hotter and circulation of air is beginning in horizontal and vertical direction of heat land thus this increase area of heat land. But in winter, the surface is not much hot enough to speed out this circulation and radiation amount is also low. So, the heat lands have smaller area in winter.

3 Conclusion

As per our study by using two satellite images of two different seasons, it is clear that temperature follows the same trend but the values change. Heat land also increases if more radiation from the sun in received. As per study, the surface temperature in the month of May is reaching around 43 °C morning while satellite cross over the city. But in daytime condition will be more worst conditions while near industrial area. It heats up and also spread heat to surroundings. As per in our study, it shows the impact of heat is also on the nearby hospital. It lies near to heat zone of the industrial area. So, a good cooling system is required for the people living in this zone, otherwise, this heat wave are not good for human being. This type of condition is at places which are between 30° of latitude where the sun wave hits at lower angle and produce more heat on the surface. On the other side, this area lies near the desert. GIS tools helps wide to investigate a wide region within limited time extent and provide more information about it.

Acknowledgements The authors would like thank for the support of Dr. Deepak Khare of IIT Roorkee and the sponsored project "Assessment of Climate change and its impact on urban hydrology: Indian Perspective [Project Number = AIC-566-WRC]".

References

1. Ghulam A, Louis S (2010) Calculating surface temperature using Landsat thermal imagery, 1–9
2. Latif MS (2014) Land surface temperature retrival of Landsat-8 data using split window algorithm—a case study of Ranchi district. Intern J Eng Dev Res 2(4):3840–3849
3. Rozenstein O, Qin Z, Derimian Y, Karnieli A (2014) Derivation of land surface temperature for landsat-8 TIRS using a split window algorithm. Sensors (Switzerland) 14(4):5768–5780. https://doi.org/10.3390/s140405768
4. Muche ME (2016) Surface water hydrologic modeling using remote sensing data for natural and disturbed lands. Kansas State University, Manhattan, Kansas. Retrieved from http://krex.k-state.edu/dspace/bitstream/handle/2097/32609/MulukenMuche2016.pdf?sequence=1
5. Li B, Chen Y, Chen Z, Li W, Zhang B (2013) Variations of temperature and precipitation of snowmelt period and its effect on runoff in the mountainous areas of Northwest China. J Geog Sci 23(1):17–30. https://doi.org/10.1007/s11442-013-0990-1
6. Rajeshwari A, Mani ND (2014) Estimation of land surface temperature of Dindigul district using Landsat 8 data. Int J Res Eng Technol 3(5):122–126. eISSN: 2319-1163, pISSN: 2321-7308

Estimation of $PM_{2.5}$ from MODIS Aerosol Optical Depth Over the Indian Subcontinent

S. L. Kesav Unnithan and L. Gnanappazham

Abstract Daily estimation of ground-level $PM_{2.5}$ ($\mu g/m^3$) at high spatial resolution is a requisite for carrying out epidemiological studies with respect to air pollution. This can be made possible by exploiting the relationship between Aerosol Optical Depth and PM to provide high spatiotemporal predictions. We have used MODIS-derived Aerosol Optical Depth (AOD) measurement for the Indian subcontinent region in Mixed-Effect Model (MEM) to calibrate the AOD–PM relationship. Ground-level $PM_{2.5}$ pollution data averaged daily were collected from 34 monitoring stations spread across the country. The MEM considers a fixed and random component, where the random components model the daily variations of the $PM_{2.5}$–AOD relationship and also derives site-specific adjustment parameters. Comparison between the observed and predicted concentrations of the $PM_{2.5}$ levels at the monitoring stations show R^2 values ranging between 0.25 and 0.667 depending upon the effect of site bias on the $PM_{2.5}$–AOD relationship.

Keywords $PM_{2.5}$ · Aerosol optical depth (AOD) · Mixed-effect model (MEM)

1 Introduction

Particulate Matter (PM) often referred to as aerosols in general, denote a collection of solid and liquid particles, of varying size, complexity, and origin suspended in air. PM which is smaller in size is most lethal in nature. It represents the most hazardous form of air pollution as these particles can easily penetrate the human lungs and bloodstream, causing lung cancer and other respiratory diseases.

S. L. Kesav Unnithan · L. Gnanappazham (✉)
Department of Earth and Space Sciences, Indian Institute of Space Science and Technology, Thiruvananthapuram 695547, Kerala, India
e-mail: gnanam@iist.ac.in

S. L. Kesav Unnithan
e-mail: slkesavunnithan@gmail.com

J. K. Ghosh and I. da Silva (eds.), *Applications of Geomatics in Civil Engineering*, Lecture Notes in Civil Engineering 33, https://doi.org/10.1007/978-981-13-7067-0_20

According to a report by Global Burden of Diseases (GBD) [1], a total of 4.2 million deaths globally were attributed to causes relating to outdoor $PM_{2.5}$ (particulate matter with size less than 2.5 μm in aerodynamic diameter) emissions in 2015, of which India alone accounted for 1.091 million deaths resulting in a net 3% Gross Domestic Product (GDP) loss. In addition, the Burden of Disease Attributable to Major Air pollution sources in India [2] finds that 75% of the number of deaths relating to air pollution is associated with rural areas. Also in recent years in Delhi, National Capital Region (NCR) has witnessed prolonged episodes lasting several days during the winter season where the $PM_{2.5}$ levels have been greater than 500 $\mu g/m^3$ signifying more than hazardous effect on human health according to the Air Quality Index (AQI) [3]. It is thus imperative to accurately map the population exposure to $PM_{2.5}$ with high spatial and temporal resolution. However, the $PM_{2.5}$ monitoring stations are limited in number, unequally distributed, and have different sampling ranges which impede the ability to accurately assess the health impact upon exposure to $PM_{2.5}$ pollution levels.

Satellite remote sensing with capabilities of high spatial coverage and resolution, in addition to repeated reliable observations, have been used to relate the Aerosol Optical Depth (AOD) with $PM_{2.5}$ measured at ground monitoring stations [4, 5]. AOD is a measure of the solar extinction coefficient over a vertical column of atmosphere. It represents the amount of sunlight that is absorbed or scattered by the aerosol particles suspended within the vertical column. AOD thus reflects as a measure to estimate $PM_{2.5}$ particle concentration at any spatial location.

Earlier studies [6] have used satellite-based AOD retrieval for $PM_{2.5}$ estimation using MODIS AOD through simple linear regression model at global and continental scale. This model assumed that AOD–PM relationship was constant with time and location and the R^2 values, which show the goodness of fit of the linear relationship, were reported from 0.3 to 0.6. This model was extended into Multiple Linear Regression [7] by considering additional meteorological parameters including humidity, temperature, wind speed, and wind direction. The corresponding R^2 values were obtained from 0.42 to 0.59 for the city to whole area spatial scales. Further, based on the vertical distribution and transmission of aerosol particles, Chemistry Transport Models (CTM) such as the Global Atmospheric Chemistry Model (GEOS—CHEM) were employed. This model reported R^2 values of 0.68 for Northern Italy, 0.41 for Canada during 2010–2012 [8]. However, Wang et al. employed CTM over urban spatial scale and obtained a daily R^2 value of 0.86 and monthly R^2 of 0.93 over the city of Montreal, Canada. Several studies have also included land use information in the form of geographically weighted regression indicating population density, urban built-up all indicative of higher probability of $PM_{2.5}$ emissions. It is thus logical to infer that the AOD–$PM_{2.5}$ relationship varies every day, and it is erroneous to assume that it remains constant throughout the study period for any given study region.

This study proposes the implementation of Linear Mixed-Effect Model (MEM) to model the AOD–$PM_{2.5}$ relationship over the Indian subcontinent. A MEM is a statistical model comprising of fixed and random parts, which allow for repeatable observations for variables under consideration. Since daily AOD–$PM_{2.5}$ relationship is modeled, similar cases have been reported as in [9, 10], this method thus offers an opportunity to assess severe pollution events within the given study region. The article is organized as follows: 'Methodology', first describes the study region detailing the geography and atmospheric conditions relevant to India, then $PM_{2.5}$ and AOD datasets used for this study domain, and the subsequent development and implementation of the Linear Mixed-Effect Model, and 'Results and Discussion' examines the trends in AOD over India, the model estimation results, the trends in estimated values of $PM_{2.5}$ and concludes with the limitations present.

2 Methodology

2.1 *Study Domain*

The spatial domain of our study included the Indian subcontinent landmass is between $8° 44'$ N and $37° 6'$ N latitude, $68° 7'$ E and $97° 25'$ E longitude. It includes 3.2 million km^2 in area and land boundary length of 15,200 km and coastline of 7517 km. India is home to the Indo-Gangetic plain (IGP), one of the largest river basins and also to highly air polluted place in the world. At 81 million h,a India's irrigated area is the largest in the world. The total forest and tree cover stands at 79.42 million ha [11] while the population density is at 382 persons per km^2. All these factors in combination are major determinants in the quality of air for human survival. The country has four seasons, winter (January–March), summer (April–June), monsoon (July–September), post monsoon (October–December), with winter recording high air pollution levels. During winter, because of low temperature and shallow boundary layer, there is the presence of large amount of aerosol concentrations near the surface and in the vertical column. Also, there is strong presence of mineral dust and sea salt during premonsoon and monsoon seasons, respectively [12].

2.2 *Datasets Used*

Ground-level $PM_{2.5}$ Monitoring Data. Daily $PM_{2.5}$ mass concentrations were obtained across 33 site locations as shown in Fig. 1. These monitoring stations at the site locations are maintained by the Central Pollution Control Board, under the National Air Quality Monitoring Programme (NAMP) in compliance with the National Ambient Air Quality Standard (NAAQS). These site locations are spread across 12 states, with 16 monitors in urban locales and 17 monitors in rural areas. The $PM_{2.5}$ concentrations have been estimated under the gravimetric method, where the mass concentrations in ambient air are calculated as the total mass of collected particles in $PM_{2.5}$ size ranges divided by the total volume of ambient air sampled, expressed in $\mu g/m^3$ [13]. The $PM_{2.5}$ data was collected from January to August 2017, the study period, totaling to 238 days.

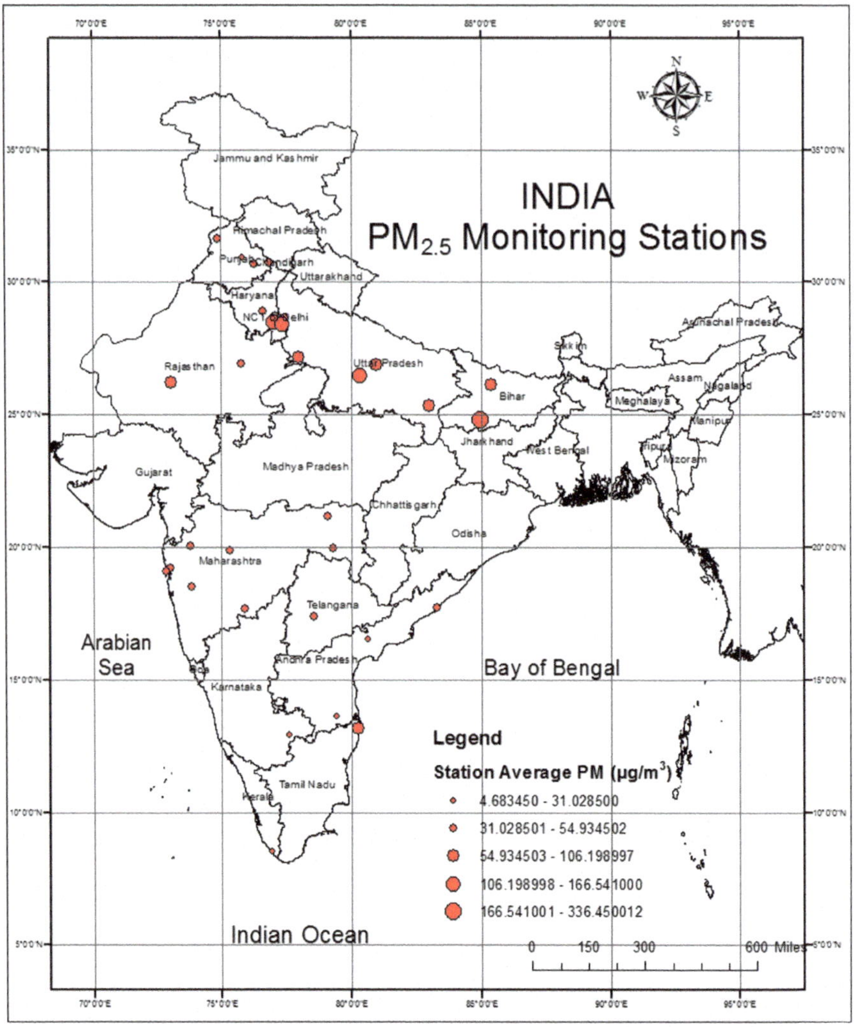

Fig. 1 $PM_{2.5}$ monitoring stations across India for the study period January–August 2017

AOD Data. Moderate Resolution Imaging Spectroradiometer (MODIS) instrument aboard the Terra satellite, part of NASA Earth Observation System (EOS) was launched in May 1999. It has a sun-synchronous, near-polar, and circular orbit of 705 km. It has a ground swath of 2330 and 10 km along the track at nadir viewing angle. The Terra satellite crosses the equator at 10.30 a.m. (descending node). It has 36 spectral bands, of which bands between 0.412 and 2.155 μm are used for retrieving aerosol properties. The Combined Deep Blue Dark Target Aerosol Optical Depth subproduct has been used from the Collection 6 MODIS Aerosol products. This subproduct is the combined result of the output of the Deep Blue and Dark

Target algorithms to estimate AOD. It is difficult to differentiate between the surface and aerosol particles in many wavelengths of the visible spectral band. In Deep Blue algorithm, the 412 nm spectral band (known as the blue band) can clearly distinguish between the bright aerosol particles and the dark surface background. The Dark Target algorithm retrieval algorithm is used for land surfaces which are mostly vegetated and not over bright surfaces like desert area. The combined product thus has more retrieval coverage over land surfaces with a 10 km-by-10 km spatial resolution [14].

2.3 Statistical Model

Model calibration was performed using AOD values of sites where $PM_{2.5}$ values are recorded. Daily data within the study period where more than two collocated observations for AOD and PM values across the entire study region were taken into consideration. A linear regression model was first employed to check the linear relationship between AOD and PM values across 33 sites of the form,

$$PM_{2.5} = \alpha + \beta * AOD \tag{1}$$

where α and β represent fixed intercept and fixed slope, respectively. In this linear regression model, the underlying assumption is that the relationship between AOD and $PM_{2.5}$ does not change throughout the study period. In effect, it is assumed that AOD and $PM_{2.5}$ have very little spatial and temporal variation. Log-transformed values for AOD were also taken for modeling with the linear regression. However, nothing substantial can be inferred from taking these log-transformed values.

In reality, AOD and $PM_{2.5}$ exhibit day-to-day variation relating to change in meteorological variables including boundary layer height, relative humidity, and also changes with surface topology and $PM_{2.5}$ pollution sources. Therefore to quantitatively model the daily variations in AOD–$PM_{2.5}$ relationship, the Mixed-Effect Model (MEM) was utilized [13]. This was implemented in the following form:

$$\begin{gathered} PM_{2.5\,(i,j)} = (\alpha + U_j) + (\beta + V_j) * AOD_{(i,j)} + S_i + \varepsilon_{(i,j)} \\ \left(U_j V_j\right) \sim N\,[00, \Sigma] \end{gathered} \tag{2}$$

where α and β denote the fixed intercept and fixed slope, U_i and V_j represent random intercept and random slope, respectively, $AOD_{(i,j)}$ represent AOD value in site i on day j, S_i represents site adjustment parameter specific to each $PM_{2.5}$ monitoring site, $PM_{2.5(i,j)}$ represents $PM_{2.5}$ value in site i on day j, and $\varepsilon_{(i,j)}$ represents the error term associated with collocated AOD and $PM_{2.5}$ values for site i and day j. The random intercept and random slope model the daily variations in the AOD–PM relationship. The random slope and random intercept are applied uniformly across all 10 * 10 km grid cells across the study region for a given day in the study period, represented by subscript j in the above equation. The S_i (~N [00, σ]) accounts for the difference in

AOD–PM relationship due to spatial variation with respect to each $PM_{2.5}$ monitoring site location. The mixed-effect model estimation with and without site bias shows the sensitivity of spatial variation affecting AOD–PM relationship. Similarly, in the implementation of this model, daily data where more than two pairs of collocated AOD–PM values are available are taken for model estimation.

3 Results and Discussion

3.1 MODIS Aerosol Optical Depth Data

The MODIS Level 2 (Collection 6) Combined Dark Target Deep Blue Retrieved AOD product was obtained for each day and Fig. 2 maps show the average month wise AOD value for a 10 * 10 km grid cell from January to August 2017. It can be observed from the maps that the Indo-Gangetic Plain (IGP) is most polluted with large presence of aerosol particles confirmed by the average AOD value of 0.8–1.6 across the states of Haryana, Delhi, Uttar Pradesh, Bihar, and northern Madhya Pradesh. There is a gradual drop in the concentration of aerosol particles across the entire region from January to March 2017. Also, there is a slight dip in concentration during May–June coinciding with the beginning of the farming season.

It can also be observed from the map that from March to May 2017, there is comparatively higher concentration of AOD (>0.6) around West Bengal, coastal Odisha, Andhra Pradesh, while the southern states of Kerala, Tamil Nadu exhibit AOD values as high as 3.23–3.42 for April 2017.

The minimum AOD concentration map shows that central part of Tamil Nadu, Karnataka record the least amount of suspended aerosol particles within the study period. It is also interesting to note that central part of Rajasthan also records AOD values in the range 0.1–0.5, even though the terrain is a desert region. Central India including parts of Maharashtra, Chhattisgarh, and Telangana record AOD values in the range 0.7–1.8 on an average daily basis. IGP constantly records alarming levels of AOD values reaching as high as 3.5, which results in hazardous conditions affecting human health. Even parts of Punjab and Haryana are prone to high aerosol loading because of the practice of burning harvested croplands in preparation for the next sowing season. This exercise even though carried out as labor-saving endeavor, ultimately ends up affecting the health of the population at large.

This daily MODIS Level 2 (Collection 6) Combined Dark Target Deep Blue Aerosol product over the Indian subcontinent is taken as input for the MEM, where corresponding $PM_{2.5}$ monitors record the $PM_{2.5}$ values against each AOD measurement.

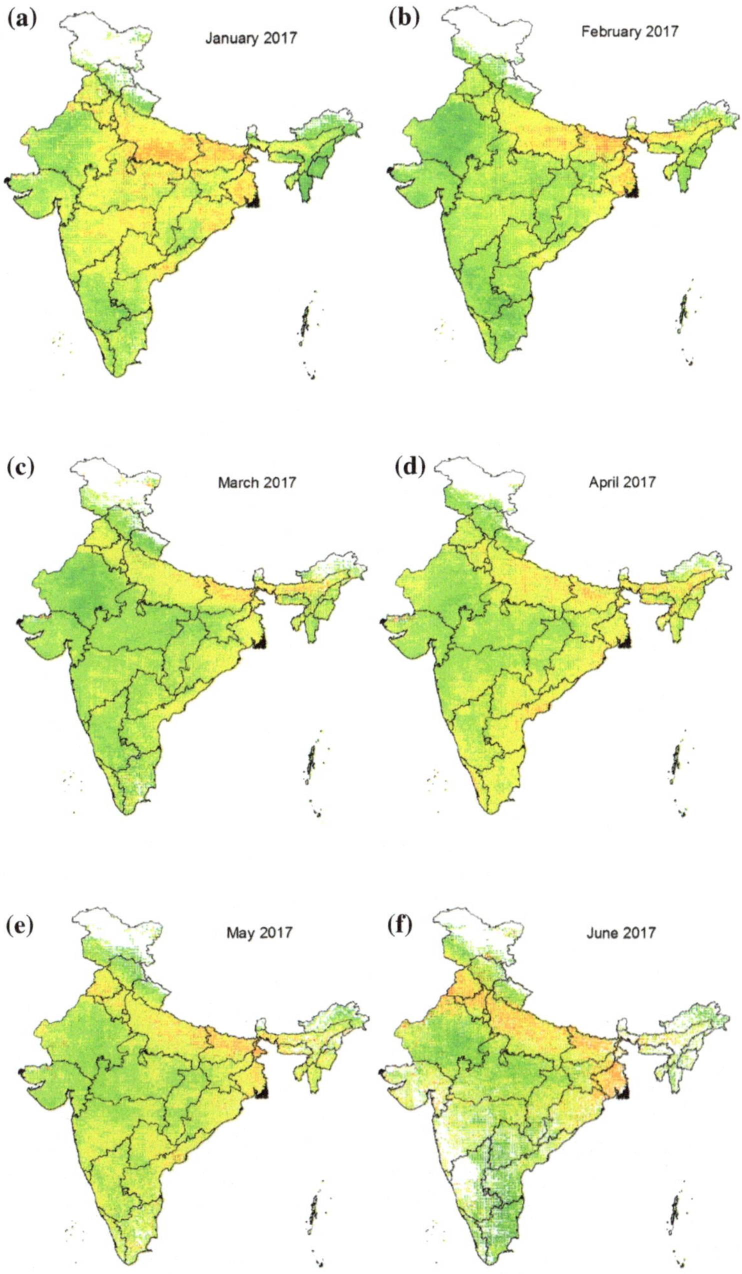

Fig. 2 Maps **a–h** each show average monthly AOD concentration from January to August 2017, while **i** and **j** depict the minimum and maximum AOD concentration, respectively, for a given 10 * 10 km grid cell over the entire study period

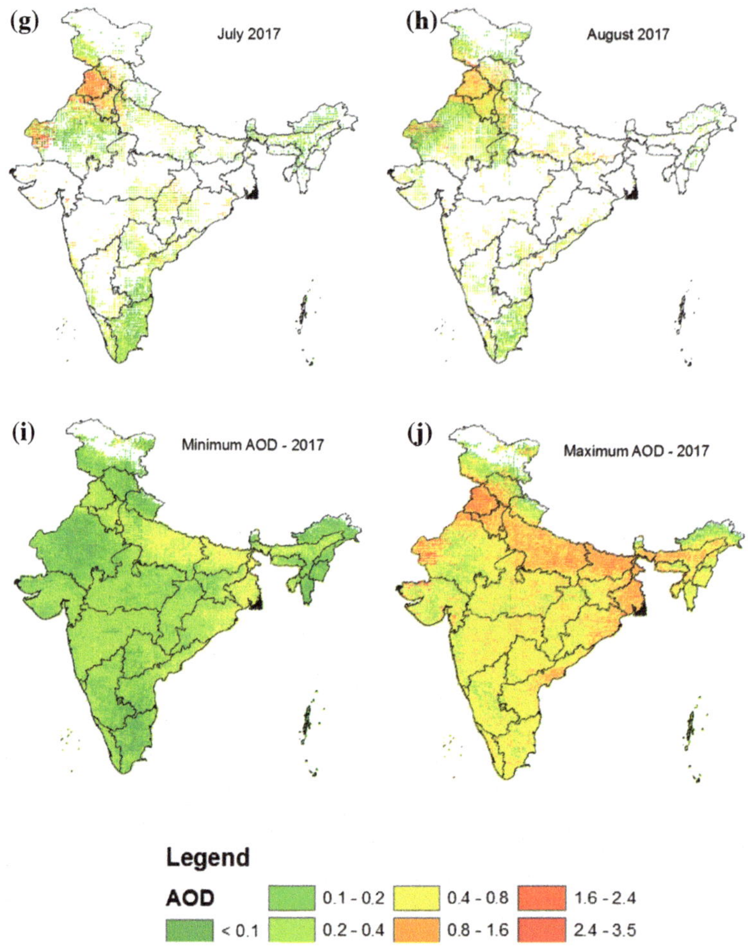

Fig. 2 (continued)

3.2 Model Estimation

The linear regression model was first used to analyze the direct relationship between AOD values and $PM_{2.5}$ concentrations. This resulted in a skewed graph (Fig. 3a) with a slope value of 0.31 and intercept value of 70. No tangible information regarding the association between the two variables can be derived as is evident from the R^2 correlation coefficient giving a value of 0.27. This confirms our conjecture that it is incorrect to assume that the AOD–$PM_{2.5}$ relationship remains constant throughout the entire study period.

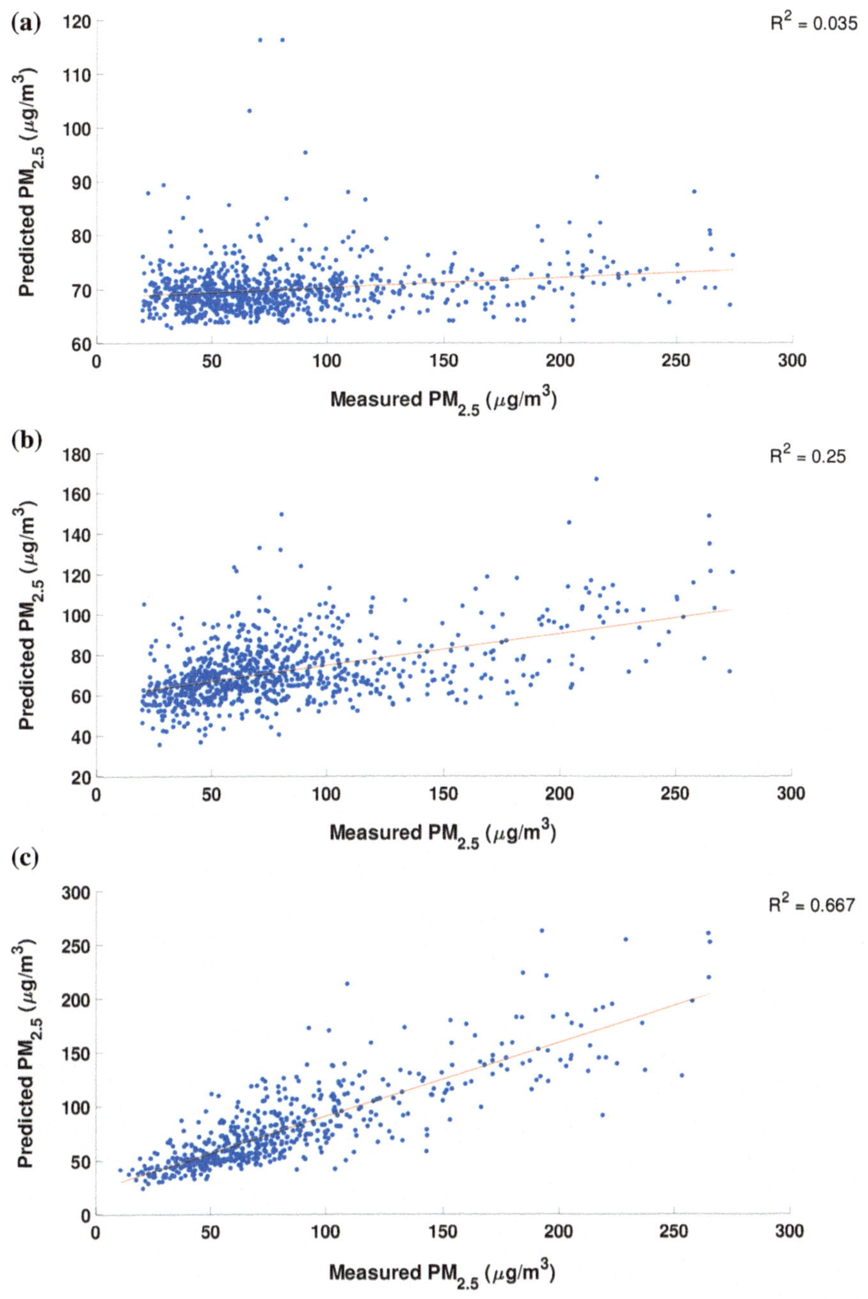

Fig. 3 **a** Linear regression estimation, **b** Mixed-effect model estimation without site bias parameter, and **c** Mixed-effect model estimation with site bias parameter taken into consideration

Table 1 Description of mixed-effect model variables

Mixed-effect model parameters	Value
Number of observations	885
Number of stations (i)	33
Number of days (j)	168
Number of fixed coefficients	2 (Slope and Intercept)
Number of random coefficients	369 (Random slope, Random intercept and Site bias)

Table 2 Statistics of random parameters of mixed-effect model with fixed slope = $2 * 10^{-4}$ and fixed intercept = 4.1474

Statistical measure	Random intercept	Random slope	Site bias
Mean	0.16936	0.00023965	0.39095
Minimum	0.07956	0.00008093	0.30092
Maximum	0.36054	0.00070962	0.50791

To overcome this deficiency, the Mixed-Effect Model was employed, which allows for the variability in AOD values throughout the epoch and the corresponding change in the resulting $PM_{2.5}$ values are observed. Here, we obtain AOD–$PM_{2.5}$ correlation separately for each day in the study period where there are two or more collocated AOD and $PM_{2.5}$ values across the region of interest. Also to be noted is that the daily intercepts and slopes were independent of the number of collocated AOD and $PM_{2.5}$ observations available for a given day across the region. Table 1 describes the different parameters associated with MEM and the corresponding values observed against each parameter. Further, the effect of spatial variation affecting the relationship between the two variables can be examined with the help of the site-specific adjustment parameter that is taken as a random intercept in the MEM. The corresponding values of fixed slope and fixed intercept obtained were statistically significant with $p < 0.0001$ and SE = 1.23 for slope and $p = 0.0023$ and SE = 1.78 for intercept. The corresponding performance of estimation was analyzed both without including the site-specific adjustment parameter and also by including it (Table 2).

The MEM estimation without site-specific adjustment parameter results in a correlation coefficient R^2 value of 0.250 (Fig. 3b), while the estimation accuracy improves when site bias is included with a higher correlation coefficient R^2 value of 0.667 (Fig. 3c). Similar R^2 values have been reported in the range 0.77–0.84 over Baltimore region, USA as in [9].

3.3 *Estimated Surface $PM_{2.5}$ Layers*

The $PM_{2.5}$ layers were generated based on the assumption that a given AOD–$PM_{2.5}$ relationship obtained for each day in the epoch would be valid for across the entire spatial extent of the region, with only the site-specific adjustment parameter accounting for deviation in correlation due to spatial variation.

From the estimated $PM_{2.5}$ layers, it is observed that the $PM_{2.5}$ trend goes through a gradual decrease in concentration across the entire study region from an average 81.24 $\mu g/m^3$ in January to 66.64 $\mu g/m^3$ in April 2017. The winter season in January results in heavy accumulation of fine particles within the country because of the shallow boundary layer as evident from [12–15]. However, there is a high incidence of estimated $PM_{2.5}$ concentration with several regions having greater than 70 $\mu g/m^3$ during May. As evident from AOD monthly maps, higher incidence of aerosol loading over IGP also resulted in higher $PM_{2.5}$ concentrations in the range 90–110 $\mu g/m^3$ on an average from January to August 2017. This is, in turn, collaborated by the average $PM_{2.5}$ concentrations greater than 100 $\mu g/m^3$ during the study period reported by $PM_{2.5}$ ground monitoring stations across Haryana, Delhi, Uttar Pradesh, and Bihar (Fig. 1).

Although the World Health Organization (WHO) standard for permissible concentration of $PM_{2.5}$ in ambient air is 10 $\mu g/m^3$, India's average $PM_{2.5}$ for the study period is much higher at 70.73 $\mu g/m^3$. Figure 4i shows that the higher extreme value of the minimum $PM_{2.5}$ concentration map ranges from 60 to 80 $\mu g/m^3$ for parts of the country including Maharashtra, Uttar Pradesh, Bihar, Jharkhand, and West Bengal, with a mean value of 63 $\mu g/m^3$ across India. These values of minimum $PM_{2.5}$ concentrations itself is recognized as unhealthy for people suffering from respiratory illness. Figure 4j depicts proximate grid cells around the IGP and central parts of Madhya Pradesh, Odisha, Chhattisgarh, and Telangana with mean value of the maximum $PM_{2.5}$ concentration of 100.17 $\mu g/m^3$ which has been known to have unhealthy to hazardous impact on healthy human according to the Air Quality Index (AQI), while the average of maximum $PM_{2.5}$ concentration stands at 82 $\mu g/m^3$ for the entire country.

However, it is visible in the maps for July and August 2017, there are large gaps in the estimated $PM_{2.5}$ values throughout Central India. This is mostly related to the corresponding gaps in AOD retrieval due to the onset of monsoon climate associated with the months of July to September. This limitation inhibits the model capacity to gainfully predict ground-level $PM_{2.5}$ concentration during these time periods which needs more input computations to be involved. In addition, the corresponding effect of temporal variables including temperature, relative humidity, wind speed and direction, and spatial variables like road features, percentage of open spaces, industries, and the interactions between these variables all impact the estimation accuracy. Therefore, upon such several auxiliary data being available, the estimation accuracy can be further improved and also the gaps in AOD retrieval can be filled by employing spatial kriging methods with help of available data which will be carried out in subsequent studies.

Despite these uncertainties, this study clearly demonstrated how AOD values obtained through satellite remote sensing data can be useful in estimating ground-level $PM_{2.5}$ concentrations with the implementation of Mixed-Effect Model. It is the first attempt to implement this model over the Indian subcontinent region on a daily basis. This model is more practical in its application since it allows for day-to-day

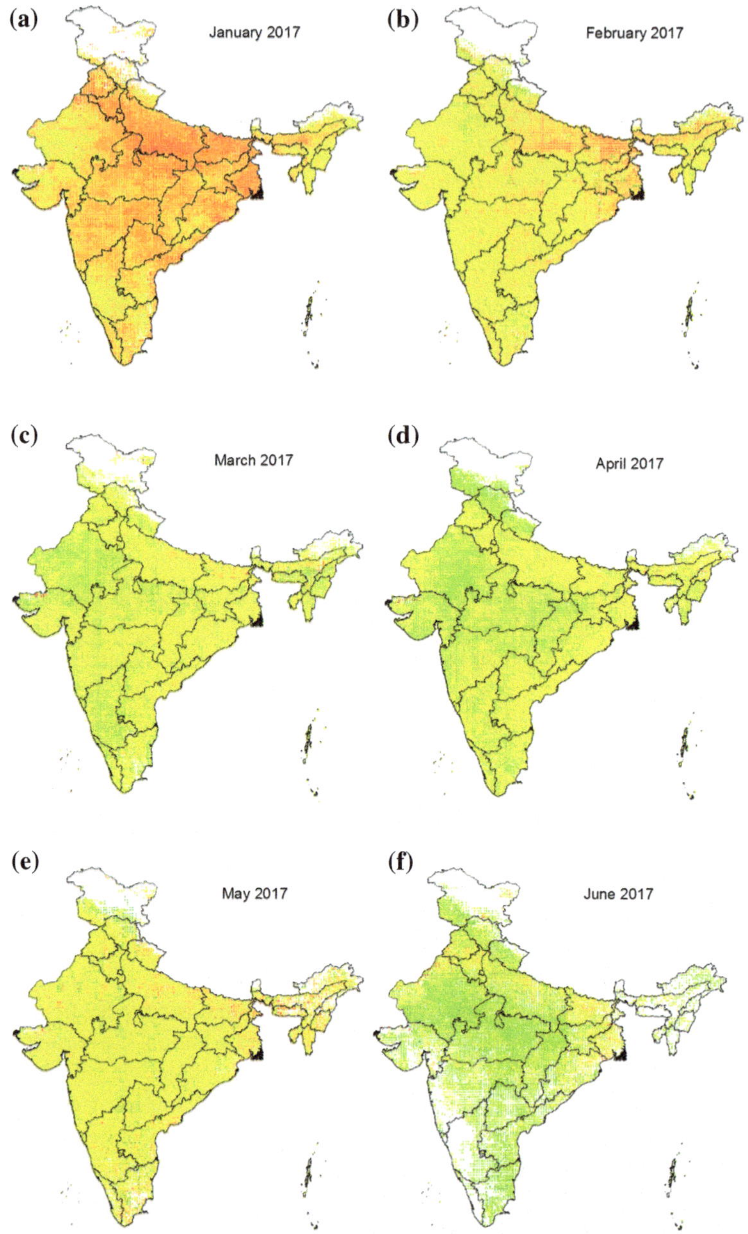

Fig. 4 Maps **a**–**h** each show estimated average monthly $PM_{2.5}$ concentration from January to August 2017, while **i** and **j** depict the minimum and maximum $PM_{2.5}$ concentration, respectively, for a given 10 * 10 km grid cell over the entire study period

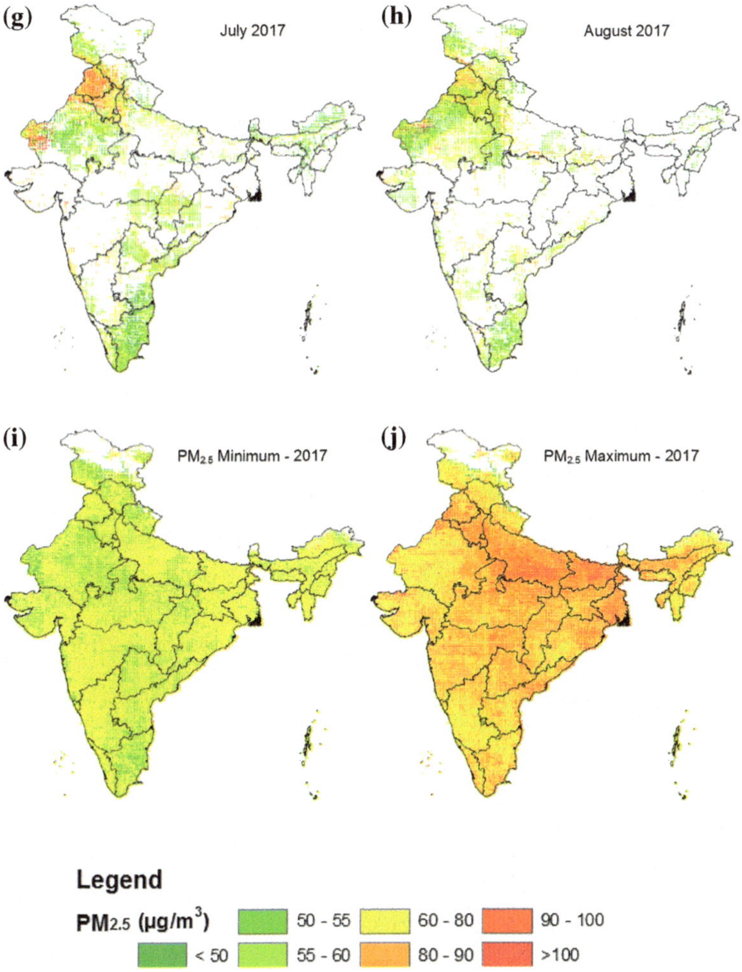

Fig. 4 (continued)

variations in the AOD–$PM_{2.5}$ relationship, meaning a distinct relationship for each day during the study period, which are mostly influenced by above mentioned spatial and temporal parameters.

A network of 33 stations across the country was installed by CPCB to monitor ground-level $PM_{2.5}$ concentrations, however, there exist significant gaps in monitoring across the central and as well as the northeastern parts of the country. Estimating $PM_{2.5}$ levels in such gaps with the satellite data which provides high spatiotemporal resolution will help to fill the gaps in monitoring the pollution levels. Outcome of such satellite data-based models to estimate air quality parameters will help the National

Air Quality Monitoring Programme (NAMP, Ministry of Environment, Forests and Climate Change, Government of India) to monitor the status and trend of the air quality, to control and regulate the pollution levels from various sources, and fixing the air quality standards for industrial and town planning.

Acknowledgements I would like to acknowledge Central Pollution Control Board (CPCB) for the free access to ground-level $PM_{2.5}$ data through the National Air Quality Monitoring Programme (NAMP), MODIS GSFC NASA for free access to MODIS Terra Aerosol Optical Depth data. I would also like to acknowledge Indian Institute of Space Science and Technology, Thiruvananthapuram for the support extended in the completion of the study as part of the master's thesis.

References

1. Global burden of disease collaborative network. Global burden of disease study 2016 (GBD 2016). Disability weights. Institute for Health Metrics and Evaluation (IHME), Seattle, United States (2017)
2. Burden of diseases attributable to major air pollution sources in India. GBD MAPS Working Group, Special Report 21. Health Effects Institute (2018)
3. Delhi's pollution levels rise again. The Hindu, dtd. http://www.thehindu.com/news/cities/Delhi/after-dip-pollution-levels-rise-again/article20355358.ece. Accessed 12 Dec 2017
4. Lee HJ, Coull B, Schwartz J, Koutrakis P (2011) A novel calibration approach of MODIS AOD data to predict $PM_{2.5}$ concentrations. Atmos Chem Phys 11:7991–8002
5. Liu, Koutrakis P, Kahn (2007) Estimating fine particulate matter component concentrations and size distributions using satellite-retrieved fractional aerosol optical depth: part 2—a case study. J Air Waste Manag Assoc 57:1360–1369
6. Wang J (2003) Intercomparison between satellite-derived aerosol optical thickness and $PM_{2.5}$ mass: implications for air quality studies. Geophys Res Lett 30
7. Di Nicolantonio, Cacciari (2010) Modis multiannual observations in support of air quality monitoring in northern Italy. Ital J Remote Sens Riv Ital Telerilevamento 43:97–109
8. Kloog I, Koutrakis P, Coull B, Lee HJ, Schwartz J (2011) Assessing temporally and spatially resolved $PM_{2.5}$ exposures for epidemiological studies using satellite aerosol optical depth measurements. J Atmosenv 45:6267–6275
9. Kloog I, Nordio F, Coull B, Schwartz J (2012) Incorporating local land use regression and satellite aerosol optical depth in a hybrid model of spatiotemporal $PM_{2.5}$ exposures in the Mid-Atlantic States. Environ Sci Technol 46:11913–11921
10. Xie Y, Wang Y, Zhang K, Dong W, Lv B, Bai Y (2015) Daily estimation of ground-level $PM_{2.5}$ over Beijing using 3 km resolution MODIS AOD. Environ Sci Technol 49:12280–12288
11. India State of Forest Report (ISFR), Ministry of Environment, Forest and Climate Change, Govt. of India (2015)
12. Kedia S, Ramachandran S, Holben BN, Tripathi SN (2014) Quantification of aerosol type, and sources of aerosols over the Indo-Gangetic Plain. J Atmos Environ 98:607–619
13. Guidelines for the measurement of ambient air pollutants, Central Pollution Control Board, National Ambient Air Quality Series: NAAQMS 36 (2012)
14. Kleidman R, Levy R, Remer L, Chu C, Mattoo S, Chu A, Martins V, Li R-R, Salustro C, Tanre D, Kaufman Y (2012) Understanding MODIS products, Goddard Space Flight Center
15. Chu Y, Liu Y, Li X, Liu Z, Lu H, Lu Y, Mao Z, Chen X, Li N, Ren M, Liu F, Tian L, Zhu Z, Xiang H (2016) A review on predicting ground $PM_{2.5}$ concentration using satellite aerosol optical depth. Atmos 7:129

Application of Geomatics for Drainage Network Delineation for an Urban City

Kaushal Kumar and Raj Mohan Singh

Abstract Traditionally, topographic maps are usually employed for the drainage analysis. However, channel network extraction and watershed delineation from topographic maps require time and expertise in cartography. The application of Geomatics (Remote Sensing (RS) and Geographic Information Systems (GIS) and other land topographic informations) is imperative for drainage network delineation. In the present study, drainage network of an urban city, Allahabad in U.P., is delineated using ArcHydro. The ASTER Global DEM (GDEM) data of 30 m resolution is processed to determine the primary natural flow routes and catchment. Hydrological analysis is performed for flow accumulation, slope, drainage path using DEM data. Results show that the maximum drainage length is 1.73 km, the longest flow path length in catchment is 2.10 km.

Keywords Drainage network · Geomatics · RS and GIS · ArcHydro · GDEM

1 Introduction

A drainage network is a linear connection of land units that can accumulate the most runoff in an area. It is difficult to determine the quantity of accumulation of runoff. However, GIS provides a platform, where the quantity of accumulation of runoff can be easily determinable [1]. Urban planners, nowadays, used different ways to deal with drainage network system, for example, headwater-tracing method [2], adaptive approach for determining stream course with heuristic information (AHI) [3], and D8 algorithm approach [4] for the outline of drainage networks using advanced DEM models of land surfaces on GIS platforms.

K. Kumar (✉)
GIS Cell, MNNIT, Allahabad 211004, UP, India
e-mail: rgi1601@mnnit.ac.in

R. M. Singh
Department of Civil Engineering, MNNIT, Allahabad 211004, UP, India
e-mail: rajm@mnnit.ac.in

J. K. Ghosh and I. da Silva (eds.), *Applications of Geomatics in Civil Engineering*,
Lecture Notes in Civil Engineering 33, https://doi.org/10.1007/978-981-13-7067-0_21

The drainage network is characterized by the relative checks wherever the upstream drainage area surpasses a specified threshold [5]. The real issue with D8 approach is that it needs depressionless DEM information [6]. The raw DEM model has sinks or depressions in the data so it must be reconditioned before processing. Sinks are the cells which have no neighbors at a low elevation and consequently, have no downslope streamflow path to a neighboring cell [7].

Some investigation needs high-resolution DEM data for better or precise outcomes. But high-resolution DEM data needs processing with good quality and quantity of computer resources and time to extricate drainage network. GIS framework provides good agreement with real-time entities. For a large basin area, number of grids in basin become too large in high-resolution DEM data, hence complexity of the system increases [8, 9]. GRASS GIS software is based on least-cost flow routing method. This software improve the speed, functionality, and memory requirements and make least-cost flow routing method more efficient for drainage network analysis for large basin [10, 11].

1.1 Study Area

The city Allahabad is at the confluence of River Ganga and River Yamuna. The geographical coverage of the city is bounded between 25° 23′ 21″ N–25° 30′ 19″ N latitude and 81° 43′ 40″ E–81° 54′ 17″ E longitude. Topographically, the city is flat in nature and the temperature is ranged between maximum 47.8 °C in summer to lowest 4.1 °C in winter season. The city also receives an average rainfall of 930 mm. The geographic land area is divided into 80 wards. The geographical location of the city is shown in Fig. 1.

2 Methodology

The flowchart diagram for the delineation of drainage network is presented in Fig. 2. Data from various sources are collected, e.g., Survey of India (SOI), DEM data (30 m) from USGS etc., Allahabad ward map from Allahabad Development Authority, Allahabad for this analysis. DEM data with topographical map were used to extract the drainage channel networks. The delineation of the channel network from digital data was carried out in the ArcGIS 10.0 environment. The detailed methodology is represented in Fig. 2.

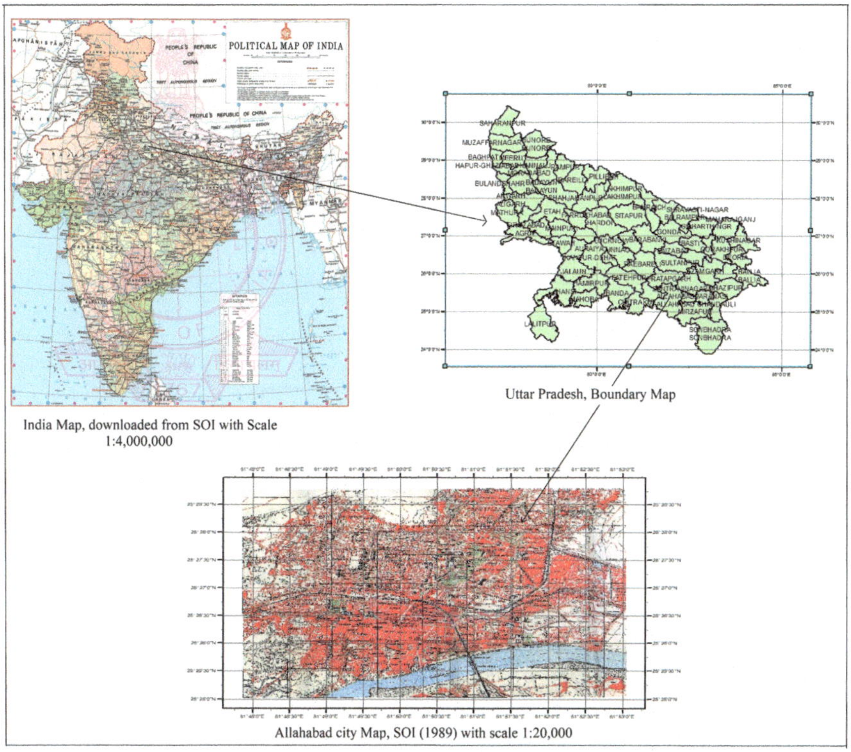

Fig. 1 Location map of the study area

3 Hydrological Analysis

The drainage channel network is delineated by utilizing DEM as input under GIS environment. ArcGIS software with the integration of Archydro tool is utilized to delineate the drainage network from DEM data [12]. The steps for the extraction of drainage channel delineation includes: preprocessing (or reconditioning) DEM, generation of flow direction, computation of the accumulation of flow, extraction of the drainage channel network. ArcHydro is working on the D8 algorithm is implemented for flow direction. A simple flowchart for extracting hydrological information and drainage network is shown in Fig. 3.

3.1 Flow Direction

For hydrological modeling, the flow direction is an important parameter. It is determined using D8 algorithm. A value ranging from 1 to 128 is assigned for each cell

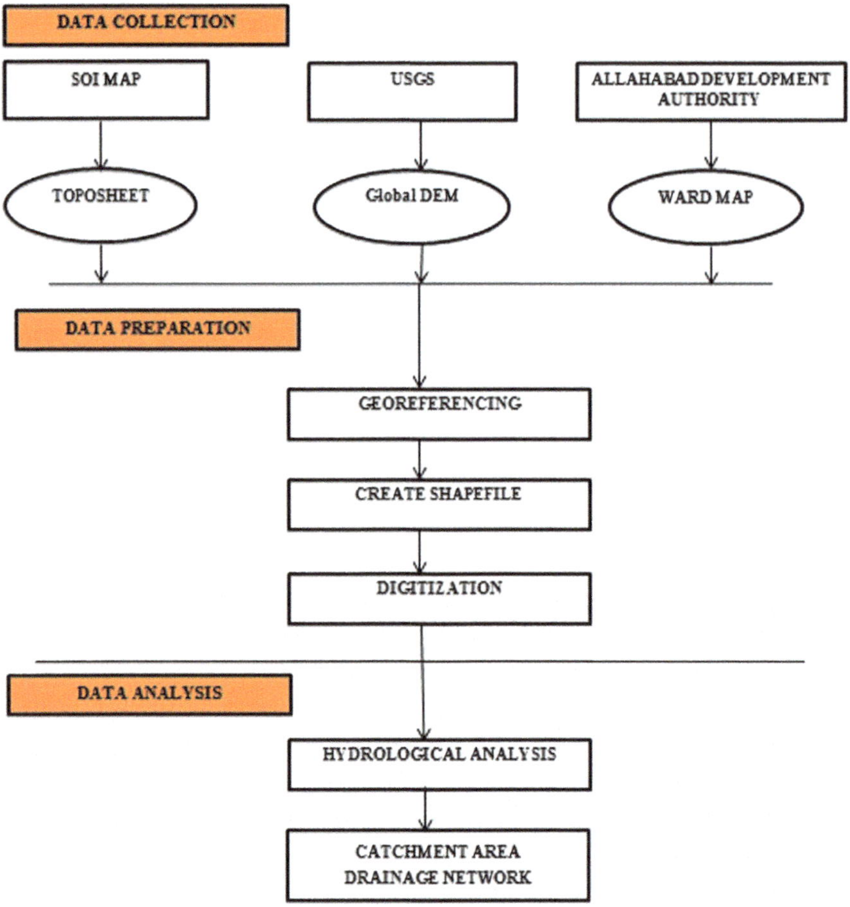

Fig. 2 Flowchart of methodology adopted

comprising the area. Flow direction is determined from higher to lower values. Flow direction calculation scheme and resulting flow direction in the study area are shown in Fig. 4 and Fig. 5, respectively.

3.2 *Flow Accumulation*

Flow accumulation, measure the additive amount of flow of water accumulated over the surface of land. This accumulated flow of water will help to the delineation drainage channel network extraction. Flow accumulation scheme and resulting flow accumulation in the study area are shown in Fig. 6 and Fig. 7, respectively.

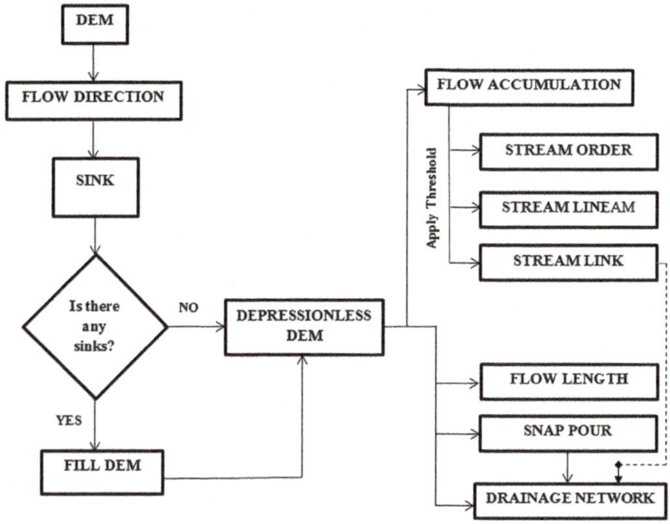

Fig. 3 Flowchart of process of extracting hydrological information from a DEM. *Source* Sitanggang and Ismail [13]

141	140	140	141	142	142	141	142
139	138	137	139	139	139	139	140
138	135	135	136	137	136	137	139
134	134	132	134	135	134	136	138
131	131	129	133	133	132	134	137
128	127	126	132	131	130	132	133
126	123	124	126	128	129	131	133
124	120	125	127	129	129	130	132

(a) DEM with elevation value

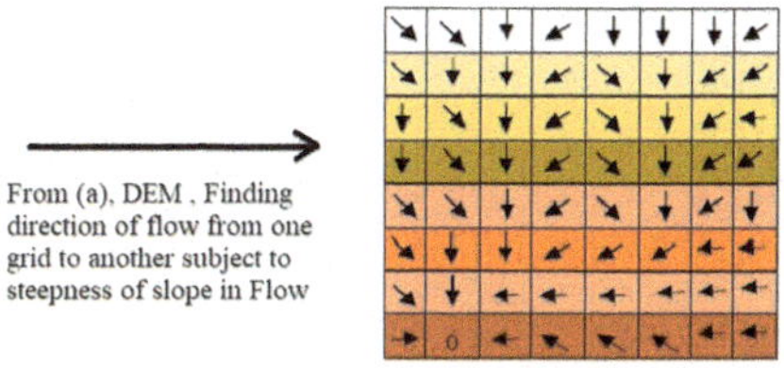

(b) Flow direction diagram

32(5)	64(2)	128(6)
16(1)	X	1(3)
8(8)	4(4)	2(7)

Fig. 4 Flow direction. *Source* Chukwuocha and Chukwuocha [14]

3.3 *Stream Delineation*

A stream network is delineated using a threshold value to cell. To understand the behavior of stream network, different values of stream threshold are applied to stream flow. The threshold value determines the number of cells contributing to stream flow. In the present study, the stream threshold (Ts) of 250, 500, 750, and 1000 are adopted. The stream delineation at different stream thresholds is shown in Fig. 8a, b, c, and d, respectively.

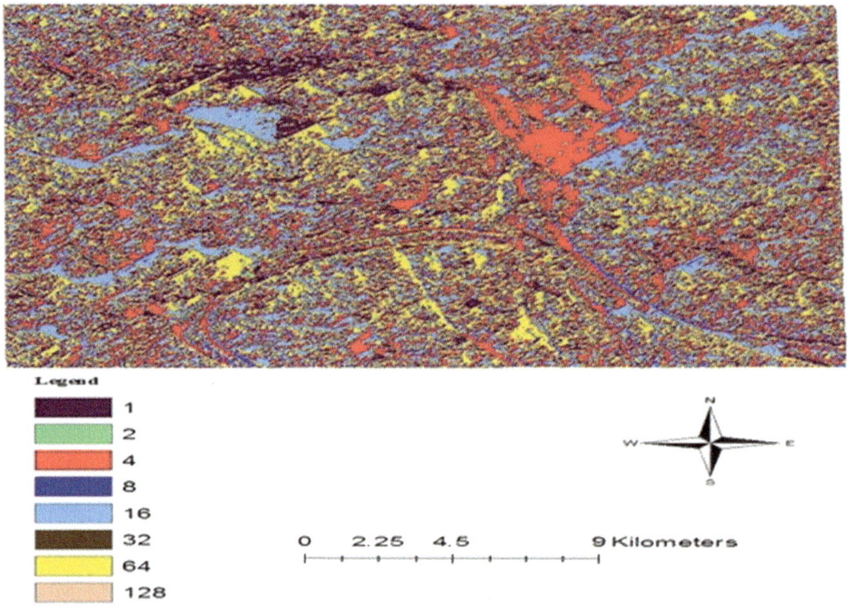

Fig. 5 Result of flow direction

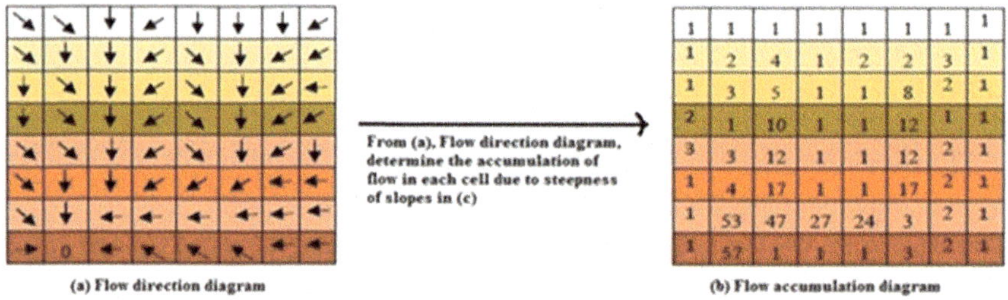

Fig. 6 Flow accumulation. *Source* Chukwuocha and Chukwuocha [14]

3.4 Drainage Network Delineation

The delineation of drainage network is carried out with flow direction and stream grid line as input. The network is extracted for city area with different stream threshold levels with the utilization of ArcGIS and ArcHydro tool under GIS framework. The different streams of different stream order contribute to mainline stream of the drainage networks which drained from different wards of Allahabad city area. The network delineated at different stream thresholds 250, 500, 750, and 1000 is shown in Fig. 9a, b, c, and d, respectively, and characteristics of the networks and its morphology are also studied.

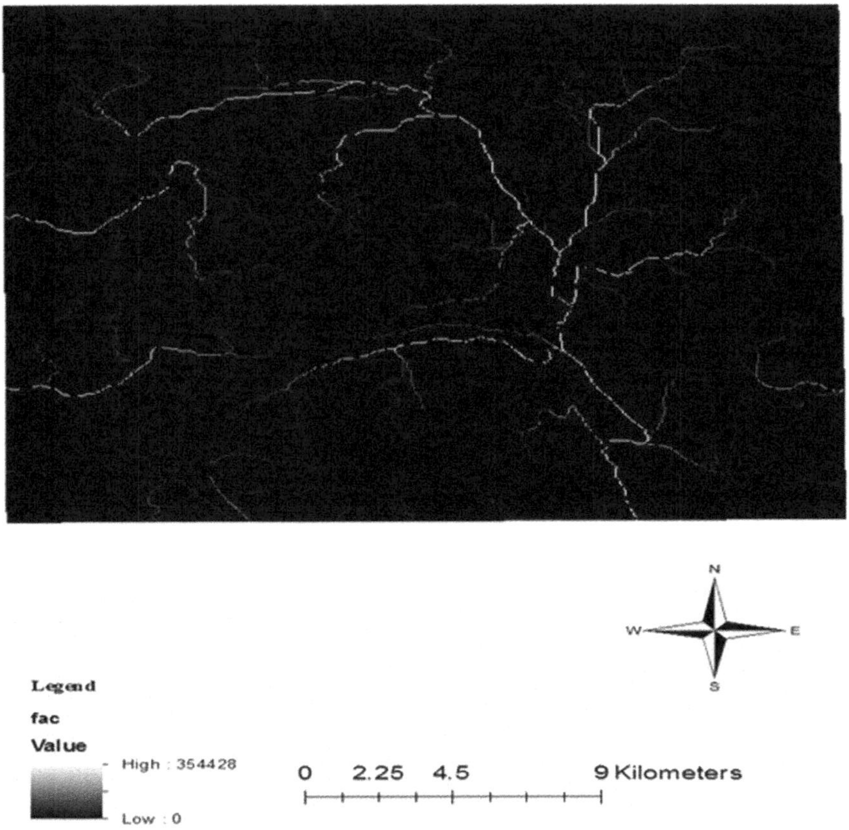

Fig. 7 Result of flow accumulation

The delineated drainage network at different stream thresholds is analyzed and summarized in Table 1.

4 Results and Discussion

The drainage network is delineated at different stream threshold values, for Allahabad city. From Figs. 8 and 9, it is clearly seen that the density of drainage network is high at low stream threshold and density of drainage network is low catchment at high stream threshold. At stream threshold 250, 500, 750, and 1000 maximum draining length are 1.15 km, 1.57 km, 1.73 km, and 1.57 km respectively, longest flow path length for catchment are 1.17 km, 1.80 km, 2.10 km, and 2.10 km, respectively, and longest flow path length for adjoint catchment are 11.79 km, 11.79 km, 11.79 km, and 11.79 km, respectively.

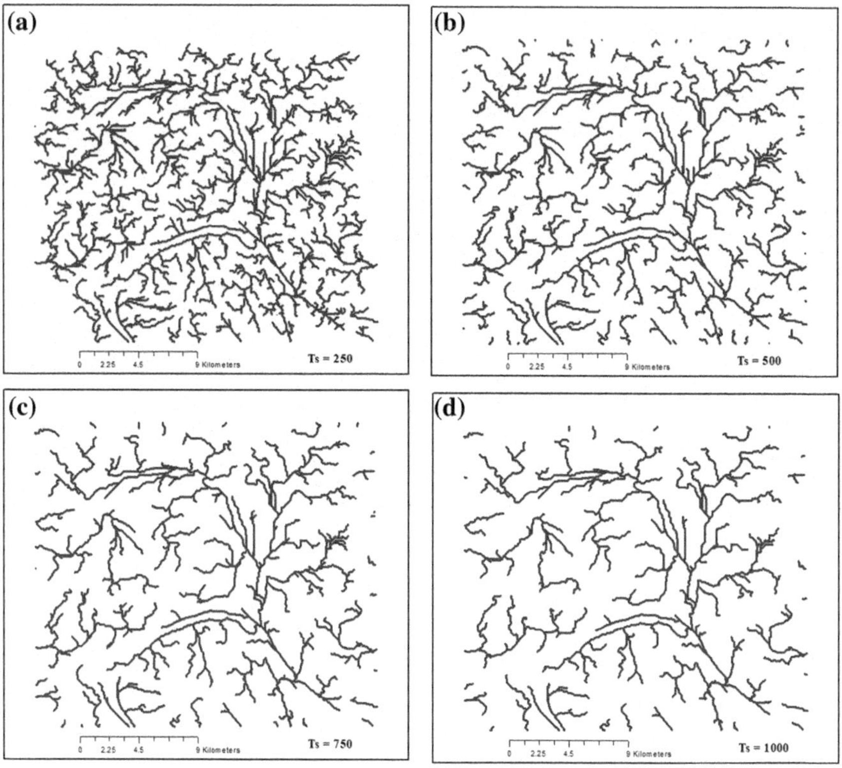

Fig. 8 Results of streamflow at different threshold values

The maximum area of catchment that contribute to the generation of drainage network at different stream threshold values of 250, 500, 750, and 1000 are 0.14 km^2, 0.31 km^2, 0.41 km^2, and 0.44 km^2 respectively, and maximum area of adjoint catchment is 11.79 km^2, 11.79 km^2, 11.79 km^2, and 11.79 km^2, respectively. The results show potential applicability of Geomatics for drainage network delineation for an urban city like Allahabad.

5 Conclusions

The natural drainage morphology is similar to the generated drainage network for the city. It is observed that longest flow path length for the contributing catchment saturate to a value (2.10 km). The longest flow length for adjoint catchment is the same in all cases even the maximum area of the adjoint catchment is the same (11.79 km) in all cases.

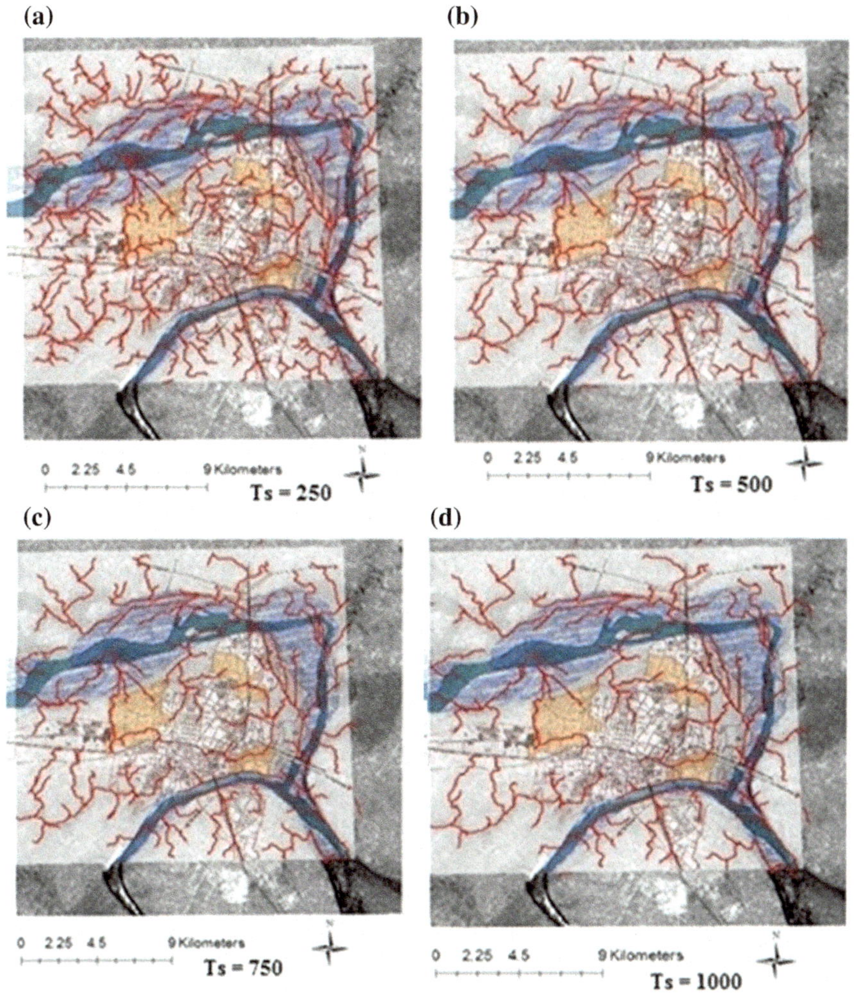

Fig. 9 Result of drainage network of Allahabad city

Table 1 Basic drainage network statistics

Stream threshold value (Ts) for cells	Maximum area of catchment (km^2)	Maximum area of adjoint catchment (km^2)	Maximum drainage length (km)	Longest flow path length for catchment (km)	Longest flow path length for adjoint catchment (km)
250	0.14	24.60	1.15	1.17	11.79
500	0.31	24.60	1.57	1.80	11.79
750	0.41	24.60	1.73	2.10	11.79
1000	0.44	24.60	1.57	2.10	11.79

References

1. Shamead et al (2014) A new automated approach to sewershed delineation for urban drainage modelling studies: a city of Toronto case study. In: International conference hydromatics
2. Lin WT et al (2005) Automated suitable drainage network extraction from digital elevation models in Taiwan's upstream watersheds. Hydrol Process
3. Sun T et al (2011) An adaptive approach for extraction of drainage network from shuttle radar topography mission and satellite imagery data. Int J Innov Comput, Inf Control 7:6965–6978
4. Mark DM, O'Callaghan JF (1984) The extraction of drainage networks from digital elevation data. Comput Vis, Graph, Image Process 28:323–344
5. Garbrecht J, Martz L (1995) The assignment of drainage direction over flat surfaces in raster digital elevation models. J Hydrol 193:204–213
6. Chorowicz J et al (1992) A combined algorithm for automated drainage network extraction. Water Resour Res 28:1293–1302
7. Martz LW, Garbrecht J (1992) Numerical definition of drainage network and subcatchment areas from digital elevation models. Comput Geosci 18747–18761
8. Mao et al (2014) An advanced distributed automated extraction of drainage network model on high-resolution DEM. Hydrol Earth Syst Sci 11:7441–7467
9. Babu GP, Sreenivas B, Rafique F, Rajesh, Importance of breaching and filling for river network extraction using high resolution DTM
10. Metz M, Mitasova H, Harmon RS (2011) Efficient extraction of drainage networks from massive, radar-based elevation models with least cost path search. Hydrol Earth Syst Sci 15:667–678
11. Mozas-Calvache AT et al (2017) Determination of 3d displacements of drainage networks extracted from digital elevation models (Dems) using linear-based methods. Int J Geo-Inf 6:234
12. USGS, Earth explorer for global DEM data. https://earthexplorer.usgs.gov/
13. Sitanggang IS, Ismail MH (2011) A simple method for watershed delineation in ayer hitam forest reserve using GIS. Bulletin Geospatial Sektor Awam
14. Chukwuocha A, Chukwuocha N (2015) Geographic information systems based urban drainage efficiency factors. FIG Working Week 2015

Assessment of Spatial Variations in Groundwater Quality of Agartala, Tripura for Drinking Employing GIS and MCDA Techniques

Santanu Mallik, Shivam and Umesh Mishra

Abstract Groundwater is an essential part of our life support system and a change in the quality of groundwater may affect human health, mapping of groundwater contamination is an important tool for groundwater protection and management. The present study figures out groundwater quality suitability for drinking by adopting Geographical Information System (GIS) and Multi-Criteria Decision Analysis (MCDA) Techniques. The Analytical Hierarchical Process (AHP) is employed to find the weightage of various parameters and reclassification along with weighted overlay tools in ArcGIS software is used. A total 18 samples of groundwater were collected from different wards of Agartala Municipality and various water quality parameters of collected samples were tested in the laboratory. It was found that only 24.24% of AMC area is having potable groundwater and 75.76% of AMC area is not potable for drinking and iron content in more than 96.25% of area is above permissible limit prescribed by BIS (2012) and WHO (2011) and for pH 22.09% of the area having value less than permissible limit.

Keywords Groundwater quality · GIS · MCDA

1 Introduction

Diversity in the level of education, awareness, poverty, practices, and socioeconomic development, especially, in the rural part of our country, access to potable drinking water has become an immense challenge which leads to poor water management. Groundwater is one of the most used raw materials for a wide range of human activity

S. Mallik (✉) · Shivam · U. Mishra
Department of Civil Engineering, National Institution of Technology, Barjala
799046, Tripura (W), India
e-mail: Coolshan02@gmail.com

Shivam
e-mail: Sonishivamsoni@gmail.com

U. Mishra
e-mail: umishra123@rediffmail.com

J. K. Ghosh and I. da Silva (eds.), *Applications of Geomatics in Civil Engineering*,
Lecture Notes in Civil Engineering 33, https://doi.org/10.1007/978-981-13-7067-0_22

and the quality of groundwater has been greatly affected by different anthropogenic activity [1]. Therefore, bundles of researches were carried out by several researchers with regard to groundwater quality and management in India [2–7]. Since the quality of groundwater is varied with location and habits of the population, therefore there is no standard method useful for drinking water management throughout India. The Geographic Information System is a very impressive tool for groundwater management and pollution study since it offers an overlay mapping techniques to identify an association between them. There are a handful of literature reported for management of water quality mapping analysis using GIS platform [8–13]. Interestingly most of the research are concentrated in few cities of India. However, very less research work focuses on the northeastern part of our country for analysis of groundwater quality. Therefore, it is forcefully brought to our attention to work with GIS platform for the above said purpose for our area of interest, i.e., Agartala, Tripura. In addition, the analytical hierarchical process method under Multi-Criteria Decision Analysis offers aggregation of geographical data and providing the best solution with regard to assigned parameters unidirectionally [14, 15]. The AHP technique analyzes the multiple datasets in a pair-wise comparison matrix and helps to find out the weightage for selecting the best alternative. The AHP not only prioritizes elements but also a group of elements as it is often necessary [16].

2 Study Area

The study area (Agartala, Tripura, India) lies between north latitude $23^{\circ}45'$ and $23^{\circ}55'$ and east longitude $91^{\circ}15'$ and $91^{\circ}20'$. The climate is commonly hot and misty with the average maximum temperature of 35 °C and the average minimum of 10.5 °C. The state receives fairly good annual rainfall around 233 cm/annum. The Agartala Municipal Council has an area of 76.50 km^2 as per census (2011). The study area is shown below (Fig. 1).

3 Material and Methods

3.1 Sample Collection and Analysis

A total of 18 samples were collected from different wards of Agartala Municipal Corporation and the station location was taken by hand GPS during August (Monsoon). Water samples were collected in polyethylene bottles (1L capacity) and stored at 4 °C before analysis. pH was measured in the site and sample was analyzed for iron (Fe), chloride (Cl^-), sulfate ($SO_4{}^{2-}$), alkalinity as bicarbonate ($HCO_3{}^-$), nitrate (NO_3^{-1}), calcium (Ca^{++}), and magnesium (Mg^{++}) as per standard methods suggested

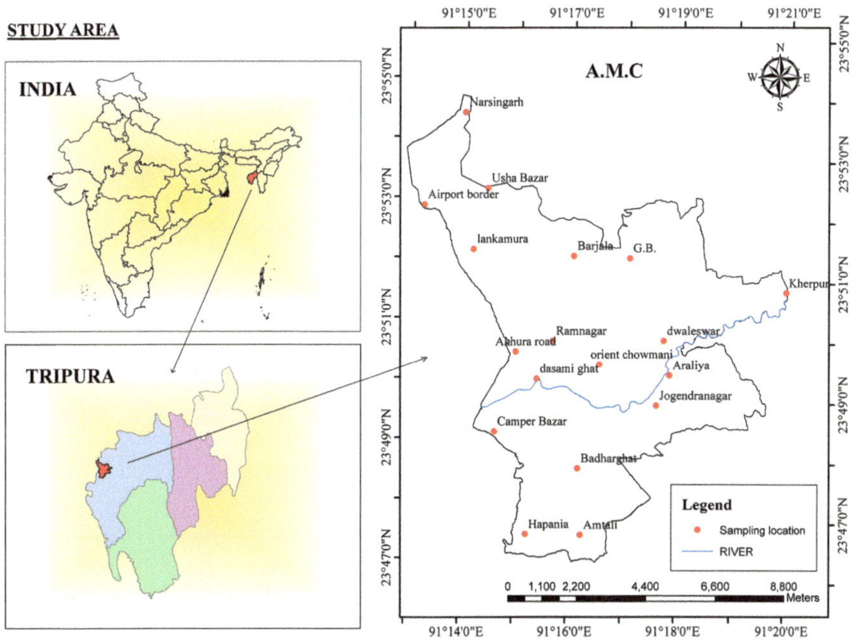

Fig. 1 Study area

by AHPA 1995 [17]. The values were compared with World Health Organization, WHO, (2011) and BIS (IS 10500:2012).

3.2 Integration of Spatial and Attribute Database

The water quality data were imported in Arc Map (10.2.2) and it was integrated with spatial data, i.e., sample location. The spatial distributed map was prepared using the Inverse Distance Weighted (IDW) interpolation method, shown in (Figs. 2, 3, 4, 5, 6, 7, 8 and 9) [17], [18]. Though there are a number of spatial modeling techniques available with reference to the application in GIS, this method is utilized because of the interpolated value predicated upon their distance from the output point, thereby generate a surface grid as well as thematic isolines. In IDW method, the weightage is the function of the distance between the point of interest and sampling station [19].

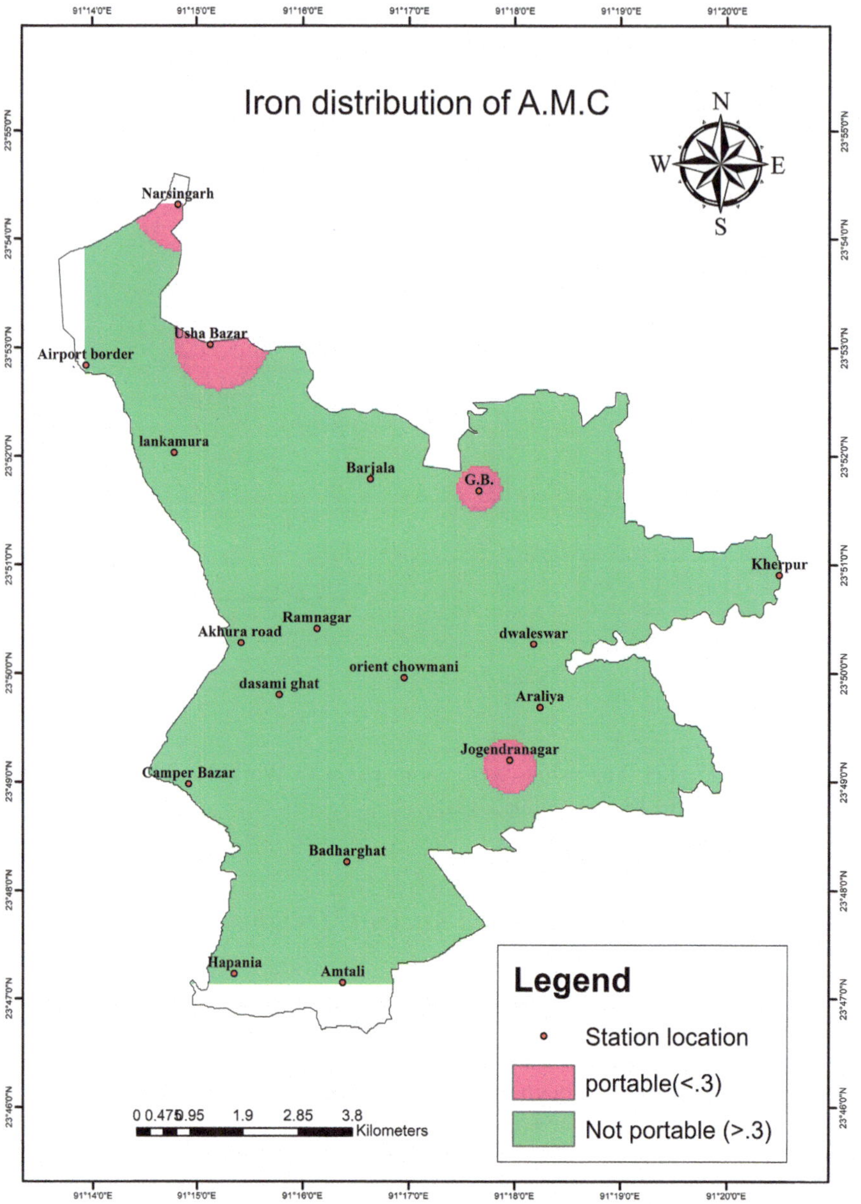

Fig. 2 Iron map

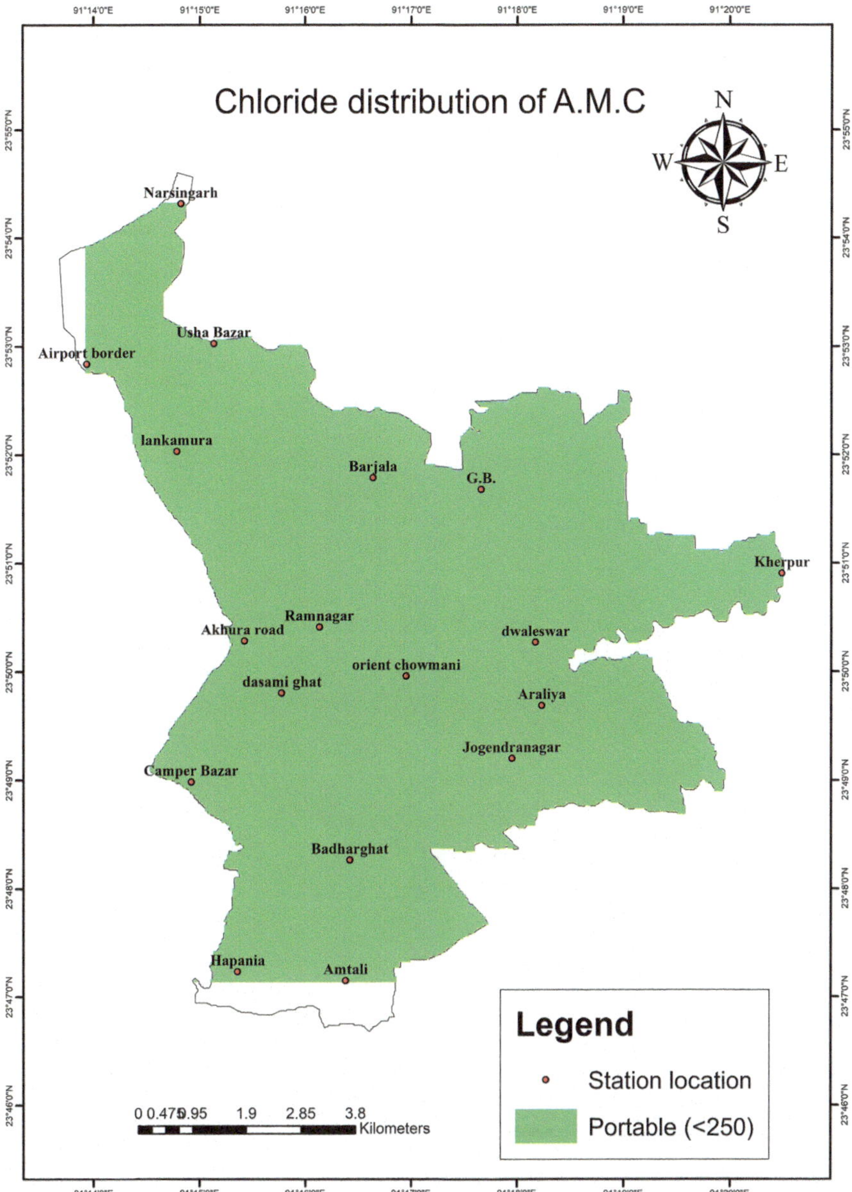

Fig. 3 Chloride map

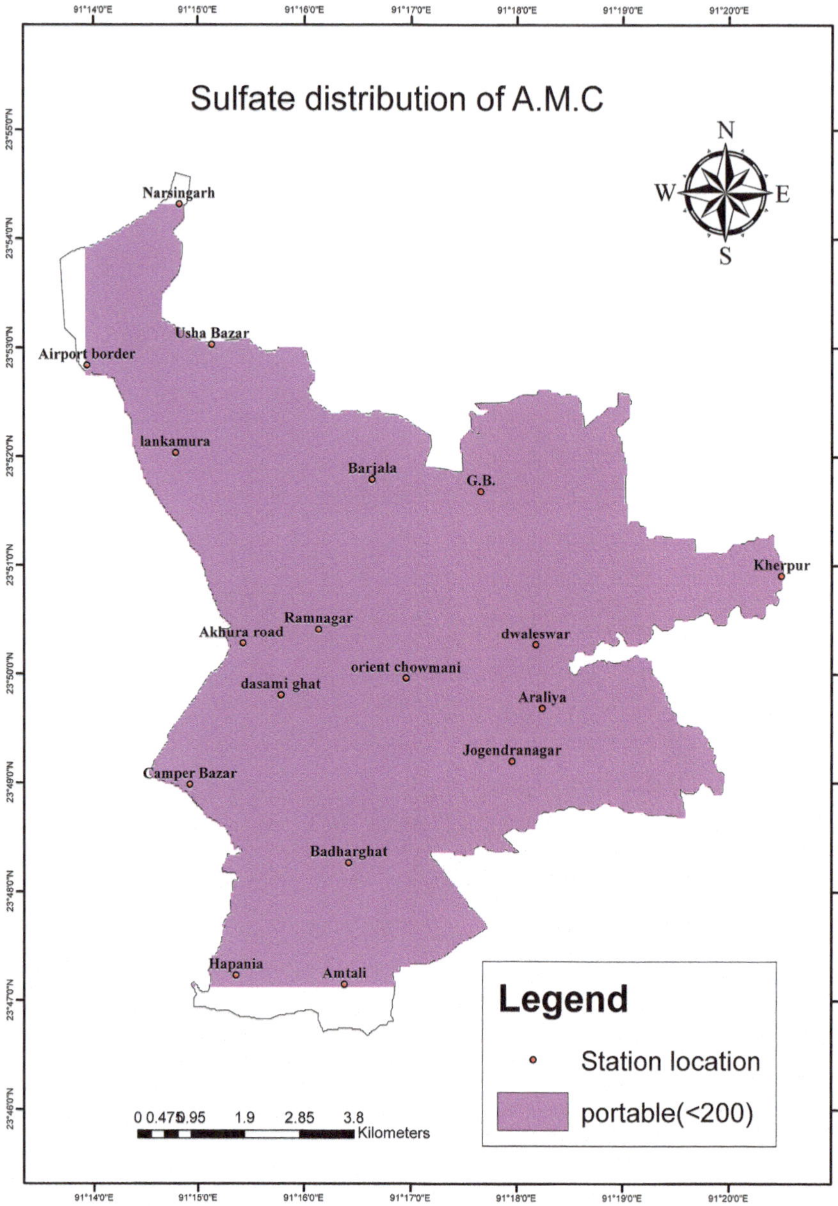

Fig. 4 Sulfate map

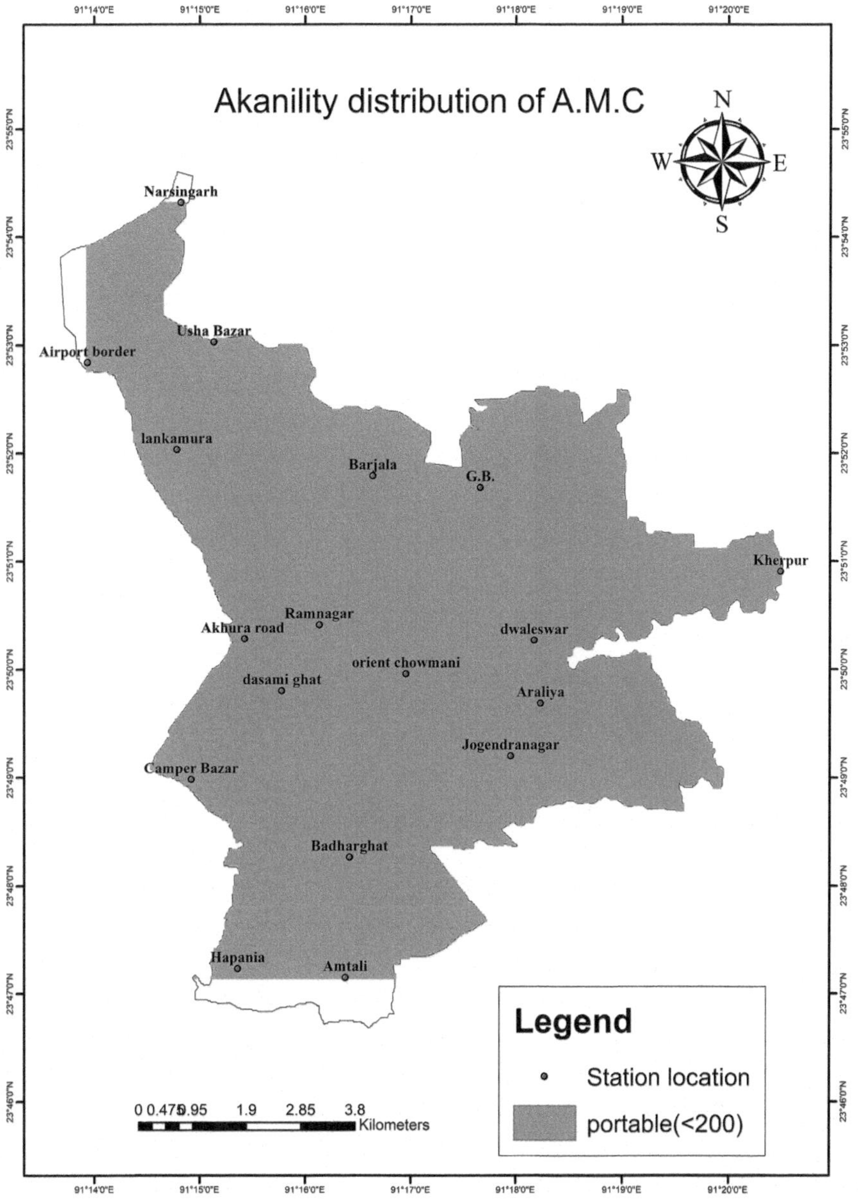

Fig. 5 Alkalinity map

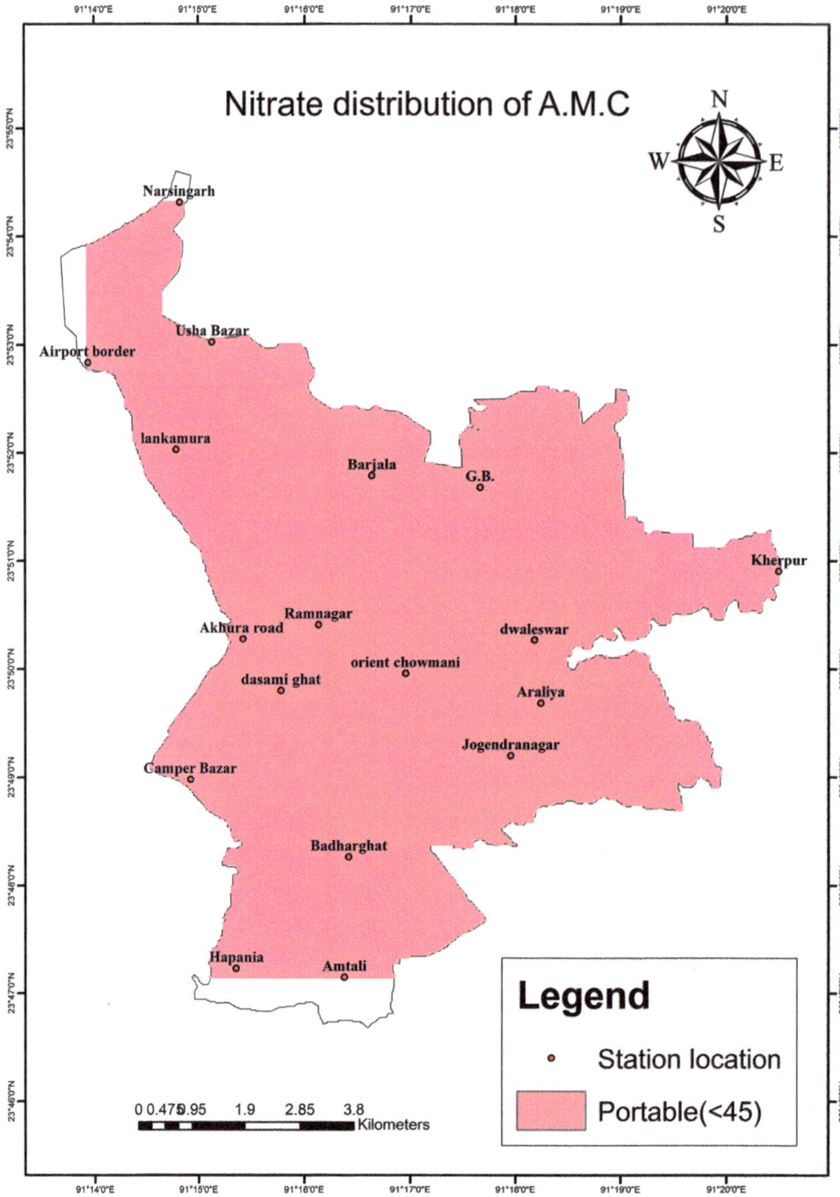

Fig. 6 Nitrate map

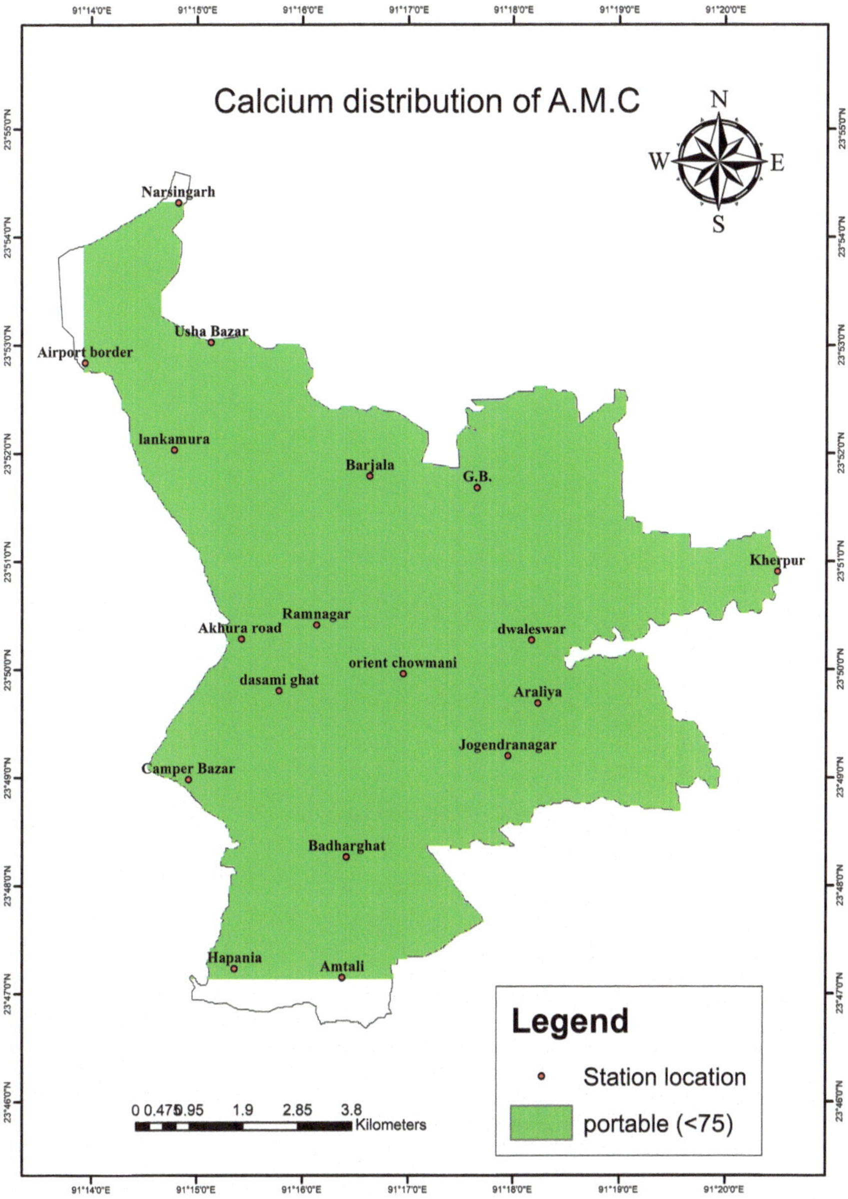

Fig. 7 Calcium map

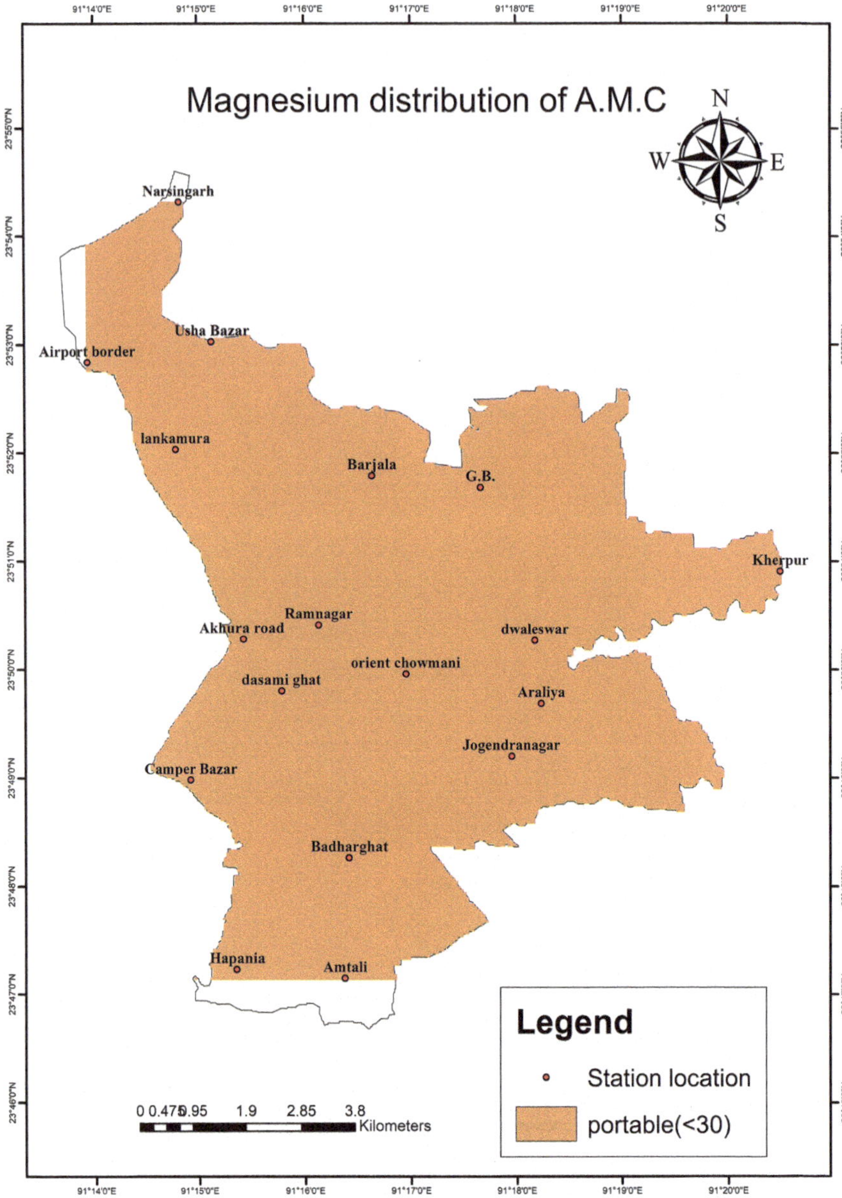

Fig. 8 Magnesium map

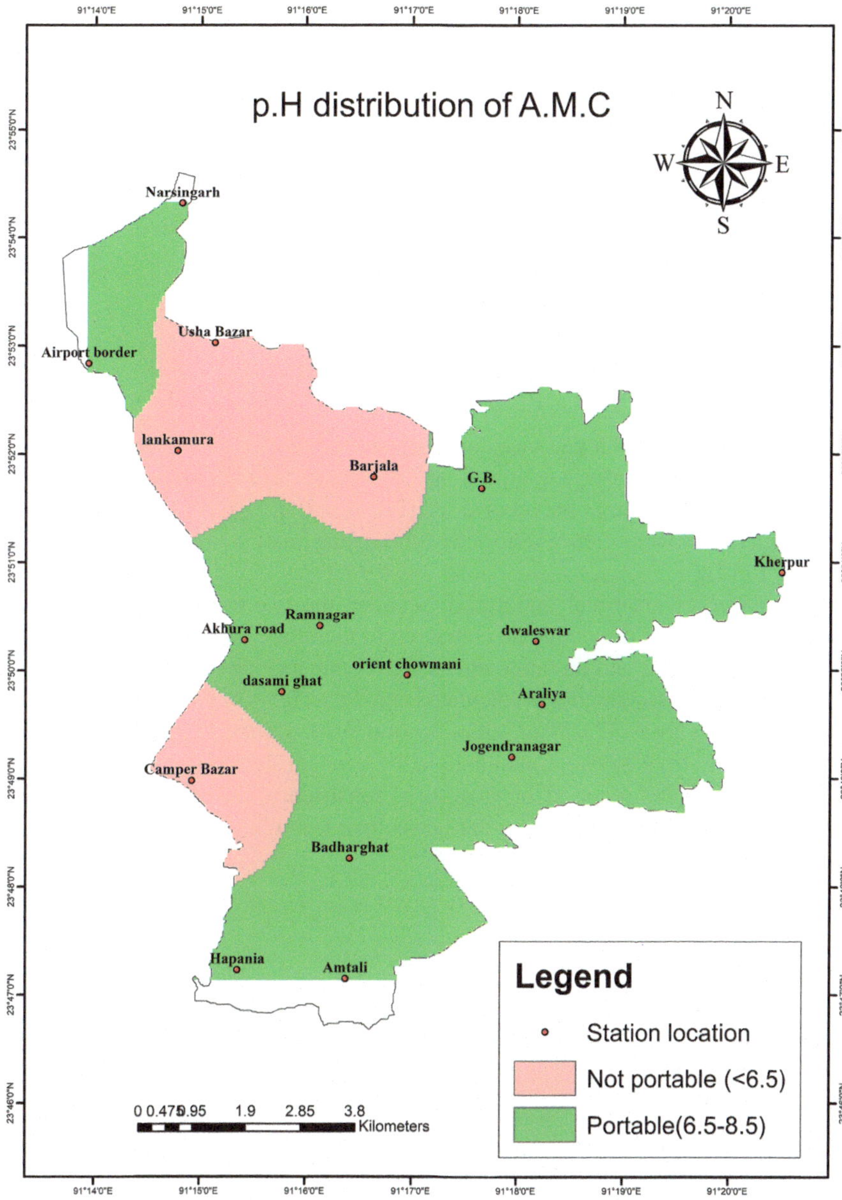

p.H distribution of A.M.C
N
W
E
S
Narsingarh
Usha Bazar
Airport border
lankamura
Barjala
G.B.
Kherpur
Ramnagar
Akhura road
dwaleswar
orient chowmani
dasami ghat
Araliya
Jogendranagar
Camper Bazar
Badharghat
Hapania
Amtali
Legend
Station location
Not portable (<6.5)
Portable(6.5-8.5)
0 0.475 0.95 1.9 2.85 3.8
Kilometers

Fig. 9 pH map

3.3 Multi-criteria Decision Analysis (MCDA)

MCDA method can give the most favorable result, where the confusion is related to choosing one option out of a cluster of options, criteria were ranked on the options reverence to the different purposes [20]. MCDA methods can be classified into either multi-attribute and multi-objective methods; these methods are mainly concerned with possibilities of joining many criteria to form optimum evaluation criteria [21, 22].

3.4 Weight from AHP for Overlay Analysis

Saaty (1978) developed a pairwise comparison method [15]. To construct a matrix AHP method gives pairwise comparisons. It takes as associate input the pairwise comparisons and produces the relative weights as output [14]. The values in accordance with relative importance were determined using Saaty's 1–9 scale shown below in Table 1 [15].

Here, 1 represents equal importance between the two criteria and 9 indicates the acute importance of one theme compared to the other. Expert suggestions from various organizations and literature are used to obtain a pairwise comparison matrix [23–26]. Criterion weighted was calculated and the weights of the different criterion were as follows iron (37.077%), chloride (4.635%), sulfate (5.285%), Alkalinity (12.312%), nitrate (9.269%), Calcium (7.415%), magnesium (5.46%), pH (18.539). Saaty (2004) [27] has reported the measure of consistency, called consistency index, (CI) which can be calculated using the subsequent formula.

$$\text{Consistency index (Cl)} = \frac{\lambda_{\text{max}} - \text{n}}{\text{n} - 1} \quad (1)$$

Table 1 Saaty's 1–9 scale of importance and their explanation

Scale of importance	Explanation
1	Equal importance
2	Weak
3	Moderate importance
4	Moderate plus
5	Strong importance
6	Strong plus
7	Very strong importance
8	Very, very strong
9	Extreme importance

Table 2 Saaty's RI value for corresponding "n"

N	1	2	3	4	5	6	7	8	9
RI	0	0	0.58	0.89	1.12	1.24	1.32	1.41	1.45

where n = number of elements (i.e., 8) and λ_{max} = average value of the consistency vector. (CR) = Consistency ratio [23]

$$\text{Consistency Ratio (CR)} = \frac{\text{CI}}{\text{RI}} \quad (2)$$

where RI = Random index. The different value of RI for the corresponding "n" is shown in Table 2 [28].

If the obtained CR value is less than 0.1, then there is an adequate amount of uniformity in the matrix formed and the value obtained is acceptable. If CR is more than or equal to 0.10, then subjective judgment must be revised [16, 22, 27, 29, 30]. The CR computed for this study is 0.01 which is less than 0.1.

4 Result and Discussions

The statistical analysis of groundwater samples for the monsoon season is presented in Table 3. It has been observed that the average value of iron is found to be 1.29 (mg/l), which is much higher than the prescribed limit given by BIS (2012) and WHO (2011). However, maximum nitrate value was found as 44.9 (mg/l) in G.B. area, which is maximum in comparison with other sampling sites. The high amount of nitrate at GB location is due to the dumping of hospital waste nearby the sample site. Moreover, this value is within the prescribed limit given by BIS (2012) and WHO (2011). This is fascinating to note that till date there is no report of nitrate found in the above experimental site. Since the maximum limit of nitrate is 45 mg/l, hence, 44.90 mg/l in GB area must be monitored routinely as it is in the alarming condition.

Table 3 Statistics of the 8 groundwater parameters used in the present study

	Min	Max
Calcium	1.0	36.0
Magnesium	0.1	12.1
Alkalinity	5.0	140.0
Chloride	0.0	15.0
Iron	0.2	3.9
Sulfate	0.0	49.3
Nitrate	0.0	44.9
pH	4.1	6.8

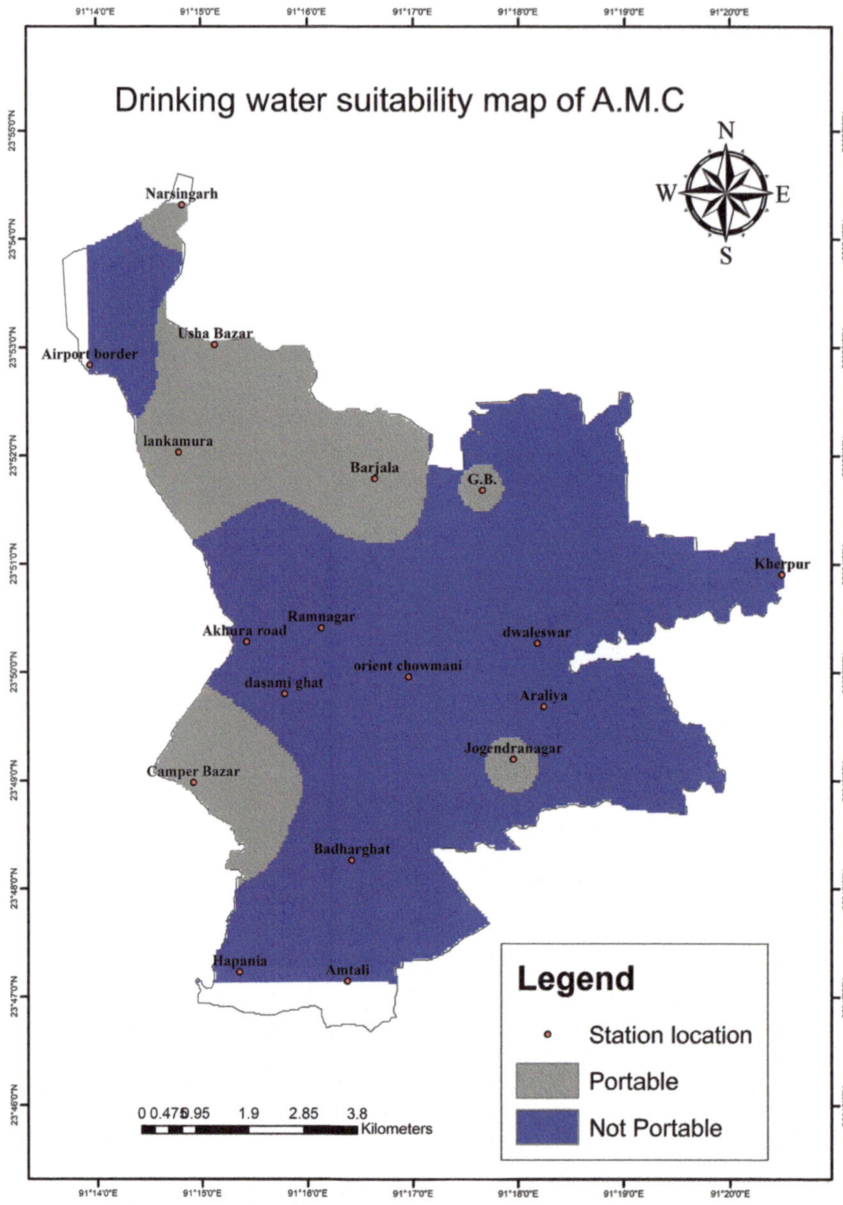

Fig. 10 Drinking water suitability map

This study illustrated the use of GIS and AHP techniques to determine the suitability of groundwater for drinking purpose for the study area using overlaying 8 hydrochemical parameter maps which are reclassified as limits prescribe by BIS (2012) and WHO (2011). This final map in Fig. 10 shows that 24.24% of AMC area is considered as portable and 75.76% of the area is found as not portable for drinking. The higher amount of iron content above permissible limit prescribed by BIS (2012) and WHO (2011) was found in 96.25% of total AMC area as shown in (Fig. 2). This may be due to the presence of red sandy lateritic soils in most part of Agartala, Tripura. The aeration and iron exchange method may be considered as a possible remedial tool for iron. On the other hand, 22.09% of A.M.C area shows a lower value of pH which is illustrated in Fig. 9.

5 Conclusions

Groundwater is a very important resource of all the living creatures on the earth. The present study presents the spatial approach toward the characterization of physiochemical parameters. In the study, it has been found that pH and iron are the two most vulnerable parameters. The nitrate near the G.B. area is almost reaching the desirable limit defined by the BIS (2012) and WHO (2011). The groundwater quality suitability map reveals a substantial amount of area comes under the non-portability zone for drinking, which must be taken care of. Consequently, the water may be treated before consumption. It is further suggested that more samples can be analyzed in various seasons to get more idea about the groundwater quality of the study area. The remote sensing can be used for change detection analysis which will be of help for planning and management of groundwater utilization and treatment.

References

1. Adhikary PP et al (2010) Assessment of groundwater pollution in West Delhi, India using geostatistical approach. Environ Monit Assess 167(1–4):599–615
2. Prasanth SS et al (2012) Evaluation of groundwater quality and its suitability for drinking and agricultural use in the coastal stretch of Alappuzha District, Kerala, India. Appl Water Sci 2(3):165–175
3. Krishan G et al (2016) Water quality index of groundwater in Haridwar district, Uttarakhand, India. Water Energy Int 58(10):55–58
4. Haldar D et al (2016) Assessment of water quality of Damodar River in South Bengal region of India by Canadian Council of Ministers of Environment (CCME) water quality index: a case study. Desalin Water Treat 57(8):3489–3502
5. Ramakrishnaiah C, Sadashivaiah C, Ranganna G (2009) Assessment of water quality index for the groundwater in Tumkur Taluk, Karnataka State, India. J Chem 6(2):523–530
6. Kumar S et al (2010) Quality assessment of potable water in the town of Kolasib, Mizoram (India). Environ Earth Sci 61(1):115–121
7. Kumar M et al (2007) A comparative evaluation of groundwater suitability for irrigation and drinking purposes in two intensively cultivated districts of Punjab, India. Environ Geol 53(3):553–574

8. Latha SP, Rao NK (2010) Assessment and spatial distribution of quality of groundwater in zone-II and III, greater Visakhapatnam, India using water quality index (WQI) and GIS. Int J Environ Sci 1(2):198
9. Babiker IS, Mohamed MA, Hiyama T (2007) Assessing groundwater quality using GIS. Water Resour Manag 21(4):699–715
10. Selvam S et al (2014) GIS-based evaluation of water quality index of groundwater resources around Tuticorin coastal city, South India. Environ Earth Sci 71(6):2847–2867
11. Machiwal D, Jha MK, Mal BC (2011) Assessment of groundwater potential in a semi-arid region of India using remote sensing, GIS and MCDM techniques. Water Resour Manag 25(5):1359–1386
12. Nas B, Berktay A (2010) Groundwater quality mapping in urban groundwater using GIS. Environ Monit Assess 160(1):215–227
13. Machiwal D, Jha MK, Mal BC (2011) GIS-based assessment and characterization of groundwater quality in a hard-rock hilly terrain of Western India. Environ Monit Assess 174(1–4):645–663
14. Malczewski J (1999) GIS and multicriteria decision analysis. Wiley & Sons
15. Saaty TL (1978) The analytic hierarchy process, NY. McGraw-Hill, USA. In: Cook WD, Seiford LM (eds) Priority ranking and consensus formation, Management Science, 24 (1980), pp 1721–1732
16. Saaty TL (1999) Fundamentals of the analytic network process. In: Proceedings of the 5th international symposium on the analytic hierarchy process
17. Association, A.P.H., Water Environment Federation. Standard methods for the examination of water and wastewater, 22, 1995
18. Robinson T, Metternicht G (2006) Testing the performance of spatial interpolation techniques for mapping soil properties. Comput Electron Agric 50(2):97–108
19. Sethy SN, Syed TH, Kumar A (2017) Evaluation of groundwater quality in parts of the Southern Gangetic Plain using water quality indices. Environ Earth Sci 76(3):116
20. Aher P, Adinarayana J, Gorantiwar S (2013) Prioritization of watersheds using multi-criteria evaluation through fuzzy analytical hierarchy process. Agric Eng Int CIGR J 15(1):11–18
21. Malczewski J (2004) GIS-based land-use suitability analysis: a critical overview. Prog Plan 62(1):3–65
22. Jhariya DC et al (2017) Assessment of groundwater quality index for drinking purpose in the Durg district, Chhattisgarh using geographical information system (GIS) and multi-criteria decision analysis (MCDA) techniques. J Geol Soc India 89(4):453–459
23. Jhariya D et al (2016) Assessment of groundwater potential zone using remote sensing, GIS and multi criteria decision analysis techniques. J Geol Soc India 88(4):481–492
24. Kumar A et al (2015) Spatial and temporal variability of surface ozone and nitrogen oxides in urban and rural ambient air of Delhi-NCR, India. Air Qual Atmos Health 8(4):391–399
25. Chowdhury A et al (2009) Integrated remote sensing and GIS-based approach for assessing groundwater potential in West Medinipur district, West Bengal, India. Int J Remote Sens 30(1):231–250
26. Yang C et al (2008) Location selection based on AHP/ANP approach. In Industrial Engineering and Engineering Management, 2008. IEEM 2008. IEEE International Conference on 2008: IEEE
27. Saaty TL (2004) Decision making—the analytic hierarchy and network processes (AHP/ANP). J Syst Sci Syst Eng 13(1):1–35
28. Kumar T, Jhariya DC (2015) Land quality index assessment for agricultural purpose using multi-criteria decision analysis (MCDA). Geocarto Int 30(7):822–841
29. Dalalah D, Al-Oqla F, Hayajneh M (2010) Application of the analytic hierarchy process (AHP) in multi-criteria analysis of the selection of cranes. Jordan J Mech Ind Eng 4(5), 2010
30. Agarwal E et al (2013) Delineation of groundwater potential zone: An AHP/ANP approach. J Earth Syst Sci 122(3):887–898

Analysis of Water Quality Parameters and Their Variation for Surface Water Using GIS-Based Tools

Rajat Chatterjee and Dilip H. Lataye

Abstract This study is aimed at evaluation of water quality parameters for the surface water taken from Futala Lake, Nagpur. The analysis of variation in 11 water quality parameters (including pH, turbidity, alkalinity, hardness, chlorides, sulphates, dissolved oxygen (DO), total dissolved solids (TDS), nitrates, chemical oxygen demand (COD) and biochemical oxygen demand (BOD)) was studied using Geographical Information Systems (GIS). The lake basin is subjected to an ecological imbalance due to idol immersions, rendering the water unusable for drinking and other purposes. Sampling was done between September and October 2017. Percentage difference for all the parameters was calculated, which showed a significant alteration from natural water quality. Kriging interpolation was used to represent the distribution of the parameters across the lake. The pre-immersion and immersion phases were compared for each parameter (change detection). The areas that are most affected and least affected spatially by idol immersion were found out and buffers were created to represent their definitive spatial extent. The areas found to be most affected by immersion pollution should be prioritized and treated more frequently during the immersion period. Lesser affected areas can be treated with lower frequency and dosages of treatment reagents.

Keywords GIS · Water quality parameter · Kriging interpolation · Buffer · Change detection

1 Introduction

'Water' is a natural resource of great importance and is the wealth of a country. Lakes act as sources of water for a variety of purposes such as irrigation, drinking, recreation, etc. In a lot of regions in India, lake water is the only dependable source of

R. Chatterjee (✉) · D. H. Lataye
Department of Civil Engineering, Visvesvaraya National Institute of Technology, Nagpur 440010, Maharashtra, India
e-mail: believe.rach@gmail.com

D. H. Lataye
e-mail: dhlataye@civil.vnit.ac.in; diliplataye@rediffmail.com

J. K. Ghosh and I. da Silva (eds.), *Applications of Geomatics in Civil Engineering*, Lecture Notes in Civil Engineering 33, https://doi.org/10.1007/978-981-13-7067-0_23

water on which almost all activities of the area are based. Further, lakes play a major role in maintaining the hydrological, ecological and environmental balance of the region. Any change in lake water both in quality and quantity will certainly hamper the development of the area. The pollutants affecting lake water can be broadly classified as decomposable organic matter, toxic chemicals and inert sediments, all of which affect the aquatic life of the lake adversely [1].

The lake considered in our study is the Futala Lake in Nagpur, Maharashtra, which is subjected to pollution from idol immersions after the festivities of Ganesh and Durga Puja. These idols are created with plaster of Paris, clay, and are adorned with paint-related items such as varnish, watercolours, etc. When the idols are immersed in the water, these metallic paints, clay, etc. dissolve slowly in water, leading to significant alteration cum degradation in the water quality. The high amount of nutrients and oxygen depletion lead to eutrophication of the lake, and death of its underwater flora and fauna. Major materials used for making idols contain Plaster of Paris (POP), i.e. gypsum (three quarters), paper and minor quantities of acids, starch and wax. Colours used are mostly chemical natured and contain heavy metals such as Cr, Cd, Pb and Hg. POP is not easily soluble in water but gradually precipitates to the lake bed over a long period of time. Metallic rods and bamboo sticks are used to hold the structure in place. The reaction of POP with water is exothermic, leading to heat production, which may burn human skin on contact. Calcium sulphate or POP is a neutral salt, thus leading to no change in pH. Calcium and magnesium concentration in water increases significantly leading to an intensification in the hardness of water. Note that, plaster of Paris ($2CaSO_4\ H_2O$) on mixing with water gives gypsum ($CaSO_4\ 2H_2O$), which is a paste that hardens over time when placed into the external mould made for the idol.

The main objectives of this study are: (i) to assess the physicochemical properties of the lake water by sampling and testing of the water quality parameters; (ii) to represent variation of these parameters using GIS-based tools of interpolation and overlay analysis; (iii) To propose a two-phase treatment scheme after demarcation of the critical regions of the lake in need of immediate treatment.

2 Study Area

The city of Nagpur, Maharashtra, found to be the centre of the country, houses a number of lakes. Of these lakes, the Futala Lake has emerged as the most desirable recreational location for the local public and tourists alike. The lake rests between the latitude 21.153°N, and longitude 79.045°E, and is spread across 60 acres. It has always been subjected to pollution due to idol immersion. The water from Futala Lake is used for irrigation, but during the period of September–October each year, the water becomes unsuitable for use.

2.1 Sampling Stations

Sampling was the first step of the study. The collection of representative water samples was done at regular intervals for water quality analysis. The focus here was solely concentrated on the impact of idol immersions on water quality. The sampling was done between September and October 2017, the period before and during Durga Puja when maximum idol immersions take place.

Water samples from three stations, strategically located at lake extremities along the periphery, were collected for analysis. Station 1, has inflow to the lake from nearby surroundings plus immersions; station 2, was at the zone of maximum immersions; and station 3, at the zone of minimum immersions. These stations represent different immersion intensities and provide an assessment of the overall surface water status within the boundary of the lake (Fig. 1).

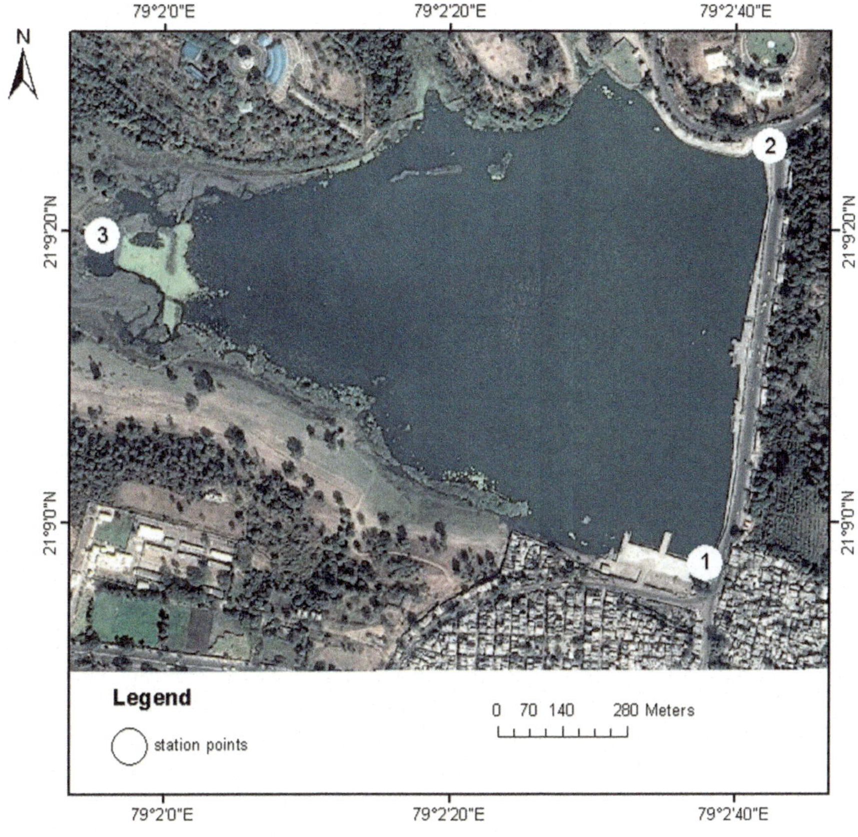

Fig. 1 Location of Futala Lake, Nagpur, Maharashtra

3 Methodology

3.1 Determination of Parameters

Samples that were collected from the three stations were analysed within 24 h of sample collection since it is the desirable period for testing. During sampling, the bottles were labelled to avoid misidentification and were rinsed with double distilled water several times, and then filled to the top to minimize the entrapment of air in water samples. Samples were stored at 4 °C in the refrigerator. The variation of 11 water quality parameters, i.e. pH, turbidity, alkalinity, hardness, chlorides, sulphate, dissolved oxygen, total solids, nitrates, COD and BOD were calculated for the period before (September 2017) and during Durga Puja (October 2017). The adverse effects of these parameters in excessive amounts have been exhaustively discussed in the literature. The experimentations were conducted in accordance with IS 3025:1987 (Methods of sampling and testing for water and wastewater) [2]. The methodology, equipment used and standards for analysis of various parameters have been listed in Table 1. After the tests were conducted, the results were compared with the standards set by the government. These standards were, IS 10500:2012 (drinking water specifications) [3] and the Environment (Protection) Rules (EPR), 1986, Schedule VI (effluent discharge standards) [4]. It was noted that COD and BOD are not specified in drinking water standards, IS 10500:2012 because they should be zero. COD and BOD values were thus referred to from EPR, 1986 and WHO drinking water standards [5].

3.2 Interpolation

The GIS interpolation function was used to depict the distribution of water quality parameters for the waters of Futala Lake. Kriging interpolation, a geostatistical interpolation technique has been found to be best suited for surface waters, as it provides a model to estimate the unknown value as a linear combination of adjacent observation values [6]. Kriging interpolation conserves the unique regional variability of the data while preparing a raster layer. Also, directional variations of parameters are best represented by this method. Inverse Distance Weighted (IDW) interpolation may also be used for the same purpose, but fails due to its deterministic approach, whereas kriging uses a probabilistic model for a parameter variation estimate [7].

The spatial analyst toolbar of ArcGIS 10.1 houses the necessary tool required for kriging. A two-dimensional approach was undertaken in data representation instead of a three-dimensional approach because the data collected was solely for the surface variation of parameters and the depth aspect was not considered.

First, a base map cum satellite image of the lake was procured from Google Earth (for the period September 2017) and imported into ArcGIS. Georeferencing was accomplished by using 6 ground control points. WGS 1984 was used as the geo-

Table 1 Methods, equipment used and standards for analysis of various parameters

S. No.	Parameter	Method of examination	Equipment	Standards (IS 10500:2012 and EPR, 1986)	
				Acceptable limit	Permissible limit
1	pH	Electrometric	pH meter	6.5–8.5	No relaxation
2	Alkalinity (mg/L of $CaCO_3$)	Titrimetric	Titration	200	600
3	Hardness (mg/L of $CaCO_3$)	Complexometric method	Titration	200	600
4	Chloride (mg/L)	Argentometric method	Titration	250	1000
5	Total dissolved solids (mg/L)	Oven drying	Hot air oven	500	2000
6	Dissolved oxygen (mg/L)	Winkler's method	Titration	6	–
7	Sulphate (mg/L)	Turbidimetric method	Spectrophotometer	200	400
8	Turbidity (NTU)	Nephelometric	Turbidity meter	1	5
9	Nitrates (mg/L)	Chromotropic acid method	Spectrophotometer	45	No relaxation
10	COD (mg/L)	Closed reflux	Digester apparatus	250 (EPR)	–
11	BOD (mg/L)	Winkler's method	Titration	3 (WHO), 30 (EPR)	6 (WHO)

graphical coordinate system. Shapefiles were created to mark the sampling stations and the lake extent. Ordinary Kriging (OK) interpolation was then used to predict the profile of variation for all 11 water quality parameters. The interpolated polygons were then clipped to the lake extent by using Data Management Tools → Raster → Clip. The reshaped rasters were then colour coded and classified based on the method of equal intervals which divides the range of attribute values into subranges of equal size. Using the generated maps, a critical zone or buffer was demarcated for each water quality parameter based on a 10% interval around the maximum value of that parameter. The critical zone was created to demarcate the region of immediate attention for treatment. Each of the parameter critical zones were then collectively analysed by using raster overlay method from the analysis toolbox. Union and Intersect functions were used for final critical zone prediction.

4 Results and Discussions

The main objectives of this study were (i) to assess the physicochemical properties of the lake water by sampling and testing of the water quality parameters [8] and (ii) to represent variation of these parameters by using GIS-based tools of interpolation and overlay analysis and come up with the critical region of the lake in need of immediate treatment. The quality parameters determined in the previous section were tabulated as shown in Table 2, and a graphical representation of the mean percentages is shown in Fig. 2. Negative values of % difference indicate fall in the magnitude of the parameter. The suitability of the surface water from Futala Lake for drinking and domestic purposes was evaluated by comparing the values of different water quality parameters with those prescribed in the standards. The higher concentration of some parameters is due to heavy pollution load due to idol immersion activities during festival season (October) resulting in deterioration of the natural water body.

4.1 *Parametric Analysis*

pH. A pH ~ 7 in the pre-immersion period indicates near neutral to slightly alkaline conditions in the lake. $[H^+]$ plays an important role in the biological processes of the lake. Variation in pH in collected water samples was from an average of 7.5 in the pre-immersion period (within acceptable limits) to 6.4 during immersion (lower than acceptable limits). The acidity is caused due to the algal bloom. This causes

Table 2 Comparison of water quality parameters before and during the immersion phase

Station	1		2		3		%Difference
Parameters	Before	During	Before	During	Before	During	
pH	7.6	6.4	7.4	5.9	7.4	6.9	−14.28
Alkalinity (mg/L of $CaCO_3$)	251.2	553	190	637	220	491	154.23
Hardness (mg/L of $CaCO_3$)	161	279	149	326	145.8	292	96.79
Chloride (mg/L)	62.7	120.3	55.1	132.9	47.6	118	124.42
TDS (mg/L)	819.4	1643	738	1824	630	1186	112.71
DO (mg/L)	5.6	2.99	6.16	2.71	6.42	4.09	−46.14
Sulphate (mg/L)	55.5	58.2	42.7	56.8	41.4	54	21.06
Turbidity (NTU)	2.7	15.2	1.6	19.3	1.2	12.1	747.27
Nitrates (mg/L)	2.4	16	1.2	28	1.7	15	1013.20
COD (mg/L)	200	745	190	925	150	665	332.40
BOD (mg/L)	14	40	12	45	9	35	242.85

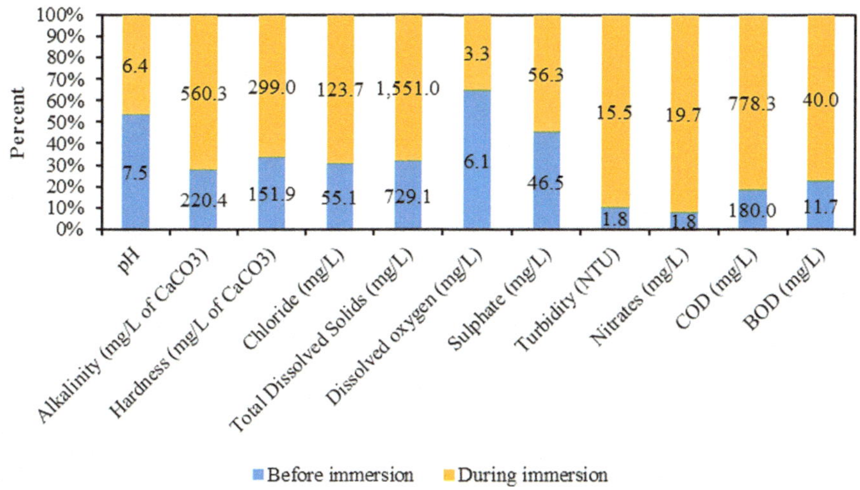

Fig. 2 Representation of mean changes in parameter values before and during immersion

the production of domoic acid in the latter half of the day. When the algae dies, the nutrients consumed by them, which include sulphates and nitrates are released back into the water, lowering the pH.

Alkalinity. Lower values of alkalinity during the pre-immersion period may be due to the usage of CO_2 by phytoplankton for growth, as well as the effect of drainage water discharged into the Futala Lake. Alkalinity increases due to the accumulation of alkali-forming metals in the lake water during immersion. Alkalinity results in a bitter taste of water and also precipitate formation with the use of chemicals. Alkalinity changes from an average of 220.4 mg/l pre-immersion (nearly acceptable) to 560.3 mg/l during immersion (not acceptable, but near to permissible), resulting in a 154% rise in its value. Kriging interpolated layers showed that while the alkalinity was least at station 2 before immersion, it spiked to the maximum value for the lake during the immersion period. Station 1 had the highest alkalinity pre-immersion due to inflow from surrounding areas. This is shown in Fig. 3a, b.

Hardness. Hardness in water is a well-documented nuisance caused mostly by dissolution of metal ions of Ca^{2+}, Mg^{2+}, and rarer Sr^{2+}, Fe^{2+}, and Mn^{2+}. Additionally, some anions (SO_4^{2-}, Cl^-, NO_3^- and SiO_3^-) expectedly exist in water. Interior lakes are predominantly affected by calcium and bicarbonate. Hardness in the collected water samples varied from 152 mg/l pre-immersion (acceptable) and 299 mg/l during immersion (not acceptable, but permissible). This is 96% increase in the values, which is nearly double of the initial values. From kriging maps, we see that pre-immersion, station 1 has the maximum hardness of water, and station 2 has one of the least values. During immersion, the scenario reverses and station 2 has most hardness, while station 1 has one of the lower values. This variation is shown in Fig. 3c, d.

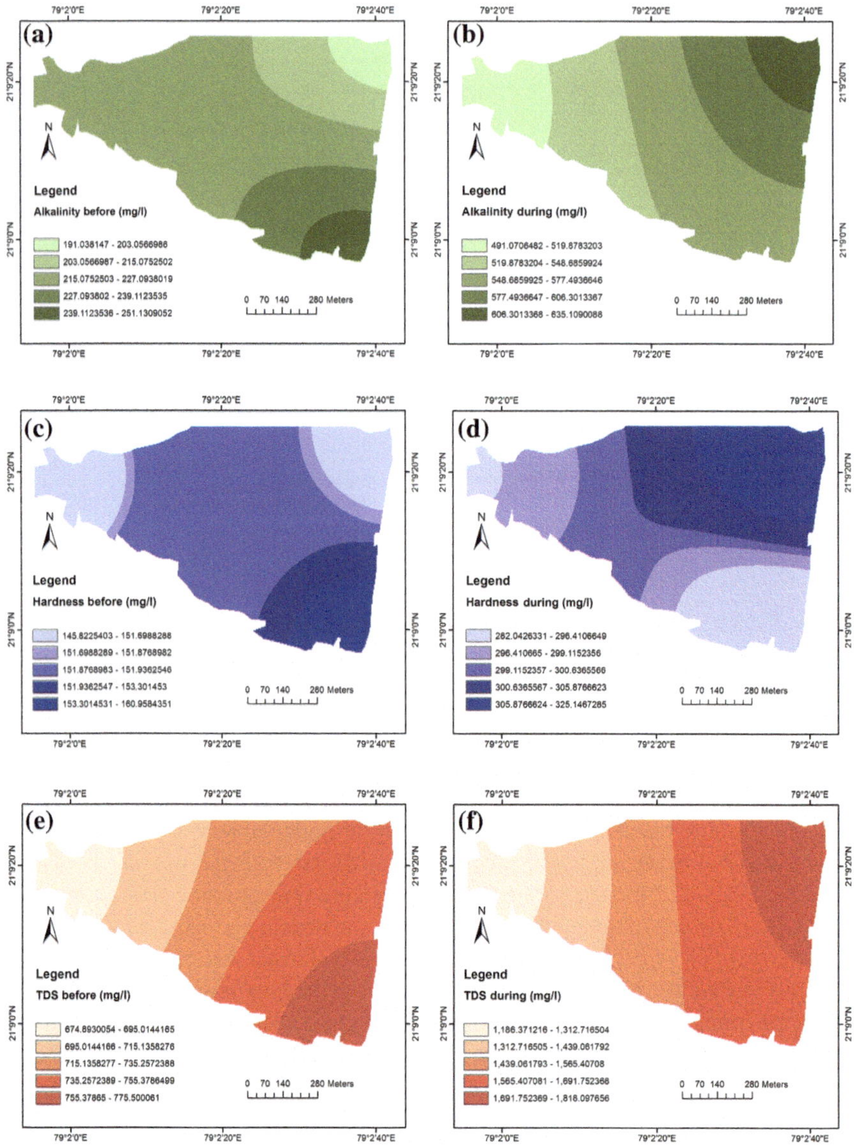

Fig. 3 Variation of parameters **a** alkalinity before immersion; **b** alkalinity during immersion; **c** hardness before; **d** hardness after; **e** TDS before; **f** TDS during; **g** DO during; **h** turbidity during; **i** BOD before; **j** BOD during; **k** COD before; **l** COD during

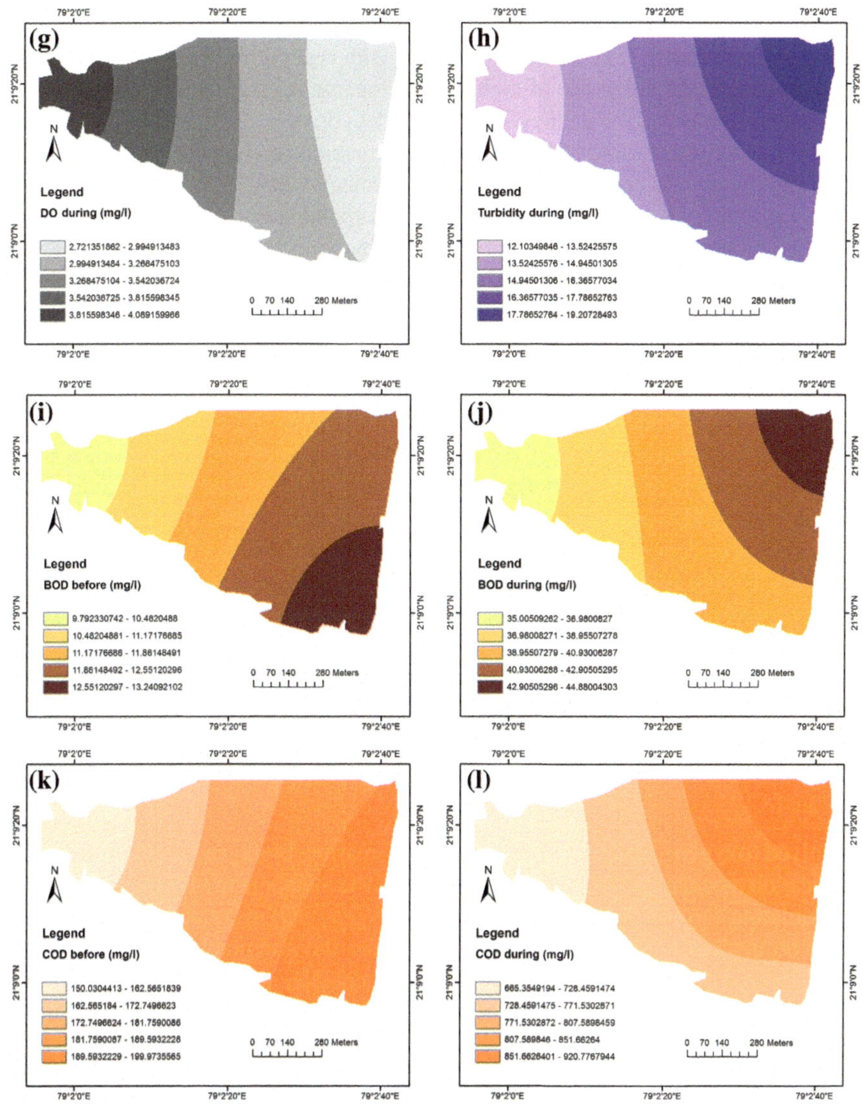

Fig. 3 (continued)

Chlorides. Chloride concentration spike seems to alter the reproduction rate of fish and upsurge mortality. It also causes problems with vegetal respiration. Chlorides in water elevate due to the intake of high organic loading from animal-related sources. In humans, it may lead to kidney failure and dehydration. Variation in chloride in collected water samples was 55.13 mg/l pre-immersion (acceptable) to 123.73 mg/l during immersion (acceptable). This is an incremental change of 124.42%, but the lake is still free from chloride contamination. Interpolated layers showed that station 1 had the highest chloride concentration pre-immersion and the least values during immersion. Station 2, had medium concentration peaked in chloride content during immersion.

Total dissolved solids. The high concentration of total dissolved solids (TDS) leads to the demise of fish by gill clogging. It increases lake temperature and blocks photosynthesis in underwater saplings. It also points to the salinity of water and is a pointer for rapid plankton growth in addition to sewage contamination, both of which were seen here. TDS chiefly consist of inorganic salts and a small amount of organic matter. TDS was found to be higher during immersion. The high values of TDS are due to deposition of inorganic salts and organic matter during immersion activities. TDS in collected water sample was 729 mg/l in the pre-immersion period (not acceptable, but permissible), and 1551 mg/l during the immersion period (lesser than the permissible limit). An average increase of 112% was observed between pre-immersion and during immersion. Interpolation showed that the TDS profile remained nearly identical before and during immersions, except for a rise in values. The values peak in the regions around station 1 and 2 and gradually decline till station 3. Lime should be added to the pond water for settling the solids. This is shown in Fig. 3e, f.

Dissolved oxygen. Oxygen is invaluable to marine biota and crucial for self-purification mechanisms of the lake. It facilitates the metabolism of all aquatic organisms utilizing aerobic respiration. Dissolved Oxygen (DO) in collected water sample was 6.06 mg/l in the pre-immersion period (near the acceptable limit of 6 mg/l), and 3.26 mg/l during immersion period (not acceptable, not permissible). An average decrease of 46% was observed between pre-immersion and during immersion, which is expected due to the increase in the loading on the lake. Kriging results showed that DO during immersion was dangerously low (2.8 mg/l) along the right shore of the lake, along stations 1 and 2. DO values gradually recover to 4 mg/l till station 3 is reached. The variation for the immersion period is shown in Fig. 3g.

Sulphates. The entire marine ecosystem requires sulphur for its normal functioning. Excess sulphur, however, serves as a bane and accelerates algal boom. Assimilation of sulphur in a marine environment takes place as sulphates but during the breakdown of carbon-based matter, sulphur is reduced to hydrogen sulphide, which is oxidized briskly. Sulphates do not limit the growth and distribution of marine wildlife. Sulphate concentration in the collected water sample was 46.53 mg/l in the pre-immersion period (acceptable) and 56.33 mg/l during the immersion period (acceptable). An average increase of 21% was observed between pre-immersion and during immersion, implying that the lake is still free from sulphate contamination. Spatially, sulphate concentration was the highest at station 1 pre-immersion. During immersion though, a constant value of nearly 55 mg/l was maintained throughout the lake.

Turbidity. Many ornamental offerings and decorations adorn the idols. From petals and leaves to clay, both organic and inorganic materials are discharged carelessly into the lake. Turbidity hits spawning beds of fish and results in the culling of larvae. The colour of the water was completely disturbed during immersions causing high turbidity. Turbidity in the collected water sample was 1.83 NTU in the pre-immersion period (not acceptable, but permissible), and 15.53 NTU during the immersion period (not acceptable, not permissible). A drastic average increase of 747.3% was observed between pre-immersion and during immersion. Station 1 was most affected pre-immersion with 2.7 NTU and station 2 during immersion with 19.3 NTU. Station 3, on the other hand, was nearly safe from turbidity pre-immersion with 1.2 NTU. The variation for the immersion period is shown in Fig. 3h.

Nitrates. Nitrates reach lakes from numerous natural sources and owing to human actions like food production, crop cultivation and careless dumping of domestic and industrial sewage. Nitrates contaminate the waters due to excessive usage of nitrogenous fertilizers in farmlands. In cities, sewage high in nitrate content, plus in our case idol immersions lead to pollution of lakes, encouraging the growth of algae and phytoplankton that further increase the nutrient in the lake leading to eutrophication. Nitrate concentration in collected water sample was 1.76 mg/l in the pre-immersion period (acceptable), and 19.67 mg/l during immersion period (acceptable), indicating that the lake is still free from nitrate contamination. As expected, kriging implied highest pollution levels for station 1 pre-immersion, but even this value was within tolerable limits. During immersion station 2 was found to have a concentration of 28 mg/l, also within acceptable limits.

Biochemical oxygen demand. The values of BOD and Chemical Oxygen Demand (COD) are comprehensive indicators for gauging the extent of fouling of water by organic wastes. According to WHO drinking water standard, BOD should not exceed 6 mg/l. Also, according to EPR 1986 effluent standards, BOD should be lesser than 30 mg/l in surface waters like a lake. BOD in collected water sample was 11.67 mg/l in the pre-immersion period (not acceptable according to WHO drinking water standards, acceptable according to EPR 1986 effluent standards), and 40 mg/l during immersion period (not acceptable according to WHO drinking water standards, not acceptable according to EPR 1986 effluent standards). This implies that the water of Futala Lake is not suitable for drinking even in the pre-immersion period, but given time, the self-purification action of the lake might have kicked in. A spike of 242.85% was observed. It has been represented in Fig. 3i, j.

Chemical oxygen demand. Toxic environments with the undoubted occurrence of biologically resistant organic substances can be estimated well by analysing the value of COD. COD in collected water sample was 180 mg/l in the pre-immersion period (acceptable according to EPR 1986 effluent standards) and 778.33 mg/l during immersion period (not acceptable according to EPR 1986 effluent standards). A rise of 332.41% was observed between the pre-immersion and immersion periods. Both BOD and COD showed similar trends during immersion season after interpolation. As earlier, stations 1 and 2 are more affected and station 3 is least affected. This has been shown in Fig. 3j, k.

4.2 Critical Zone Determination

From the study of parameters, it was clear that (i) pre-immersion: water can be made drinkable after treatment for alkalinity, TDS, turbidity, BOD; and (ii) during immersion: water is not drinkable and needs rigorous treatment for pollution. Treatment needed for pH, alkalinity, hardness, TDS, DO, turbidity, BOD.

It was found on visual inspection that while station 3 was the least affected of all stations, which was to be expected due to least number of immersions there. Station 2 seemed to be most affected due to immersions, with its effects reaching up to station 1 that is already affected by inflows from surrounding areas. Station 1 seemed to be most affected in the pre-immersion phase. A concrete region within the lake boundaries was to be determined to study the extent of the lake that was severely affected by pollution. The parameters rendering water unsuitable during the immersion period were considered in preparing a critical zone around the most affected parts of the lake. For the determination of critical zone, the maximum value of required parameters was considered from previous experimentation and interpolation results. A 10% range around the maximum parametric values was found and demarcated spatially for each layer in GIS using buffer analysis. The intersection tool was used to overlay these demarcated buffers upon each other determine the most critical zone of the lake affected by highest pollution. The region obtained has been shown in Fig. 4. This region is in immediate need of treatment and treatment should be commenced even as the immersions are taking place. This phase of treatment was called Phase 1. The area around station 2 was found to be most affected. The overlay analysis was further extended to find the area that is in lesser danger than Phase 1 but needs to be given priority in treatment after the immersion season is completed. Union tool was used for this purpose, and the region that emerged is proposed to be treated in Phase 2 of the treatment programme. It is shown in Fig. 5. Both stations 1 and station 2 are affected in this case. It was seen that nearly 25% of the lake was affected in Phase 1 and 60% in Phase 2.

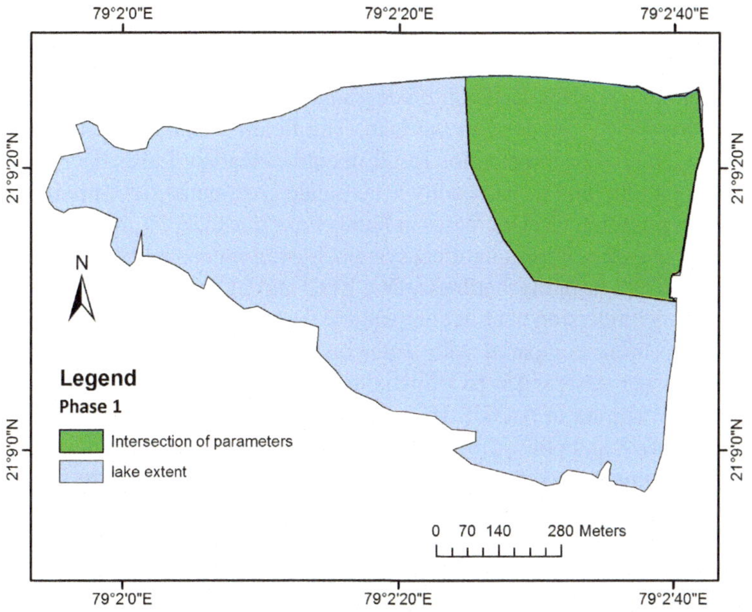

Fig. 4 Overlay analysis showing phase 1

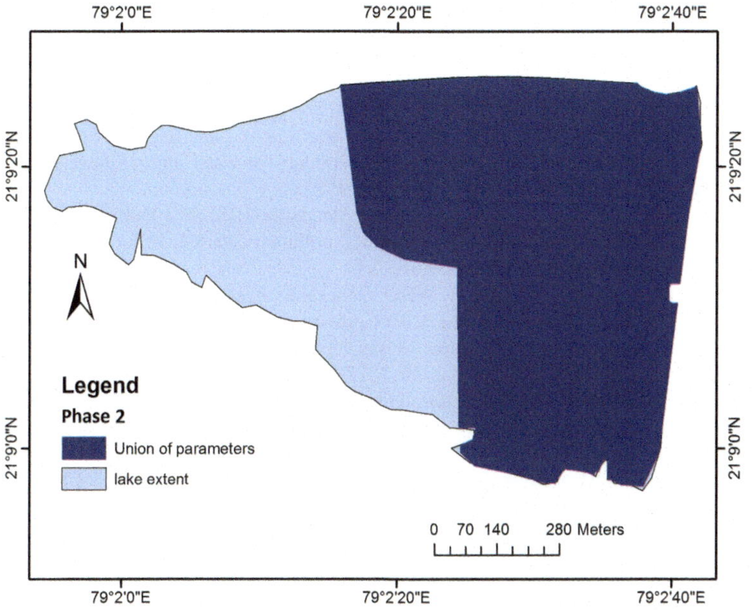

Fig. 5 Overlay analysis showing phase 2

5 Conclusion

The input of biodegradable and non-biodegradable substances deteriorates the water quality and enhances silt load in the lake. The floating materials released during idol immersion after decomposition result in eutrophication. Lakes have a very low self-cleaning ability and hence readily accumulate great quantities of pollutants.

The water pollution level increases in Futala Lake due to idol immersions, causing adverse effects on the entire aquatic ecosystem. Eleven water quality parameters were analysed and TDS, turbidity, hardness, DO, BOD and COD have shown a significant increase during immersion of idols compared to the pre-immersion period. Parameter concentrations were compared with water quality standards prescribed by Indian pollution control boards and WHO. Spatial trends for the parameters were examined using ordinary kriging in ArcGIS 10.1. Critically polluted regions of the lake were found and phase 1 and phase 2 of treatment were determined. Treatment with alum and lime is recommended. A reduction in idol size is suggested with the use of paper pulp, natural gum, food colours and natural coir of trees. These idols are lightweight, close to traditional beliefs, have lesser metal content and easily disintegrate in water. POP idols take a very long time to disintegrate (15 days–3 months), but the eco-friendly idols begin disintegration in roughly 3 h.

References

1. Puri PJ, Yenkie MKN, Rana DB, Meshram SU (2015) Application of water quality index (WQI) for the assessment of surface water quality (Ambazari Lake). Eur J Exp Biol 5(2):37–52
2. Methods of sampling and test (physical and chemical) for water and wastewater. Bureau of Indian Standards (BIS), IS 3015 (various parts) (1987)
3. Drinking water specifications. Bureau of Indian Standards (BIS), IS 10500: 2012 (2012)
4. General standards for discharge of environmental pollutants, Part-A: Effluents. Schedule—VI. The environment (protection) rules (1986)
5. Guidelines for drinking-water quality. WHO, Geneva, Switzerland (2004)
6. Murphy RR, Curriero FC, Ball WP (2010) Comparison of spatial interpolation methods for water quality evaluation in the Chesapeake Bay. J Environ Eng 136:160–171
7. Koko S, Irvine KN, Jindal R, Thondara R (2017) Spatial and temporal variations of dissolved oxygen in Cha-Am municipality wastewater treatment ponds using GIS kriging interpolation. J Water Manag Model 25:C427
8. Sener S, Sener E, Davraz A (2017) Evaluation of water quality using water quality index (WQI) method and GIS in Aksu River (SW-Turkey). Sci Total Environ 584–585:131–144

Assessment of Surface Water Quality Using GIS: Case of Tapi Basin, Surat, Gujarat, India

Divya Lad, Mehali Mehta and Manisha Vashi

Abstract Quality and quantity of water bodies is of prime importance for sustainable development. Tapi basin is the lifeline of Surat city, as it fulfills the majority of its water requirement. Due to the impact of urbanization and industrialization, the water quality of the Tapi basin has been adversely affected. The present study is focussed on assessing water quality of Tapi basin, with the aims: (i) to give an overview of present water quality of Tapi basin, (ii) to determine spatial dissemination of water quality parameters by means of BOD, COD, chloride and TDS, and (iii) to generate map for river water quality in the study area by using GIS. Water quality of the basin was assessed by taking samples from nine locations through the flow path of basin within the assessment area and an interpolation technique, that is, Inverse Distance Weighting (IDW), is used to obtain the spatial dissemination of water quality parameters of the basin. The results indicated that the quality of water is more deteriorated on the downstream side as compared to that of the upstream side. From the assessment of water with GIS and IDW, it is concluded that despite urbanization and industrialization, the quality of surface water is predominately average to unsuitable.

Keywords Inverse distance weighting (IDW) · Water quality · GIS · Tapi river

1 Introduction

Water is the most valuable gift to mankind and the soul of nature. Water is the energetic source that is used for various purposes such as irrigation, drinking, industrial cooling, fish production, energy generation, and several others. Freshwater is only

D. Lad (✉) · M. Mehta · M. Vashi
Department of Environmental Engineering, Sarvajanik College of Engineering and Technology, Dr. R. K. Desai Marg, Opp. Mission Hospital, Athwa Lines, Surat 395001, India
e-mail: divyalad19@gmail.com

J. K. Ghosh and I. da Silva (eds.), *Applications of Geomatics in Civil Engineering*, Lecture Notes in Civil Engineering 33, https://doi.org/10.1007/978-981-13-7067-0_24

just 0.3–0.5% of the total water available on the earth. Rivers are the main resource of water supply for different applications containing industry, drinking, and agriculture. For that reason, observing the quality of these sources owing to the current drought and rural and urban development is one of the main tasks in the arena of the environment [1].

The river water is an essential component playing a vital role in country's development. In India almost 70% of the surface water has been contaminated as domestic and industrial wastewater is discharged into natural sources such as lakes, streams as well as rivers [2].

The quality of surface water is a very sensitive and life-threatening problem in many nations. As people become more aware of the significance of raw water quality to aquatic life and drinking water quality to public health, it is highly desirable to assess the quality of surface water [3]. Surface water resources are dynamic in nature and due to the enlargement of urbanization, industrialization and irrigation activities, it is essential to monitor and protect this important resource [4]. For that water quality assessment is necessary for the safeguard of surface water resources.

2 Study Area

The area under study is the basin of river Tapi basin, which passes through district Surat, Gujarat. The Tapi basin is everlasting tropical basin and one of the main river mouths on the Gulf of Khambhat (Fig. 1). The basin is situated at geographic location Latitude 21° 40′ N and Longitude 72° 40′ E. Its stretch from source to mouth is around 720 km and approximately 1,650,000 km^2 area of a catchment. The river originates in the state of Madhya Pradesh and after its way through Maharashtra; Tapi basin terminates in Gulf of Khambhat of the Arabian Sea. The river along with its way takes industrial wastewater discharged from urban cities like Surat and many chemical and fertilizer industries, both inside and outside the Hazira industrial zone, which originates the main source of pollution into the basin [5].

3 Methodology

3.1 Sample Collection and Analysis

The water quality of the basin was assessed by taking samples from nine locations in November 2017 through the flow path of basin within the assessment area as shown in Fig. 2. Inverse Distance Weighting (IDW) is used to acquire the spatial dissemination of water quality parameters of the basin.

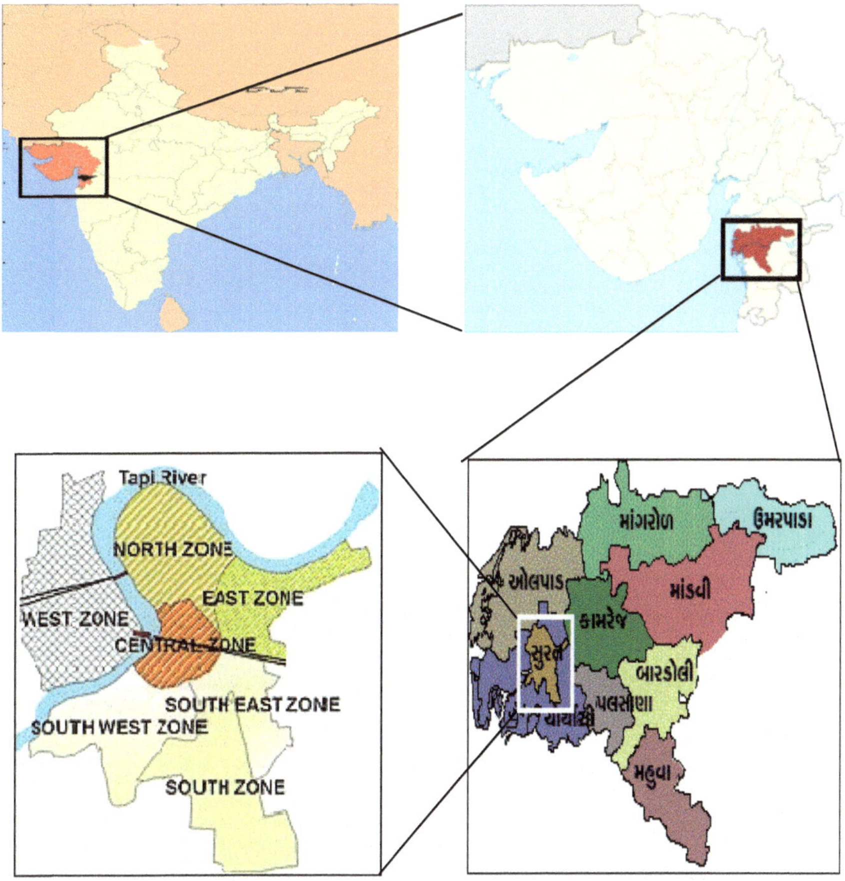

Fig. 1 Map of study area

The samples were analyzed using a standard procedure (APHA1989) [6]. Winkler's method is used for the analysis of BOD. The titrimetric method is used to determine the COD value. The chloride ion concentration is also determined by the titration method. A gravimetric process (APHA 1989) is used for determination of TDS. BOD is varied between 79 and 511 mg/l. COD is ranging from 200 to 5000 mg/l and Chloride is between 141.8 and 930.7 mg/l. TDS is varied between 500.4 and 31956 mg/l.

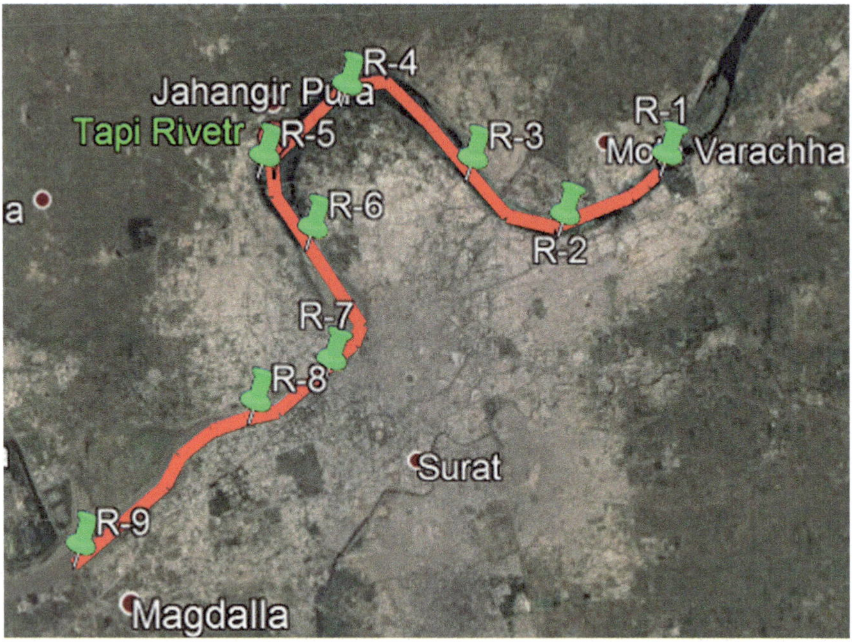

Fig. 2 Sampling location and sampling numbers of Tapi basin in the study area

3.2 GIS Analysis

Using GIS methods with ArcMap, the various thematic maps of BOD, COD, Chloride and TDS concentrations have been generated. The Inverse Distance Weighting (IDW) method has been used as the most appropriate methods and desired parameters were interpolated using these methods. Spatial distribution maps for BOD, COD, Chloride, and TDS have been created for Tapi basin. Drinking water is traced according to the Indian Standard (ISI 1983) and the proposed methodology flowchart is followed up to improve the water quality of the Tapi basin classification map from thematic layers (Figs. 3, 4, 5, 6 and 7). Water quality classification is crucial to assess the suitability for domestic, agriculture or industrial uses.

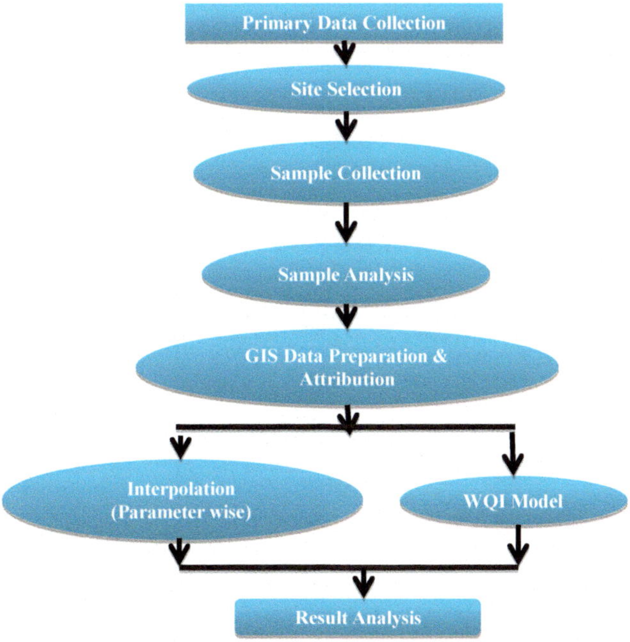

Fig. 3 Flow diagram of the proposed methodology

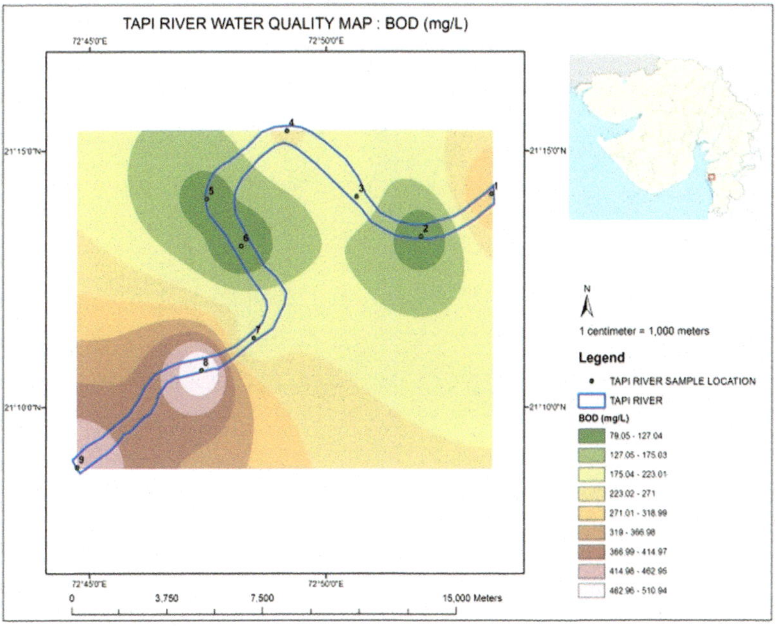

Fig. 4 Distribution of BOD levels in Tapi Basin

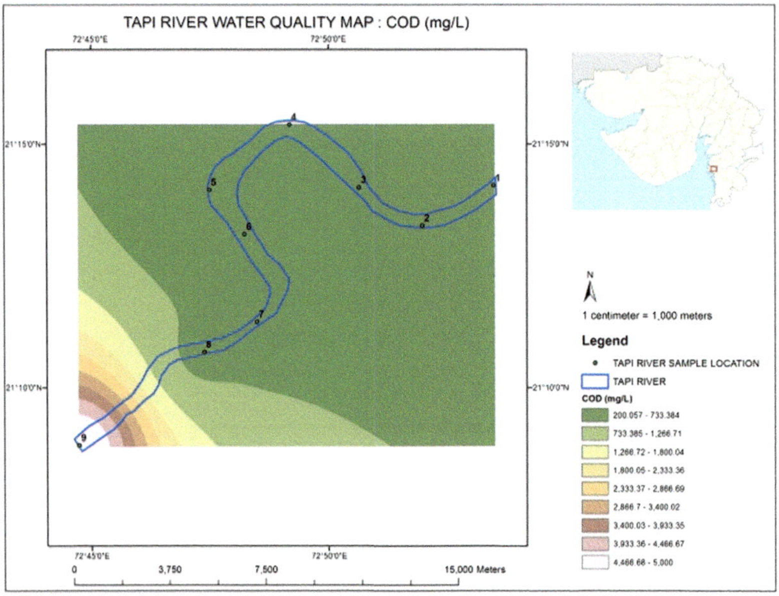

Fig. 5 Distribution of COD values in Tapi Basin

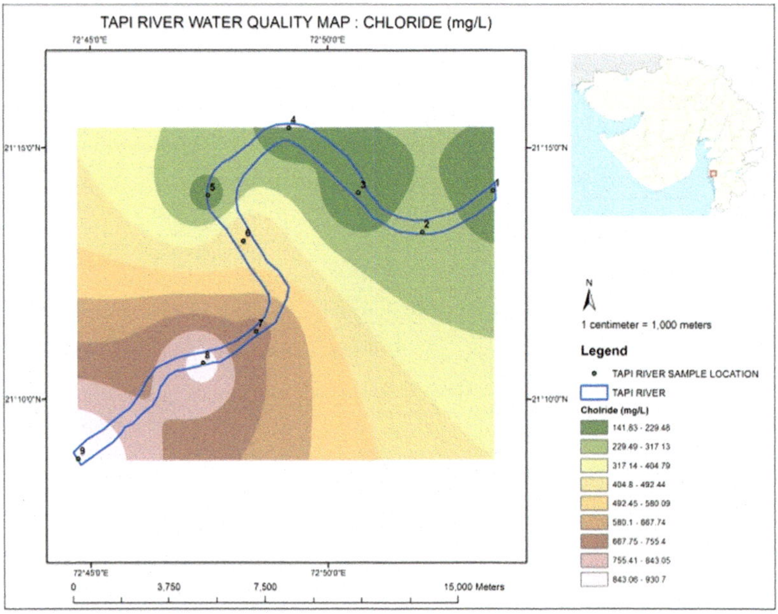

Fig. 6 Distribution of Chloride values in Tapi Basin

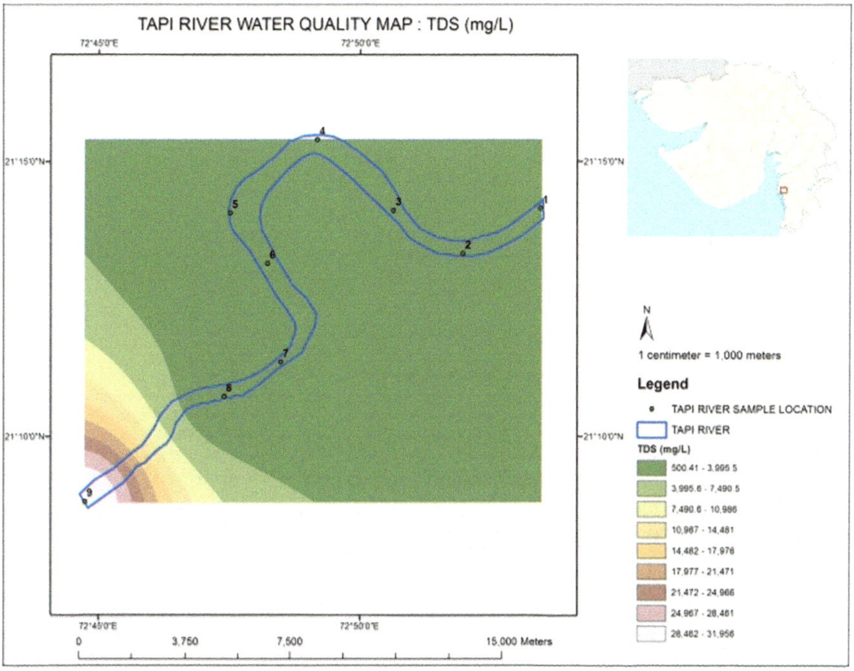

Fig. 7 Distribution of TDS values in Tapi Basin

4 Results and Discussions

In the present study, nine water samples are collected from nine different locations in November 2017 to evaluate the water quality of the Tapi basin. The analytical results of the physicochemical characteristics of the river waters are presented in Table 1 with a basic statistical summary.

The water of the Tapi basin in the study area was found to unsuitable for the direct drinking purpose. It can be treated to meet Indian drinking water standards as per IS: 10500. From Table 1, we can see that the BOD value is ranging between 79 and 511 mg/l, the value of COD is varied between 200 and 5000 mg/l, chloride is ranging between 141.8 and 930.7 mg/l and TDS is ranging between 500.4 and 31956 mg/l.

BOD value is exceeding the permissible limit at all the sampling points as per IS: 10500. COD is very high at point R-9 as 5000 mg/l and COD value is exceeding the permissible limit at all the sampling points except point R-5. Chloride is higher at the point R-9 as 930.7 mg/l which is exceeding the permissible limit as per IS: 10500 and COD remained under the permissible limit at points R-1, R-2, R-3, R-4, R-5, and R-6. TDS value is lower at the point R-4 as 500.4 mg/l which is under the permissible limit and higher at the point R-9 as 31956 mg/l which is exceeding the permissible limit.

Table 1 Physical and chemical parameters of the Tapi Basin

Sr. no.	Location point	BOD (mg/l)	COD (mg/l)	Chloride (mg/l)	TDS (mg/l)
1	R-1	320	320	177.25	541.2
2	R-2	90	301	254.6	520.6
3	R-3	183	280	141.8	536.4
4	R-4	230	280	177.25	500.4
5	R-5	109	200	194.97	507.2
6	R-6	79	440	496.3	542
7	R-7	195	554	679.86	648.9
8	R-8	511	639	875.1	2164.89
9	R-9	454	5000	930.7	31956

The BOD value is low at point R-6 and high at point R-8 as shown in Fig. 4. The distribution map of COD is indicated that the value of COD is higher at point R-9 as shown in Fig. 5. The chloride is higher at point R-9 and lower at point R-3 as shown in Fig. 6. Figure 7 shows that the TDS is not exceeding the permissible limit except point R-9.

5 Conclusion

From the study of water quality assessment considering BOD, COD, Chloride, and TDS, river water cannot be used for drinking purpose directly. Water quality of Tapi basin should be assessed using GIS technology. Four parameters have been selected to represent water quality in the Tapi Basin. Application of GIS technologies in water quality studies has a great role in the environmental management and assessment studies of Rivers which are easier and cheaper than the conventional method. The above study indicated that the quality of water is more deteriorated on the downstream side as compared to that of the upstream side. From the assessment of water with GIS and IDW, it is inferred that due to urbanization and industrialization, surface water quality is predominately average to unsuitable. From the observations during sampling, it was noted that Tapi basin has led to slum developing in urban areas near the river banks disposing the wastewater into the river which is deteriorated the quality of the river basin.

References

1. Mir A, Piri J, Kisi O (2017) Spatial monitoring and zoning water quality of Sistan River in the wet and dry years using GIS and geostatistics. Comput Electron Agric 38–50

2. Ujjania NC, Dubey M (2013) Water quality and pollution status of Tapi river, Gujarat, India. Int J Pure Appl Zool 261–266
3. Sener S, Sener E (2017) Evaluation of water quality using water quality index (WQI) method and GIS in Aksu River. Sci Total Environ 131–144
4. Chatterjee R, Tarafder G, Paul S (2010) Groundwater quality assessment of Dhanbad district, Jharkhand, India. Bull Eng Geol Environ 137–141
5. Kumar N, George B, Kumar RN (2009) Assessment of spatial and temporal fluctuations in water quality of a tropical permanent estuarine system—Tapi, West Coast India. Appl Ecol Environ Res 7(3):267–276
6. Samantray P, Mishra BK (2009) Assessment of water quality index in Mahanadi and Atharabanki rivers and Taldanda Canal in Paradip area, India. J Hum Ecol 153–161

Investigate Groundwater Quality Parameters for Accomplishing Demand of Bhimrad of Surat City

Manisha Desai and Jayantilal Patel

Abstract Water, the basis of analysis for the chemical unknown substance is the essential need of every locality. However, be the important part of human life it is nowadays found in scares due to rapid growth of urbanization and industrialization. The quality of water is usually described by its physical, chemical and biological properties. It is essential to investigate groundwater quality for new urban area Bhimrad of Surat city to accomplish public needs. For this investigation, Groundwater samples were collected from different seven places in the study area during the pre-monsoon and post-monsoon period of the year 2015. Experimental work was carried out for determination of different parameters of water quality. GIS is the best tool for the analysis of water quality assessment. Each of the groundwater samples was examined for physical and chemical parameters like Color, Odor, Turbidity, TDS, pH, Chloride, Hardness, and EC using standard procedures. By comparing all these parameters with the Indian standard it is found that this groundwater quality is not suitable for domestic/drinking purposes. As per GIS investigation, Groundwater quality should be recovered and practiced.

Keywords Groundwater · GIS · Chemical parameters · TDS (Total dissolved solids) · EC (Electric conductivity)

1 Introduction

Groundwater is being the most vital renewable which is extensively distributed and an essential source of water throughout the world. The quality of water is a significant issue that is directly related to human prosperity [1]. Groundwater resources are dynamic so by the expansion of irrigation activities, industrialization and urbaniza-

M. Desai (✉) · J. Patel
Civil Engineering Department, Sardar Vallabhbhai National Institute of Technology, Ichchanath, Surat, Gujarat, India
e-mail: mpvashi@gmail.com

J. Patel
e-mail: jnp@ced.svnit.ac.in

J. K. Ghosh and I. da Silva (eds.), *Applications of Geomatics in Civil Engineering*, Lecture Notes in Civil Engineering 33, https://doi.org/10.1007/978-981-13-7067-0_25

tion, and other factors, it is important to monitor and protect this important resource [2]. Pollution Groundwater may cause poor quality of drinking water. Water quality impost consists of evaluating water quality. The changes in the specific physical, chemical and biological properties of water may cause damage to human and aquatic life [3].

Much research has been done on Geographic Information Systems (GIS) that is a powerful tool for the assessment of water quality [4]. GIS software has greatly benefited natural resources and environmental issues, including groundwater. A typical example of GIS application in groundwater is a site suitability analysis that manages site inventory data to estimate the vulnerability of groundwater. Potential pollution from non-point source pollution, groundwater movement modeling, and modeling of solutes transport and leach out and creation of space by combining groundwater quality assessment models with spatial data decision support systems [5]. This study of groundwater quality map, query, and analyzing the data is done by ArcGIS software.

The purpose of this study is as follows: (1) To provide an overview of the current groundwater quality of study area, (2) clear spatial distribution of water quality parameters, such as pH, TDS, alkalinity, total hardness, residual chlorine, EC and turbidity, (3) that were used to study the quality of groundwater in the area for drinking GIS and Geomatics.

2 Study Area

In the present paper, the investigation of groundwater is focused on the case of Bhimrad, a new urban area of Surat city. Bhimrad is situated on southwest zone of Surat Municipal Corporation. The area of South-West Drainage zone is about 2608 ha and is covered by following drainage boundaries; North: River Tapi, East: South Drainage Zone, South: Mindhola Basin, West: Arabian Sea.

Bhimrad is covered by following boundaries; North: Tapi Basin (10 km away), East: South Drainage Zone, South: Adjoining creek (Moti Khadi) and Industrial zone (Sachin GIDC—8 km away, Bamroli GIDC—9 km away, Pandesara GIDC—9 km away), West: Adjoin State Highway No. -5, Arabian Sea (12 km away). This is shown in Fig. 1.

3 Methodology

In this research, the methodology is described for Groundwater quality assessment status which water quality parameter is analyzed. For that, the samples were collected from seven locations of the study area during pre-monsoon and post-monsoon period of the year 2015. Experimental work should be carried out for that collected samples for analyzed concentration of different parameter of water quality. Each of the

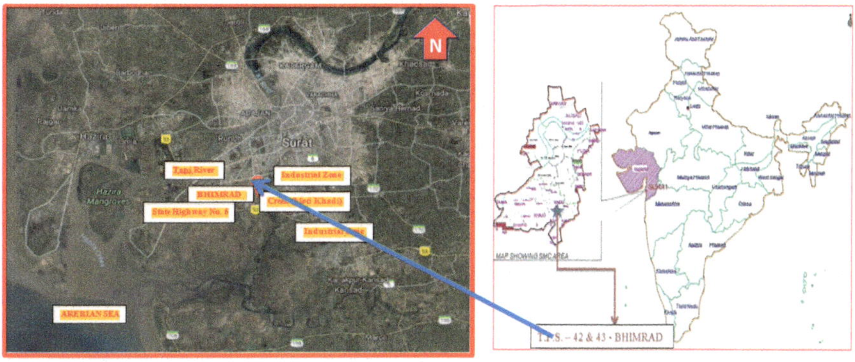

Fig. 1 Location of the study area with boundaries

Table 1 Permissible values of water quality parameters

Sr. no.	Parameters	Permissible value
1	pH	6.5–7.5
2	TDS (mg/l)	500
3	Total hardness (mg/l)	300
4	Chloride (mg/l)	250
5	EC (electric conductivity) (μs/cm)	0–750

Source (BIS 1991) 10500, Reference no. 7

groundwater samples was analyzed for physical and chemical parameters like Color, Odor, Turbidity, TDS, pH, Chloride, Dissolved Oxygen, Hardness, and Alkalinity using standard procedures. Following seven locations were selected for collecting groundwater samples (Fig. 1); (1) Aakash Era, (2) Halpati Niwas, (3) Aakash Avenue, (4) Ashirwad Enclave (Near Khadi), (5) Santosh Height (Near Khadi), (6) Aastha Flats (Near Khadi), (7) Swagat Residency (Near Khadi). This is shown in Fig. 2 and Drinking water standards as per BIS are shown in Table 1.

3.1 pH

The pH rate determined is based on a well-defined scale as that of temperature. The pH of the water cannot be identified by its physical parameter but is measured in terms of concentration. The acidity or alkalinity of the water body falls in the range from 0 to 14 along the logarithmic scale. On the pH scale value above seven indicate acidic nature and below it indicates the basic nature while seven is the neutralization point. The aquatic life is affected by an increase or decrease in the pH of water. Solubility, the toxicity of chemicals, and heavy metals are also affected by the pH

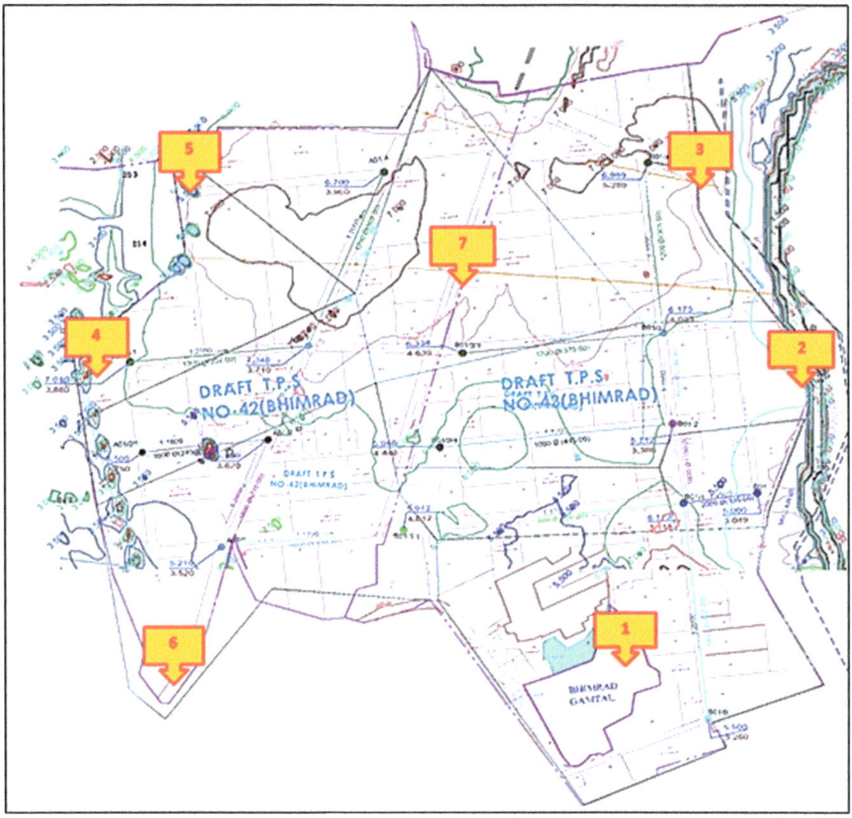

Fig. 2 Location of collecting groundwater samples

value of water. Most aquatic animals prefer a pH range of 6.5–9.0, although some can live in water outside the pH range.

There are many influences that can disturb the pH value of the water, including natural and artificial water. Most natural changes are due to interaction with surrounding rocks (especially carbonate form) and other materials. The precipitation caused by acid rain and wastewater or mining emissions fluctuates with the pH value. In addition, the concentration of carbon dioxide can affect the pH value.

3.2 *Turbidity*

Turbidity is the visual clarity of the water. The physical appearance of the turbid water depends upon the cloudy, dark or other colors. The clarity of the water is reduced by suspended solids and dissolved colored materials which produce opaque, turbid or muddy appearance. Indicators of water quality are based on transparency and estimated total suspended solids in water which serves the common measurement of turbidity.

Turbidity is a measure of how the particles suspended in water affect the transparency of the water. This is an important indicator of sediment and erosion levels. Usually during and after the rain will increase dramatically, which will lead to sediment into the river. Increased turbidity also increases the water temperature, reduces dissolved oxygen, prevents light from reaching aquatic plants, reduces its ability to photosynthesis, and damages fish gills.

3.3 *Total Dissolved Solids (TDS)*

All the inorganic and organic substances contained in the liquid, ionized or particulate are the measures of TDS. The operational definition is that the two microns sized filter paper allows very small particles to get through it. Total dissolved solids are usually discussed on freshwater systems as the salinity includes some of the ions that relate to the TDS definition. The main application of TDS is to study the water quality of streams, rivers, and lakes, although TDS is usually not considered to be a major contaminant, it is generally used as a sign of visual features of drinking water A comprehensive indicator of the existence of a wide range of chemical contaminants.

The vital cradle of TDS is the agricultural and residential runoff, clay-rich landscape, industrial or sewage treatment plant, soil pollution, and point source water pollution emissions. The most common chemical constituents are calcium, phosphate, nitrate, sodium, potassium and chloride, the presence of nutrient runoff, general rainwater runoff and application of road melting salt in snowy weather. The chemical substance may be a cation, an anion, a molecule or an agglomerate above the molecular order of magnitude, as long as the soluble microparticles are formed. TDS is more exotic and harmful factors are caused by surface runoff pesticides. The weathering and dissolution of rocks and soils are the main factors which are responsible for naturally occurring TDS. The United States has established a secondary water quality standard of 500 mg/l to provide portability of drinking water.

3.4 *Total Hardness*

A simple definition of water hardness is the amount of calcium and magnesium dissolved in water. Hard water that dissolves minerals, calcium and magnesium are very high. Hardness is caused by calcium and magnesium compounds and various other metals. The general criteria for water classification are: 0–60 mg/l, because calcium carbonate is classified as soft; 61–120 mg/l, moderately hard; 121–180 mg/l; more than 180 mg/l is difficult.

In hard water, soap reacts with calcium to form a "soap scum". Hard water causes less foaming which gives rise to the need for more soap or detergent to clean, whether it is hand, hair or laundry. Many industrial and domestic water users care about the hardness of their water. When the hard water is heated, for example, in a domestic water heater, a solid deposit of calcium carbonate can be formed. This scale can reduce the life of the equipment, increase the cost of heating water, reduce the efficiency of electric water heaters, rust the pipeline.

3.5 Alkalinity

Alkalinity refers to the ability to neutralize the acidic nature of the aqueous solution of water. This is actually the performance of the buffering capacity. The buffer is a solution that can be added without changing the available H+ ion concentration (without changing pH). It essentially absorbs excess H+ ions and protects water from pH fluctuations. The presence of calcium carbonate or other compounds such as magnesium carbonate will allow carbonate ions to enter the buffer system. Alkalinity is usually associated with hardness, since the main source of alkalinity is usually carbonate (limestone), mainly $CaCO_3$. If $CaCO_3$ actually accounts for most of the alkalinity, the hardness in $CaCO_3$ is equal to the alkalinity. As the hard water contains metal carbonate (mainly $CaCO_3$), it has high alkalinity. In contrast, soft water usually has low alkalinity and cushioning unless the carbonate is associated with sodium or potassium that does not contribute to hardness. Therefore, in general, soft water is more prone to pH fluctuations than acid rain or acid contamination.

Alkalinity is a measure of how much acid can be added to the liquid without causing a great pH change. A higher alkalinity in surface water can relieve acidic wastes such as acid rain to prevent harmful pH changes in aquatic organisms. In order to protect aquatic organisms, the buffer capacity should be at least 20 mg/l. If the alkalinity is naturally low (20 mg/l below), it can be reduced by 25%.

3.6 Chlorides

Chlorides are minerals produced by the combination of chlorine and metal. Some common chlorides include sodium chloride (NaCl) and magnesium chloride ($MgCl_2$). Chlorine salts being amphoteric is highly toxic, which is commonly used as a disinfectant. It is combined with a metal such as sodium to make it a necessity of life (nontoxic compound). Plant and animal life requires a small amount of chloride in normal cell function.

The public drinking water standard requires a chloride content of no more than 250 mg/l. The standard for the protection of aquatic organisms is long-term exposure to 600 mg/l and short-term exposure to 1200 mg/l.

Chlorides can disintegrate metals and affect the taste of food products. Chloride contents are altered at different stages. Therefore, for industrial and water use chloride content is kept at a high level.

4 Results and Discussions

Samples were collected from seven locations in pre-monsoon and post-monsoon months during the year 2015 and different quality parameters like Electrical Conductivity (EC), Total Dissolve Solids (TDS), pH, Chlorine (Cl) and Total Hardness (TH) were found by laboratory experiments. Experimental results and comparison with standard values with GIS analysis are shown Tables 2, 3, 4, 5, 6, 7, and 8 as well as in Figs. 3, 4, 5, 6, 7, 8, 9, 10, 11, and 12.

The above all results and GIS maps/analysis shows that the groundwater compares with Indian Standard for drinkable water as well as pre-monsoon and post-monsoon, it is clear that groundwater quality in the study area is not suitable for public use, it needs to be improved. This result also shows that the groundwater quality is diminutive to improve in post-monsoon due to natural infiltration through previous layers of the study area. Due to urbanization, the impervious layers are supplementary than previous layers and infiltration rate is curtailed. Than better option to infiltrate rainwater into the ground is artificial recharge of groundwater so it may be improved groundwater quality as well as quantity.

Table 2 Pre-monsoon sample test results-2015

Sr. no.	Collection date	Source/location	TDS (mg/l)	pH	Chlorine (Cl) (mg/l)	Total hardness (TH) (mg/l)	EC (μs/cm)
1	05-31-2015	Aakash Era	4513	7.74	0.2	32	4865
2	05-31-2015	Halpati Niwas	3416	8.25	0.4	46	4583
3	05-31-2015	Aakash Avenue	3127	8.11	0.3	48	4320
4	06-01-2015	Ashirwad Enclave (near Khadi)	2896	7.99	0.3	82	4249
5	06-01-2015	Santosh height (near Khadi)	3910	7.92	0.1	74	4106
6	06-01-2015	Aastha flats (near Khadi)	4126	8.1	0.4	24	4598
7	06-01-2015	Swagat residency (near Khadi)	3189	7.47	0.3	40	4320

Table 3 Post-monsoon sample test result-2015

Sr. no.	Collection date	Source/location	TDS (mg/l)	pH	Chlorine (Cl) (mg/l)	Total hardness (TH) (mg/l)	EC (μs/cm)
1	11-23-2015	Aakash Era	4486	7.68	0.22	31	4723
2	11-23-2015	Halpati Niwas	3187	8.2	0.37	45	4478
3	11-23-2015	Aakash Avenue	2956	8.05	0.24	42	4219
4	11-23-2015	Ashirwad Enclave (near Khadi)	2619	7.84	0.1	80	4015
5	11-24-2015	Santosh height (near Khadi)	3649	7.62	0.11	73	3998
6	11-24-2015	Aastha flats (near Khadi)	3973	8	0.3	22	4476
7	11-24-2015	Swagat residency (near Khadi)	2646	7.59	0.23	41	4213

Table 4 Comparative analysis of EC

Sr. no.	Source/location	EC (μs/cm)		
	Indian standard value as per IS: 10500	Standard value	Pre-monsoon value	Post-monsoon value
1	Aakash Era	750	4865	4723
2	Halpati Niwas	750	4583	4478
3	Aakash Avenue	750	4320	4219
4	Ashirwad Enclave (near Khadi)	750	4249	4015
5	Santosh height (near Khadi)	750	4106	3998
6	Aastha flats (near Khadi)	750	4598	4476
7	Swagat residency (near Khadi)	750	4320	4213

Table 5 Comparative analysis of TDS

Sr. no.	Source/location	TDS (mg/l)		
	Indian standard value as per IS: 10500	Standard value	Pre-monsoon value	Post-monsoon value
1	Aakash Era	500	4513	4486
2	Halpati Niwas	500	3416	3187
3	Aakash Avenue	500	3127	2956
4	Ashirwad Enclave (near Khadi)	500	2896	2619
5	Santosh height (near Khadi)	500	3910	3649
6	Aastha flats (near Khadi)	500	4126	3973
7	Swagat residency (near Khadi)	500	3189	2646

Table 6 Comparative analysis of pH

Sr. no.	Source/location	pH		
	Indian standard value as per IS: 10500	Standard value	Pre-monsoon value	Post-monsoon value
1	Aakash Era	7	7.74	7.68
2	Halpati Niwas	7	8.25	8.2
3	Aakash Avenue	7	8.11	8.05
4	Ashirwad Enclave (near Khadi)	7	7.99	7.84
5	Santosh height (near Khadi)	7	7.92	7.62
6	Aastha flats (near Khadi)	7	8.1	8
7	Swagat residency (near Khadi)	7	7.47	7.59

Table 7 Comparative analysis of Cl

Sr. no.	Source/location	Chlorine (mg/l)		
	Indian standard value as per IS: 10500	Standard value	Pre-monsoon value	Post-monsoon value
1	Aakash Era	0.2	0.2	0.22
2	Halpati Niwas	0.2	0.4	0.37
3	Aakash Avenue	0.2	0.3	0.24
4	Ashirwad Enclave (near Khadi)	0.2	0.3	0.1
5	Santosh height (near Khadi)	0.2	0.1	0.11
6	Aastha flats (near Khadi)	0.2	0.4	0.3
7	Swagat residency (near Khadi)	0.2	0.3	0.23

Table 8 Comparative analysis of TH

Sr. no.	Source/location	Total hardness (mg/l)		
	Indian standard value as per IS: 10500	Standard value	Pre-monsoon value	Post-monsoon value
1	Aakash Era	300	450	430
2	Halpati Niwas	300	580	570
3	Aakash Avenue	300	520	510
4	Ashirwad Enclave (near Khadi)	300	810	790
5	Santosh height (near Khadi)	300	740	710
6	Aastha flats (near Khadi)	300	470	460
7	Swagat residency (near Khadi)	300	490	450

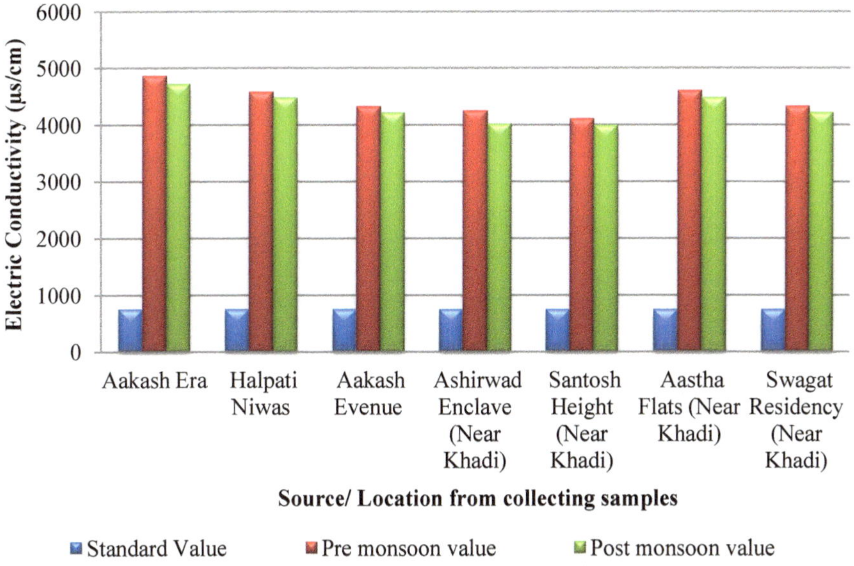

Fig. 3 Comparative analysis of EC

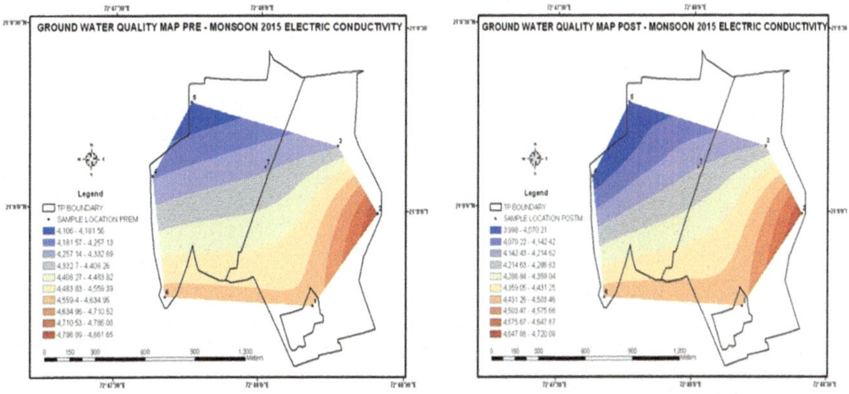

Fig. 4 Test result GWQ concentration of EC for 2015 pre- and post-monsoon

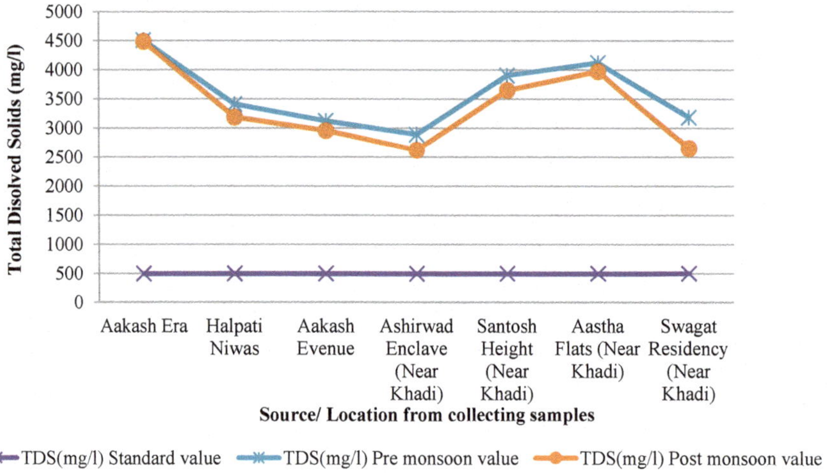

Fig. 5 Comparative analyses of TDS

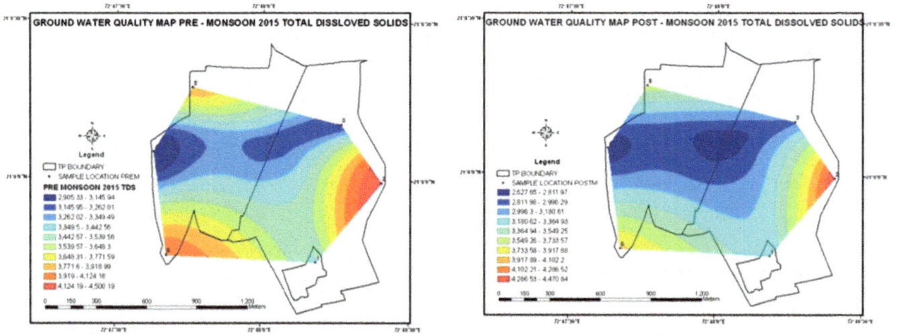

Fig. 6 Test result GWQ concentration of TDS for 2015 pre- and post-monsoon

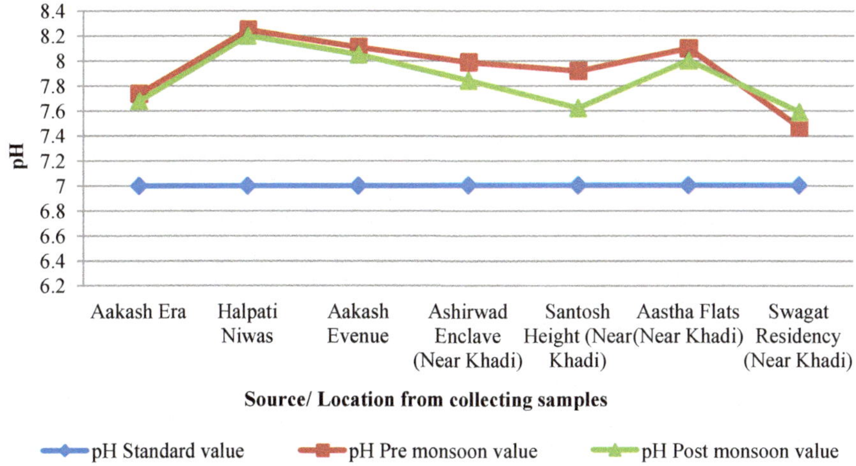

Fig. 7 Comparative analysis of pH

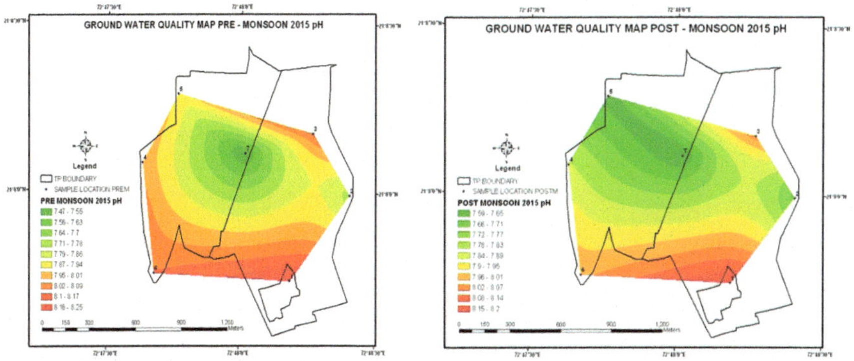

Fig. 8 Test result GWQ concentration of pH for 2015 pre and post-monsoon

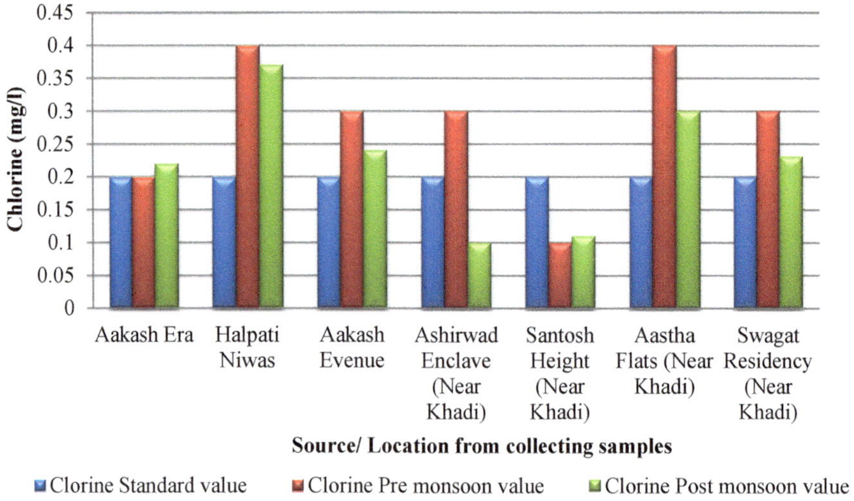

Fig. 9 Comparative analysis of Cl

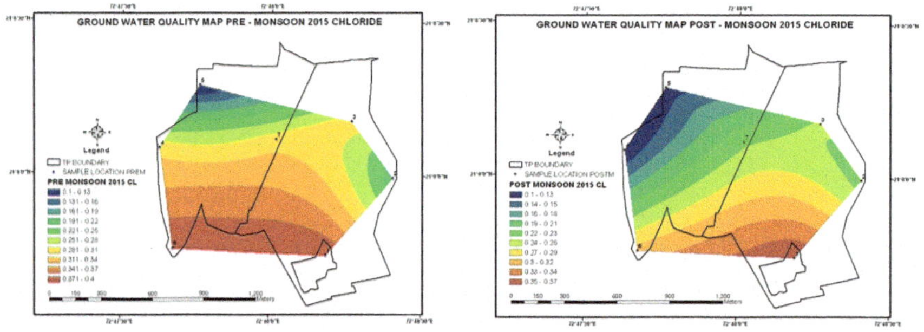

Fig. 10 Test result GWQ concentration of Cl for 2015 pre- and post-monsoon

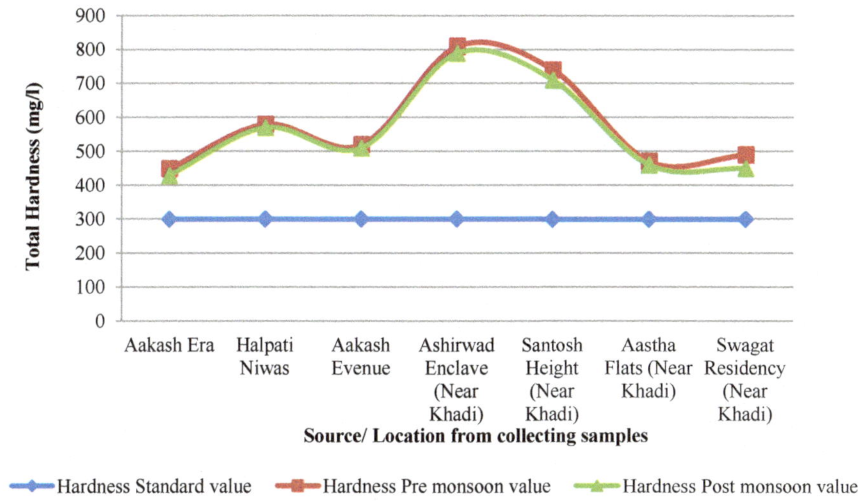

Fig. 11 Comparative analysis of TH

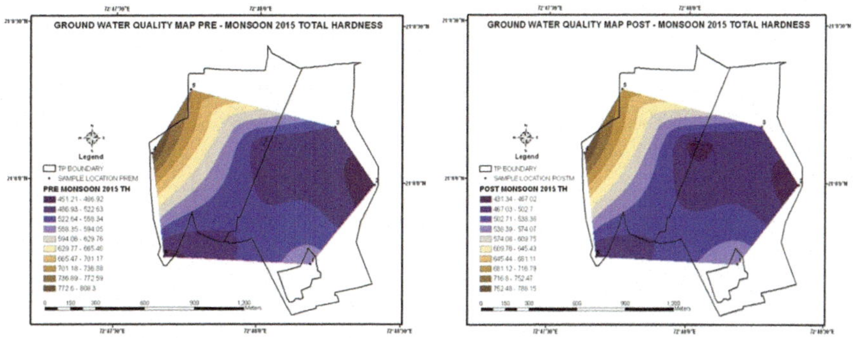

Fig. 12 Test result GWQ concentration of TH for 2015 pre- and post-monsoon

5 Conclusion

From the investigations of the groundwater quality of the study area, it is finally concluded that the location of the study area, as well as rapid industrialization and urbanization of the city day by day the groundwater quality, is contingent. Also, the demand for drinking water is increasing due to urban growth of the study area and only natural source of the river cannot accomplish the future demand of the usable water. So, groundwater is the best source and GIS is the best tool to investigate and fulfill future demand. From the results, it shows that existing groundwater quality is not usable for domestic/drinking water purposes as per Indian standard (IS: 10500).

References

1. Mahalingam B, Bhauso R, Deepali M, Jayashree P (2014) Assessment of groundwater quality using GIS techniques: a case study of Mysore City. Int J Eng Innov Technol (IJEIT) 117–122
2. Nas B, Berktay A (2010) Groundwater quality mapping in urban groundwater using GIS. Environ Monit Assess 215–227
3. Jebastina N, Arulraj GP (2017) GIS based assessment of groundwater quality in Coimbatore District, India. J Environ Anal Toxicol
4. Krishan G, Singh S (2016) Water quality assessment in terms of water quality index (WQI) using GIS in Ballia District, Uttar Pradesh, India. Environ Anal Toxicol
5. Jeihouni M, Toomanian A, Shahabi M, Alavipanah SK (2014) Groundwater quality assessment for drinking purposes using GIS modelling (case study: City of Tabriz). Int Arch Photogramm Remote Sens Spat Inf Sci 163–168

Part VI
Geomatics in Transportation Engineering

Evaluating Transit-Oriented Development Using a Sustainability Framework: A Case Study of Bhopal

Rupali Khare, Vasanta Govind Kumar Villuri, Devarshi Chaurasia and Supriya Kumari

Abstract Globally, TOD is gaining popularity as an admitted tool to implement Smart Growth and Sustainable Development. TOD includes high density, compact, mixed-use type of development around the transit station, suitable for bicycle and pedestrian users in order to encourage the use of the nonmotorized vehicle and public transit system. The objective of this study was to build up a strategy for measuring the performance of TOD in Bhopal against selected five criteria, including 1. Built environment; 2. Social environment; 3. Travel behavior; 4. Natural environment; 5. Economic Development; and other performance indicators to set up the structure for a database required to attempt performance estimation. And, it evaluates the sustainability of the TOD based on various criteria. Therefore, this study has set up a pattern that future analysis can be measured against.

Keywords Sustainability · Urbanization · Smart growth · Transportation

1 Introduction

As the cities are experiencing rapid growth, the level of urbanization is rising much faster than its total population. Bhopal is competing with the fastest growing city in India. The urban population of Bhopal was nearly 1.79 million (2011) and will reach 3.5 million in 2030 [1]. Urbanization has led to the horizontal growth of cities

R. Khare (✉) · V. G. K. Villuri
Indian Institute of Technology (Indian School of Mines), Dhanbad, Jharkhand, India
e-mail: rupali.16dr000204@me.ism.ac.in

V. G. K. Villuri
e-mail: govindvilluri@gmail.com

D. Chaurasia
School of Planning and Architecture, Bhopal, India
e-mail: dchaurasia@spabhopal.ac.in

S. Kumari
Central University of Jharkhand, Ranchi, India
e-mail: ksupriya120@gmail.com

J. K. Ghosh and I. da Silva (eds.), *Applications of Geomatics in Civil Engineering*, Lecture Notes in Civil Engineering 33, https://doi.org/10.1007/978-981-13-7067-0_26

thus creating the problem of urban sprawl. This has resulted in increase of trip length and higher use of private vehicle, deterioration in the environment, congestion, low quality of life, loss of neighborhood, traffic accidents, poor market shares of transit due to economically unfavorable settlement structure, dangerous condition for bicyclists and pedestrian, and increased demand of infrastructure. To address this issue, internationally, many cities have transit system like BRTS, MRTS, etc., to cater the growing travel demand [2]. Public transportation is one of the most important components of the urban transport system, particularly in the pace of rapid urbanization in Bhopal city. Today in the city, the usage of Bus Rapid Transit System (BRTS) is progressing step by step looking for cost-effective transit solutions and for environmental concerns. But the problems faced by the users to use this system are poor accessibility status, lack of integration of parking facilities, and pedestrian unfriendly. For effective planning, it is quite essential to understand the barrier to smart growth. A sustainable strategy is required to maintain the economic growth and alleviate the problems arising due to the growth. Globally, Transit-Oriented Development (TOD) is achieving popularity as a tool to implement Smart Growth and Sustainable Development. It refers to a form of urban design that achieves moderate to high density, mixed-use, pedestrian friendly type of development close to transit service. Hence, TOD has become a popular planning response to the impacts on urban growth. Sustainable development looks to make an urban environment which augments economic improvement and social value, while limiting the effect of negative externalities on the natural environment (see Fig. 1). From a land use and transport point of view, the need in such manner is to decrease automobile dependency through the improvement of mixed land use and compact urban communities with a variety of travel choices concentrated on walking, bicycling, and public transport [3–7].

Fig. 1 Sustainable development [8]

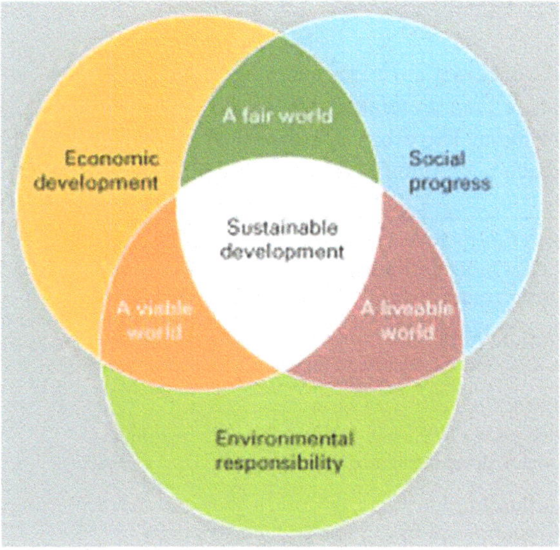

While finding a framework for estimating land use and transport outcomes, it becomes hard to categorize indicators utilizing the three essential categories of sustainable development (social, environmental, and economical) since numerous indicators cross limits [3]. This method evaluates five aspects of TOD outcomes, including 1. Built environment; 2. Social environment; 3. Travel behavior; 4. Natural environment; and 5. Economic Development.

ArcGIS is a useful tool to analyze large amounts of information through different layers and provide a structured, traceable, and flexible analysis. To analyze the level of service (LOS) of TOD, it is important to have a base map and attribute data corresponding to different areas in the base map. ArcGIS could help us understand the characteristics of each area (each census tract, specifically) and thus determine which area is most suitable for TOD.

This study is a response to the requirement for an assessment of Indian attempts at creating TOD. It follows the approach of TOD in India and its progress throughout the years [9].

The framework created for assessing these TOD designs is produced after reviewing a number of literature surveys, drawing particularly on the work done by Renne et al. [3]. The first section comprises of a literature review of writing on TOD from worldwide encounters with the identification of indicators on which the TODs can be evaluated on their TODness, and the present scenario of integration of land use and transportation. We evaluate the case study of Bhopal with respect to the identified indicators to comprehend the present situation of TODness in India.

2 Literature Review

There are many ways to define TOD with the perspective of new urbanism. TOD is pedestrian-friendly, compact, and mixed-use developments within walking distance of a transit station and a core commercial area [9–12].

TODs are introduced both to help public transport and to use the development opportunities that such a framework may give. A few examinations have recommended that coordinated land use arrangements around public transport stations can prompt increment in ridership. They also increase the financing of public transport.

After some time, TOD has come to procure the significance of planned development around a transit. TOD is not only any development close to transit. It is an improvement that expands "area proficiency" so people can walk, cycle, and utilize public transport and decrease the utilization of private vehicles; gives a mix of housing, occupations, shopping, and recreational choices; offers some opportunities for people and private sectors, and for both new and existing residents and provide quality of life [13].

Likewise, the transportation center point ought to be situated in the core of the area, inside a 400 m, or 10-min walk from residents. This center location increases the significance of transit in the group and in all the areas (Fig. 2). TOD contains a mix of commercial, residential, and institutional development built to encourage a

Fig. 2 Transit-oriented development

transportation center point and to support nonmotorized vehicle versatility alternatives, for example, cycling and walking, inside the group. A TOD zone could envelop a range of radius as 0.5 miles or 1 mile from a transit station [9, 14].

TOD coordinates land use and transport planning and intends to create planned sustainable urban development focuses, having walkable and decent communities with high-density mixed land use. The residents can approach green and open spaces and parallelly travel services are productively utilized. TOD expands the accessibility of the transit stations by making pedestrian and Non-Motorized Transport (NMT)-friendly framework that advantages a vast number of residents, and in this manner adding more ridership to the transit service and enhancing the monetary and money-related feasibility of the framework. Since the transit neighborhood has mixed land use, where the travel stations are either starting point (house) or goal (work), the utilization of the travel framework would streamline traffic in the peak hour movement of both directions in the corridor [15].

3 Need for TOD in Bhopal

As the cities are encountering fast development, travel frameworks like metro rail, BRTS, and so forth are being implemented to take into account the developing travel demand. Therefore, it is good to practice TOD for every such city which has a current mass transit system or are experiencing unsafe access, lack of traffic calming, poor transit quality, lack of multi-model integration, lack of planned parking facility, high on street parking encroaching NMT infrastructure, and lack of last mile connectivity [16]. This study fits their objective in finding those areas where transit may be required but is not present sufficiently.

4 Study Area

The study area is a part of Bhopal city, the capital of Madhya Pradesh, which is situated in Central India (see Fig. 3). The total population of Madhya Pradesh is about 7.26 crores having the sixth rank in India. In population wise, it is the second largest city of Madhya Pradesh. The study area is northeast part of the city having coordinates 77.28–77.55 E longitude and 23.122–23.333 N latitude and an average altitude of 427 m from Mean Sea Level (MSL). There is five transit stations (bus stop) taken in the study of the Bhopal transit station, as shown in Fig. 4a. The transit core area is 400 m radius and transit neighborhood is 800 m radius surrounding the transit station, as shown in Fig. 4b.

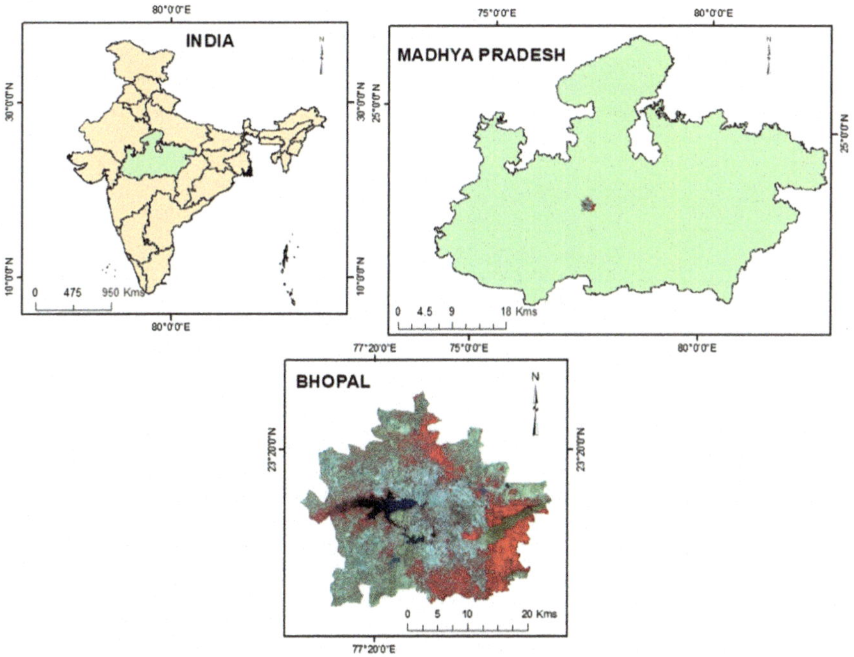

Fig. 3 Bhopal study area

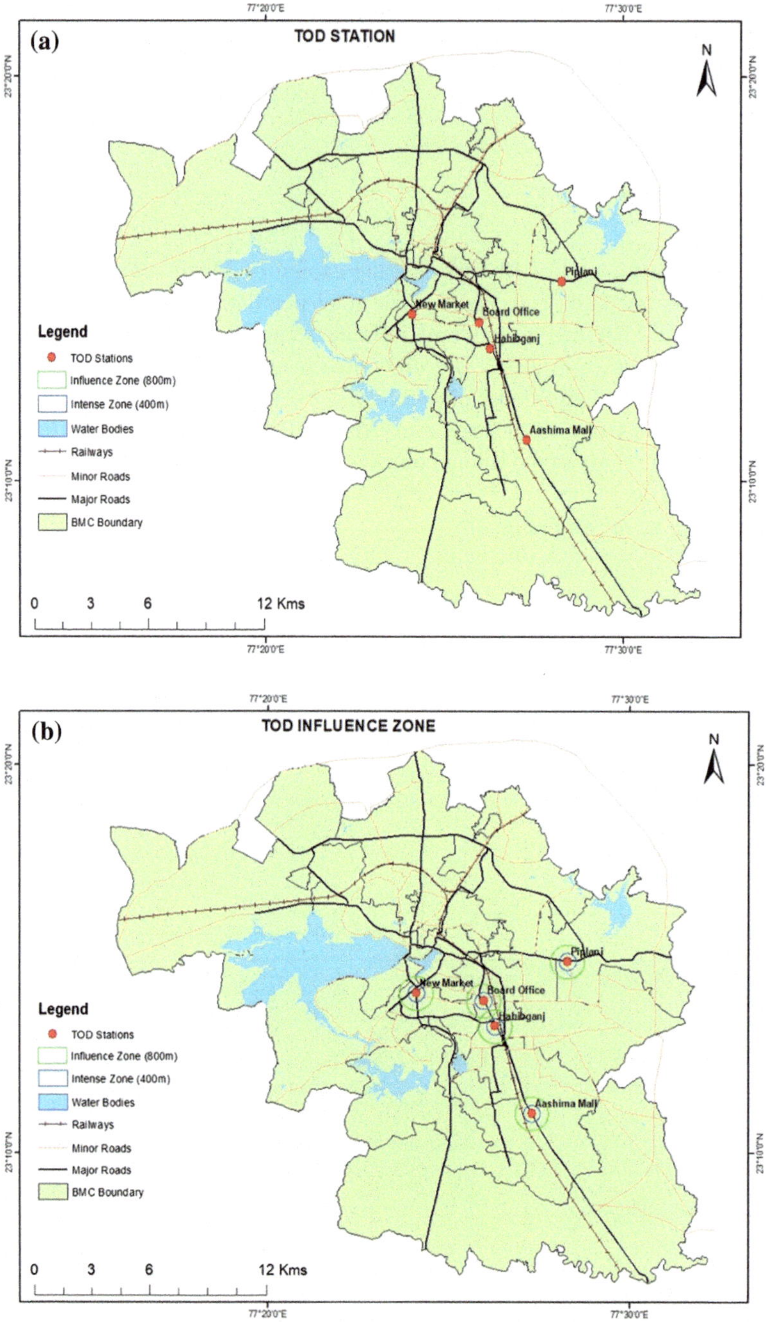

Fig. 4 TOD stations and its surroundings

5 Methodology

The goal of the study was to develop a method to evaluate the level of services of TOD. For this, it was necessary to understand the indicator availability and set up standard information accordingly. The study includes 10 transit stations. As presented in Fig. 4a, b, they were chosen based on the identified built types and land uses along the proposed transit corridors, the amount of station area land within the BMC boundary lines, the presence of developable vacant land, variety of existing building and land use types, proximity to other transport infrastructure, etc. [13].

Also, covering entire Bhopal city was considered to analyze the coverage of urban and suburban areas. The ten stations selected were: 1. Piplani—Moderate density, compact, mixed-use, and mixed income suburb; 2. Board Office—compact, horizontal as well as vertical type of mixed use, walkable, commercial area, and educational hub of the city; 3. Habibganj—gives connectivity to Bhopal from other places through railways; 4. Aashima mall—developing high-density residential area with a high level of mixed use of land; and 5. New Market—center hub for the shopping. After finalizing the station, the next step is to examine the existing situation of the particular area and decide on the appropriate indicator. In the literature, various indicators were found useful for evaluating the performance of TOD [6, 14–21]. Based on that literature, we mainly focused on five categories for sustainability of TOD: 1. Built environment; 2. Social environment; 3. Travel behavior; 4. Natural environment; and 5. Economic Development. The 43 indicators under these categories are:

1. Category: Built Environment

    1. Population density/housing density
    2. Commercial density
    3. Access to station/stop
    4. Location Accessibility provided by the stop/station
    5. Interchange
    6. Routes
    7. Parking spaces that are shared
    8. Parking spaces for bicycle and locker provided
    9. Walkable and cyclable path
    10. Pedestrian amenities
    11. Quality of streetscape
    12. Mixed use of land
    13. Land use land cover classification

2. Category: Social Environment

    1. Residential diversity
    2. Information display system
    3. Street light
    4. Basic amenities

5. Safety
6. Number of affordable housing units
7. Land use diversity
8. Amount of crime
9. No. of locality institution/organization/club
10. Last Mile connectivity

3. Category: Travel Behavior

1. Vehicle km traveled per household
2. No. of trips per day, by mode, per household
3. Method of journey to work(resident)
4. Method of journey to work (employee)
5. Method of journey to others(visitor)
6. Average daily commuting time and distance(residents)
7. Average daily commuting time and distance (employee)
8. No. of high frequency, line haul, and local public transport service available
9. Integration of services both spatially & time table
10. Number of vehicles per household 1. Auto traffic speed and volume
11. No. of traffic calming features available
12. Ped shed
13. Pedestrian activity count

4. Category: Natural Environment

1. Percentage of green and natural spaces
2. Parks and playfields
3. Vacant land parcel

5. Category: Economic Development

1. Employment level
2. Property value
3. Tax earning
4. Home ownership and rental

After reviewing the literature, we found that many of the surveys include only a few indicators for measuring these criteria. Table 1 presents the most useful ones and Table 2 the percentage of easy to collect. First, we tried to collect data from secondary source after that we do effort for primary data collection. Our primary data gathering efforts appeared as a site visit and TOD household survey. The site visit tried to gather indicators from field visit while the household survey goal to gather information and perception of people living within the study areas. A total of 1,260 households were arbitrarily chosen over the TOD areas and answered a questionnaire survey. The limitation of this methodology is time-consuming, because the accuracy of the work depends mainly on the data collected. Primary data are collected by doing manually survey. Finding secondary data is difficult because we need location dependent information within the 800 m boundary area around the transit station and people are not aware that they do not give much response. As many problems are encountered during data collection, merely, the most ideal approach to test a methodology for estimating TOD success is actually to collect data from the primary source, which act as a benchmark to track future development [15].

Table 1 Indicators rated very useful for TOD by at least 50% of the respondents

Indicators	Percentage as "very useful"	Category
Population/housing density	71	Built environment
Mixedness of residential with other land use	68	Built environment
No. of buses, auto, minibus, or service connecting to transit station	70	Travel behavior
Quality of transit	68	Travel behavior
No. of traffic calming features	61	Travel behavior
Pedestrian activity count	67	Travel behavior
Access to station	68	Built environment
Safety	70	Social environment
Parking space available for resident	60	Built environment
Public perception	69	Social environment
Land use diversity	62	Built environment
Pedestrian amenities	56	Built environment
Quality of streetscape	69	Built environment
Employment level	55	Built/economic development
Property value	48	Economic development
No. of transit boarding	70	Travel behavior
Shared parking space	58	Built environment

Table 2 Indicators rated very easy to collect for TOD by at least 50% of the respondents

Indicators	Percentage as "easy to collect"	Category
Percentage of natural and green space	67	Natural
Vacant land parcel	67	Natural
Property value	65	Economic
Mixedness of residential with other land use	54	Built environment
Population/housing density	69	Built environment
Access to station	68	Built environment
Pedestrian activity count	67	Travel behavior
Vehicle use	54	Travel behavior
No. of buses, auto, minibus, or service connecting to transit station	76	Travel behavior
Quality of transit	71	Travel behavior
No. of traffic calming features	61	Travel behavior
Safety	61	Social environment
Parking space available for resident	60	Built environment
Public perception	50	Social environment
Location accessibility provided by the stop/station	62	Built environment
Street light	62	Social environment
Land use diversity	52	Built environment
Pedestrian amenities	56	Built environment
Quality of streetscape	59	Built environment
Employment level	55	Built/economic development
No. of affordable housing	68	Social environment
Property value	58	Economic development
No. of transit boardings	70	Travel behavior
Shared parking space	58	Built environment
Amount of crime	61	Social environment

6 Results

The study identified five categories, including 1. Built environment; 2. Social environment; 3. Travel behavior; 4. Natural environment; and 5. Economic Development. All the indicators are taken from these categories. Here, we present the result of identifying and collecting indicators within each category. Table 3 shows the indicators measures with data collection source of each category included in the study of Bhopal.

Table 3 Recommended measure and indicators with data sources to evaluate TOD for Bhopal

Measure	Indicator	Possible data source
Urban density	1. Population density/housing density 2. Commercial density	DPI, CENSUS DATA
Transit system	1. No. of transit boarding 2. No. of bus, auto, minibus, or service connecting to transit station	BCLL, BMC, RTO
Travel behavior		Survey
Vehicle use	1. Vehicle km traveled per household 2. No. of trips per day, by mode, per household 3. Method of journey to work (resident) 4. Method of journey to work (employee) 5. Method of journey to others (visitor)	survey
Trip length	1. Average daily commuting time and distance(residents) 2. Average daily commuting time and distance(employee)	Survey, BMC
Transit quality	1. No. of high frequency, line haul, and local public transport services available 2. Integration of services both spatially and time table	Survey, site visit, BCLL, BMC
Vehicle ownership	1. No. of vehicles per household	RTO office, survey
Traffic	1. Auto traffic speed and volume 2. No. of traffic calming features available	BMC, site visit, Traffic police
Pedestrian accessibility	1. Ped shed 2. Pedestrian activity count	DPI, Site visit, BMC
Accessibility	1. Access to station/stop 2. Location accessibility provided by the stop/station 3. Routes (no. of connection to different routes)	Site visit, survey, DPI

(continued)

Table 3 (continued)

Measure	Indicator	Possible data source
User friendliness	1. Information display system 2. Street light 3. Last mile connectivity 4. Basic amenities 5. Safety (security at station/building frontage)	Site visit, DPI, public transport authority
Parking	1. No. of spaces for shoppers only 2. No. of spaces that are shared 3. No. of spaces for bicycle or lockers provided	Site visit, survey, DPI, traffic police
Public perception	1. How would you rate your town as a place to live? 2. Do you feel the transit station area is more or less attractive now compared to years ago? 3. Is it more or less no. of public transports available to travel comparative to years ago? 4. Does the downtown seems more or less safe now compared to years ago? 5. Does downtown offers better or worse shopping now compared to year ago? 6. Does the downtown clean and well maintained now compared to years ago? 7. Does the walkable path are in the good condition now compared to year ago?	Survey
Diversity	1. Residential Diversity 2. Land use diversity	LULC map, Google Earth, BMC
Social quality	1. No. of affordable housing units 2. Amount of crime 3. No. of neighborhood organizations	Department of housing and works, real estate agents, local government., census data, BMC
Walkability or cyclability	1. Walkable path 2. Cyclable path 3. Mixing of residential with other land use 4. Quality of streetscape	Site visit, DPI, BCLL, BMC, LULC Map
Economic development	1. Employment level 2. Tax earning 3. Property value 4. Home ownership and rental	Census data, survey
Natural environment	1. Percentage of green and natural space 2. Parks and playfields 3. Vacant land parcel	Site visit, Survey, T&CPO, LULC Map

6.1 Built Environment

In this section, we talk about the human surrounding facilities made for the human activity, including building, transportation, and recreation.. Table 3 shows the indicators decided for the study with the ideal information accumulation source. For the success of TOD, urban planning plays a vital part how territory is created, how it encourages occupants with the end goal of parking availability, parking spots, strip development, walking path, essential amenities, and so on. Density concentrates and intensifies activities near frequent transit and supports a strong demand for transit service. Figure 7 shows the density graph of the TOD area.

6.1.1 Land Use

Encourage a mix of land uses around frequent transit node to create a complete neighborhood. Collocating jobs and housing create short commutes within the region. And to reduce peak crowding and spread level demand throughout the day [13]. Figure 5a–e shows the land use by Landsat 7 data and Google Earth image of TOD area. Figure 6 shows the analysis of mixed land use of the TOD area. Find the board office station and the new market station has high mixed use above 0.5, while Aashima mall has the least score for mixed land use.

6.2 Travel Behavior

Travel behavior study gives an idea about how people use transport and what people do. The survey questions, in this case, are designed in such a way that we can achieve the traveler's behaviors, attitude, and the gaps between them in relation to the sociological and environmental impacts of travel. Table 3 lists the measures, indicators, and data collection sources for measuring the travel behavior of people living in TOD areas. As discussed, the TOD household survey (Table 4) provided a much larger sample across the station area. The questionnaire asked residents how they use public transport (Table 5), how long it takes to reach the transit station (Table 6), and how they go for shopping (Table 7). The survey also collected the number of vehicles per household and TOD household people perception about transportation (Table 8).

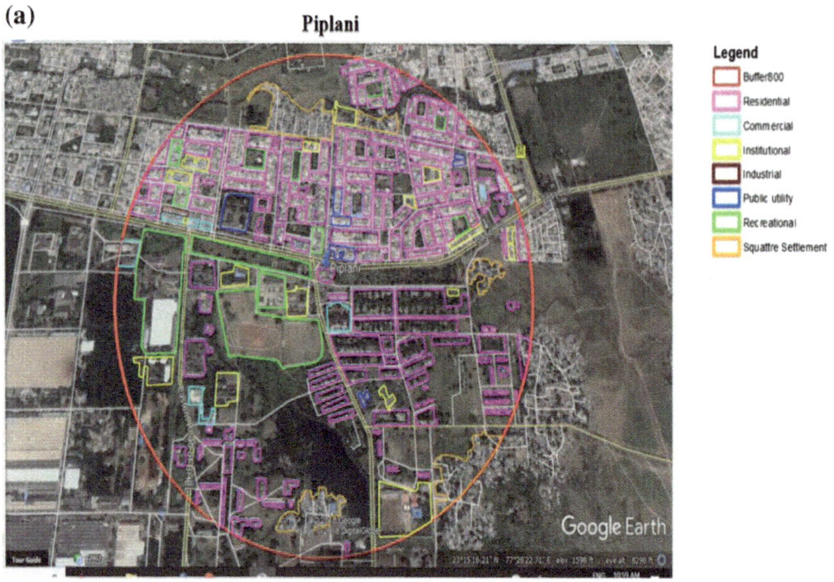

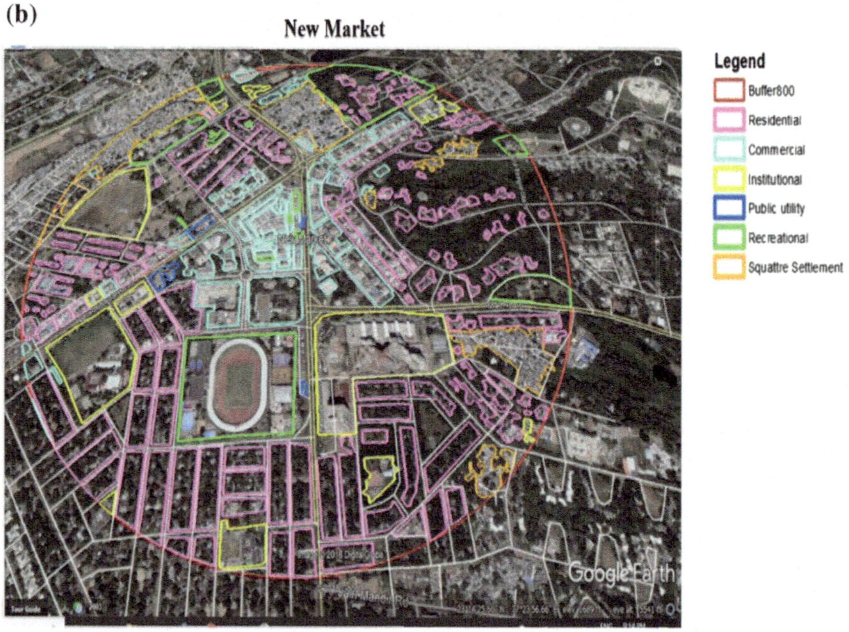

Fig. 5 **a**–**e**—Land use by using Landsat 7 data and Google Earth image of TOD area Piplani, new market, Habibganj, board office, and Aashima Mall, respectively

(c)

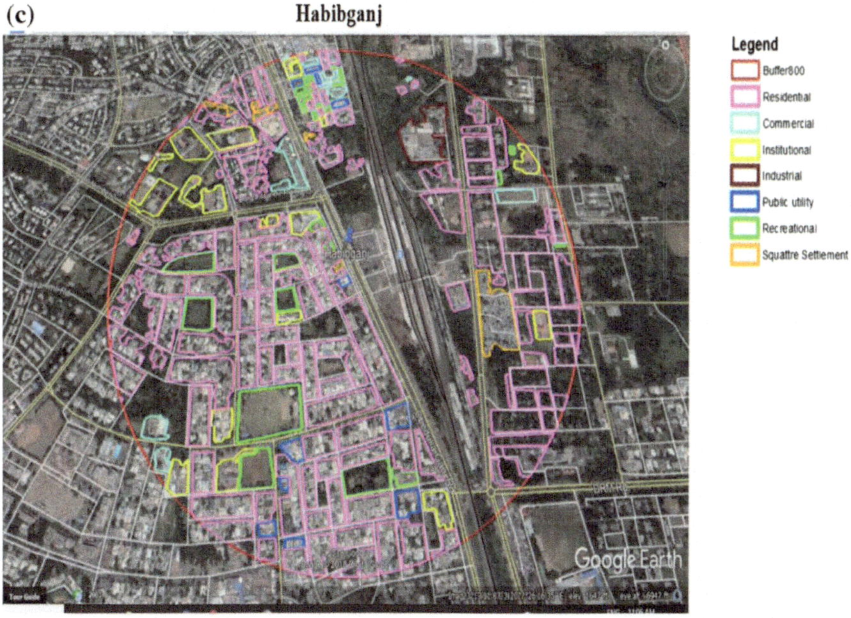

(d)

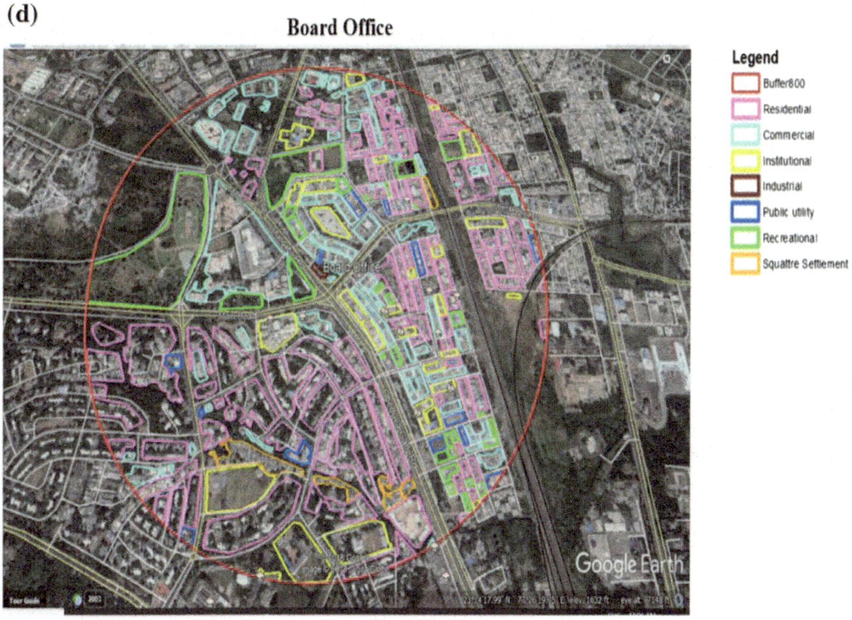

Fig. 5 (continued)

Fig. 5 (continued)

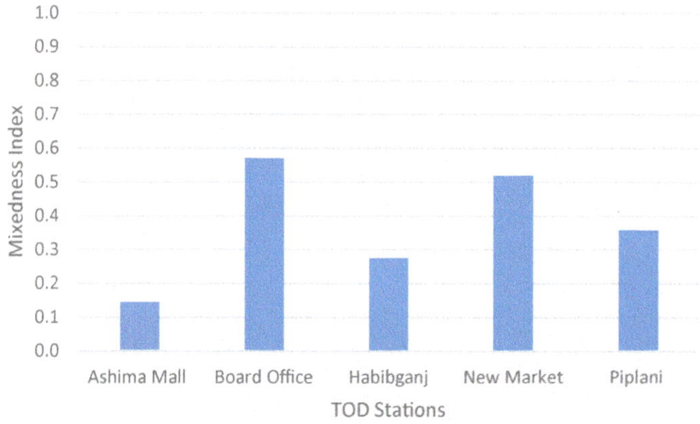

Fig. 6 The analysis of mixed land use within five transit stations from Google Earth image

6.3 *Economic Development*

Sustainable economic growth is an economic development that endeavors to fulfill the necessities of human, however, in a way that maintains common assets and the earth for future generations [4, 22]. Therefore, straightforwardly worried about increasing the material standard of living of the poor at the grassroots level, which can be quantitatively measured in terms of increased real income, educational service,

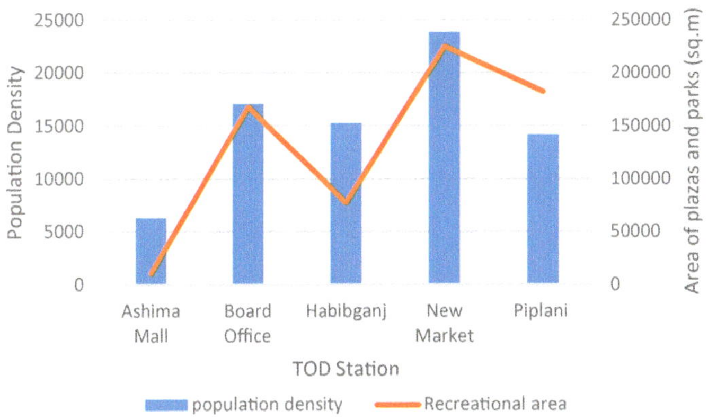

Fig. 7 Density map of the study area

Table 4 Travel behavior of TOD station residents from household survey

TOD Performance indicator		Piplani	Board Office	Habib Ganj	Ashima Mall	New market
		Vehicle use				
Vehicle km traveled per household		38.50	24.10	22.74	32.12	23.44
Trips per day, per household (%)	Motorized vehicle	49.2	24.4	42.5	41.2	49.6
	Public transport	20.3	38.4	18.9	28.5	20.4
	Walk or cycle	30.5	37.2	38.6	30.3	30
Method of journey (residents) (%)	Motorized vehicle	40	39	68	52	49
	Public transport	37	20	17	28.2	45
	Walk or cycle	23	41	15	16.8	6
Method of journey (Employee) (%)	Motorized vehicle	34	51.3	37.2	33	49
	Public transport	39	39.2	43.7	30	28
	Walk or cycle	23	18.5	20.1	37	23

(continued)

Table 4 (continued)

TOD Performance indicator		Piplani	Board Office	Habib Ganj	Ashima Mall	New market
Method of journey (Visitors/others) (%)	Motorized vehicle	59.3	40.7	57.9	55.1	63
	Public transport	30.6	44.2	30	39.2	29.2
	Walk or cycle	10.1	15.1	12.1	5.7	7.8
		Trip length				
Avg. daily commuting time and distance (Residents) (%)		10.841	8.321	10.761	11.240	8.133
Avg. daily commuting time and distance (Employees) (%)		12.841	6.231	10.321	15.297	8.211
No. of high-frequency public transport services available (%)		80	80	80	60	80
Integration of services both spatially and time (%)		0.9	0.9	0.9	0.3	0.9
No. of vehicles per household (%)		1	1.59	1.79	1.47	1.36
Ped shed (%)		60	80	70	70	71
Pedestrian activity count (%)		50	80	60	60	70
No. of traffic calming features (%)		0.6	0.9	0.7	0.2	0.9
Last Mile Connectivity (%)		0.5	0.5	0.4	0.3	0.5

Table 5 Frequency of public transport usage from the TOD household survey

Use of public transport	Percent
5 days per week or more	32.4
1–4 days per week	24.3
1–3 days per week	23.1
Less than 2 days per week	9.1
never	11.1
N = 540	

Table 6 Walking distance to the nearest train station from the TOD household survey

Walking time to reach nearest transit station	Percent
Less than 5 min	24.5
5-10 min	60.1
10–20 min	8.2
More than 20 min	0.9
Don't know	6.3
N = 540	

Table 7 Mode choice for shopping and commute tips from the TOD household survey

Mode	Shopping trips	
	Survey respondent (%)	Survey respondent family (%)
Motorcycle or two wheeler	44	39
Public transport	41	43
Walk and bicycle	7	7
Automobile and taxi	8	11
N = 450		

health care, transportation, affordable housing, etc. TOD encourages a high level of mix use of land and diversity so that all people get benefited. Table 3 shows the indicators with data collection source taken for the study. Table 9 shows the expensive ranking of the study area among all selected TOD area new market and board office found most expensive according to stakeholders. From the site visit and survey, we find out the housing tenure of TOD household area, as shown in Table 10 and the employment rate of the TOD household (Fig. 8).

6.4 Social Environment

Social life has developed a framework for social sustainability which dimensions include amenities and infrastructure, social, and cultural life and space to grow. Table 3 shows suitable indicators with possible data collection source included in the study. Here, we discuss the quality of life of people living in TOD areas. The perception of the people is shown in Table 12. Table 11 shows their neighborhood from various stockholders, ranking the property value according to the location nearby transit type.

Table 8 Transportation opinion from the TOD household survey

S. no.		Strongly disagree	Slightly disagree	Neutral	Slightly agree	Strongly agree
1	My locality is well served with public transport	0.56	0.84	10.1	28.2	60.3
2	Traffic is not a major issue in the area	18.7	30.6	26.5	15.3	8.9
3	The main reason I live here is to be near the transit	0.8	2.9	35.2	40.8	20.3
4	The neighborhood is easy to walk around	16.6	20.5	31.1	22.6	9.2
5	Footpaths are in good condition	1.7	16.9	47	20.2	14.2
6	It is easy to cross the street	1.3	9.4	30.3	36.3	22.7
7	I feel safe from traffic while walking	2.1	7.1	60.3	20.4	10.1
8	Drivers give way to pedestrians crossing the road	10	17.6	27.3	30.2	14.9
9	I can easily walk to the transit station from my house	0.7	2.3	28.5	28.2	40.3
10	Walkable path to the train station is user friendly	1.4	14.1	24.3	27	33.2

Table 9 Ranking of the selected TOD areas in Bhopal city by stakeholders (1 meaning most expensive and 5 meaning least concerned)

Area	Rank
Piplani	3
New market	1
Board office	1
Aashima Mall	4
Habibganj railway station	2

Table 10 Housing tenure of TOD household area by field survey

TOD performance indicator	Piplani	Board office	Habib Ganj	Aashima Mall	New market
Fully owned	42	12	53	32	26
Being purchased	25	17	15	20	21
Rented	25	54	30	38	43
other	8	17	2	10	10

Table 11 Land value near transit stations in Bhopal discussed by stakeholders

S. no	Location	Property	400 m around transit station	800 m around transit station	Increase in property value	Decrease in property value	No effect in property value
1	BRTS	1. Commercial property	1	2	yes		
		2. Residential property	2	3	yes		
2	Railway station	1. Commercial property	1	2			yes
		2. Residential property	2	3			yes
3	Proposed metro station	1. Commercial property	1	2	yes		
		2. Residential property	2	3	yes		

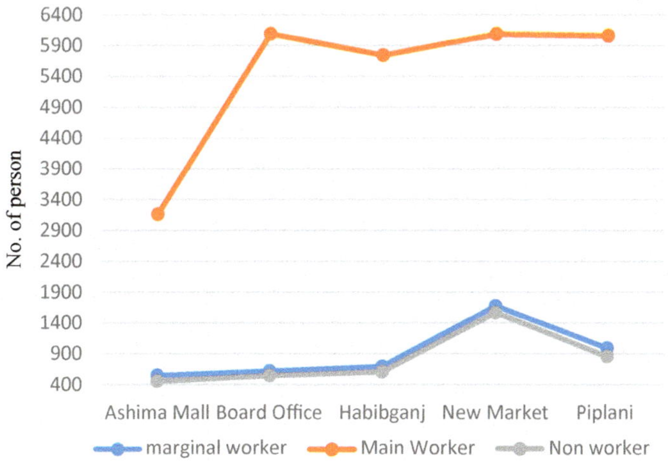

Fig. 8 Number of jobs in the TOD area as shown in [22] and field survey

Table 12 Neighborhoods opinion from the TOD household survey

	Strongly disagree	Slightly disagree	Neutral	Slightly agree	Strongly agree
My locality is a good place to live	0.8	1.9	13.3	47.3	36.7
Compared to other parts of Bhopal, my locality is a better place	2.8	6.4	7.1	44.2	39.5
My locality is well planned and clean	0.79	1.68	22.03	34.3	41.2
In crime count my locality is better than other parts of Bhopal	1.2	6.2	49.6	19.3	23.7
I feel safe and secure in my locality even at night time	0.76	1.1	31.04	20.6	46.53
My locality is a attractive place, nice to be here	8.2	13.3	40.2	9.9	28.4
My locality is well coupled with shopping center for my day to day shopping	0.93	2.4	30.1	26.57	40.3
We can do weekly shopping in the locality center	0.7	1.3	4.1	40.2	53.7
My locality has community feeling	2.9	16.4	49.1	11.2	20.4
My locality is a peaceful place free from traffic and noise to live in	4.8	10	35.4	29.4	20.4
My locality is well equipped with recreation facilities	2.7	13.3	24	43	17

Table 13 Sustainable natural environment indicators value for selected TODs area based on land use land cover classification from satellite image and Google Earth image

TOD performance indicator	Piplani	Board office	Habib Ganj	Aashima Mall	New market
Percentage of green and natural spaces	9.62	7.25	13.76	5.77	24.91
Parks and playfields	21.04	18.52	8.73	3.01	19.80
Vacant land parcels	57.07	54.99	56.13	83.07	43.48

6.5 Natural Environment

The objective of environmental sustainability is to monitor natural assets and to create substitute sources of energy while decreasing contamination and harm to nature [23]. For environmental sustainability, we present indicators with possible data collection source (Table 3). Our findings (Table 13) revealed that among all selected TOD stations, new market has large green and natural spaces and parks playfield found more in Piplani, whereas Aashima mall has more development potential because of largely vacant land availability. From the site visit and the household survey, approximately, 120 houses found Piplani has less amount of recreational areas.

7 Conclusion

This paper examines the requirement for an assessment of Indian endeavors at delivering TOD. It traces the approach of TOD in Bhopal and its improvement throughout the years. It displays complete literature of TOD about indicators, measures, and data source. The purpose of this study is to summarize the current knowledge about the factors which influence individual travel behavior in Bhopal. The focus is mainly on indicators selected for Bhopal under the sustainability framework criteria. At last, the paper closes with recommendations to make TODs more comprehensive. The areas chosen for the study are different from a specific degree concerning TOD potential. Piplani is one of the most vehicle subordinates and is reasonably worked out. The potential for changing Piplani into a TOD is pretty low. Board office and new market are mostly completed TOD. Aashima mall is developing so that it supports TOD. It also has less need for change. Other station precincts, such as Habibganj, may have more development potential. Mostly, the parking and last mile connectivity problem are seen in all areas. Only autorickshaw available for last mile connectivity, which charges high. The mixed-use ratio of station precincts Habibganj station and Aashima Mall station is less.

8 Recommendation

Once more, the objective of this investigation was not to make a particular methodology to quantify TOD achievement utilizing a sustainability framework. The objective was to begin a discourse. Future examinations ought to investigate which indicators are ideal, what number of are required, and how to best dissect the data once it has been gathered. The sustainable development and transit-oriented development have turned out to be very well known with organizers crosswise over most of the urbanized zones, particularly in Australia and North America. Now, in India also, most of the studies concentrate on city level study and to intensely be a single part of TOD achievement. This study is independent of a particular aspect of TOD and it is about a collection of data by which we do further study on the different aspects of TOD. In Indian cities, area-based development, city renewal, and city extensions are the plus point and the reason why this would be the perfect time to consider and implement policies that facilitate TOD.

References

1. Madhya Pradesh District Census Handbook Directorate of Census Operations Madhya Pradesh
2. National Transit Oriented Development (TOD) Policy
3. Renne JL (2009) Evaluating transit-oriented development using a sustainability framework: lessons from Perth's network city. In: Planning sustainable communities: diversity of approaches and implementation challenges, pp 115–148
4. Banister D, Pucher J, Lee-gosselin M (2007) Making sustainable transport politically and publicly acceptable: lessons from the EU, USA and Canada. Regulation, pp 1–28
5. Abane A, Abelson D, Acevedo J, Van Acker V, Aditjandra P, Agrawal A, Aguiléra A, Akyelken N, Aldred R, Algers S, Allen J, Amoh-Gyimah R, Anable J, Anderson W, Andersson DE, Anderton K, Andrey J, Antipova A, Ardila-Gomez A, Arentze T, Attard M, Avineri E, Axhausen K, Bae C, Baier S, Bailey K, Baird A, Barr S, Filho MNMB, Batterby S, Baum-Snow N, Beecroft M, Behnen T, Ben-Edigbe J, Bergqvist R, Bertolini L, Birtchnell T, Black J, Blainey S, Bliemer M, Blumenberg E, Bocarejo JP, Bonham J, Bonilla D, Bono F, Boschmann E, Boussauw K, Bowen J, Brand P, Brand C, Broach J, Browne M, Buckwalter D, Budd L, Buddhavarapu P, Buehler R, Buliung R, Burger M, Burghouwt G, Burgoine T, Burke M, Button KJ, Caiado G, Caldas M, Calvo F, Campos J, Canters F, Xinyu J, Cao, Cappuccilli J-F, Cardozo O, Cariou P, Carrasco J-A, Carstensen T, Cary K, Celik H, Ceylan H, Chan CK, Chang D, Chapman L, Charlton C, Chatterjee K, Chen C, Chen C-L, Yu B, Zhenhua C, Chen CA, Cherrett T, Chiou Y-C, Chorus C, Chow A, Chow ASY, Chun Y, Ciari F, Cidell J, Clark S, Clement S, Coffin A, Comtois C, Cools M, Corcoran J, Cox P, Curl A, Currie G, Dablanc L, Dai D, Daigle J, Daley B, Datta J, Davison L, Day K, De Jong M, De Langen P, De Montis A, Deka D, Dekker M, Delbosc A, Dell'olio L, Delmelle E, Delmelle E, Dennis N, Derrible S, Floridea BD, Ciommo D, Dickinson J, Dijst M, Ding J-F, Dobruszkes F, Docherty I, Dodson J, Dresner M, Drummond H, Ducruet C, Dumbaugh E, Duncan M (2014) J Trans. Geogr 34:307–310
6. Renne JL, Chandra L, Tippett B, Kolapalli S (2007) Measuring the performance of transit-oriented developments in Western Australia planning and transport research Centre Sponsored by the Western Australian Department for Planning and Infrastructure and the Public Transport Authority of Western Australia

7. Renne J (2005) Transit-oriented development in Western Australia: Attitudes, Ostacles and opportunities. Transit Oriented Development: Making It Happen, 2005, Perth, Western Australia, Australia
8. Smart Map Bhopal. https://smartmapbhopal.city/bhopalsmartmap/
9. Joshi R, Joseph Y, Chandran VM, Darji V (2017) Transit-oriented development: lessons from Indian experiences, pp 1–31
10. Lyu G, Bertolini L, Pfeffer K (2016) Developing a TOD typology for Beijing metro station areas. J Trans Geogr
11. Hamre A, Buehler R (2014) Commuter mode choice and free car parking, public transportation benefits, showers/lockers, and bike parking at work: evidence from the Washington, DC Region. J Public Trans 17:67–91
12. Transportation Institute, M.: MNTRC Report 12-67 The Impact of Transit-Oriented Development on Social Capital
13. Bhatt C, Paradkar P, van de Fliert N (2012) Station area planning: a guide to planning and implementing transit oriented development in Indian cities
14. JPT 9-5 Cervero
15. Renne JL. 6 evaluating transit-oriented development using a sustainability framework: Lessons from Perth's Network City
16. Ottawa BC, Bhopal OC. Introducing Bhopal
17. Singh YJ (2015) Measuring transit-oriented development (TOD) at regional and local scales—a planning support tool
18. Belzer MH (2002) Technological innovation and the trucking industry: information revolution and the effect on the work process. J Labor Res 23:S.375–S.395
19. Singh YJ, Flacke J, van Maarseveen MFAM (2015) Measuring tod over a region using GIS based multiple criteria assessment tools Pengwei he measuring TOD over a region using GIS based multiple criteria assessment tools. SPA J Plan Archit 19:970–706
20. Lavallee D (2005) The effect of a life development intervention on sports career transition adjustment. Sport Psychol 19:193–202
21. Performance-based transit-oriented development typology guidebook (2010)
22. Census of India Madhya Pradesh. Series-24 (2011)
23. Conserve Energy Future. https://www.conserve-energy-future.com/what-is-environmental-sustainability-and-sustainable-development.ph

Application of Geospatial Technology in Planning and Acquiring Land for Proposed Roads Under a Master plan: A Case Study of Sultanpur Lodhi Local Planning Area

Lakhvir Singh, Simerjit Kaur, Sana, Ramandeep, Harmanpreet, Balwan, Aman Kumar Balihar, Preeti Kashmira, Tapti Baskey, Reenu Sharma, Ajay Mathur and Brijendra Pateriya

Abstract A general problem faced by many Indian cities is the criss-crossing of major roads through the centre of the city. The present study focuses on one such city in Punjab, Sultanpur Lodhi. Sultanpur Lodhi is a religious centre in Punjab of Sikhs attracting large number of tourists throughout the year. Due to the location of Gurudwaras in congested city and crossing of major road connecting Firozpur to Jallandhar through the city, the traffic becomes a huge problem. During festive seasons, the traffic diverts to the internal roads causing chaotic conditions. The present study focuses on the master plan development of the area with special regard to the proposed ring road to ease the traffic volumes. Geospatial technology using high-resolution satellite data has been used to prepare the master plan of Sultanpur Lodhi, which includes a 12 km long proposed ring road with a right of way of 150 feet.

Keywords Cadastral · Geospatial · Existing land use · Proposed land use

1 Introduction

In most of the cities in India, congestion of roads due to vehicular traffic has become a chronic problem with practically no solution coming from urban planners or municipal corporations and other related government bodies. A slow-moving traffic adds up to cost of travel due to the reduction in productivity, late delivery of goods and material adds up to cost of delay, in case of emergency the slow-moving traffic is fatal.

The various root causes of traffic congestion can be attributed to narrow roads, which are unable to handle the traffic of the area, roads dotted with potholes, rough patches affecting traffic flow, obstructions and encroachments, accident, breakdown

---

L. Singh (✉) · S. Kaur · Sana · Ramandeep · Harmanpreet · Balwan · A. K. Balihar · P. Kashmira · T. Baskey · R. Sharma · A. Mathur · B. Pateriya
Punjab Remote Sensing Centre, Ludhiana, India
e-mail: singhlakhvir707@gmail.com

J. K. Ghosh and I. da Silva (eds.), *Applications of Geomatics in Civil Engineering*, Lecture Notes in Civil Engineering 33, https://doi.org/10.1007/978-981-13-7067-0_27

of vehicle, bad weather, non-availability of parking space. In addition, the increase in purchasing capacity of the middle class has resulted in the increase in number of vehicles in the city.

Many solutions may be possible to ease congestion in the city, which includes traffic management, improving road infrastructure, odd–even formula imposed in Delhi, congestion charges in central London, public awareness to use bicycles or share cars to working places.

However, traffic congestion in many parts of Indian cities can be attributed to highways passing right through the city. For example, Roorkee, the highway from Delhi to Haridwar passes right through the congested part of the city with hardly any capacity to broaden the road. The problem exaggerates during Kumbh Mela leading to traffic blocks and inconvenience.

The present study focuses on one such similar city in Punjab—Sultanpur Lodhi, which is a historic and religious place of Sikh religion dotted with many Gurudwaras in the city which is more or less very congested. The city is also criss-crossed by a highway from Firozpur to Jallandhar. The traffic on this road is very heavy as it comprises not only of cars but trucks and buses also. One way to ease congestion in Sultanpur Lodhi is to restrict entry of heavy vehicles in the city during the daytime but this would have a cascading effect on the economy.

The way out is to have a ring road around the Sultanpur Lodhi city so that the fast-moving heavy traffic does not enter the city but is channelized through the ring road. The present paper focuses on ways and means of finalizing the alignment of such a ring road through the master plan. A master plan is a legal document for 20 years, which reflects the proposed land use to which a parcel of land can be used for. The master plan affords a legal way of acquiring land for proposed roads.

2 Study Area

The study area comprises of Local Planning Area (LPA) of Sultanpur Lodhi for which the master plan was conceived. It comprises of the city part and 11 villages totalling an area of 33.55 km^2 (Fig. 1). The city is dotted with many Gurudwaras and is criss-crossed by road connecting Firozpur to Jallandhar (Fig. 1).

3 Methodology

The master plans in Punjab are being prepared by the Punjab Remote Sensing Centre in conjunction with Punjab Urban Development Authority (PUDA) using Geospatial technology. The master plan is a legal document, which reflects the proposed land use to which a parcel of land can be used for. The process of preparing the master plan involves first to map the Existing Land Use (ELU) to have an understanding of inventory of Land use/cover of the Area of Interest (AOI) more commonly called

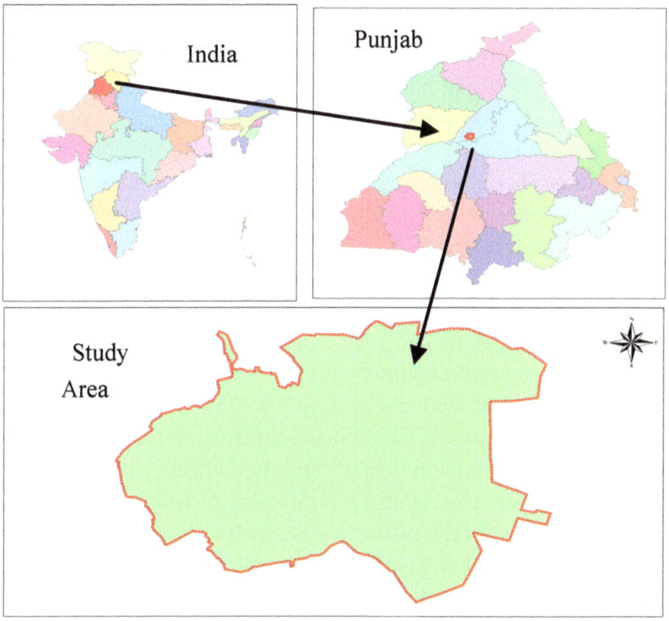

Fig. 1 .

as Local Planning Area (LPA). Based on ELU master plans also called as Proposed Land Use are generated.

The LPA of Sultanpur Lodhi as provided by PUDA in the form of a list and the blueprint map was first delineated on Cartosat-1 satellite data. The AOI comprises the outer area comprising of villages and the core area comprising the congested town part of the Sultanpur Lodhi. A detailed mapping using QuickBird data was carried for the core part of the city with the outer villages mapped to lesser details using Cartosat-1 data.

The Cadastral maps of the villages falling in LPA were procured from the State Revenue department by the office of Deputy District Town Planner, Kapurthala and these maps were scanned and registered with Cartosat-I data in ARC Map GIS s/w to demarcate village boundaries.

The areas of the village polygon as demarcated in GIS were cross-checked with the revenue areas as a measure of accuracy. In addition, the revenue details such as musteel and khasra of the villages were also demarcated and polygons populated with Musteel no and khasra no.

The existing land use and land cove features map were demarcated on the satellite data.

The features like roads, rails, built-up, forest, drains, etc. were delineated from Cartosat-1 data. The attributes of some of the land cover features could be discerned from satellite data but mostly the attributes of land use features, in particular were collected by officials of PUDA from the field. The database was edited in GIS based on inputs from field and features populated to generate with attribute data to generate ELU. After the preparation of ELU on 1:10,000 scale using Cartosat-1 data, the draft base map for densely populated built-up areas of Sultanpur Lodhi town (core areas) was also prepared on 1:5000 scale using QuickBird satellite data of 0.6 m spatial resolution in a similar fashion.

After number of deliberations (keeping the ELU as the basis) by the officials of the urban body, a draft proposed land use plan (master plan) is framed, which is posted at many locations including the Deputy Commissioner office in Sultanpur Lodhi for 30 days for calling any objections to the Draft Master plan from the public. The public concerns were deliberated and genuine concerns addressed and the master plan is finalized. This is then placed before the state government for notification.

The master plan is a legal document for 20 years, which reflects the proposed land use to which a parcel of land can be used for. Many times, the public would not consult the master plan or take official permissions from authorities and come up with constructions not allowed on the parcel of land. In addition, change in land use over the one given in the master plan is also warranted at times for developmental work. For this, the applicant has to take prior permission for change in land use (CLU) for any developmental works other than those prescribed for the parcel of land on the master plan if the said change is permissible. In the absence of any such information on the public domain, many a construction came up in utter disregard to the master plan in force. This may result in haphazard development and affect the planning for which the master plan was in effect. For instance, land parcels earmarked for ring road may be used for any other purpose by the public and when the ring road project is to be executed, the very purpose of the master plan would be defeated.

However, to ward off the above problem, the master plan was overlaid with the cadastral information and the information was made available on the web. This helped the public to understand if the change in land use sought by them can be undertaken as per master plan or as per change in land use policy of the government. However, with the overlay of cadastral information on the master plan and hosting the information on the web, the applicant can check the proposed land use khasra wise (for each parcel of land as per cadastral information) online. The information of master plans may be accessed up to killa level on the digital map by selecting the same through a dropdown menu which helps to select the village, musteel no and finally to killa level.

4 Results and Discussion

One of the concerns in acquiring land for transportation is the absence of master plans in major parts of the country. As master plans is a legal document, and reserves land for planning including transport network which may include widening of the existing roads or reserving land for the ring road to ease the traffic in the city. For, Sultanpur Lodhi, ELU map was generated using remote sensing data (Fig. 2) and Table 1.

Table 1 Statistics of various land use/cover categories in the Sultanpur Lodhi LPA

Sr. no.	Land use	Area in hect.	Percentage of total area
1	Residential	166.92	5.13
	Residential (urban)	107.19	
	Rural residential	55.53	
	Mixed land use (residential + commercial)	4.2	
2	Commercial	49.15	1.51
	Retail shopping/marriage places/Cinema	17.21	
	Wholesales, godowns, warehousing	31.94	
3	Industrial	19.86	0.61
	Small scale light and service industry	19.86	
4	Public and semi-public	56.04	1.72
	Govt/semi govt/public offices	8.04	
	Educational and research	19.78	
	Medical and health	2.1	
	Social, cultural and religious	19.51	
	Cremation and burial grounds	2.28	
	Electric substation	4.33	
5	Utilities	6.54	0.2
	Waterworks	0.24	
	Solid waste/dumping/sanitary land filling	1.29	
	Sewerage treatment plant	5.01	
6	Recreation	3.58	0.11
	Playground, stadium, sports complex	2.91	
	Parks and gardens (public open spaces)	0.67	
7	Rasportation and communication	95.86	2.94
	Roads	54.46	
	Railway line and siding	38.29	
	Bus terminus	0.48	
	Railway station	2.63	
8	Rural and agriculture	2857.68	87.7
	Agriculture	2570.39	
	Forest	142.82	
	Water bodies	63.82	
	Plantation and Orchards	9.19	
	Vacant land	71.46	
9	Special areas	0.13	0.08
	Fort/heritage building	0.13	0.08
	Total	3255.76	100

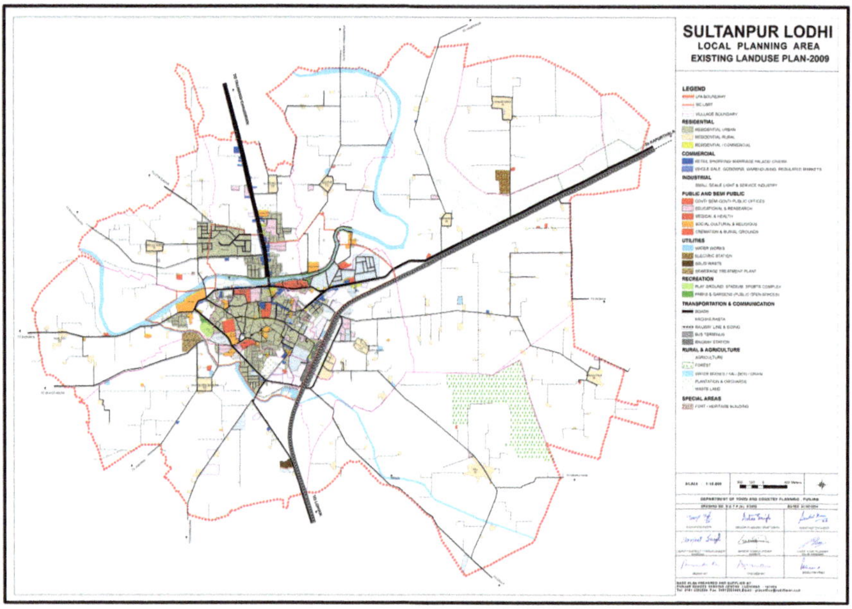

Fig: 2 Existing land use of Sultanpur Lodhi LPA

Table 2 Statistics of various land use/cover categories in the master plan

Sr. no.	Proposed land use	Urbanizable area*		Total LPA, Sultanpur Lodhi town	
		Area in hect.	% age	Area in hect.	% age
1	Residential	829.8	59.2	873.95	26.86
2	Commercial	24.6	1.75	24.6	0.75
3	Mixed land use	207.8	14.83	207.8	6.39
4	Industrial	129.95	9.26	129.9	3.98
5	Public and semi-public	24.8	1.76	24.6	0.75
6	Utilities	9.3	0.66	13.4	0.41
7	Recreational areas	73.7	5.25	297	9.13
8	Traffic and transportation	102.35	7.29	133.68	4.1
9	Rural and agricultural			1550.83	47.63
	Total	1402.3	100	3255.76	100

Based on the ELU map, the draft master plan was finalized which was refined by incorporating the concerns of the public and the final master plan was notified by the Government of Punjab vide letter No (Fig. 3 and Table 2).

The master plan focused on the transport requirement of Sultanpur Lodhi. In order to provide relief to the town roads and keeping in view the existing roads and

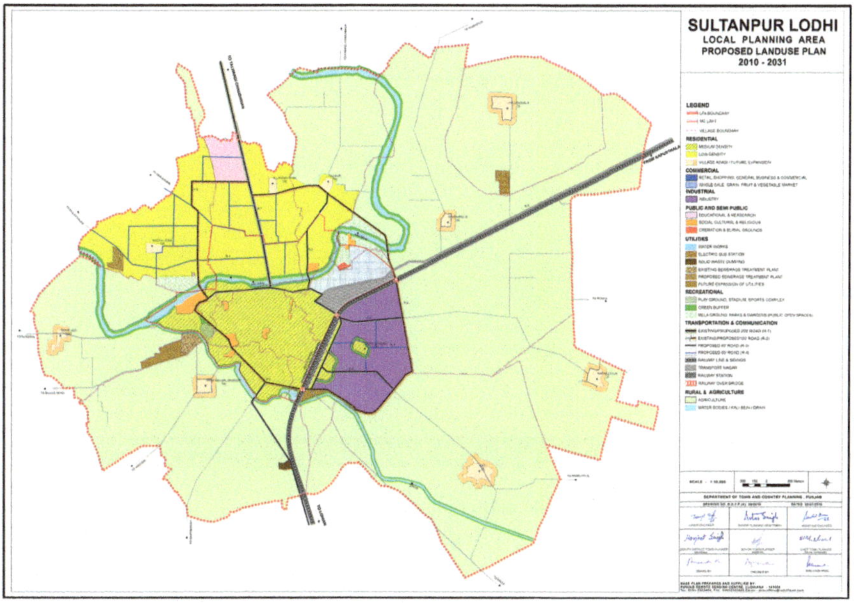

Fig. 3 Master plan of Sultanpur Lodhi

the increased volume of traffic in future, the concept of ring radial road has been proposed. The ring road is proposed to be developed on the existing revenue rasta and other roads wherever possible. The existing roads have been adopted as radial roads. Efforts have been made to follow the existing roads wherever available. The proposed ring road is expected to free Sultanpur Lodhi from the heavy traffic passing through the town. The road hierarchy proposed in the master plan includes R-1 (200 feet wide), R-2 (150 feet wide), R-3 (100 feet wide), R-4 (80 feet wide), R-5 (60 feet wide) with no construction zone of 5 mts on either side of the road.

However, to keep the public well informed of the master plan, the cadastral information was overlaid on the master plan of Sultanpur Lodhi and the data hosted on the website (http://www.pbhousing.gov.in/) (Fig. 4).

The website has been inaugurated by Hon'ble Chief Minister of Punjab, on 1 September 2017. The overlay of cadastral information on the master plan and hosting the information on the web, the applicant can check the proposed land use khasra wise (for each parcel of land as per cadastral information) online. The information of master plans may be accessed up to killa level on the digital map by selecting the same through a dropdown menu which helps to select the village, musteel no and finally to killa level. This would help to reserve the land allocated for ring road or widening of roads as per master plan and acquiring land as per master plan becomes legally viable.

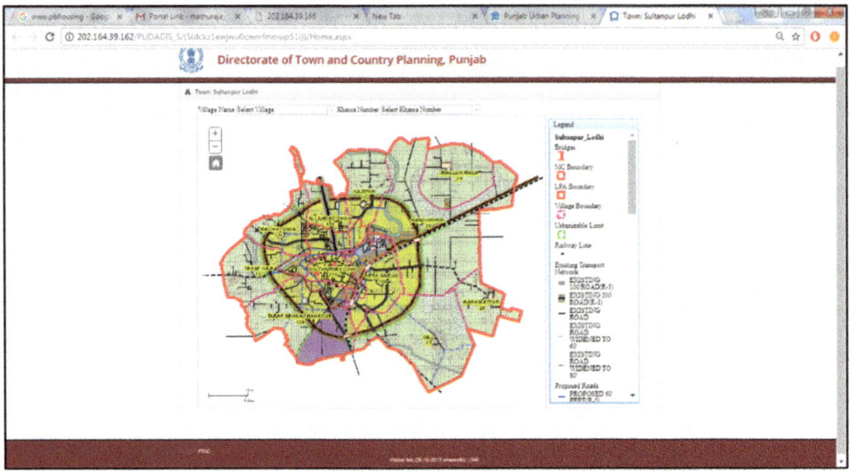

Fig. 4 Cadastral information with the master plan

5 Conclusion

The geospatial technology was used extensively to prepare the existing land use plan, the basis of finalizing the master plan of Sultanpur Lodhi. The satellite data provides truthful information of the earth's surface. The existing land use plan of the LPA was quickly finalized using satellite data, which is not possible with the traditional manual methods. The time lag would have with the traditional method changed the land use of the area by the time master plan was formulated and defeat the very purpose. The satellite data provided a quick means of preparing the master plan including that is related to road network. The ELU provided a means to develop the existing roads and link the existing roads with the proposed ring roads. Further, it helped in aligning the ring road through areas not located on forests, built-up, avoided rivers, avoided religious places to achieve an alignment practically and economically viable. The legalization of the alignment of ring road or widening of the existing road through the master plan notification makes land acquisition an easy proposition.

References

1. https://www.dda.org.in/planning/master_plans.htm
2. http://www.puda.gov.in/
3. http://www.pbhousing.gov.in/

Route Analysis of Hyderabad City Using Geomatics Application—A Case Study

Bipin Chand Pandey

Abstract Surveying technology is enhancing the process of old method of collecting geospatial data, through coming new hardware and software. Previously, the data collection and handling both are very tedious work, but nowadays, both are going to be easier and faster for a trained person. If data quality is good as per user demand and makes the availability of data is easy to some extent. So, it will be a new era for real analysis of ground in any field. Compared to other parts of the world, urbanization in India is rapidly growing. At present, due to urbanization and population growth, urban transportation system and its management is a challenging task for all over the world. However with the development of a proper methodology and its systematic approach, transportation problems can be resolved. Network Analyst tool provides a network-based spatial analysis tool for solving complex routing problems. Network Analysis data can be shared through the server to improve data analysis and its productivity. Route problems can be solved using a topologically correct network dataset. The study of Hyderabad City area is basically done to keep in mind about the problem faced mostly in the unprecedented developing urban areas, where the population is growing blindly in an unsystematic manner due to socio-economic issues. If the transportation system are not maintained properly and the current ground reality of transportation are not accessed quickly, then it may be difficult to rearrange it systematically in future for any growing city and also make a challenge for coming smart city.

Keywords Geospatial data · Network · Analysis · Diminishing · Unprecedented · Transportation

B. C. Pandey (✉)
Indian Institute of Surveying and Mapping, Survey of India, Hyderabad, India
e-mail: bcppandey@gmail.com

J. K. Ghosh and I. da Silva (eds.), *Applications of Geomatics in Civil Engineering*,
Lecture Notes in Civil Engineering 33, https://doi.org/10.1007/978-981-13-7067-0_28

1 Introduction

In topographical map, we are always seeing the most important and useful linear feature like roads. Roads are an essential part of transportation in any country, which creates a link between source to destination places, it may be village to village, city to city, state to state or from one country to another country.

Due to rapid growing urban infrastructure, it is a necessity of our society that we should follow a proper and systematic transportation system. In GIS technology, Network Analysis is a common tool, which is used in large scale for improving the transportation system.

Town and urban planners always designed a best road network, so that it should optimize the transportation safety, cost, land use and other environmental issues. The basic problem in transportation is to find out the optimal path between different locations on a network.

Here for our study, Arc Map is used as the main application software. In the Arc Map Network Analyst extension—Dijkstra's algorithm is used for finding shortest paths in route network.

The classic Dijkstra's algorithm solves the shortest path problem. To find a shortest path from a starting source location '***a***' to a destination location '***b***'. It picks the nearby unvisited node (vertex) and calculates the distance for each nearby node (vertex), and updates the nearer distance if smaller. The whole algorithm works until solving the nearest path from '***a***' to '***b***'.

Hyderabad is the largest city and capital of the southern state Telangana (India). The estimated population of Hyderabad is 8.7 million (approx.) in 2014, which makes it the fourth most populated city in India. It is a rapidly growing urban metro city with a huge road infrastructure, so in this study, it has been made to analyse the road network of the surrounding Hyderabad City (Telangana, India) area.

2 Objective

The study of this paper is to develop and create a best possible route network in metro cities. It is very helpful for future city/urban planning and healthy city development.

3 Data

3.1 Documents/Data Used

- Hyderabad Guide Map of edition 2014 (Survey of India).
- Soft Copy of Hyderabad Guide Map of edition 2000.
- Survey of India OSM Sheet No. E44M 6, 7, 10, 11 of Scale 1 : 50,000.

- Imageries downloaded from Google Earth Pro for digitization/updation.

3.2 *Hardware/Software Used*

- A high configuration computer system to carry out the project.
  (Processor-i3, Hard Disk-500 GB, RAM-2 GB, Latest Monitor Screen with keyboard and mouse).
- HP Design jet 500PS Plotter to take a hard copy of imageries/Map/necessary plots.
- Arc Map software available/new version, Erdas Imagine 10.1, Microstation V8.
- Internet connection for downloading imageries (if latest imagery not available for updation).
- All software tool of Microsoft office available/new version.

4 Study Area

The study and analysis area of this paper is the surrounding Hyderabad City, Telangana India, and the area is lying between latitude of 17 : 15 : 00 and 17 : 37 : 00 (approximately) and longitude of 78 : 15 : 00 and 78 : 43 : 00 (approximately). It covers an area of 2290 Km^2. The area falls in Survey of India OSM Sheet No. E44M 6, 7, 10 and 11.

5 Methodology

Proper planning was done before the initialization of this study. Geospatial data has prepared using software Microstation V-8 and Arc GIS for specified area.

Geo-referenced Hyderabad city Guide Map and update with the help of Mosaic and Geo-referenced Google imagery. After updation of vector data, quality control and validation were also completed for the whole vector data. Then prepared GIS layer like road and railway. Finally, Route Network Analysis has been done of different parts of the study area.

The study was executed in a number of systematic stages as shown in the workflow diagram:

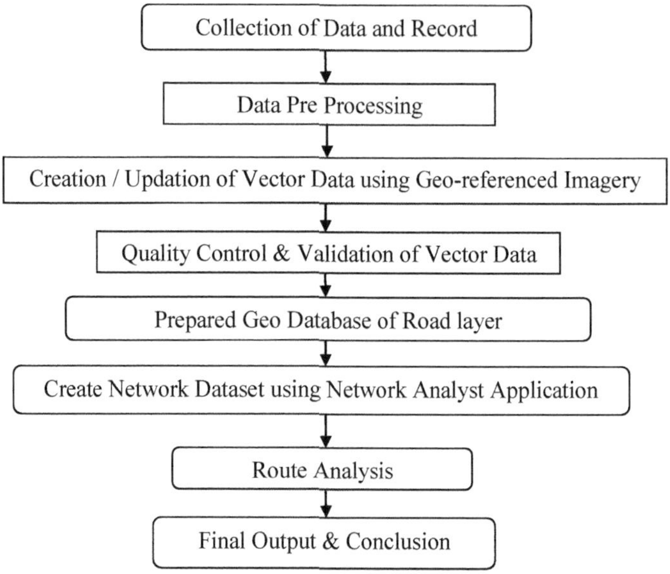

5.1 *Network Analysis*

A network is a system, where elements are connected through points and line on junction. It helps to find out the location and the possible routes of different places.

Analysis is a systematic examination of something in detail in order to understand it better. Network analysis is a method of solving network problem using network connectivity.

Networks include many features like highways connecting to cities, streets are interconnected to each other at street junctions, sewer and water lines that connect to residential, commercial and industrial sectors.

Connectivity is important in order to travel over the network. Network elements, such as edges and junctions, must be interconnected properly to allow navigation over the network. These elements have their spatial properties to control navigation on the network. To do network analysis first, a network dataset was created with the help of topologically validated spatial network database. It is used to find the shortest path between the source and destination.

5.2 *Network Dataset*

Spatial data from Geographic Information System database is being used more and more in transportation planning due to the convenient structure. They provide for entering, viewing and manipulating spatially oriented data.

An application of GIS in the traffic safety area has been limited mostly to visual representation of accident locations. Network dataset is built mainly from two GIS data layers. They are major road network which is captured as line features, and junctions and important landmarks that are captured as point features. These two are playing a prominent role in keeping the network alive at all times.

Road network is properly connected in GIS with junctions and important landmarks. Necessary attribute data such as name of roads, length of roads, name of the junctions and important landmarks have been given as input in network dataset.

5.3 *Route Analysis*

Route analysis can be found out quickly, after selection and giving proper impedance input during analysis. Shortest path is only where least cost with lowest impedance. To find out best route, any cost attribute can be used as the impedance. If the impedance is time, then best route is quickest route, but if impedance is time attribute with live, then best and quickest route is dependent upon the date and time of the day or historical traffic.

5.3.1 Route Analysis Layer

All the inputs, parameters, and results of a route analysis stored in route analysis layer. A route analysis layer can be created from the Network Analyst toolbar—**Network Analyst → New Route** (Fig. 1).

Figure 2: The route analysis layer appears in the Table of Contents as a composite layer named Route. There are five feature layers—Stops, Point Barriers, Routes, Line Barriers, and Polygon Barriers, has default symbology, it can be modified using Layer Properties dialog box.

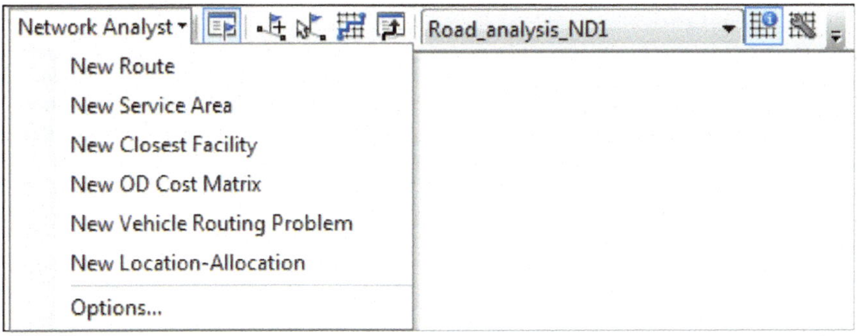

Fig. 1 Network analyst applications

Fig. 2 Route (table of contents)

5.3.2 Route Analysis Classes

The route analysis layer is classified into different network analysis classes—stop class and stop properties, route class and route properties, point, line and polygon barrier.

Stops Class

Stop class stores the network locations that are used as stops in a route analysis. The Stops layer has four default symbols: located, unlocated, errors, time violations (Fig. 2).

Stop Properties

In network analysis, stop properties are only available when a start time is defined or time windows are enabled. Time window is an attribute table which has information about start time, end time, arrival time, and drive time for different routes.

Routes Class

The Routes class gives the output of routes only. It gives the output when the route analysis is complete. Once the best route is found, it is displayed in the Network

Fig. 3 Routes after analysis

Analyst window. The Routes class stores the resulting routes. Symbology can be changed from its Layer Properties dialog box (Fig. 3).

Routes Properties

The route properties are ObjectID, Name, FirstStopID, LastStopID, StopCount, Total_[Impedance], TotalWait_[Impedance], TotalViolation_[Impedance], Start-Time, EndTime.

5.3.3 Point, Line, and Polygon Barriers

Point, Line and Polygon barrier are basically used for diversion of the traffic. It temporarily restricts the route network. If it is compulsory to add barriers, then barriers are allowed by the network analyst to restrict particular place (it may be point or line or polygon barrier) as per ground reality.

5.4 Route Analysis Parameters

Route Analysis parameters are set on the Layer Properties dialog box (Fig. 4).

5.4.1 Impedance

For finding out the best and shortest route, any cost attribute can be selected as impedance. Impedance plays an important role to select best possible routes during network analysis.

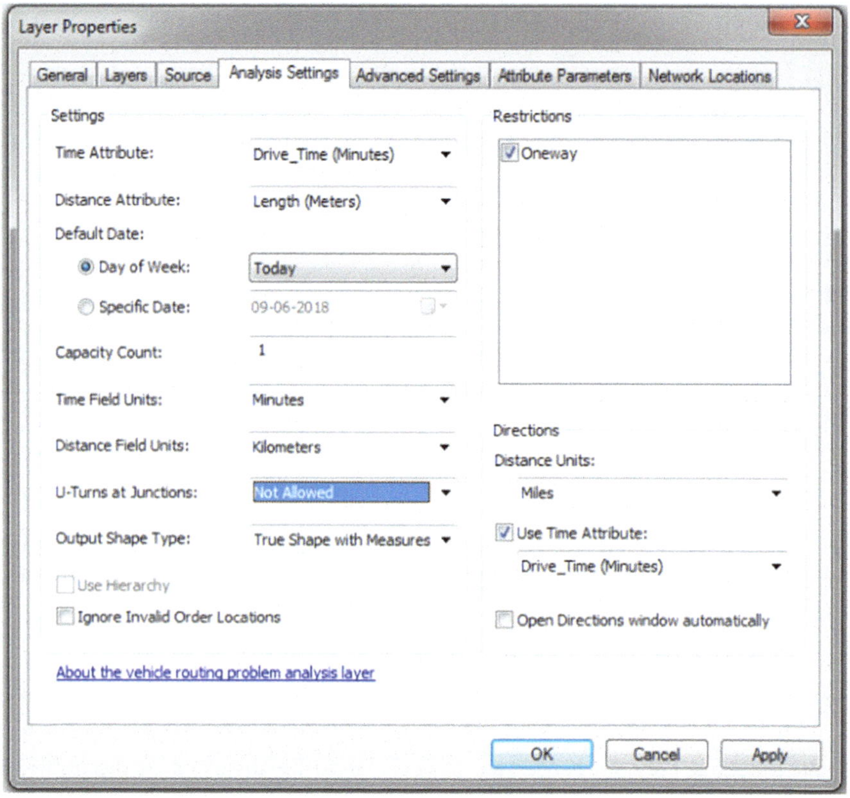

Fig. 4 Layer properties window (analysis settings)

5.4.2 Restrictions

At the time of solving network analysis restriction attribute can be selected. The restriction may be caused on any route due to point, line or any polygon barrier. It attributes are like One-way, U-turn, Height limit, Weight limit, avoid certain roads or bridges, etc. Mostly, restrictions cause roads to be prohibited. As per suitability restriction, attribute can be set during network analysis.

It depends upon the network analyst solver that U-turns can allow everywhere or nowhere as per ground reality, or only at dead ends or only at intersections and dead ends. U-turns allows that vehicle can turn around at a junction and double back on the same street.

5.4.3 Directions

Directions properties can be set at the time of creating new network dataset, set the units for display length and time attribute, to open directions automatically after the generation of a route. If display of direction property not required, unselect the direction window button  (Fig. 5).

6 Observation

Nowadays with developing urbanization, the systematic and well planned road network system can play an effective role to customize the transportation system in a planned way. Therefore, it is necessary that the Agencies/Organizations, those who are maintaining the geospatial data of whole country, if that geospatial GIS layer of different features are available with free from topological error and linked with GPS services and authenticated web server then it will be helpful to detect and analyse many transportation and other problems in an organized way.

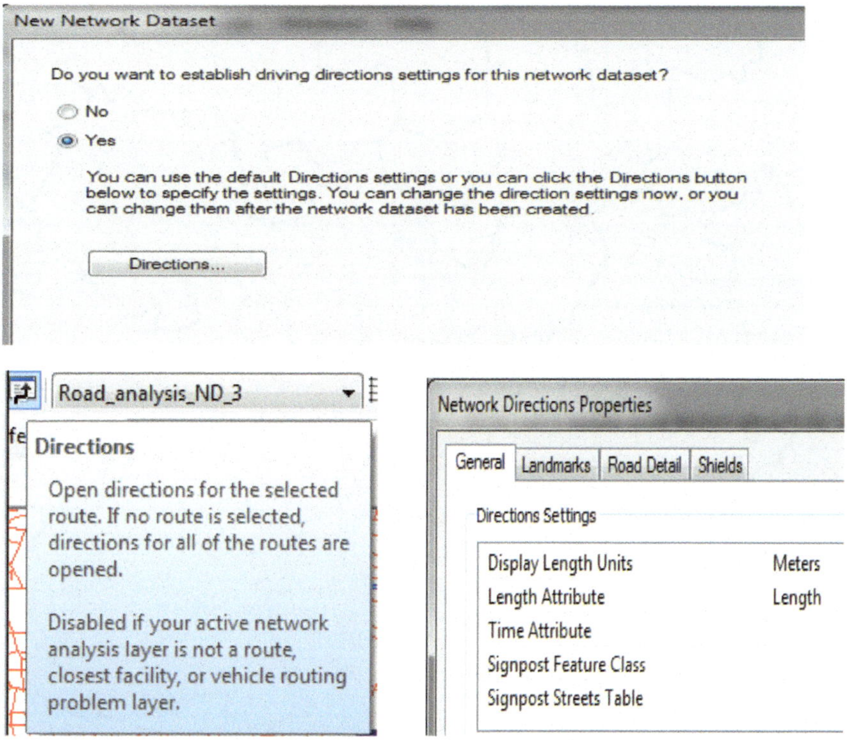

Fig. 5 Network directions setting windows

7 Scope of Paper

- Route analysis will be very useful for Crime Route analysis, and is also used to solve vehicle routing problem.
- Geo coding can also be done. For example, we can find out the details of every house by telephone no.s/mobile no. It is easy for tax collection and development funds.
- We can also add and analyse another layer into this geodatabase like Water Pipe line, Sever Line, Gas Pipe Line and underground electrical lines, etc.

8 Final Output

The outputs of Route analysis are as follows (Figs. 6, 7, 8, 9, 10, 11 and 12).

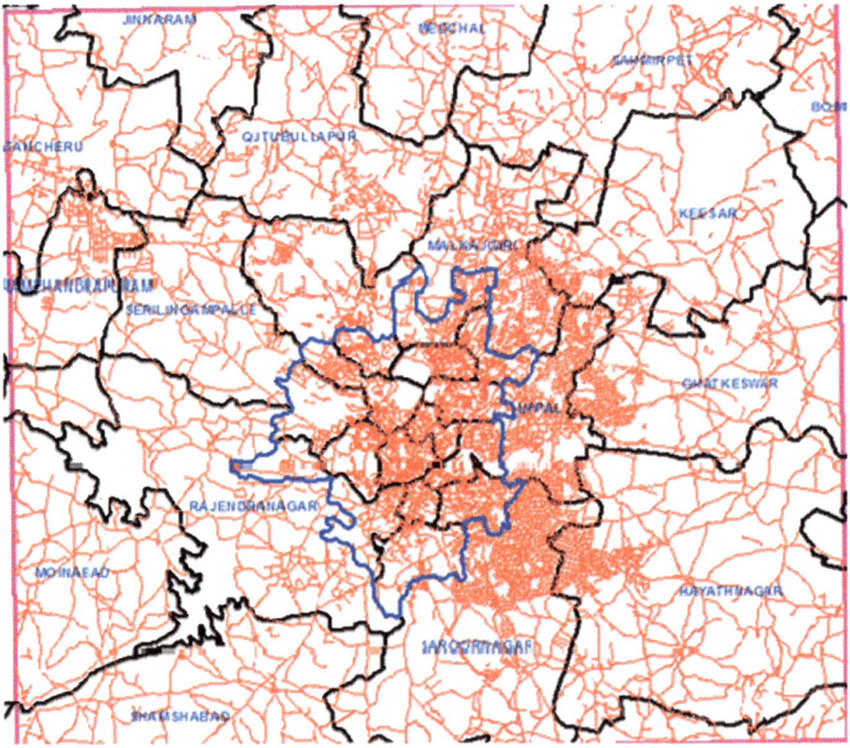

Fig. 6 Study area with mandal boundary

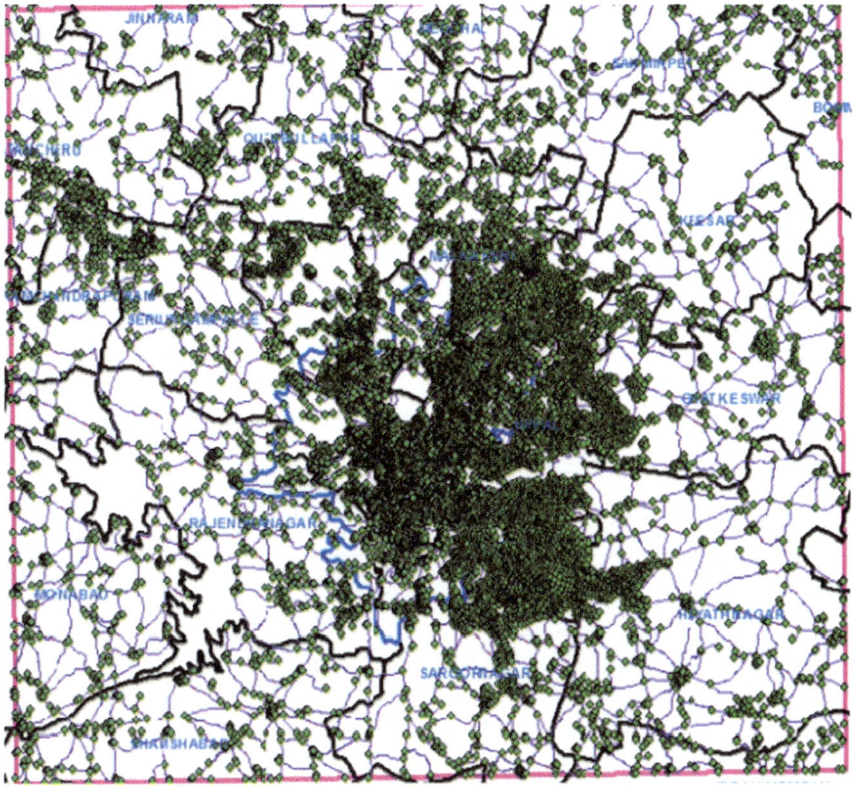

Fig. 7 Network dataset output

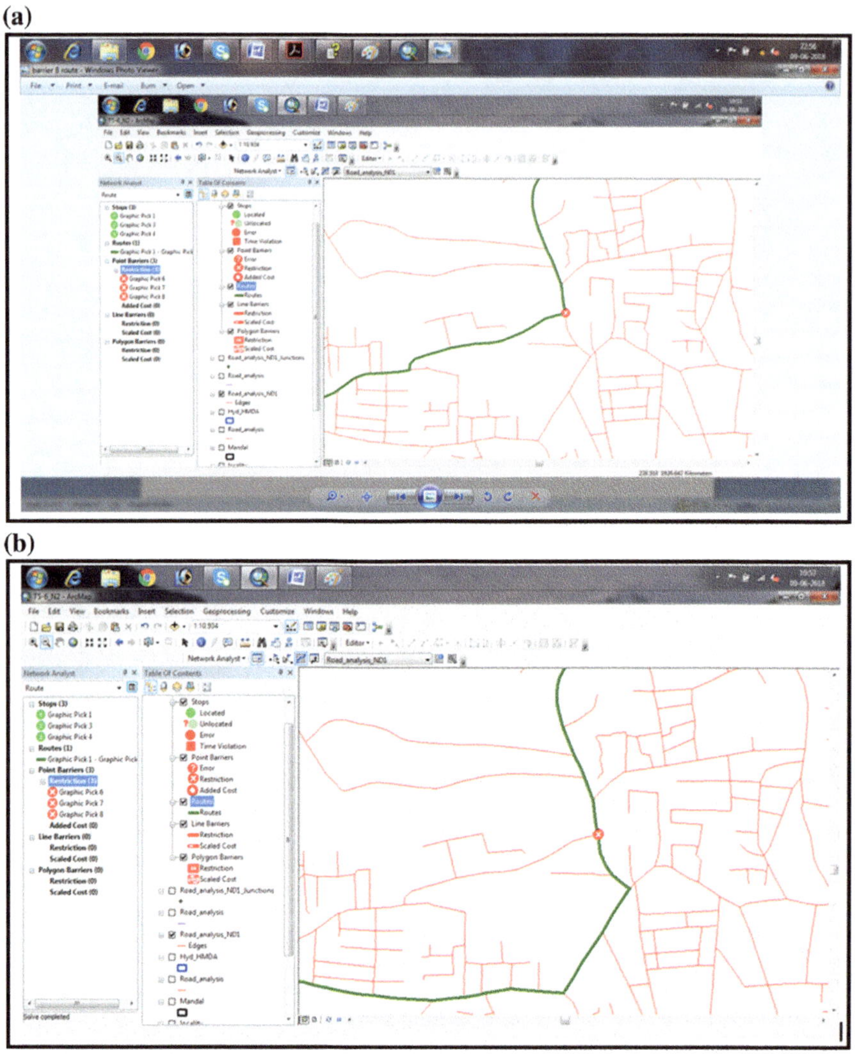

Fig. 8 Point barrier output

Fig. 9 **a** Line barrier output. **b**, **c** Polygon barrier output

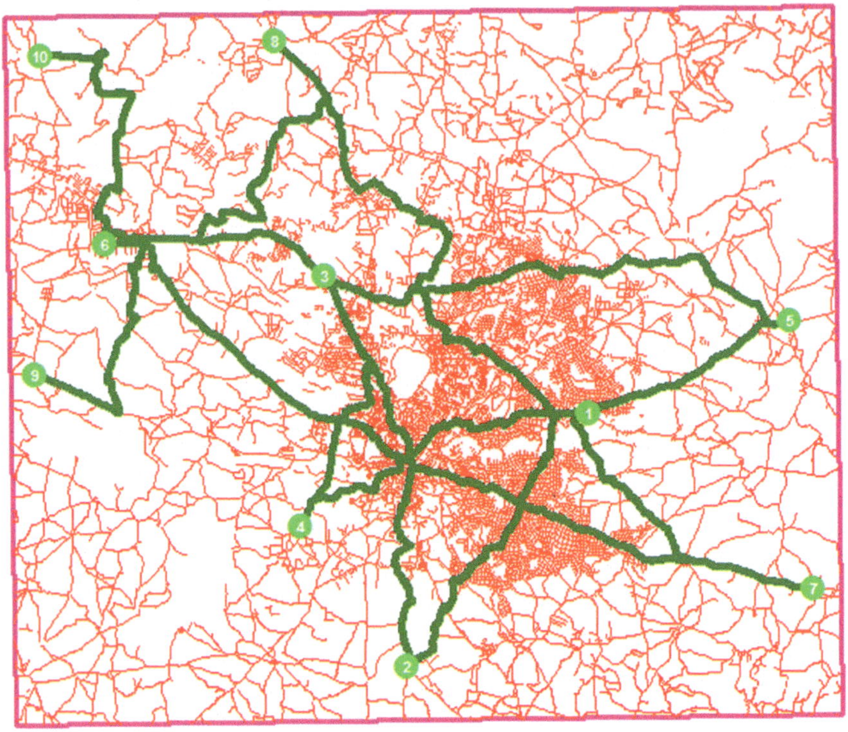

Fig. 10 Different routes network

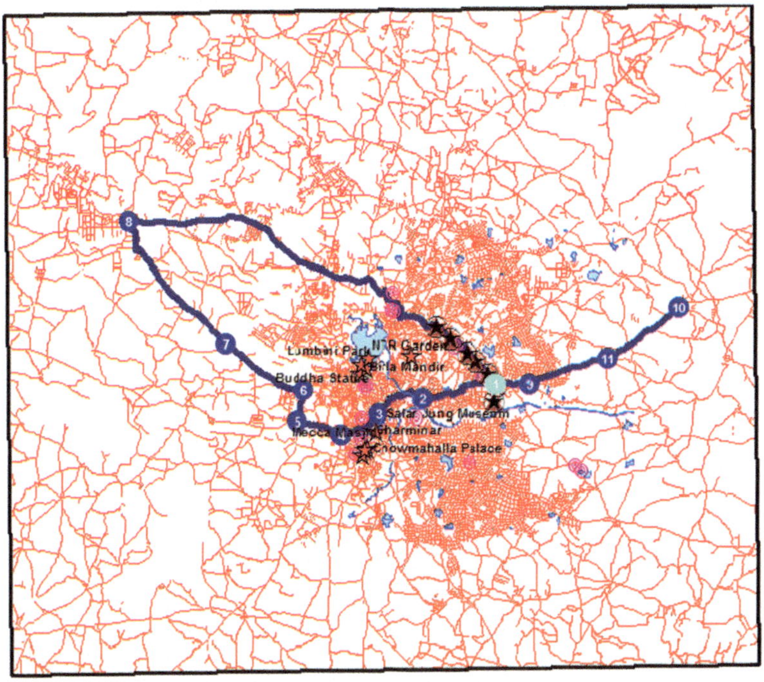

Fig. 11 Route layout in sequence (from 1 to 10)

12: 6936.1 m Turn left to stay on Primary 760.0 m 14 min

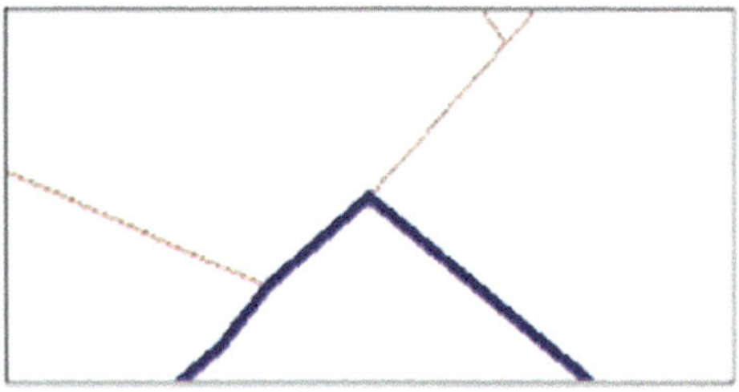

13: 7696.1 m Turn right on Primary and immediately turn right on Primary 291.1 m 5 min

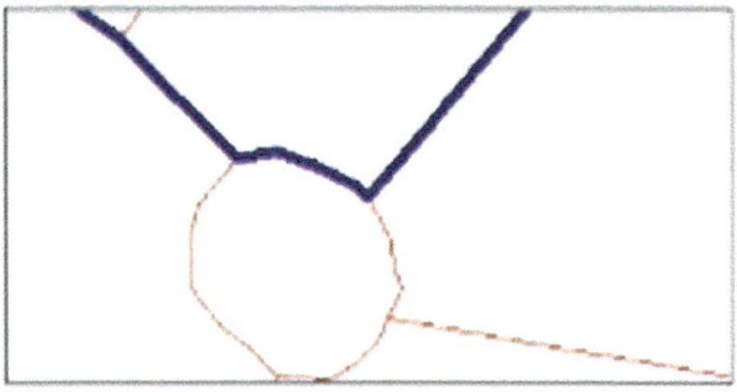

14: 7987.2 m Turn left to stay on Primary 709.2 m 13 min

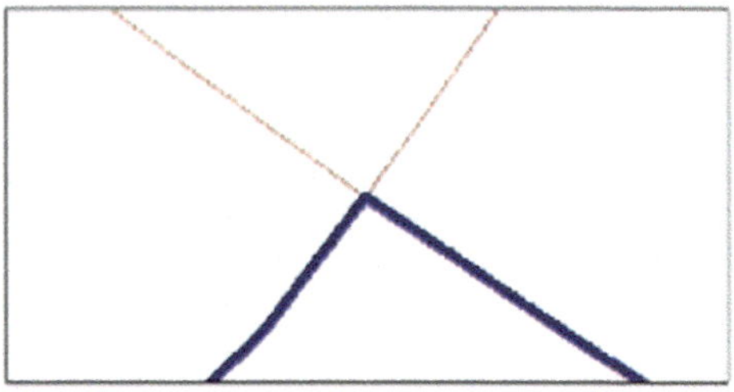

Fig. 12 Route with direction distance and drive time

9 Conclusion

In this study and analysis, it demonstrated about the network analyst applications of new route. Shortest path analysis has given the optimal path between two destinations, when there is some obstruction in route. When this data is web-enabled along with GPS services, then online queries can be made all over the world. On having topologically correct geospatial data from any part of the country/world and along with proper impedance parameter, this study can be expanded on a large scale.

References

1. Network Analyst Tutorial Arc GIS desktop
2. Surveyofindia.gov.in
3. Googleearth.com

Identification and Removal of Accident-Prone Locations Using Spatial Data Mining

Rashmi A. Mestri, Ravindra R. Rathod and Rahul Dev Garg

Abstract Road accidents are responsible for a great amount of morbidity and mortality, hence, public health safety on transport channels is a major concern in the present scenario. The present study shows a method for identification and removal of accident-prone locations on roads (hot spots). The proposed methodology includes road network analysis followed by hot spot analysis using Kernel Density Estimation (KDE), buffer operation of Geographic Information System (GIS) and spatial data mining techniques. Various causes for occurrences of accidents are further correlated with identified hot spots. For the present study, accident records on internal roads of an engineering institute have been used. The results reflect that the junction points and low visibility at turns are the main causes behind the occurrences of accidents at identified hot spots. The obtained results can be further used by institute administration for implementing accident prevention measures and removal of hot spots.

Keywords Road accident · Road network analysis · Hot spot · Kernel Density Estimation (KDE) · Buffer operation

1 Introduction

Road accidents resulting in injuries and fatalities to people around the world and hence, it is globally recognized as a serious public health problem. In the year 2016; 480,652 accidents were reported which caused 150,785 deaths and injuries to 491,624 persons [1]. It has been also observed that nearly 37% of total accidents took place at

R. A. Mestri (✉) · R. R. Rathod
Walchand College of Engineering, Sangli, Maharashtra, India
e-mail: rashmi.mestri07@gmail.com

R. R. Rathod
e-mail: rrr.wce@gmail.com

R. D. Garg
Indian Institute of Technology Roorkee, Roorkee, Uttarakhand, India
e-mail: garg_fce@iitr.ac.in

J. K. Ghosh and I. da Silva (eds.), *Applications of Geomatics in Civil Engineering*, Lecture Notes in Civil Engineering 33, https://doi.org/10.1007/978-981-13-7067-0_29

junctions as they are points of conflicts [1]. Mostly, road accidents are caused due to driver's mistakes, a few of these causes behind road accidents are listed as follows:

i. Human factors such as driver's behavioural factors [2, 3], visibility range of driver [4] and visual obstruction due to obstacles, etc.
ii. Rapid increase in number of vehicles running on roads.
iii. The improper interactions between the pedestrians and the vehicles.
iv. The improper road designs with turns, intersection and banking of roads.
v. Pavement surface characteristics [5] like friction, unevenness, etc.
vi. Environmental conditions (fog, rain, dusk, etc.).

The focus of the present study is to present a methodology to identify the accident-prone locations (hot spots) using spatial data mining methods and finding the causes behind them and proposing a solution to remove them. Spatial data mining discovers previously unknown but useful patterns from the spatial database. A better understanding of the spatial distribution of the road accidents plays a very important role in addressing road safety problems. Hence, in recent time spatial data mining on road accidents has received a great attention. The results of data mining approach for accident data helps to explore occurrence of accident patterns and predict future behavioural patterns for effective decision making to reduce accidents.

The main objective of the present study is to analyse the distribution of the accidents by means of road network analysis and hot spot analysis using spatial data mining. Road network analysis helps to solve many transportation engineering problems like routing, finding shortest path, etc. whereas with the help of GIS, it is very useful to analyse distribution of spatial as well as non-spatial data for solving such problems.

2 Related Work

Nowadays, it is observed a tendency to apply spatial data mining methods to detect locations with higher number of accident occurrences known as hot spots. The spatial attributes like location of accidents on roads, road designs and weather as well as non-spatial attributes such as traffic flow, human factors which contribute to road accidents are studied. The studies carried out for analysis of road accidents can be classified in technical and non-technical work.

Human factors [2–4] (old age, disability, fatigue, alcoholism), obstructed view due to trees or buildings and sometimes driver's/rider's misinterpretation of the oncoming vehicle due to inclination cause the road accidents. Such factors also have been studied to determine causes and influence on the road accidents.

Several studies have been carried out to find the distribution of road accidents with spatial data mining techniques and to identify the unusual concentration of accidents in areas, i.e. point pattern analysis [6, 7]. Areas with higher concentration of accident events are known as hot spots. Hot spot identification is a process of finding locations on roads that suffer from high risk of accidents. Hot spots are identified based on some precise selection criteria such as collision frequency and severity of collision. Hot spot analysis [8] is normally carried out in four phases namely identification of hot spots, ranking of hot spots, profiling of hot spots and selection of region/sites of interest. To minimize the number of road accidents, it is important to know the location/region where the concentration of road accidents occurs.

Major studies have used Kernel Density Estimation (KDE) [9–14] method to estimate the accident occurrence intensity for identification of hot spots. KDE represents the set of accident occurrences and creates density levels to give the intensity of the hot spots. Nearest Neighbour distance analysis [11, 15] has been also carried out to determine whether the accidents are clustered or not based on the distances between neighbouring accident events pertained to locations.

Local method of spatial autocorrelation [16, 17] has been used to identify statistically significant clusters of accident locations. Moran's I [14, 16, 17] method is popularly used to indicate the level of spatial autocorrelation at local scale and evaluates the pattern of accidents into clustered, dispersed or random manner.

Various other studies include spatio-temporal analysis [6, 17] carried out to identify hot spots and temporal causes behind accidents and also analyses the impact of seasonal variations on the accidents.

3 Methodology

The proposed methodology for the present study is shown in Fig. 1. The road network has been created and used for road network analysis for identification of weighted road segments. The identification of hot spots has been carried out using KDE and buffering followed by correlation of causes behind accidents to identified hot spots. Obtained results are used to propose solutions for hot spot removal.

3.1 Creation of Road Network

Road network is a system of interconnected road segments which are mainly used for transportation purpose. Road network consists of intersections, turns, vehicle lanes, footpath, etc. This road network is used to carry out spatial analysis for identification of hot spots.

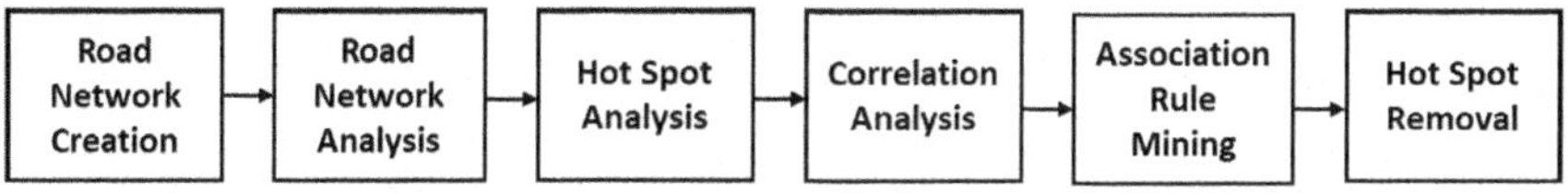

Fig. 1 Proposed methodology

3.2 Road Network Analysis

Roads interconnections and their spatial relations, generate a specific network structure. The road network analysis allows us to solve problems like finding the most efficient travel route, creating route directions, finding nearby facilities or services.

In the present study, internal roads are assigned weightage on the basis of road geometry and traffic characteristics such as width, length, number of lanes, number of linked roads and traffic flow on the road. On the basis of assigned weights, important roads are selected against the threshold weight and considered for hot spot analysis. Figure 2 shows the block diagram for road network analysis and Table 1 shows the weight assignment.

3.3 Hot Spot Analysis

Important road segments are identified using road network analysis and used to perform hot spot analysis. Hot spot analysis consists of two phases as in Fig. 3:

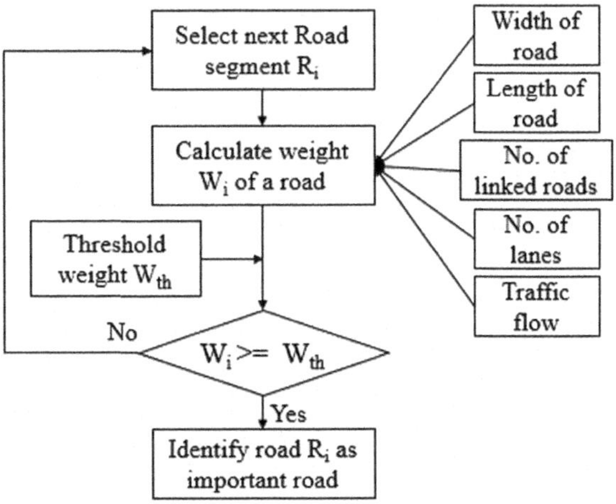

Fig. 2 Road network analysis

Table 1 Road weight calculation

Road width	Road length	No. of lanes	Traffic flow	No. of linked roads	Road weight
Wider	Long	≥2	High	>5	5
Wide	Long	2	High	2–5	4
Narrow	Medium	1	Medium	2	3
Narrow	Short	1	Medium	2	2
Narrower	Short	1	Low	1	1

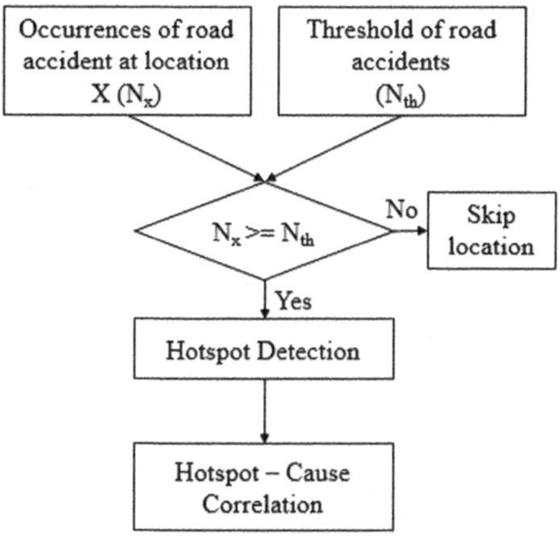

Fig. 3 Hot spot analysis

i. Hot spot detection, i.e. detection of spatial clusters of road accidents with high risk
ii. Correlation of hot spot with the cause behind hot spot.

Kernel Density Estimation. Hot spots are detected by using KDE technique [9, 10] which explores spatial patterns of road accidents. In this technique; a circular surface defined by a kernel function Eq. (1) is positioned over every accident location that results in an individual density surface.

$$f(x, y) = \frac{1}{nh^2} \sum_i^n K\left(\frac{d_i}{h}\right) \tag{1}$$

where

h is bandwidth around location (x, y),
n is the number of observations,
K is function of kernel density,
d_i is the distance between the location (x, y) and ith observation.

A grid of cells is then placed over the study area and total kernel density is calculated by adding the overlapping density surfaces resulted from each accident location.

The bandwidth around accident location affects the KDE. The bandwidth needs to be selected wisely as smaller bandwidth will generate a finer mesh density estimates with finer spikes and a larger bandwidth will produce a smoother density estimate but with less variability.

Vector Spatial Analysis Using Buffer Operation. In GIS, the buffer is a zone around a spatial feature. Buffering [18] is a process of creating such buffer zones around target feature, within a predefined distance from the feature. These resulting buffers may get automatically merged where they overlap. The larger the buffer zone distance, larger the spatial features fall into buffer zones.

Buffering is also used to identify hot spots by creating buffer zones around road accident locations (spatial point data) with predefined distance, so the number of accident occurrences within merged buffer zone is considered for hot spot identification by comparing them with some threshold value. Like KDE method, selection of distance to create buffer zones plays important role for density evaluation.

3.4 Correlation Analysis

A correlation analysis allows to search for relationship between the hot spot location and the cause of the accident. The causes for accidents at hot spots [19] can be deadly curves on roads [12, 20], obstructed view due to trees or buildings at intersections, inclination of the roads, vehicle speed factors, population density within the area, etc.

3.5 Association Rule Mining

A data mining technique that extracts interesting rules between various attributes in a large data set is known as association rule mining [21] which is based on market basket analysis. Association rules $A \rightarrow B$ holds in transaction set (where A and B are sets of items) with support *s*, where *s* is the percentage of transactions that contain $A \cup B$ and confidence *c*, where *c* is the percentage of transactions containing A that also contain B. An analysis of accident type has been carried out by finding frequent itemsets which satisfy minimum support threshold value and by generating association rules using Apriori [21, 22] algorithm.

3.6 Hot Spot Removal

Accident causes are correlated with identified hot spots and possible solutions are proposed in order to remove the hot spots. Following solutions are proposed for hot spot removal.

i. Revised road planning in terms of road design.
ii. Suggesting alternate safer road.

iii. Installing traffic control measures like speed breakers and appropriate sign-boards.

4 Results and Discussion

The proposed methodology has been used on road network of Walchand College of Engineering (WCE) and reported accidents in campus. The institute campus has a campus area of around 100 acres having around 3.27 km total run length of internal road network, around 50–100 four wheelers, 400–500 two wheelers, hundreds of cyclists and around 2500–3000 peoples.

It has been observed that most of the accidents occurred at junction points and turns. Very few accident occurrences are detected on straight roads. Road network analysis is performed followed by hot spot analysis to identify hot spots.

4.1 Road Network Analysis

Open Source Software—QGIS is used to carry out hot spot analysis on accident data collected for Walchand College of Engineering (WCE), Sangli.

Road network analysis is performed on road segments which resulted in the following weight assignments as shown in Table 2. Road network analysis identified four road segments having weights greater than the predefined threshold weight (W = 3) out of which two road segments Road_1 and Road_2 are within the campus. Figure 4 shows different internal road segments with identification numbers from Road_1 to Road_7 connected to two external roads (a state highway and a city road) with assigned weights showed with different widths and accident occurrences are plotted as points (red circles). Table 2 shows the reported 18 occurrences of accidents in study area.

4.2 Hot Spot Analysis

Hot spot analysis is carried out to identify hot spots using KDE and buffer analysis method. The results detected two hot spots as shown in Fig. 5. Accident densities estimated using KDE is represented with heat map in Fig. 5a and buffering resulted in eight clusters out of which two clusters containing accident occurrences greater than threshold (3) are identified as hot spots shown in Fig. 5b.

Table 2 Accident records at study area

Sr. No.	Accident location	Road segment	Road width	Road length	Traffic flow	No. of lanes	No. of links	Weight	Accident type
1	Back gate	Road_8	Narrow	Medium	Medium	1	4	4	Injury
2	Staff quarter	Road_2	Narrow	Medium	Medium	1	5	3	Injury
3	Front gate	Road_1	Wide	Long	High	2	5	4	Injury
4	Font gate road	Highway	Wider	Long	High	2	7	5	Fatal
5	Front gate	Road_1	Wide	Long	High	2	5	4	Injury
6	Director house	Road_1	Wide	Long	High	2	5	4	Injury
7	Lipton	Road_1	Wide	Long	High	2	5	4	Injury
8	Cafeteria corner	Road_2	Narrow	Medium	Medium	1	5	3	Fatal
9	Cafeteria	Road_2	Narrow	Medium	Medium	1	5	3	Injury
10	Chemistry lab	Road_3	Narrow	Medium	Low	1	3	2	Injury
11	Rector office	Road_1	Wide	Long	Medium	2	5	4	Injury
12	Front gate	Road_1	Wide	Long	High	2	5	4	Injury
13	Cafeteria corner	Road_2	Narrow	Long	Medium	1	5	3	Injury
14	Back gate	Road_8	Narrow	Long	High	1	4	4	Injury
15	Mini CCF	Road_3	Wide	Long	High	1	3	2	Injury
16	SLI-MRJ road	Highway	Wider	Long	High	2	7	5	Fatal
17	Main parking	Road_1	Narrow	Long	Medium	2	5	4	Injury

4.3 Correlation Analysis

Causes behind hot spots identified in the present study are obstructed view at junction points and high traffic flow. These are further elaborated as follows:

i. T-Junction:
   The entrance of campus at north which is connected to external state highway having high traffic flow which forms T-junction. Due to difference in traffic rate on both road and due to T-junction; frequency of road accidents is higher at this location.
ii. Obstructed view at sharp turns:
   Inadequate sight distance (viewing angle) due to sharp turns causes low intersection visibility at these locations which resulted in accidents. Sharp turns having acute angle lowers the vision of driver even though drivers pay attention to road

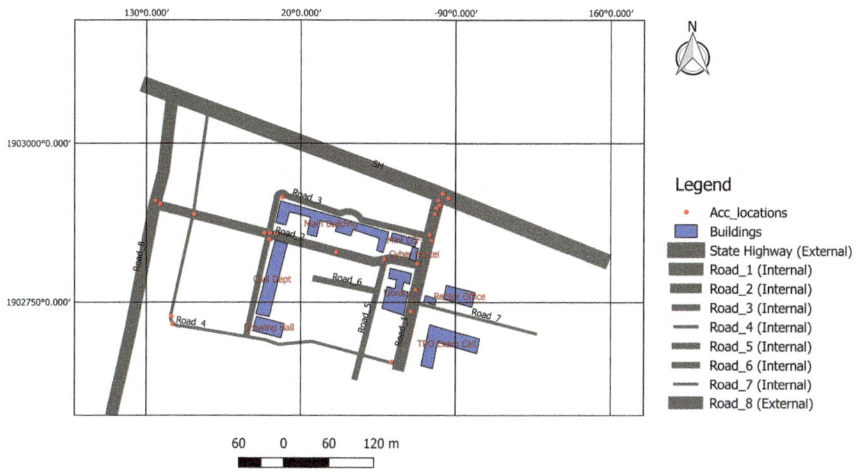

Fig. 4 Accident occurrences in study area

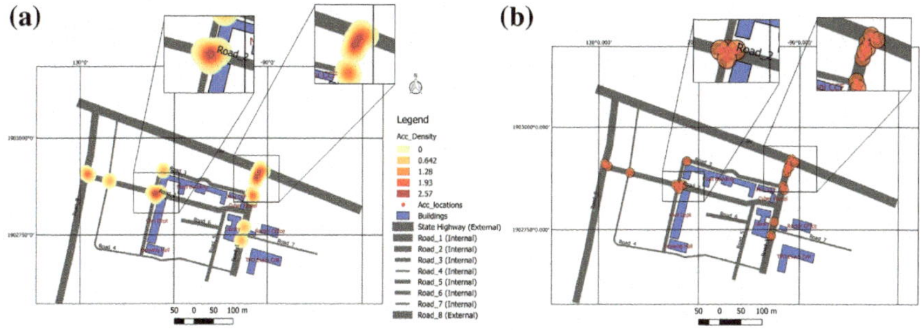

Fig. 5 Hot spot analysis **a** using KDE, **b** using buffering

Table 3 Association rules

Rule No.	Rule body
1	{Wide Road, Long Length, High Traffic Flow} → {Injury}
2	{Narrow Road, Medium Traffic Flow} → {Injury}
3	{Highway, Wider Road, High Traffic Flow} → {Fatal}
4	{Long Road, Medium Traffic Flow, Links 3–5} → {Injury}

but may misjudge the danger of forthcoming curves. Traffic flows from all directions at intersecting road segments increases the probability of the road accidents.

4.4 Association Rule Mining

Apriori algorithm of association rule mining produced following 4 best rules as shown in Table 3, with support 40% and confidence 50%, indicate the type of accident either injury or fatal. These rules imply that the type of accident is associated with road geometry and traffic flow. Highways are having more chances of fatal accidents while internal roads have injured accident occurrences.

4.5 Hot Spot Removal

To remove these identified hot spots, the following measures are suggested:

i. Removal of manmade obstacles which obstructs driver's view, increasing angle of curves on roads to increase visibility of road to driver.
ii. Installation of reflectors and proper sign boards at junctions with obstructed view, so that driver can have proper perception of oncoming vehicles.
iii. Limiting speed of vehicles by constructing speed breakers and installing sign board at proper locations.

5 Conclusions and Future Scope

The present study identifies important road segments using road network analysis based on the various parameters and the spatial distribution of road accidents using KDE method. Two hot spots are detected with hot spot analysis. Main causes have been identified as low visibility at sharp turns and at junction points and lack of speed control mechanisms on straight road segments. Based on the identified causes and suggested solutions for removal of hot spots, administrations can adopt corrective measures to remove the accident-prone locations identified in road network.

Analysis of curves on roads to find angle of curvature will be the motivation for future research to predict the risk of accident occurrence at turns. If the vision angle is smaller and it is accident-prone location; some alert system can be designed as preventive measures. Application of methodology in the present study to identify and remove hot spots on other road types like regional road, state roads and highways and the comparison of the results will be a significant avenue for future research.

References

1. Govt. of India (2017) Ministry of Road Transport Highways, Road Accidents In India—2016
2. Petridou E, Moustaki M (2000) Human factors in the causation of road traffic crashes. Springer 16(9):819–826
3. Issa Y (2016) Effect of driver's personal characteristics on traffic accidents in Tabuk city in Saudi Arabia. J Transp Lit 10(3):25–29
4. Kim Y (2001) Effects of driver, vehicle, and environment characteristics on collision warning system design
5. Suryakanta (2015) 4 Must consider characteristics of a pavement surface—CivilBlog.Org. http://civilblog.org/2015/09/07/4-must-consider-characteristics-of-a-pavement-surface/. Accessed 22 Aug 2017
6. Brockmann D, Hufnagel L, Geisel T (2006) Data mining and knowledge discovery handbook
7. Mennis J, Guo D (2009) Spatial data mining and geographic knowledge discovery-An introduction. Comput Environ Urban Syst 33(6):403–408
8. Moons E, Brijs T, Wets G (2008) Hot spot analysis: improving a local indicator of spatial association for application in traffic safety. In: Computational science and its applications—ICCSA, pp 221–231
9. Anderson TK (2009) Kernel density estimation and K-means clustering to profile road accident hotspots. Accid Anal Prev 41(3):359–364
10. Thakali L, Kwon TJ, Fu L (2015) Identification of crash hotspots using kernel density estimation and kriging methods: a comparison. J Modern Transp 23(2):93–106
11. Ali G, Famili A (2017) GIS-based spatial analysis of urban traffic accidents: case study in Mashhad, Iran. J Traffic Transp Eng (English Ed) 4(3):290–299
12. Mohaymany AS, Shahri M, Mirbagheri B (2013) GIS-based method for detecting high-crash-risk road segments using network kernel density estimation. Geo-spat Inf Sci 16(2):113–119
13. Loo BPY, Yao S, Wu J (2011) Spatial point analysis of road crashes in Shanghai: a GIS-based network kernel density method. In: Proceedings—2011 19th international conference on geoinformatics, Geoinformatics 2011
14. Aghajani MA (2017) Applying GIS to identify the spatial and temporal patterns of road accidents using spatial statistics (case study: Ilam Province, Iran) accide. Transp Res Procedia 25:2131–2143
15. Shariff SSR, Maad HA, Halim NNA, Derasit Z (2018) Determining hotspots of road accidents using spatial analysis, Indonesian. J Electr Eng Comput Sci 9(1):146–151
16. Deshpande N, Chanda I (2011) Accident mapping and analysis using geographical information systems. Int J Earth Sci Eng 4(6):342–345
17. Prasannakumar V, Vijith H, Charutha R, Geetha N (2011) Spatio-temporal clustering of road accidents: GIS based analysis and assessment. Procedia—Soc Behav Sci 21:317–325
18. de Smith MJ, Goodchild MF, Longley PA (2015) Geospatial analysis: a comprehensive guide to principles, techniques and software tools, 5th edn. The Winchelsea Press, Winchelsea, UK

19. Vanlaar W, Yannis G (2006) Perception of road accident causes. Accid Anal Prev 38(1):155–161
20. Wang C, Quddus MA, Ison SG (2013) The effect of traffic and road characteristics on road safety: A review and future research direction. Saf Sci 57:264–275
21. Han J, Kamber M (2006) Data mining concepts and techniques, 2nd edn. Elsevier Inc.
22. Agrawal R, Shafer JC (1996) Parallel mining of association rules. IEEE Trans Knowl Data Eng

Study of Pedestrian Movement in Correlation to the Transportation Infrastructure and Land Use/Land Cover in a Fast Developing Indian City

M. S. Mukesh and Y. B. Katpatal

Abstract The cities in developing countries are growing at a faster rate, which has ultimately generated stress on the transportation networks within them. During transportation planning, less attention is paid for pedestrian movement during the design of any street functionaries. So the study focuses on understanding the pedestrian movement in a developing country like India w.r.t., rate of development. Intersections located along the Nagpur inner ring road were considered and based on the trends in the development rate of built-up, study intersections were selected. The video data were acquired and Pearson's correlation coefficient and ANOVA analysis have been carried out to find the correlation between development and the pedestrian movement. From the analysis, it is observed that the development rate and infrastructure are directly related to the pedestrian movement at the intersection.

Keywords Pedestrian behavior · Pedestrian movement · Development rate · Built-up density · GIS

1 Introduction

The transportation functionaries design mainly concentrates on the vehicle and geometric aspect of parameters concerning to mostly operating speed and safety of vehicles and less concern is shown toward the pedestrian safety, comfort, and convenience. The street functionaries' designs are generally oriented without any prominence to a pedestrian who share the same road space with the vehicles, due to the complex nature of pedestrian behavior. The pedestrian problems are more serious where either infrastructure is absent or poorly maintained.

M. S. Mukesh (✉)
Research Scholar, Transportation Engineering, Visvesvaraya National Institute of Technology, Nagpur, India
e-mail: mukeshms060@gmail.com

Y. B. Katpatal
Civil Engineering Department, Visvesvaraya National Institute of Technology, Nagpur, India

J. K. Ghosh and I. da Silva (eds.), *Applications of Geomatics in Civil Engineering*, Lecture Notes in Civil Engineering 33, https://doi.org/10.1007/978-981-13-7067-0_30

At the intersection, the interaction between the vehicle and pedestrians are high, where it is controlled by traffic signals. Less Prominence is given to pedestrian safety during road design and are usually neglected during design in the developing countries. But to less prominence pedestrian are less bound to traffic rules there by causing more fatalities at the intersection as efficiency of the traffic signal is less. Currently, with the high rate of development, urban sprawl has gained importance and the behavior of different economic classes moving to different distances from the Central Business District (CBD) has become important.

In the recent years, the researchers have found that more number of pedestrians are involved in fatal crashes and have a larger share in traffic fatalities [1]. Commuters are in a state of high-risk while crossing the road in the urban areas [2]. Pedestrian design aspect requires data regarding the route they choose [3]. Pedestrian attitude toward using the facilities is an important factor which needs to be considered during planning [4, 5]. The pedestrian behavior has a greater importance in the designing and planning aspect of the urban region [6, 7]. Studies have been carried out to find the pedestrian and flow characteristics along the sidewalks and walkways [7]. Previously, the research work has been done to find the pedestrian interaction [5, 8–11] at the intersection and also to find out the behavior of the pedestrian gender and age, [9, 12] waiting time [13, 14], risk [4], crossing behavior [15], economic behaviour [6, 16].

Only few research works have been carried out to examine the micro parameters associated to the pedestrian movement in relation with the land use/land cover or to be precise the built-up. This paper mainly concentrates on finding the relationship that exists between the pedestrian movement, rate of change in built-up, and availability of infrastructure for cities in developing counties like India.

2 Methodology

2.1 Selection of Intersections for the Study

The trend in development is an important criterion which is driving the present development system in Indian city. The rate of development in the region provides information regarding nature, class, sociality, and other parameters related to the land use/land cover of that region [17]. High-resolution satellite data for the years 1997, 2007, and 2017 of Nagpur was considered and it was digitized into two classes: one built-up and the other non-built-up (Fig. 1).

All the major intersections located along the inner ring road of Nagpur were considered for the present study (Fig. 2). A buffer zone of about 2 km was created by considering the walking speed of 1.2 m/s (IRC 1985) with a service time of 30 min. The built-up data for different years was extracted from the classified set of images at the intersection.

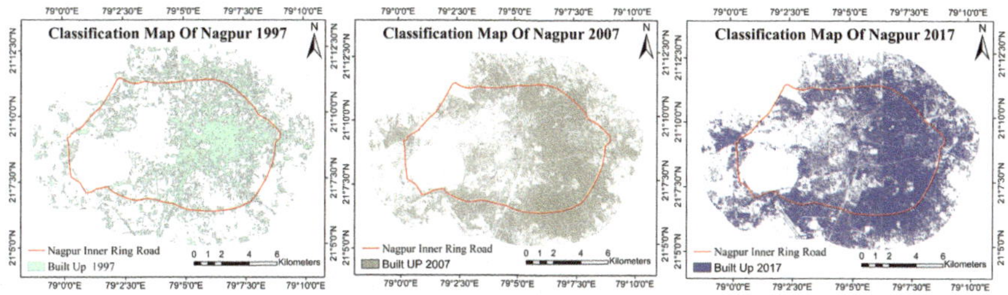

Fig. 1 Built-up classification map of Nagpur city for 1997, 2007 and 2017

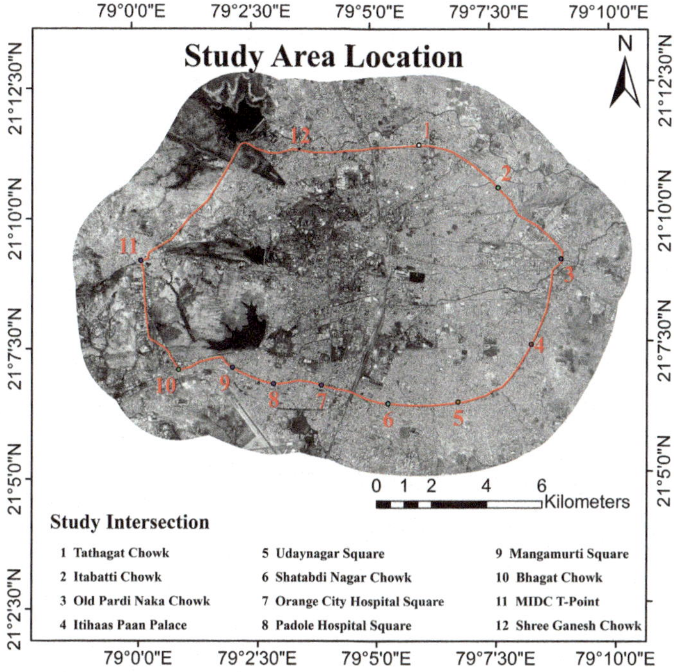

Fig. 2 Nagpur city satellite image with study intersection

From these sets of data, the percentage development for different years was determined. Then, the rate of change in built-up was determined by considering two consecutive years. From the available trend in the graph, four locations were selected; first set consists of MIDC T-Point and Tathagat Chowk where the built-up development rate is gradual and the other set consists of Udaynagar square and Padole hospital square where there is the sudden rise and saturating built-up (Fig. 3).

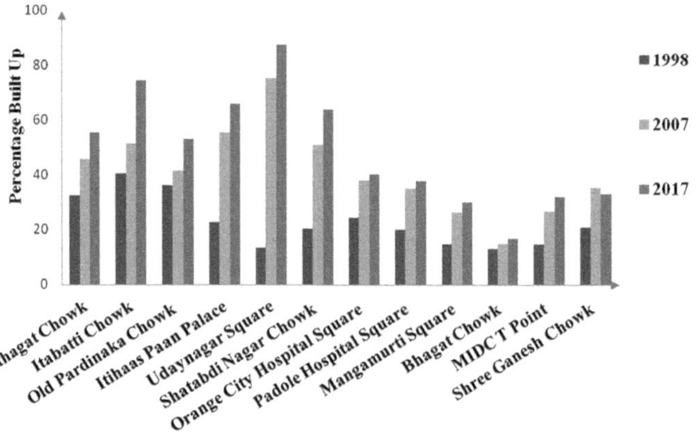

Fig. 3 Percentage built-up density at intersection for different years

2.2 *Human Movement Behavior*

Human movement is so uncertain that it is mostly influenced by factors like physical environment and operational environment. Hence, the pedestrian movement is analyzed by using Pearson's correlation coefficient and ANOVA test statistical methods. The dependency of the factors like signal break, crossing type, crossing time and crossing location with respect to age, gender, group, development rate, and the available infrastructure was determined using SPSS 24.0 software. These tests have been carried out at a confidence interval of 95%. The dependence of the factors is discussed in results.

3 Data Collection

High definition video data were collected from NMC (Nagpur Municipal Corporation) at 4 signalized intersections located along the inner ring road of Nagpur, India. For the purpose of data collection, only those cameras were selected which were oriented toward the main ring road and data from the intersecting road were not considered as the volume of movement of the vehicle and cause of accident on those roads were very less in comparison to the mainstream flow. Pedestrian movement along the intersection was examined by a direct observer. From the video recording, the information regarding the pedestrian signal break, crossing type, crossing time and crossing location were available with respect to age, gender, and group size. Crossing type was classified into risk averse and risk-taking. The pedestrian usage of mobile phone while crossing, haphazard movement and negligence toward vehicular movement behavior were considered to be the risk-taking factors. The crossing time was classified into slow, normal, and fast movers, with respect to IRC 1985 Standards

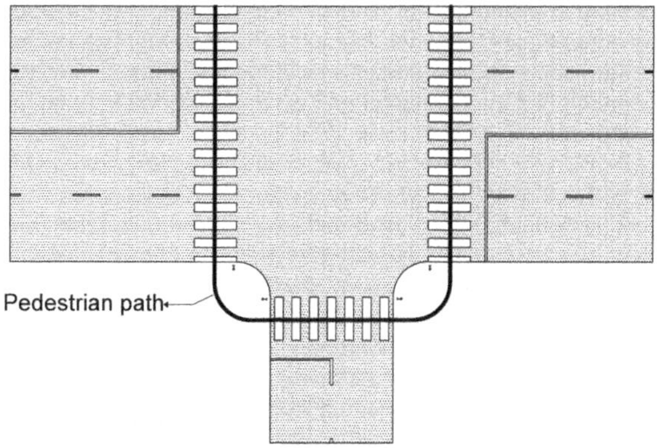

Fig. 4 Ideal crossing path at the intersection

of pedestrian normal cruising speed of 0.95 m/s and were compared. The crossing locations have been differentiated into the proper crossing and improper crossing (Fig. 4).

4 Result

From the quantity of data obtained, it was observed that study locations with gradual rate of development have more pedestrian movement than the locations with a very high accelerated rate of development.

4.1 For the Gradual Increase in the Built-up Density

From the result, it is observed that the persons with more age (approx. more than 50) cross the road with less risk and the time required to cross the road also increases. Males have a tendency to cross using the proper path than the females observed. With the increasing group size of pedestrians, the time required to cross the road also increases. But, the other factors like the signal break are found to have no correlation with the parameters like age, gender, and group size (Table 1).

4.2 For the Sudden Rise in the Built-up Density

Even in case of sudden rise in built-up densities, old age persons cross the road with less risk and the time required to cross the road is more. Gender is not an influential characteristic on the pedestrian movement behavior in a location having

Table 1 Statistical results for factors affecting pedestrian movement for location having a gradual rise in built-up density

Factors		Pearson's correlation coefficient		ANOVA analysis		Remark
		Pearson coeff.	Sig	F Value	Sig	
Age	Signal break	0.037	0.445	14.154	0.000	Not significant
	Crossing type	−0.102	0.032	7.245	0.001	Significant
	Crossing time	0.336	0.000	33.804	0.000	Significant
	Crossing location	0.031	0.521	6.897	0.001	Not significant
Gender	Signal break	−0.065	0.171	1.882	0.171	Not significant
	Crossing type	0.014	0.770	0.085	0.770	Not significant
	Crossing time	−0.003	0.945	0.005	0.945	Not significant
	Crossing location	0.111	0.020	5.483	0.020	Significant
Group size	Signal break	−0.070	0.142	2.160	0.142	Not significant
	Crossing type	−0.028	0.554	0.351	0.554	Not significant
	Crossing time	0.093	0.050	0.3804	0.050	Significant
	Crossing location	−0.064	0.180	1.805	0.180	Not significant

[a]Statistical test performed at 95% confident interval

a sudden rise in built-up density. As group size of the pedestrian increases, the time required to cross the road increases. Percentage of violation of signal break and the improper crossing location is more in the study area, even though it does not have any implication on the age, gender, or the group size (Table 2).

4.3 *For Combined Dataset*

By considering the combined dataset, the dependency factors of the infrastructure and the built-up density were found. It is evident from the result that the signal break, crossing time, and crossing location are depending on availability of infrastructure, i.e., with availability of infrastructure the signal break and risk-taking nature is reduced, and road crossing behavior is improved at intersection. With the increase in built-up density, pedestrians choose the proper crossing location at intersections while crossing (Table 3).

Table 2 Statistical results for factors affecting pedestrian movement for location having sudden rise in built-up density

Factors		Pearson's correlation coefficient		ANOVA analysis		Remark
		Pearson coeff.	Sig	F Value	Sig	
Age	Signal break	0.053	0.648	0.776	0.464	Not significant
	Crossing type	−0.189	0.002	5.183	0.006	Significant
	Crossing time	0.266	0.000	10.097	0.000	Significant
	Crossing location	−0.090	0.142	1.097	0.337	Not significant
Gender	Signal break	−0.087	0.463	0.554	0.463	Not significant
	Crossing type	0.105	0.086	2.971	0.086	Not significant
	Crossing time	0.027	0.658	0.197	0.658	Not significant
	Crossing location	0.060	0.329	0.985	0.329	Not significant
Group size	Signal break	−0.048	0.683	0.168	0.683	Not significant
	Crossing type	−0.047	0.444	0.587	0.444	Not significant
	Crossing time	0.165	0.042	3.521	0.042	Significant
	Crossing location	0.007	0.908	0.013	0.908	Not significant

[a]Statistical test performed at 95% confident interval

Table 3 Statistical results for a combined set of location

Factors		Pearson's correlation coefficient		ANOVA analysis		Remark
		Pearson coeff.	Sig	F Value	Sig	
Infra structure	Signal break	0.197	0.000	28.495	0.000	Significant
	Crossing type	−0.088	0.008	7.075	0.008	Significant
	Crossing time	0.005	0.889	0.019	0.889	Not significant
	Crossing location	−0.205	0.000	39.521	0.000	Significant
Built-up density	Signal break	0.003	0.929	0.008	0.929	Not significant
	Crossing type	0.004	0.910	0.013	0.910	Not significant
	Crossing time	0.020	0.541	0.375	0.541	Not significant
	Crossing location	0.129	0.000	15.210	0.000	Significant

[a]Statistical test performed at 95% confident interval

5 Discussion

The study was conducted to find out the dependency of pedestrian movement with the available infrastructure and the change in built-up density. During setting a buffer at the intersections, a constant radial buffer was considered wherein service area was not taken into account due to change in the road density in the region over a period of two decades. The change in built-up density at any location depicts the nature of the people settling in the region. The sudden rise in the built-up density suggests new identification of resources and economically rich people moving to settle in these regions, while less change in built-up indicates that the availability of resources is limited. Heterogeneous pedestrian group and the cyclist movement at the intersection were kept out of the study. During data extraction, the pedestrians crossing the road near the stipulated zebra crossing were also considered as proper crossing.

The result clearly depicts that parameters like the crossing time and the crossing type depend on the nature of the surrounding location. Pedestrian characteristic like gender is an independent parameter having no effect on the behavior of pedestrian movement. The signal break and crossing location are factors which neither depends on age, gender, nor the group size but instead, these are the factors depending on the infrastructure and the built-up. With the availability of infrastructure, the pedestrian behavior is improved thereby minimizing the interaction with the vehicles.

The result is confined to two types of development patterns and other patters were not taken into consideration during the study. During the study, the interdependency of the independent pedestrian characteristics is not taken into account. The study is confined to cities having only one Central Business District (CBD) and the development is radial and outwards.

References

1. Kharola P, Tiwari G, Mohan D (2010) Traffic safety and city public transport system: case study of Bengaluru, India. J Public Transp 13(4):63–91
2. Lassarre S, Papadimitriou E, Yannis G, Golias J (2007) Measuring accident risk exposure for pedestrians in different micro-environments. Accid Anal Prev 39(6):1226–1238
3. Zacharias J (2001) Pedestrian behavior pedestrian behavior and perception in urban walking environments. J Plann Lit 16(1):3–18
4. Holland C, Hill R (2007) The effect of age, gender and driver status on pedestrians intentions to cross the road in risky situations. Accid Anal Prev 39:224–237
5. Wang J, Fang S (2008) Pedestrian-vehicle conflict observation and characteristics of road section. J Tongji Univ 36(2):503–507
6. Hana V (2014) Economic development and individual and social behaviour. Procedia—Soc Behav Sci 109:1116–1119
7. Laxman KK, Rastogi R, Chandra S (2010) Pedestrian flow characteristics in mixed traffic conditions. J Urban Plan Dev 136(1):23–33
8. Guo JY, Bhat CR, Copperman RB (2007) Effect of the built environment on motorized and nonmotorized trip making substitutive, complementary, or synergistic? Transp Res Rec 1–11
9. Hamed MM (2001) Analysis of pedestrians behavior at pedestrian crossings. Saf Sci 38(1):63–82

10. Marisamynathan, Perumal V (2014) Study on pedestrian crossing behavior at signalized intersections. Traffic Transp Eng (English Edition)
11. Pei Y-L, Feng S-M (2007) Study on hazard degree of pedestrian crossing based on traffic conflict. J Harbin Inst Technol 39:285–287
12. Van Der Molen HH (1981) Child pedestrian's exposure, accidents and behavior. Accid Anal Prev 13(3):193–224
13. Tarawneh MS (2001) Evaluation of pedestrian speed in Jordan with investigation of some contributing factors. J Saf Res 32(2):229–236
14. Tiwari G, Bangdiwala S, Saraswat A, Gaurav S (2007) Survival analysis: pedestrian risk exposure at signalized intersections. Transp Res Part F 10(2):77–89
15. Evans D, Norman P (1998) Understanding pedestrians road crossing decisions: an application of the theory of planned behaviour. Health Educ Res Theory Pract 13:481–489
16. Cao XY, Mokhtarian PL, Handy SL (2009) The relationship between the built environment and non-work travel: a case study of Northern California. Transp Res Part A: Policy Pract 43
17. Basiago AD (1999) Economic, social, and environmental sustainability in development theory and urban planning practice, vol 19. Kluwer Academic Publishers, Boston, pp 145–161

Remote Sensing and GIS-Based Analysis to Envisage Urban Sprawl to Enhance Transport Planning in a Fast Developing Indian City

N. P. Anoona and Y. B. Katpatal

Abstract Rapid changes and sudden shootouts of development are occurring in most of the major cities in developing countries, leading to a requirement for mass transit facilities. Envisaging Urban Sprawl helps in finding out the future needs and critical locations of development. This paper validates the use of Shannon's entropy in finding out the sprawl in Nagpur, one of the fast developing cities of India, with the help of remote sensing data and GIS tools. In the present study, the spatial pattern of urban sprawl has been worked out based on location factors. Remotely sensed data obtained from Landsat and IRS LISS-III sensors were utilized. The entropy values show that the city is spreading outward following the transportation corridors. The observed growth pattern can be used for the future development plans of the city to mitigate some of the environmental impacts and provide sustainable growth.

Keywords Urban sprawl · Shannon's entropy · GIS · LU/LC

1 Introduction

Urban areas in India are growing at an alarming rate [1]. The number of million-plus cities in India has increased from 23 in 1991 to 35 in 2001 and still growing due to rapid urban growth, which results in problems like urban sprawl, unregulated ribbon development, inadequate infrastructure, and deterioration of the quality of living. This has also induced stress in the planning process, which needs to be properly projected commensurate to the enlarging urban areas [2].

The land use and land cover play a prominent role in planning and design of the transportation network in the urban areas, and when scientifically done, in accordance

N. P. Anoona (✉)
Transportation Engineering, Visvesvaraya National Institute of Technology,
Nagpur, India
e-mail: anoonanp@gmail.com

Y. B. Katpatal
Civil Engineering Department, Visvesvaraya National Institute of Technology,
Nagpur, India

J. K. Ghosh and I. da Silva (eds.), *Applications of Geomatics in Civil Engineering*,
Lecture Notes in Civil Engineering 33, https://doi.org/10.1007/978-981-13-7067-0_31

to LU/LC, it is the most effective [3]. The problem arises when accurate planning of transit is not achieved [4]. Envisaging Urban Sprawl helps in finding out the future requirements and possible locations of development [5]. Proper planning of transportation can lead to a well-planned city as the city usually grows around the transportation corridors [6].

The measure of urban sprawl is the essential component in the planning process, and vast research has been carried out in finding out the factors causing urban sprawl, monitoring of urban sprawl and results of urban sprawl [7]. With the advancement of technologies, various methods have been evolving to monitor spatial changes. Rapid and easy access to remotely sensed data and rapid analysis using possibilities of Geographical Information System (GIS) have analyzed spatial phenomenon like urban sprawl easier [8, 9].

Shannon's entropy, which gives the measure of the degree of spatial concentration or dispersion of a geophysical variable, has been used to quantify urban sprawl in the past years [7, 10]. The integration of Remote sensing and GIS with landscape matrices such as Shannon's entropy is one of the most reliable techniques so far to determine the Urban sprawl.

In this study, land use/land cover classification map has been used to extract built-up, and this result is compared with the entropy values.

2 Study Area

Nagpur is the thirteenth largest city in India, and the Nagpur Metropolitan Region is an urban agglomeration coming under Million Plus UA/City [11]. The city has a population of 2.81 million with a growth of 0.062 million per year [12], and the district population was 4,653,171 [11]. It is also a major commercial and political center of Central India. Nagpur being the geographical center of India, have many major highways passing through the city which has influenced the development of logistics industry. The Multi-model International Cargo Hub and Airport (MIHAN) shows the potential of Nagpur in the logistics sector.

The high rate of population growth is leading to the expansion of urban agglomeration beyond the city boundaries. This rapid growth has hence presented challenges for the planned growth of the city and its fringes.

Hence, for better management of growth and allied provision, the Nagpur Metropolitan Area (NMA) was formed comprising of 721 villages with an area of 3567.37 km^2 [13]. NMA (Fig. 1) is located between the North Latitudes 20° 45′ 26″ N and 21° 39′ 37″ N and the East Longitudes 78° 41′ 16″ E and 79° 34′ 38″ E. The study of Urban sprawl, its direction, momentum, and rate in NMA can help the authorities in the planning of infrastructure development and mitigation processes.

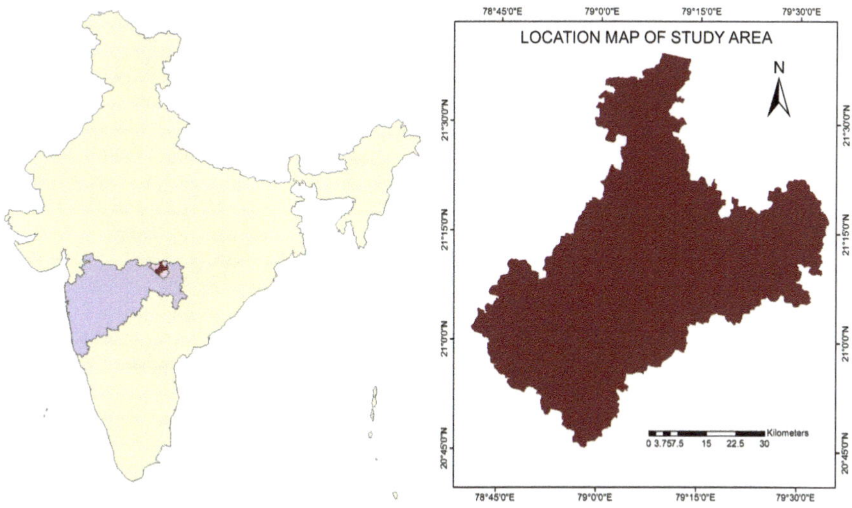

Fig. 1 Location of study area, NMA

3 Methodology

3.1 Data Collection

Landsat MSS, TM, ETM+, OLI, and TIRS sensor data obtained from United State Geological Survey (USGS) in GeoTIFF format, IRS LISS-III from ISRO Bhuvan and Google earth images were used in the present study. The images of the study area for the years 1995, 2005, and 2015 were obtained. The NMA map was obtained from the Nagpur Improvement Trust. Demographic data were collected from [11].

3.2 Image Classification

Land use classification is the primary requirement to understand a dynamic spatial phenomenon like urban sprawl. As the images were obtained from various sources, initially the coordinates were defined. Images were rectified, and the cloud cover was removed using ArcGIS and bought to standard false color composite for easy identification of various land use classes.

Unsupervised classification of images was done using Iterative Self-Organizing Data Analysis Technique Algorithm (ISODATA). Classified images were then compared with the Google Earth images and FCC images and the built-up areas were identified. The images were then classified using Maximum Likelihood Classification method into a binary map of built-up and non-built-up area. The reclassification

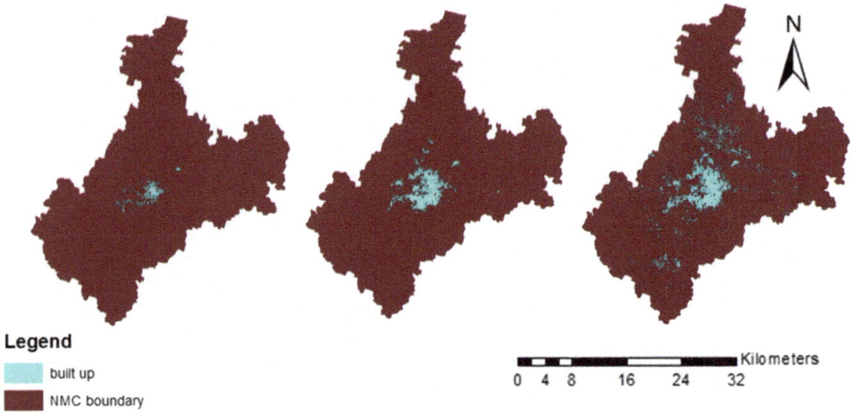

Fig. 2 Built-up area of 1995, 2005, and 2015, respectively

of images was done to extract the built-up area from the binary image by spatial analysis tool, which helps us to reclassify the values of the raster (Fig. 2). The extracted built area was then converted to vector format to enable corrections. Corrections were done in comparison with high-resolution Google earth images by selection of random samples all over the study area.

3.3 Urban Sprawl Assessment

The urban sprawl is the dispersion of a spatial phenomenon (built-up) over unoccupied tracts of land. Thus, the study of sprawl requires an understanding of different landscape matrices like Shannon's entropy.

Shannon's entropy is the measure of the degree of spatial concentration or dispersion of a geophysical variable among n number of zones [14–16]. As sprawl is the dispersion of special variable of impervious area or built-up, Shannon's entropy can be used to quantify urban sprawl [17, 18].

Shannon's entropy is given by the equation

$$\mathrm{H_n} = -\sum \mathrm{P_i}\log(1/\mathrm{P_i}) \quad (1)$$

where $\mathrm{P_i} = \frac{x_i}{\sum_i^n x_i}$ is the probability of the phenomenon (built-up) in the *i*th zone, $\mathrm{x_i}$ is the observed value of the phenomenon in the *i*th zone, and n is the total number of zones. The value of entropy ranges from 0 to log(n). If the spatial distribution is concentrated in one zone, the value will be lowest, i.e., closest to zero. Conversely, if there is even distribution of the phenomenon, the entropy value will be closest to log(n).

Relative entropy is used to scale the entropy value into a range between 0 and 1. Relative entropy is given by

$$\mathrm{H}_{\mathrm{n}}' = \sum_{i}^{i} \frac{p_i \log \frac{1}{p_i}}{\log(n)} \tag{2}$$

The degree of urban sprawl can be indicated using relative entropy by examining whether the land development is dispersed or compact. Larger values, i.e., values closest to 1 indicate the occurrence of urban sprawl.

From the city center, the "Zero Mile" considered as the geographical center of India, 1 km buffer zones were created. Thus, 58 buffer zones were created to cover the NMA. The study area was further clipped with each buffer as the extent to find out the area of built-up in each zone.

The magnitude of change of sprawl can be obtained by finding out the temporal variation of the entropy values as entropy measures the distribution of the geophysical phenomenon. The difference of entropy between time t and t + Δ can be used to indicate the magnitude of change of urban sprawl.

$$\Delta H = H_n'(t + \Delta) - H_n'(t) \tag{3}$$

The change in entropy helps us to find out whether the direction and momentum of land development. The overall methodology has been shown in Fig. 3.

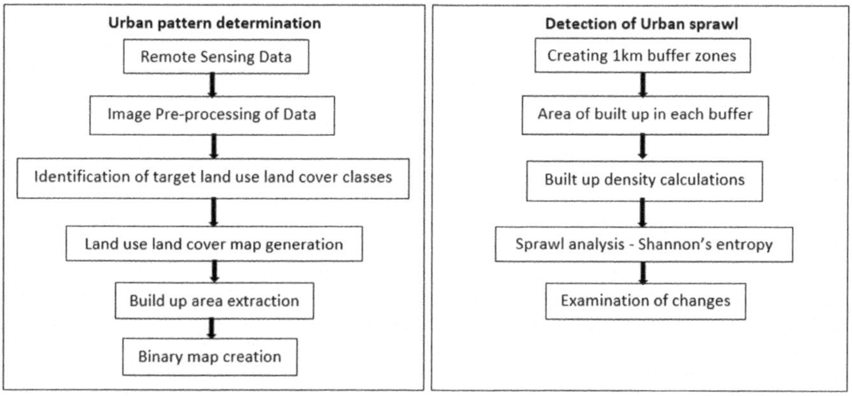

Fig. 3 Flow chart of methodology

4 Results and Discussions

Binary images of built-up and non-built-up areas obtained by the reclassification of supervised classified images are shown in Fig. 4. The figure shows variation in built-up areas of two decades. The central portion of the city was sparsely filled until the year 1995. Even though the increase in built-up area was high in the period 1995–2005, there is no sprawl in the city. The density of built-up areas in the inner zones has significantly increased. During 2005–2015, the next 10 years, the increase in built-up was only 40 km^2, but as the central portion of the study area is already filled, the built-up spread more in the Northern direction and comparatively less in the Western and Southern direction.

Table 1 shows that there is a sudden hike in built-up of 187.32 km^2 during the period 1995–2005 and there is only an increase of 37.672 km^2 in the next decade. Now the question arises that what is the present problem if the rate of growth has decreased. The problem is that as the central city has already been developed to the maximum extent in the first decade, the built-up is spreading outward converting the vegetation and pervious classes. This causes an ecological imbalance within the study area.

The Shannon's entropy values of 1995, 2005, and 2015 support that there is sprawl in the city. The relative entropy values are 0.508 (1995), 0.479 (2005) and 0.757 (2015). In the year 1995 and 2005, the entropy values are lesser compared to 2015 indicating that the built-up is compact in 1995 and 2005. There is a significant sprawl during 2005 and 2015 as the entropy value is closer to one, which is also visible by the visual examination of the binary image of 2015.

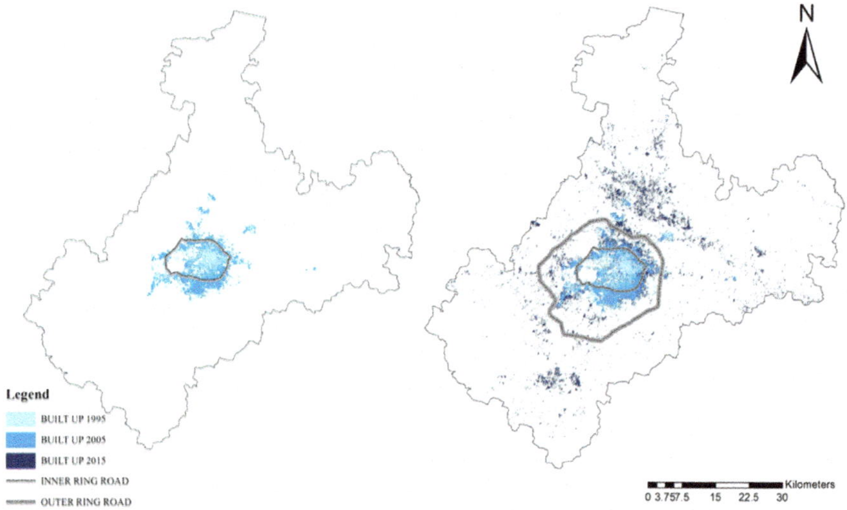

Fig. 4 Variation of built-up area 1995–2005 and 1995–2015, respectively

Table 1 Changes in built-up area, density, and growth rate over 20 decades

Year	Built-up area (km^2)	Built-up density (%)	Decades	Change in built-up (km^2)	Built-up area growth rate
1995	37.569	0.973			
2005	224.885	5.822	1995–2005	187.316	4.986
2015	262.557	6.797	2005–2015	37.672	0.168

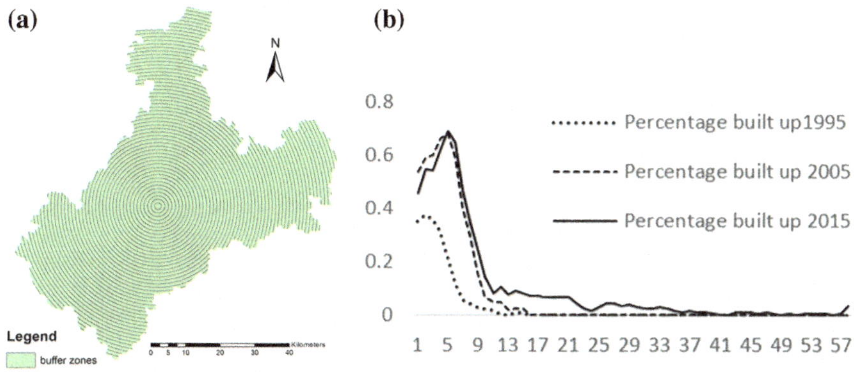

Fig. 5 **a** Circular buffers created at a 1 km interval from the center of the city. **b** The variation of percentage of built-up over time in each buffer zone

The variation in percentage built-up over two decades helps us in identifying the regions of sprawl or unplanned development. Figure 5 shows that, in the period from 1995 to 2005, the built-up was concentrated in the first 17 zones, which cover a radial distance of 17 km from the center. There is almost no or zero built-up in further zones. But in the case of 2015, 68% of the total built-up were concentrated in the first 17 km radial distance whereas 32% is spread outside this 17 km buffer zone, indicating a high rate of sprawl. It may be mentioned here that the outer ring road was planned after 2005 and constructed by 2015. The planning and construction of the major roads thus show a pronounced control over the urban sprawl.

The change in entropy was calculated to find out the magnitude of change in urban sprawl. From 1995 to 2005, there is a negative difference in entropy of −0.0291 indicating that increase in built-up is in a compact manner. But in the next decade, the change in entropy is 0.278 which indicates that there is an increase in sprawl.

Shannon's entropy values helped to understand the degree of urban sprawl. The overall assessment shows that the integration of GIS and entropy gives us accurate results of urban sprawl.

5 Conclusion

In this study, Shannon's entropy was used to find the urban sprawl whose results coincide with that of Land Use/Land Cover change of the NMA region. The study gives the direction, momentum, and rate of the urban sprawl of the region, which can be used in the development process. As the transportations corridor influences the areas of built-up, proper planning of transit facilities in fast developing cities has become more significant. Planned transportation facilities drive the systematic urban growth required for ensuring sustainable development.

References

1. Chadchan J, Shankar R (2012) An analysis of urban growth trends in the post-economic reforms period in India. Int J Sustain Built Environ 2012:36–49
2. Yeh AGO (1991) The development and applications of geographic information systems for urban and regional planning in the developing countries. Int J Geogr Inf Syst 5:5–27
3. Mallupattu PK, Reddy S, Reddy J (2013) Analysis of land use/land cover changes using remote sensing data and GIS at an urban area, Tirupati, India. Sci World J
4. Estrada M, Roca-Riua M, Badiaa H, Robustéa F, Daganzob CF (2011) Design and implementation of efficient transit networks: procedure, case study and validity test. Procedia Soc Behav Sci 11–135
5. Terando AJ, Costanza J, Belyea C, Dunn RR, McKerrow A et al (2014) The southern megalopolis: using the past to predict the future of urban sprawl in the Southeast U.S. PLoS ONE 9(7):e102261
6. Sustainable Urban Transport in the Developing World (2015) Pojani, Dorina., Stead, Dominic. Beyond Megacities Sustain 7:7784–7805
7. Ozturk D (2017) Assessment of urban sprawl using Shannon's entropy and fractal analysis: a case study of Atakum, Ilkadim and Canik (Samsun, Turkey). J Environ Eng Landscape Manage 25(3):264–276 (2017)
8. Duran Z, Musaoglu N, Seker DZ (2006) Evaluating urban land use change in historical peninsula, Istanbul, by using GIS and remote sensing. Fresen Environ Bull (8a):806–810
9. Hegazy IR, Kaloop MR (2015) Monitoring urban growth and land use change detection with GIS and remote sensing techniques in Daqahlia governorate Egypt. Int J Sustain Built Environ 4(1):117–124
10. Yeh AGO, Li X (2001) Measurement and monitoring of urban sprawl in a rapidly growing region using entropy. Photogr Eng Remote Sens 83–90 (2001)
11. Census of India. http://censusindia.gov.in/
12. India population (2018). http://indiapopulation2018.in/
13. Nagpur Improvement Trust (2015) Nagpur metropolitan area development plan 2012–2032. Draft Dev Plan Report
14. Theil H (1967) Statistical decomposition analysis. North-Holland, Amsterdam
15. Thomas RW (1981) Information statistics in geography, concepts and techniques in modern geography 31
16. Theil H (1972) Economics and information theory. North-Holland, Amsterdam
17. Jat MK, Garg PK, Khare D (2011) Monitoring and modelling of urban sprawl using remote sensing and GIS techniques. Int J Appl Earth Obs Geoinform 26–43
18. Rahman MT (2016) Land use and land cover changes and urban sprawl in Riyadh, Saudi Arabia: an analysis using multi-temporal Landsat data and Shannon's Entropy Index. Int Arch Photogram Remote Sens Spat Inf Sci XLI-B8:1017–1021

Application of GIS in Road Information System—An Experience with State Highways of West Bengal

Sudipta Pal

Abstract In last few years, application of GIS technology in highway and transportation engineering field increased substantially in India. In the present study, GIS has been used to develop a user-friendly road information system for State Highway Network in state of West Bengal, India. Objective of the study is to prepare a "Base map" to have all the traffic, road, and structure features mapped into a GIS platform. This map will provide the basic platform for all spatial features of the road and structure assets with an integrated database for storing attributes of the data features and screening and ranking of developmental requirements and to prioritize resource allocation for such development. The present paper demonstrates with the help of data collected from State Highways of West Bengal that GIS can be used as a useful tool for developing an interactive road information system.

Keywords Road information system · GIS · Layers · Attributes · Land use

1 Introduction

People working in many different fields use GIS technology for various purposes. In last few years, application of GIS and GPS technology in highway and transportation engineering increased manifold in India. Road information system aims to provide detailed information on road profile which would be accessible both by public and policy-makers.

A few research papers on application of GIS technology in various areas are reviewed for the present study. A paper [1] provides an introduction to Geographic Information Systems (GIS) and a research framework for information systems researchers. The paper summarizes the main GIS features, functions, and capabilities, including a research framework for GIS. In a book [2], Geographical Information Systems (GIS) is described as computer-based technologies for handling geograph-

S. Pal (✉)
RITES Ltd., Highway Division-Kolkata, 56, C.R. Avenue (4th Floor), Kolkata 700012, West Bengal, India
e-mail: spalonline@rediffmail.com

J. K. Ghosh and I. da Silva (eds.), *Applications of Geomatics in Civil Engineering*, Lecture Notes in Civil Engineering 33, https://doi.org/10.1007/978-981-13-7067-0_32

ical data in digital form in order to capture, store, manipulate, analyze, and display diverse sets of spatial or georeferenced data. In a research paper [3], authors discussed in detail how a spatial analyzing system can be used for urban expansion and land management. In another work [4], it is demonstrated that GIS can be used as a very useful tool for real estate management system and 3D visualization. Another paper [5] highlights the need to implement GIS as a decision support system in road maintenance and development with example of roads in Ghana under budget-constrained situation.

Government of West Bengal being concerned about the overall development of the state intends to develop a sustainable road network system which drives economic growth. Recognizing the immediate needs of upgrading the road transport system, the State government has initiated a Strategic Option Study (SOS) for State Highways. This study was intended to identify high-density traffic corridors and roads with various strategic importance in order of priority compatible with the present-day traffic needs and future demands. This will establish the need for improvement, upgradation, and interconnectivity of state highway network on priority basis. As a first step, baseline data was collected related to traffic, road assets, and formulate a Geographic Information Systems (GIS)-based base map for screening and ranking of developmental programs of state highways. GIS will provide the basic platform for all special features of the road and structure assets with an integrated database for storing attributes of the data features.

GIS database developed under this study will serve as a central repository to keep departmental records regarding past maintenance of the roads, fund allotted from various schemes, road assets, and historic traffic data. This GIS database shall have different levels of access control starting from public/ordinary road user to highest level of state administration.

2 Study Sections

Study corridors consist of 19 State Highways of West Bengal, i.e., SH1, SH2, SH3, SH4, SH4A, SH5, SH6, SH7, SH8, SH9, SH10, SH10A, SH11, SH11A, SH12, SH12A, SH13, SH14, and SH15 having total length about 4000 km. Summary of the project corridors as per work order is furnished below. Details within 100 m width were collected along each project corridor (Table 1).

3 Objective of the Study

The objective of the proposed “GIS-based digital map” is to facilitate state government for effective decision-making in selection of the corridors with maximum traffic potential and identify the type of development needed in order of priority. The objective of the work is to establish the following:

Table 1 Summary of state highway corridors in West Bengal

Sl No	SH No	Length (km)	Districts
1	SH-1	151	North 24 Parganas, Nadia, South 24 Parganas
2	SH-2	323	Bankura, Hooghly, North 24 Parganas
3	SH-3	260	Nadia, North 24 Parganas, South 24 Parganas
4	SH-4	466	Purulia, Bankura, Paschim Medinipur, Purba Medinipur
5	SH-4A	39	Purulia
6	SH-5	376	Purulia, Bankura, Paschim Medinipur, Purba Medinipur
7	SH-6	266	Birbhum, Burdwan, Hooghly, Howrah
8	SH-7	289	Birbhum, Mursidabad, Burdwan, Hooghly, Paschim Medinipur
9	SH-8	292	Purulia, Bankura, Burdwan, Nadia
10	SH-9	251	Burdwan, Bankura, West Medinipur
11	SH-10	173	Malda, Dakshin Dinajpur
12	SH-10A	63	Dakshin Dinajpur, Uttar Dinajpur
13	SH-11	251	Birbhum, Murshidabad, Nadia
14	SH-11A	65	Murshidabad
15	SH-12	352	Darjeeling, Jalpaiguri
16	SH-12A	217	Jalpaiguri, Alipurduar, Cooch Behar
17	SH-13	150	Birbhum, Burdwan, Hooghly
18	SH-14	226	Birbhum, Burdwan, Nadia
19	SH-15	242	Burdwan, Hooghly, Howrah

(i) Creation of GIS-based "Central Repository" of road data, which can be used by the state administration for decision-making;
(ii) To assist in e-governance initiatives;
(iii) To provide facility for carrying out route and network analysis;
(iv) Location reference of the "State Highway Road Network" and integrated road data;
(v) Platform to graphically review, view, and analyze the road data;
(vi) Framework for collection of road data and updating "GIS-based digital map" database on a sustainable basis.

4 Methodology

A systematic approach was taken to develop GIS-based map and road information system. Methodology adopted for the study has been discussed under following broad heads.

4.1 Establishment of DGPS Control Points on the Ground

Control points were established in every 10 km interval through dual frequency DGPS. These control points were connected to established BM of Survey of India (SOI)/GTS benchmark. Reduced levels of each 100 m interval of the road have been shown with road centerline. Submeter (<1.0 m) accuracy was ensured for (X, Y) and 10 cm is considered as maximum permissible limits for Z value with respect to absolute values with GTS benchmarks and relative levels between two points on road centerline. Closed loops consisting of a network of triangles, connecting these control points were set out. The triangles were well formed and preferably not too acute nor obtuse, with sufficient redundancy so that a baseline could be confirmed by observations from multiple control points. The coordinates for the entire section have been taken with reference to single arbitrary grid in metric system and WGS 84 reference frame on UTM projection system.

The control point locations have been identified considering following criterion:

(i) Clear of HT/LT lines;
(ii) Free from multipath problems associated with tall features in the vicinity;
(iii) Free from foliage;
(iv) Open to sky with a clear view of the horizon;
(v) Close to or on the permanent structure such as bridge, culvert, etc. If such features were not available, pillars were erected specially for this purpose.

4.2 Mapping of the Project Corridor Through RTK/DGPS

After establishment of GCP, survey team comprising one base station with two movable rovers collected project features within 100 m corridor width like road center with RL, carriageway, shoulder, utility line, crossroad, river, canal, junction, religious place, KM post, traffic sign, landmark, ROBs, etc. Georeferenced photographs of important objects like bridges, culverts, etc. have been taken and linked with the feature ID.

4.3 Georeferencing of the Satellite Imagery

The satellite imageries were imported into .img format and georeferenced with respect to the GCPs collected. Registration of the satellite image is performed in ERDAS environment. Resampling was done by using "Nearest neighborhood" algorithm.

- Projection systems used are UTM, Datum: WGS-84, Zone: 45 N.
- Geometric model: Polynomial and Polynomial order: 1.

All the information collected from secondary sources are converted to georeferenced and brought to the same spatial framework that of satellite imagery.

4.4 Link and Section IDs and Identification of Structures

Project corridors have been divided into number of different links as well as homogeneous sections. Links are selected based on their connectivity with other roads and other road features but the homogeneous sections have been denoted based on the traffic pattern. Each link and section has been assigned a unique ID as per the following:

- Link ID begins with L if it has no overlapping SH/NH: L14-01, L12-02, etc., where 14 and 12 stand for SH number;
- Link ID begins with N if it has overlapping NH: N12-01, N12-02, etc.;
- Link ID begins with S if it has overlapping SH: S12/10-01 (e.g., in the overlap section of SH-10 with SH-12, based on the maintenance records SH-12 continues);
- Section ID has been denoted as SH12-01, SH-12-02, etc., based on their intensity of the traffic on each link, where 12 stands for SH number.

Each structure was assigned with a unique ID for ease of use to identify the details individually in each section to avoid confusion. Demarcation of different types of structures as Major Bridge—MJB, Minor Bridge—MNB, Causeway—CWAY, Slab Culvert—SC, Rail Over Bridge—ROB, etc. For example, third Major Bridge on Link L12-01 will have structure ID: L12-01-MJB-03.

4.5 Preparation of GIS Database of State Highway Network

Data collected on Carriageway, Medians, Traffic islands, Footpath, Right of way of roads, Rivers, Culverts, Bridges, ROBs, RUBs, and Flyovers were converted to GIS format and the same has been overlaid on the satellite imagery. GIS database of district boundaries, state boundary, and international boundary along the state highway has been created from georeferenced and spatially adjusted administrative boundary maps.

4.6 Mapping of the Land Use from Satellite Imagery

Digitization of all the visible land features was carried out into point, polyline, and polygon layer from the satellite imageries along the state highway. ArcMap was used for mapping purpose. The SOI toposheets were used as a reference for delineation of reserved forests, landmarks, roads and railway networks, water bodies, rivers, canals, etc. Separate layers for each of the feature were created for each of the features.

4.7 Data Integration and Conversion

The entire field data (spatial and nonspatial components) related to inventory, traffic, and other field surveys collected on the project corridors were integrated with the GIS database. The spatial components were converted into shape file and overlaid on the base map. Based on the field data necessary corrections have been applied on the base map. Then the nonspatial components were introduced into the base map by joining/linking the attributes with the spatial features using unique id.

Total 32 layers and 115 attributes have been included in the GIS database. Uniformity regarding data type, field length, etc. was strictly maintained. Legends have been used for landmark point features like School, Bank, Temples, Hospital, Railway Station, etc. Different legends were used with specific color code for representing various land use patterns.

4.8 Quality Check of the Data

Following quality checks have been performed on the individual layers to identify the errors:

- (i) Correctness of the layers w.r.t digitized feature,
- (ii) Completeness,
- (iii) Matching of all the features with the image and maps,
- (iv) Snapping of different features with other features as applicable,
- (v) Continuity of various features such as rivers, canals, roads,
- (vi) Missing features,
- (vii) Dimensional accuracy—over capture, shape errors, displacement,
- (viii) Mismatched edges,
- (ix) Correctness of the data entered as nonspatial attribute.

5 Road Information System (RIS)

In road information system, road data, and information would be obtained by searching the concerned Road Name, District, Block, Division, Road Category, etc. Road profile information generally includes typical photograph and profile Information like Carriageway Type and Width, Repair Works, New Construction, Pavement composition, Traffic data, etc. In the present study, GIS-based road information system has been developed with the objective of creating up-to-date digital database for entire State Highway Network of West Bengal. It will help policy-makers to identify the problem associated with road development, keeping records of maintenance, and proper resource allocation. Provision of access control system has been kept for various levels of user starting from common road users to highest level of state administration. Using this web-based GIS map one can easily find out various utility services like school, college, government offices, hospitals, religious places, etc. In addition to standard tools like distance and area measurements, quarry-based services have also been provided to list out various physical features like bridges, culverts, and ROBs along selected road length. Traffic data, engineering details of structures, and carriageway along with typical photographs have been incorporated in GIS-based map as attributes. Tools like shortest path and network analysis can be used during natural disaster and other emergency situations.

After thorough checking of the GIS map features and the GIS database, it has been converted into ".shp" format, which is universally accepted GIS format and the same is compatible with all the GIS software and these are superimposed into the Google Maps for their land use and efficiency of presenting the data to National Informatics Centre (NIC), New Delhi for web publishing. NIC developed application software for web-based GIS map and road information system for entire state highway corridor of West Bengal. Typical snapshots from developed web-based GIS map have been shown in following Figs. 1, 2, 3, and 4.

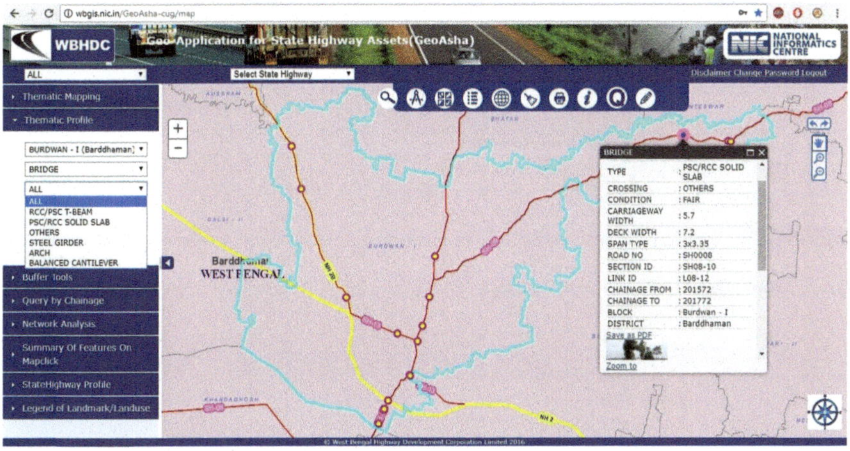

Fig. 1 Thematic profile of road sections in Burdwan district with typical bridge details

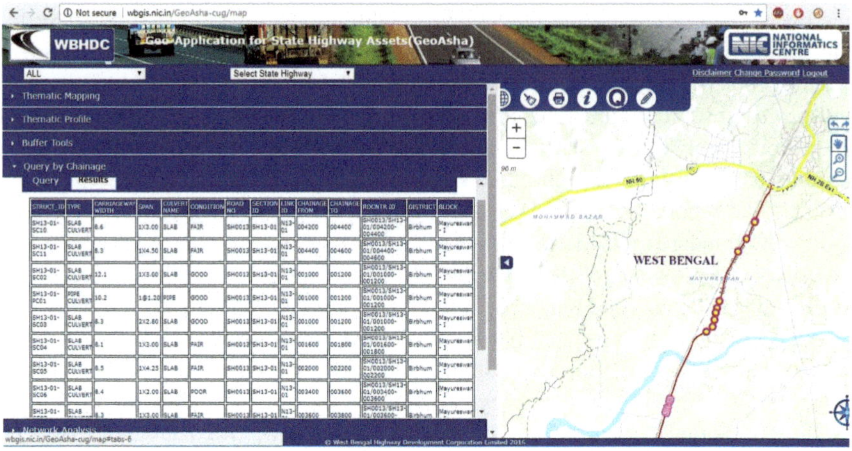

Fig. 2 Quarry by chainage with typical tabular display

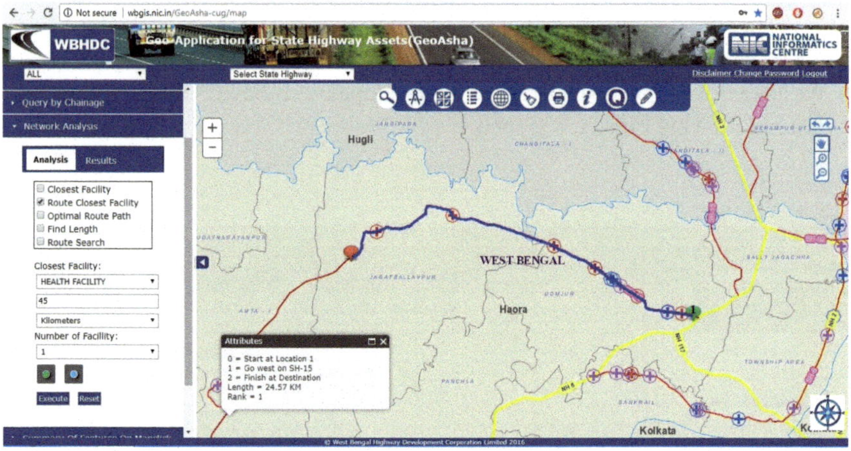

Fig. 3 Network and shortest path analysis

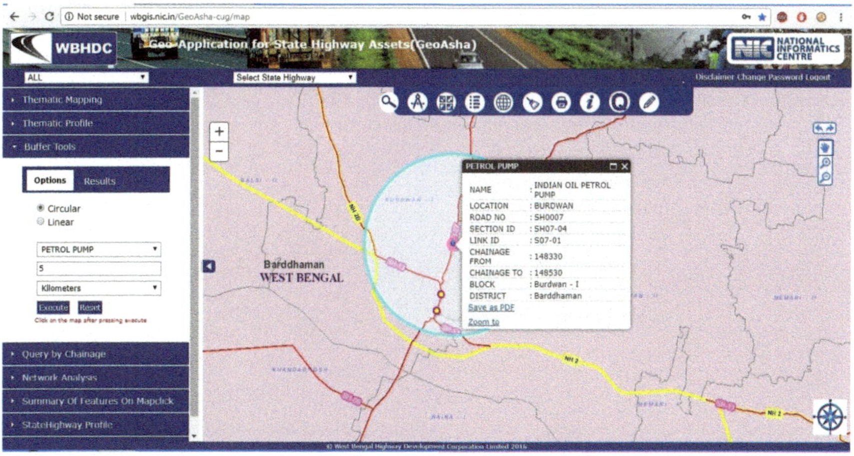

Fig. 4 Buffer analysis for searching utility services within a distance

Acknowledgements Field data collected by RITES Ltd for Strategic Option Study (SOS) initiated by West Bengal Highway Development Corporation (WBHDC) is used as basic input for present research paper. I thankfully acknowledge the contribution of Shri J. R. Sarkar of AARVEE associates in the study. National Informatics Centre (NIC), New Delhi is the agency developing application software for web-based GIS maps and road information system. Few snapshots from web GIS developed by NIC are included in the present paper. I convey my sincere thanks to RITES Ltd and WBHDC for giving me permission for using project data as basic inputs for this paper.

References

1. Mennecke BE, Crossland MD (1996) Geographic information systems: applications and research opportunities for information systems researchers. In: Proceedings of the Hawaii international conference on system sciences (HICSS-29)
2. Burrough PA, McDonnell RA (1998) Principles of geographical information systems. Oxford University Press
3. Yang F, Zeng G, Du C, Tang L, Zhou J, Li Z (2008) Spatial analyzing system for urban land-use management based on GIS and multi-criteria assessment modeling. Prog Nat Sci 18:1279–1284
4. Lin C, Meng L, Pan H (2001) Application and research on GIS for real estate. In: 22nd Asian conference on remote sensing. Singapore, 5–9 November
5. Acquah PC, Fosu C (2017) Implementation of geographic information system application in the maintenance management of roads in Ghana: a case study of roads in Kumasi Metropolis. Am J Geogr Inf Syst 6(3):90–102

An Intelligent Gas Pipeline Route Alignment System

Suraj Sawant, Roshan Kumar and Rupendra Kumar

Abstract Pipeline is one of the premier means of transportation for petroleum products. Optimizing the cost, time of construction and environmental considerations for pipeline route, depends on the determination of optimal route for pipeline alignment. In this paper, an intelligent system capable of identifying optimal (based on cost, engineering criteria, and environmental issues) land corridor for alignment of natural gas pipeline from source to destination within onshore has been proposed. Accessing results by analyzing large datasets is achieved through spatial multi-criteria decision-making strategy. A preliminary implementation of the proposed system is being done to identify land corridor for alignment of gas pipeline from Shahdol (Latitude 23 18′ 07″ N and Longitude 81 21′ 24″ E in Madhya Pradesh, India) to Phulpur (Latitude 25 33′ 03″ N and Longitude 82 05′ 18″ E in Uttar Pradesh, India). Slope, crossings (road networks, rail tracks, rivers etc.), forests, vegetation and costs have been considered as criteria for identification of optimal land corridor. For implementation and analysis of the proposed work, Landsat 8 images, having spatial resolution 30 m, are used for analysis purpose. The study also made use of state boundary, road network, river network, elevation, and land cover type data. Data layers for analysis have been created using distance criteria. Finally, different layers of criteria have been merged using Analytical Hierarchical Process (AHP) algorithm to de ne two cost layers the Cost Distance Layer and Cost Direction Layer. Finally, depending on guiding criteria, the least cost path from source to destination station has been identified as the optimal land corridor for alignment of gas pipeline. To make the system more intelligent is the future scope of this work.

Keywords Route alignment · Analytic hierarchy process (AHP) · Automated system

---

S. Sawant (✉) · R. Kumar · R. Kumar
Geomatics Engineering, Indian Institute of Technology Roorkee, Roorkee, Uttarakhand, India
e-mail: suraj.t.sawant@gmail.com

J. K. Ghosh and I. da Silva (eds.), *Applications of Geomatics in Civil Engineering*,
Lecture Notes in Civil Engineering 33, https://doi.org/10.1007/978-981-13-7067-0_33

1 Introduction

Pipeline is one of the premier means of transportation for petroleum products. Optimizing the cost, time of construction and environmental considerations for pipeline, depends on identifying the optimal path. Thus, Route Alignment is an important step in Pipeline projects. Pipeline route alignment is always a challenge because of the di cult terrain [1]. However, route alignment is usually carried out manually utilizing existing techniques. Such methods are not optimal in nature as it cannot take care of all the factors together that affect the path alignment. The route alignment process requires consideration of technical, economic, environmental and sociological issues. There is a need to identify well-defined routing criteria and sub-criteria in a sensitive context. The routing decisions are implemented according to defined fixed rules, hence it requires uncertainty to be considered. An intelligent pipeline route alignment system may resolve all these limitations. The proposed system will identify an optimal route for alignment of pipeline from a source to a destination station lying onshore. Identification of the pipeline route requires taking decisions based on location [2].

2 Related Work

Traditionally, land suitability analysis has been used to identify possible routes. First, the physical criteria for the study were identified. Physical criteria usually include zoning, land use, ownership, and soils, to name a few. These criteria were depicted on transparent maps which were overlaid, and the routes with the most desirable features were identified. This approach was suggested by [3], among others, as an alternative to an economic cost–benefit analysis which ignored the costs and benefits of the effect of the project on the fronts of natural and social environments. Land suitability mapping is performed at various measurement scales, the lowest, the nominal scale, is best represented by the Gestalt method [4], in which the study area is implicitly subdivided into homogeneous regions and classified according to their suitability for various land uses. This method demands the decision maker to be very familiar with the study area, which is not the case always. Results represented are in Gray shades, from light to dark. This approach has some limitations, like, it is mathematically incorrect to perform addition at the ordinal scale, and the measurement of factors at the ordinal scale is subjective. The methods described by Hopkins utilize manual overlays, the process is easily computerized. As the map layers are themes, a Geographical Information System can be used to perform the map overlays. This work introduced the application of GIS in Pipeline Route Identification system [5] prepared Pipeline Risk maps for study area by identifying appropriate risk factors. Optimized the pipeline route to avoid environmentally sensitive areas and also minimized exposure to seismic activity areas [6] model can be modified to identify the important criteria and compute the weights for linear engineering structures, such as pipelines, water lines, roads, channels, railways, and energy transfer lines. The accuracy obtained was dependent on the data used, hence not acceptable. In, [7], region of interest is modeled by constraint-based suitability analysis. The resolution algorithm used, produces a minimized number of solutions, which makes it feasible for the decision maker to easily identify the solutions respective to the best adjustments in decision makers point of view.

This approach considers only two criteria, which restricts the best available path. Search Algorithm used cannot be used for complex terrain [8] focused on the communication among the Decision Makers (DMs) throughout the Delphi process and SWOT analysis. PROMETHEE-GDSS, developed in this study, the DMs decomposed the process into manageable steps and integrated the results to arrive at a consistent solution to follow organizational goals and objectives. This system does not identify the route but supports the Decision Makers to take decision [9] suggests that importance of this method lies in the ability to embed more knowledge into the pathfinding process and to account for multiple linkages and considerations between the proposed project and the various environmental, social, technical and economic factors it will interact with. It is potentially time consuming and technically complex to implement [10] used the Genetic algorithm to identify the optimal pipeline route. As the use of classic static penalty approach is not adequate, especially in more complex scenarios, it is important to associate these criteria to more advanced constraint-handling techniques [11] developed an automated analytical solution and designed a tool to identify the most cost-efficient path for a pipeline between two known terminal points across a heterogeneous terrain. This work used mathematical techniques for optimal route computation.

Proposed solution is terrain specific [12] developed a multi-objective model for the combined natural gas network and electricity network. It takes into account the uncertainty. The main objective of the model was to optimize investment cost and production cost of the system while considering the network security factor. Elitist Non-Dominated Sorting Genetic Algorithm-II (NSGA-II) is implemented to compute the objective functions. Decision makers may use a fuzzy decision-making technique according to preference to select the final optimal solution [13] stresses the use of high-quality data, which is especially a big problem. It used Technique for Order of Preference by Similarity to Ideal Solution TOPSIS method for multi-criteria evaluation. Another significant point is the identification of well-defined factors and subfactors in a sensitive context.

Thus, to analyze geospatial decision problems, an optimized, intelligent and integrative use of information, interdisciplinary knowledge and continuous learning is required. Thus, the objective is to develop a system which will address the onshore pipeline routing alternatives and identify the optimum alignment corridor.

3 Proposed Methodology

The first step involved in the methodology is selection of routing criteria affecting the pipeline route. Once the route alignment criteria and factors affecting route alignment are finalized, data gathering step is carried out. Data gathering phase depends on the routing criteria and sub-criteria, as the finalization of required information depends on routing criteria. The process of data gathering involves collection of topographic maps, satellite imagery, digital elevation model (DEM) etc. for the selected study area. Once the required data is gathered, it needs to be preprocessed for making it suitable for further analysis. One of the examples of preprocessing of data is stack layering of the satellite images of study area, so as to stack the various bands of satellite image together. Then the stack layered images are mosaicked with other satellite images, which constitutes the study area. Next step is to create knowledge base. Combination of reclassification of datasets and derived datasets constitutes to knowledge base. The classification of dataset changes/modifies cell values to alternative values using a variety of methods. This is done in order to transform the different types of data formats into a similar data type. When a dataset undergoes a transformation, it is called as derivative dataset. Knowledge base contains the different types of data which is readily accepted by next steps of methodology. Suitability analysis is a geo-spatial phase used to identify the aptness of a selected location for a specific use. The basic purpose of suitability analysis is to check whether the proposed location is apt or not for the activities being planned. Suitability is identified by systematic, multi-factor analysis of the various aspect of the land cover. Output of suitability analysis is a map representing areas from high to low suitability. Suitability analysis is divided in the following steps to simplify the process. It consists of creating Routing criteria datasets, Reclassification, and Accumulated cost dataset. The input layers are transformed to cost distance layer and cost direction layer, which constitutes to cost weighted computation step. In cost weighted computation, all the criteria are considered and its cost in terms of distance is calculated. Next step is route analysis which finds the quickest, shortest, or even the most economical route, depending on the impedance chosen to solve for. Any cost attribute can be used as the impedance when determining the optimal route. One of the most important step and contribution of the proposed methodology is use of automated techniques and automatic reasoning. Use of intelligent technique is proposed in the methodology to enhance and automate the gas pipeline route alignment process. To make the system intelligent enough, the learning is carried out. Learning tasks will be accomplished by the feedback given to the existing database. Learning occurs when the database is given immediate feedback about the value of each action. Feedback will be in terms of rewards and punishments to the system after a sequence of actions. This will make easier for the system to infer based on the previous history. The steps of the proposed methodology are represented in Fig. 1.

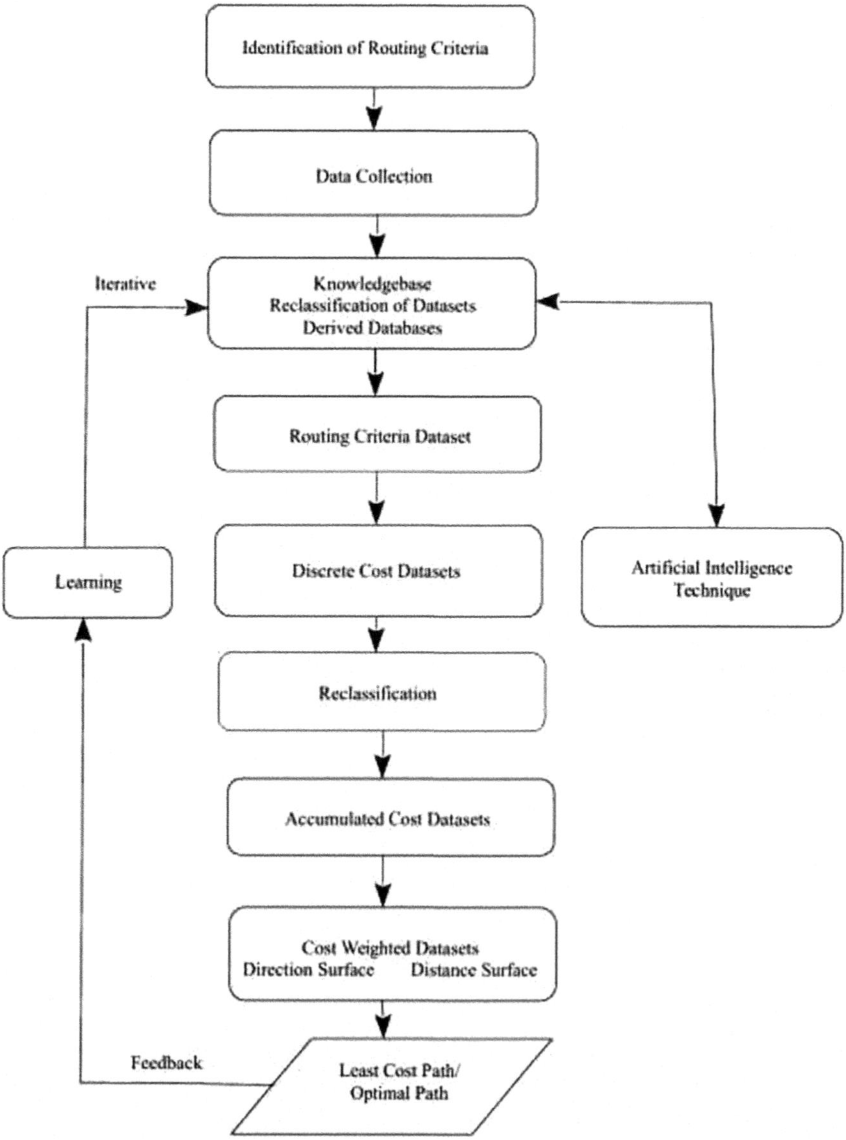

Fig. 1 Proposed system architecture

4 Experiment

A preliminary implementation of the proposed methodology is being done to identify land corridor for alignment of gas pipeline from Shahdol (Latitude 2318′ 07″ N and Longitude 8121′ 24″ E in Madhya Pradesh, India) to Phulpur (Latitude 2533′ 03″ N and Longitude 8205′ 18″ E in Uttar Pradesh, India). Map of study area is shown in Fig. 2. On the whole, the terrain in between station consists of small mountains, hills, plateau and plane region. Slope, crossings (road networks, rail tracks, rivers etc.), forests and vegetation have been considered as criteria for identification of optimal land corridor.

For implementation and analysis of the work, Landsat 8 satellite multi-spectral images are used. The resolution is 30 m. While considering the satellite image tiles for the creation of study area dataset, the cloud cover is kept less than 4%. The satellite data used provides sufficient spectral resolution and dense earth materials are acquired by the sensor, which is enough for analysis purpose of the proposed work. Table 1 summarizes the details about the satellite image taken for study area. The required state boundary shape files, road network shape le, river network shape le, elevation data in grid le, and land cover data le in grid format are taken. The details are mentioned in Table 1. This data is used to analyze the criteria mentioned above. The input layers are transformed to the same coordinate system, i.e., Geographic Coordinate System (GCS). A map projection is a methodical modification of the latitudes and longitudes of locations from the sphere surface or an ellipsoid into locations on a plane. All input dataset can only be viewed together at its correct coordinates if it converges with the correct map projection as well as transforming in the same measures. It is most important process before proceeding to suitability analysis as any kind of analysis can only take place accurately when all the theme

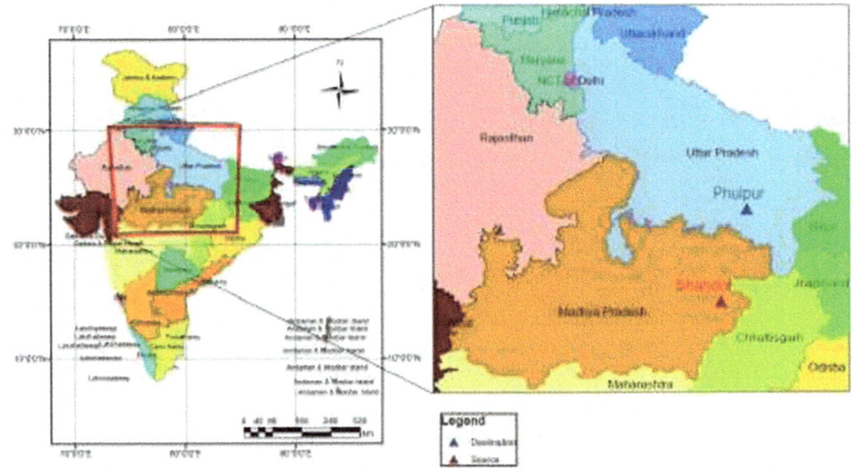

Fig. 2 Map of study area

Table 1 Study area details

Sr. No.	Sensor	Producer	Area image	Resolution	Altitude
1	L8	OLI/TIRSUSGS	Madhya Pradesh and Uttar Pradesh, India	30 m	705 km

Table 2 Data to be used in route identification

Name	Description	Format	Resolution
Administrative areas	Country outline and administrative subdivisions	Vector (area)	–
Roads	Roads	Vector (line)	–
Railroads	Railroads rivers, canals, and lakes	Vector (line)	–
Inland water	Separate les for line and area feature	Vector (line and area)	–
Elevation	Digital elevation model	GeoTIFF	2.5 m

Table 3 Assigned weights

Criteria	Vegetation	Slope	Water ways	Rail network	Road network
Vegetation	1	3	4	6	7
Slope	0.333333	1	2	3	5
Water ways	0.25	1	2	5	4
Rail network	0.166667	0.333333	0.5	1	3
Road network	0.142857	0.2	0.25	0.333333	1

layers are combined in the exactly same map projection system with the same unit (Table 2).

Analytical Hierarchy Process (AHP) [14] a technique of multi-criteria analysis is used to create accumulated cost layer. In order to assign the priorities amongst the criteria of pipeline routing, weights are assigned to the criteria after confirming with the expert. Table 3 shows the rankings allotted to the respective routing criteria as the expert has given highest priority to vegetation.

This shows that the approach used here is to reduce the environmental loss. The second highest priority assigned is to Slope. Ranking is assigned as per the intensity of importance. The ranking is subjective in nature. Change in the ranking will result in the different optimal path. The weights assigned to criteria as per intensity of importance is shown in Fig. 3.

The AHP weights obtained from the calculations as per the ranking of criteria after applying the steps in AHP are shown in Table 4. Highest ranking is assigned to vegetation, then slope, waterways, and rail and road network. So, the route obtained will consider the importance of this factor and will be abide to this constraint. The Consistence Index (CI) obtained for above priorities is 0.049747 and Consistence Ratio (CR) is 4.441667.

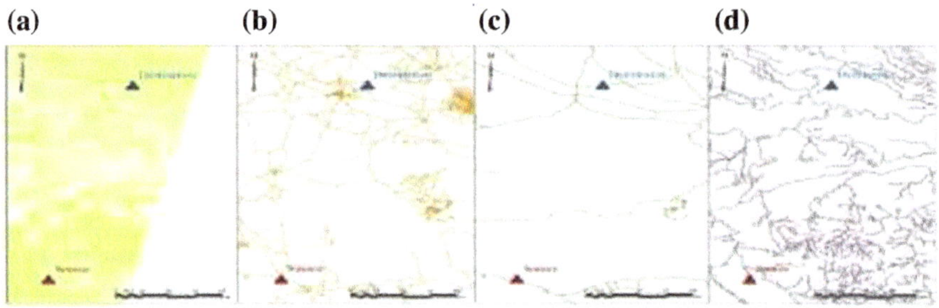

Fig. 3 Criteria layers a Vegetation. b Road network c Railway network. d Water streams

Table 4 Ranking

Criteria	Vegetation	Slope	Water ways	Rail network	Road network	Priority vector
Vegetation	0.528302	0.596026	0.516129	0.486486	0.35	0.495388779
Slope	0.176101	0.198675	0.258065	0.243243	0.25	0.225216777
Water ways	0.132075	0.099338	0.129032	0.162162	0.2	0.144521528
Rail network	0.144521528	0.066225	0.064516	0.081081	0.15	0.089974538
Road network	0.075472	0.039735	0.032258	0.027027	0.05	0.044898378

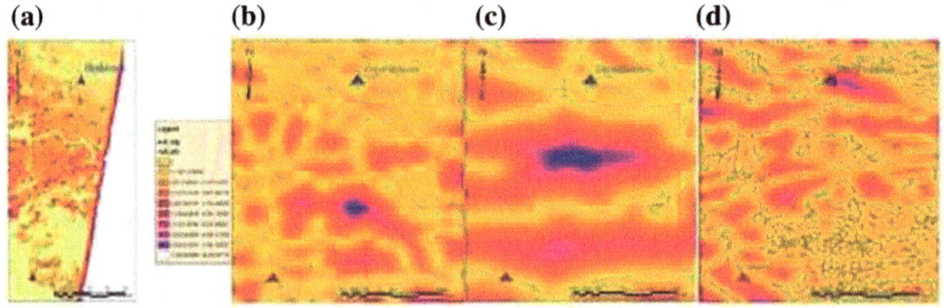

Fig. 4 Cost layers a Vegetation. b Road network. c Railway network. d Water streams

For generating discrete cost layers, Euclidean Distance is computed. To create cost layers a source location and a routing criteria layers are required. Cost of traversing through each pixel is computed in cost dataset. Figure 4 represents the cost layer datasets.

To determine the contribution of every routing criterion to the cost of generating optimal path, it is required to bring the criteria layers within the same range of interval values. Thus, reclassification is used to change a value ranges in a specified range of integers. It sets a common scale from 1 (highly preferred) to 10 (least preferred) for rating each interval within a dataset. Higher the values, higher are the cost of construction and lower the values, lower is the cost of construction. Therefore, the values are given as per the level of expected construction cost, for example 1 is assigned where there is no or very less vegetation, 10 is assigned to higher degree of vegetation and 1 is assigned to distance further away from existing roads, rails, vegetation, water bodies, 10 is assigned to distance close to the routing criteria mentioned above [15].

Once the common scale is applied to the raster datasets through reclassification to generate cost datasets, an accumulated cost surface is generated after combining and weighing cost datasets. Weights of the datasets are assigned as per the importance or influence of the routing process. Percentage assigned to datasets always must be equal to 100 percent. The higher the percentage assigned to a particular dataset, the more influence it has in the final outcome (Fig. 5).

To analyze cost distance, the creation of cost surface is important. In cost weighted layer the value of each pixel represents the cost per unit distance of crossing that cell, necessarily, it is not the physical distance traveled. The costs are finalized on basis of the routing criteria like the area being traversed is considered for its land use, natural vegetation, the slope, road network, water streams, and rail network. The accumulated cost layer is shown in Fig. 6.

After generating accumulated cost surface, cost-weighted distance dataset is used to create two datasets, first is distance surface and second is direction surface. A distance surface specifies the distance to the nearest source from every pixel in the dataset. A direction surface leads the path to the destination. Generated cost distance surface and cost direction surface are shown in Fig. 7.

To create the least cost path, cost distance surface, direction surface and location of destination point is required. As the distance and direction are known, the algorithm tries to compute cost and the minimum cost is selected. The path is generated until destination point is not reached. By applying this procedure, the path which is computed from source to destination points is called as least cost path and this is the optimal path. The least cost path is represented in Fig. 8.

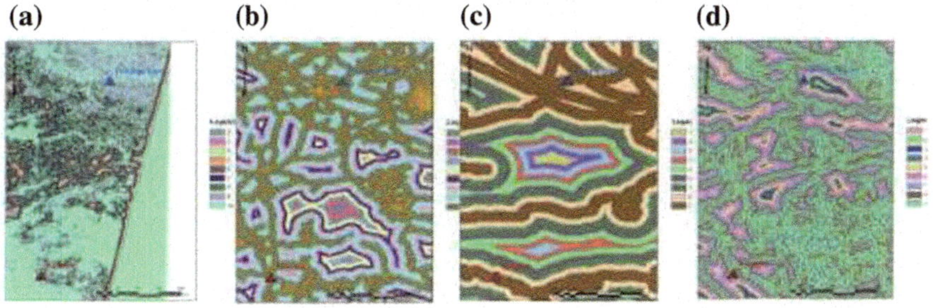

Fig. 5 Reclassification **a** Vegetation. **b** Road network. **c** Railway network. **d** Water streams

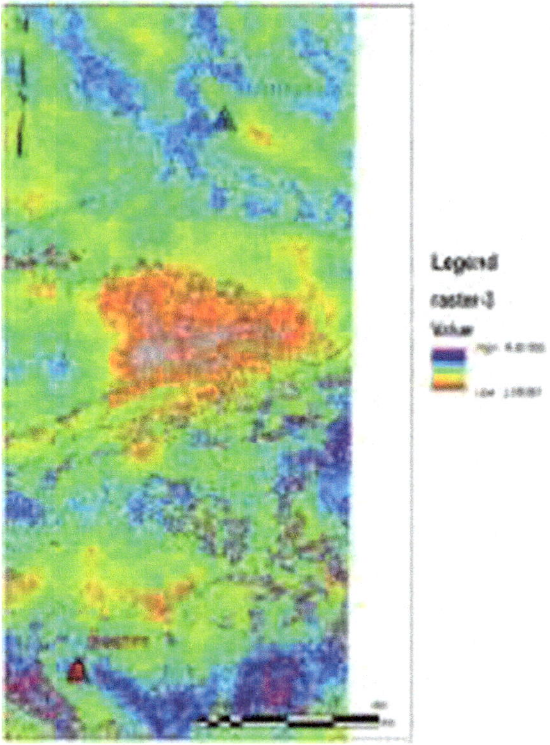

Fig. 6 Accumulated cost surface

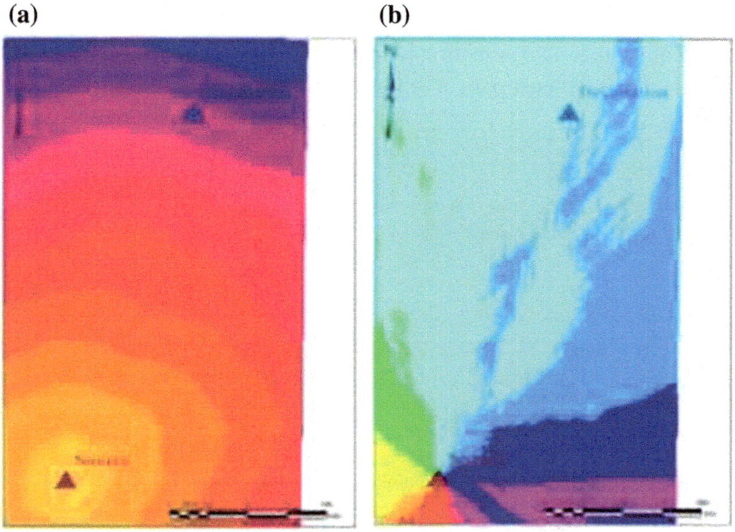

Fig. 7 **a** Cost distance surface. **b** Cost direction surface

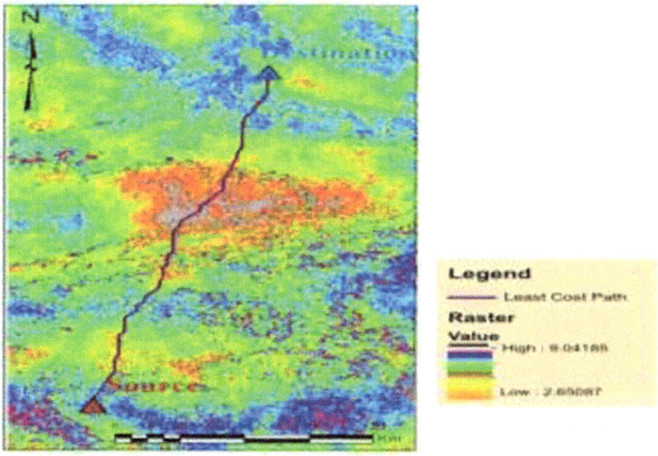

Fig. 8 Least cost path

5 Result and Analysis

The result of the optimal routing of the land corridor of pipeline subjected to the constraint of vegetation area, slope, waterways, rail network, and road network is shown in Fig. 9. The length of generated optimal route is 274 km.

This path avoids the vegetation area and hence reduces the environmental loss. From the analysis of this study, the alignment of the generated path is found to be acceptable as it satisfies the provided routing criteria. The obtained path crosses less steep areas, hence, the cost of construction and need of pump stations is reduced. The obtained optimal path fulfills the top priorities which are the environment safety and construction cost in the pipeline route identification process.

6 Conclusion

Identification of routing criteria is done successfully, to achieve acceptable results. The results are obtained on the basis of the priorities assigned to routing criteria. The pipeline route generated with vegetation as highest priority crosses minimum of the vegetation areas resulting in environmental safety. The obtained path passes through most of the at areas which directly results in the reduction of pipeline setup cost. The approach proposed in this paper for obtaining optimal path is simple, more precise and handy approach as compared to previous path selection methods.

This approach is still not in its final version as it lacks the uncertainty management. The developed system lacks most of the intelligence of the proposed system and thus, making the developed system more intelligent is the future scope of this work.

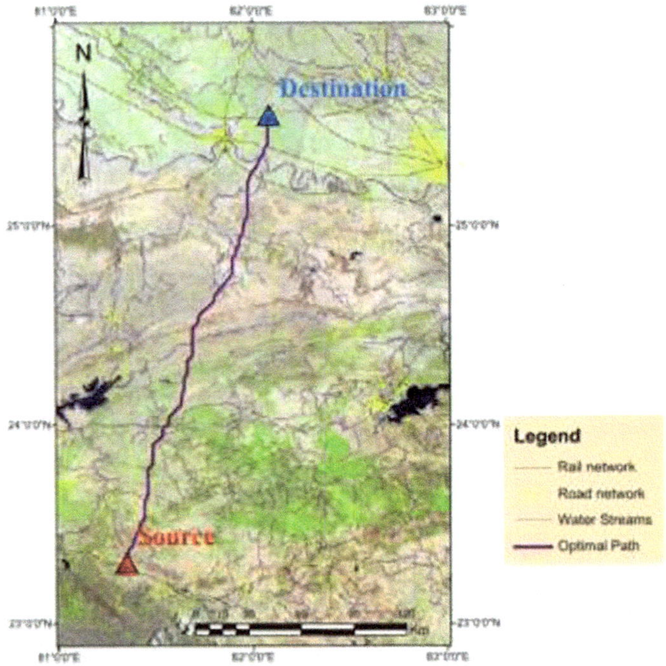

Fig. 9 Optimal path

References

1. Malczewski J, Ogryczak W (1995) The multiple criteria location problem: 1. A generalized network model and the set of efficient solutions. Environ Plan A 27(12):1931–1960
2. Gamarra A (2014) Gis suitability modeling to support a pipeline route selection
3. McHarg I, Mumford L (1969) Design with nature. American Museum of Natural History, New York, pp 7–17
4. Hopkins LD (1977) Methods for generating land suitability maps: a comparative evaluation. J Am Inst Plan 43(4):386–400
5. Byron S et al (2010) Risk-based pipeline routing improves success probability. Oil Gas J 108(24):58
6. Yildirim V, Yomralioglu T (2011) Nabucco pipeline route selection through turkey comparison of a gis-based approach to a traditional route selection approach. Oil Gas-Eur Mag 37(1):20–24
7. Aissi H, Chakhar S, Mousseau V (2012) Gis-based multicriteria evaluation approach for corridor siting. Environ Plan B: Plan Des 39(2):287–307
8. Tavana M, Behzadian M, Pirdashti M, Pirdashti H (2013) A promethee-gdss for oil and gas pipeline planning in the Caspian Sea basin. Energy Econ 36:716–728
9. Seel K, Dragan M, Coulombe-Pontbriand M, Laird CS, Campbell C (2014) A spatial multicriteria analysis process to optimize and better defend the pipeline route selection process. In: 2014 10th international pipeline conference, American Society of Mechanical Engineers, V001T04A004-V001T04A004
10. de Lucena RR, Baioco JS, de Lima BSLP, Albrecht CH, Jacob BP (2014) Optimal design of submarine pipeline routes by genetic algorithm with different constraint handling techniques. Adv Eng Softw 76:110–124

11. Asad FH (2015) Cost effective pipeline route selection: an exact analytical solution. J Intell Transp Urban Plan 3(4):116–127
12. Hu Y, Bie Z, Ding T, Lin Y (2016) An nsga-ii based multi-objective optimization for combined gas and electricity network expansion planning. Appl Energy 167:280–293
13. Yildirim V, Yomralioglu T, Nisanci R, Colak HE, Bediroglu S, Saralioglu E (2017) A spatial multicriteria decision-making method for natural gas transmission pipeline routing. Struct Infrastruct Eng 13(5):567–580
14. Saaty TL (1980) The analytic hierarchy process: planning, priority setting, resources allocation, vol 281. McGraw, New York
15. Matori AN, Lee H (2009) Suitability analysis of subsea pipeline route using gis. PETRONAS Technol J 2(1)

Part VII
Miscellaneous

The Digital Cadastral Map/Layer Generation and Conclusive Titling of Land Parcels Using Hybrid Technology (Aerial/High-Resolution Image (HRSI) and DGPS and ETS Survey) Adopted by Govt. of Odisha Under Digital India Land Record Modernization Programme (DILRMP), Govt. of India—The Technical Challenges and Solutions

P. K. Parida, M. K. Sanabada and Sandeep Tripathi

Abstract To bring efficacy in survey, creation, and updation of Land Records within a short time span with perfection and accuracy compared to the old method of survey and record preparation, the High-Tech Survey involving High-Resolution Satellite/Aerial Orthoimage, Differential Global Position System (DGPS) and Electronic Total Station (ETS) has been initiated in the State under the umbrella of Digital Land Record Modernization Program (DILRMP), a Govt. of India flagship program. The Odisha Special Survey and Settlement Act, 2012 has been enacted by the Government of Odisha to facilitate such Survey. The four costal Districts and five Western Districts of the State were taken in the first phase using HRSI and Aerial Survey Methodology, respectively. Odisha Space Applications Centre (ORSAC) is the technical collaborator to Revenue and Disaster Management Department (R&DM), Govt. of Odisha for this activity. A Standard Operating procedure (SOP) is prepared to facilitate the revenue officials in the field to overcome the technical challenges faced during execution.

P. K. Parida (✉) · M. K. Sanabada · S. Tripathi
Odisha Space Applications Centre, Bhubaneswar, India
e-mail: pkparida@yahoo.com

M. K. Sanabada
e-mail: m_sanabada@yahoo.co.in

S. Tripathi
e-mail: sandeeptrip.ifs@gmail.com

J. K. Ghosh and I. da Silva (eds.), *Applications of Geomatics in Civil Engineering*, Lecture Notes in Civil Engineering 33, https://doi.org/10.1007/978-981-13-7067-0_34

Keywords Land records · High-tech survey · High-resolution satellite/aerial orthoimage · DGPS · ETS · DILRMP · Odisha survey and settlement act, 2012 · Standard operating procedure (SOP)

1 Introduction

Till date, cadastral maps were being prepared as per the provisions of the Odisha Survey and Settlement Act, 1958, which recommended Survey and Settlement Operation in the State through four methods of Survey namely, (a) The Theodolite Traverse, (b) Prismatic Compass Traverse, (c) Plain Table Traverse, and (d) Chain Triangulation Method. These processes are resource hungry (Time, Cost, and Manpower). Besides, accuracy levels have been very low in those traditional methods of survey as well as such survey methodologies are not capable of surveying the area having slopes beyond 10° and 30°, respectively. The plot area calculation and preparation of Records of Right (RoR) was also time consuming. With the advancement of Science and Technology, the Survey and Settlement operation in the State will enable the survey beyond 30° slope with the adoption of Hi-Tech Methodology and with the help of Total Ground Method using Electronic Total Station (ETS), Differential Global Positioning System (DGPS) or Hybrid Method using Aerial Photography and High-Resolution Satellite Imaging (HRSI) as well as Ground Truthing with ETS and DGPS.

National Land Records Modernization Programme (NLRMP), presently renamed as Digital Land Records Modernization Programme (DILRMP) envisages deployment of such modern equipment and methodology to bring efficacy in survey and creation and updation of Land Records with shorter time span with perfection and accuracy compared to the old method of survey and record preparation. The Odisha Special Survey and Settlement Act, 2012 has been enacted by Govt. for such Hi-Tech Survey to minimize the time span without compromising quality, transparency, and grievance redressal with the involvement of lesser manpower. Revenue and Disaster Management Department, Government of Odisha has also come out with a Technical Manual for Cadastral Survey using Modern Technology and a Standard Operating Procedure (SOP) for the revenue officials to implement the project in a time bound manner by adopting the Modern Technology under DIRLMP. The Technical Manual and the SOP in PDF format is placed at Odisha Government Website under Revenue and Disaster Management Department for easy access. The Revenue and Disaster Management Department, Government of Odisha initiated four Coastal Districts (Ganjam, Khordha, Cuttack, and Keonjhar) and five Western Districts (Sundargarh, Deogarh, Sambalpur, Sonepur, and Bolangir) following High-Resolution Satellite Orthoimage and Arial Ortho Photographs clubbing DGPS/ETS Survey wherever, obscurity found in High-Resolution Orthoimages as well as Aerial Ortho Photographs. This Paper covers the complete road map of two villages namely Malidihi and Susarsanpur of Subdega Tehasil under Sunadargarh District and the technical challenges found during the process execution and actions taken to sort out the issues by the authorised team including Scientists ORSAC and Revenue Officials within the frame work of High-Tech Survey are highlighted.

District and the technical challenges found during the process execution and actions are taken to sort out the issues by the concerned stake holders.

2 Objective

The objective of this study is to share the process of mapping for preparation of Digital Cadastral Layer and generation of Record of Rights (ROR) following High-Tech Methodologies as under Digital Land Record Modernization Programme and also share the actions taken to sort out the technical challenges faced at the different stages of execution of the project.

3 Study Area

Two of the villages, namely Malidihi and Sudarsanpur under Sudega Tehasil of Sundargarh District are taken to showcase the execution of the methodology. The bounding coordinates of the two villages are given below.

Malidhi Bounding Coordinates:

Upper Left Co-Ordinate: 84°05′29.958″E 22°14 10.727″N
Lower Right Co-Ordinate: 84°06′27.315″E 22°13 15.226″N

Sudarsanpur Bounding Coordinate:

Upper Left Co-Ordinate: 84°06′18.301″E 22°14 19.836″N
Lower Right Co-Ordinate: 84°07′08.781″E 22°13 30.239″N

4 Stake Holders

The following Government Departments are the Stake Holders in executing the Project and their roles are given below.

i. **Government in Revenue and Disaster Management Department**: Responsible for taking decisions regarding the areas to be surveyed and methodologies to be adopted for survey in a particular area. The overall supervision of High-Tech Survey Operation is being conducted by the Administrative Department.

ii. **Director, Land Records, and Surveys**: Director, Land Records, and Surveys, Odisha is responsible for selection of private vendor and for guiding the survey operation entirely and issuing executive instructions/clarifications from time to time for smooth operation of High-Tech survey operation.
iii. **Joint Director, Survey and Map Publication, Odisha**: He works under the direct supervision of Director, Land Records, and Surveys, Odisha. He is responsible planning up to Tehasil level and monitoring of the survey and settlement operation being done in close contact with the Vendor, ORSAC, and the District Administrations.
iv. **Odisha Space Applications Centre (ORSAC)**: It acts as the technical partner in the matter of taking decisions involving technical issues during survey operations. ORSAC is responsible for quality checking all digital deliverables submitted by the private vendor during various stages of survey operation.
v. **National Informatics Centre (NIC)**: NIC is another technical partner and responsible for development of digital Record of Rights (ROR) format, development of application module for generation of different forms required during different stages of survey operation, hoisting the digital ROR, and preservation and maintenance of record of rights.
vi. **District Administration**: The role of district administration is for planning and successful implementation of field survey activities, preparation and publication of RoR with the involvement of grassroot level of Revenue Field Functionaries. On behalf of the District Administration, it is the responsibility of the Additional District Magistrate (ADM) to take steps for creation of awareness among general public about survey operation using Modern Technologies, to conduct training programs for the capacity building of the grassroot level Revenue Officials, regular monitoring of survey and settlement processes in all Tehasils and bridging the administration loop holes during the entire process of High-tech Survey operation. Tehasildar plays the pivotal role in executing the entire activities of survey settlement and preparation of map and preparation of map and RoR at field level. The Revenue Inspectors, Assistant Revenue Inspectors, and Amins working under the leadership of the Tehasildar shall be responsible in giving shape to the entire activities of survey, settlement, and preparation of map and RoR in close association with the agency personnel and the public.
vii. **Public**: Unlike other Acts, the Odisha Special Survey and Settlement Act, 2012 mandates certain responsibilities upon general public for successful implementation of High-Tech survey operation. The act provides for submission of self-declaration by landowners over the landed properties owned by them. The survey activities in field level require active participation of general public for preparation accurate land records. The general public must be made aware of different facets of High-tech survey operation for their active participation during the process and for gaining their confidence.

5 Data, H/W, S/W, and Instruments Used

A. The following High-Resolution Satellite Data and Aerial Photographs were used for the four coastal districts (Ganjam, Khordha, Cuttack and Keonjhar) and five Western Districts (Sundargarh, Deogarh, Sambalpur, Bolangir, and Sonepur) of Odisha, respectively.

   i. World View II, Panchromatic Stereo Data having 0.5 m spatial resolution.
   ii. Aerial Stereo Photography Multispectral Data having 0.09 m spatial resolution.

B. High-End Work Stations with Online Storage System were used to process Photogrammetric Blocks and High-End Desktops are used for the vectorization process.
C. Leica Photogrammetric Suite with the latest version was used for the Photogrammetric job, AutoCAD was used for Digitization and ARC GIS 10.5 Version was used for vectorization and GIS works.
D. Differential Global Positioning System (DGPS) and Electronic Total Station of Trimble Make were used for the survey of obscured area and generation of Ground Control Points (GCPs).

6 Methodology

The stepwise process flow under the methodology with the technical challenges faced during execution and the solutions arrived is given below.

1. The first necessary requirement before the commencement of survey using high-tech technology was the establishment of Ground Control Points (GCP). Once these GCPs were established and monumented, the coordinates of the GCPs were recorded through DGPS observation along with simultaneous observations taken using Survey of India (SOI) reference GCPs and benchmarks in order to transfer latitude, longitude, and altitude from the reference points of SOI following triangulation principle and the network adjustment of the GCPs to distribute the errors equally among all the GCPs. Each GCP was numbered following a standard numbering scheme. The High-Resolution Satellite (HRSI) Images/Aerial Photograph of the selected districts requisitioned after the establishment of GCP network. So that the imprints of the GCPs were available in the satellite image/aerial photographs. These were used during photogrammetric process to generate orthophoto/image as well as the reference points during the high-tech survey. The GCPs established with 16 km × 16 km, 4 km × 4 km, and 1 km × 1 km grid intervals were named as Primary Control Points (PCP), Secondary Control Points (SCP), and Tertiary Control Points (TCP), respectively. The Auxiliary Control Points (ACP) established depending upon the need during the high-tech survey for the two villages.

2. The raw image/photo was then converted into orthoimages following photogrammetric block adjustment process in the laboratory in order to make it suitable for survey operation as the raw image contained undulations of earth and the same was removed through photogrammetric block adjustment. The photogrammetric block adjustment used the stereo raw image/photo (stereo image/photo: a pair of photograph/image of the parcel of the land taken from two viewing angles) as well as the GCPS to generate Digital Elevation Model and orthophoto/image in the laboratory. The work was carried out under the supervision of Scientists of Odisha Space Applications Centre (ORSAC), the technical collaborator of Revenue and Disaster Management (R&DM) Department for the High-Tech Survey Project. The orthoimages were thus relief corrected and the ortho-rectified photographs, thus reflected the ground realities in terms of dimensions/areas truly within the RMS error limit.
3. The orthoimages were validated by ORSAC following tie-line measurements taken on the orthoimages for the district in the laboratory and the corresponding ground measurements were taken using Electronic Total Station (ETS)/Differential Global Positioning System (DGPS) for those points in the field. A comparison of orthoimage distance and field measurement distance were made and it was confirmed that the differences was less than 20 cm.
4. Since scanned cadastral maps were static (which could not be converted to any scale), they were first digitized to make it dynamic using CAD Software (.DWG format) so that the digitized AutoCAD files could be transferred to other formats like .shp vector format for using the same in the GIS environment. As per NLRMP/DILRMP guidelines, the AutoCAD files (.DWG files) were initially transposed to meter scale by multiplying suitable scale factor in AutoCAD environment.
5. All map sheets of two revenue villages were first digitized and mosaiced (joining of sheets) to create the complete Village mosaics in the AutoCAD environment. In cases of Gharabari clusters where it was shown in larger scale either in separate cadastral map sheet or in the same sheet at different locations, the said portion in the Cadastral Map usually shown as blank. The said block sheet digitized and inserted in the appropriate blank place in the cadastral map while mosaicking.
6. The transposed meter scaled seamless mosaiced AutoCAD file (.DWG file) for two villages converted to vector format, i.e., .shp file with all the layers like road, canal, forest, plot parcels, etc. The vector layers geo-referenced with the orthoimage/photo by taking the coordinates of the common unchanged features in both ortho images, as well as in the vector layers so that the points distributed throughout the village symmetrically and the number of points selected was more than 3, as first-degree polynomial transformation algorithm was utilized for georeferencing otherwise the number of common points would have been more as per degree of polynomial equation following the thumb rule as $2n + 1$, where n is the polynomial degree. The transformation of meter scaled vector to the orthoimage followed the best fit method taking correctly the common points identified both in the vector, as well as in the orthoimage and the Root

Mean Square error was minimum in order to get the best fit of the vector in the orthoimage. Thus, the identification of common points was the most vital process in the georeferencing of the cadastral vector where the common points selected were mostly the cross section of narrow roads, bund cross section of parcels, canal cross section, etc., of the village which had not been changed in the field over the period of time and was clearly visible accordingly both in the orthoimage as well as in the cadastral vector layers. The geo-referenced cadastral vector projected to a local projection, i.e., Universal Transverse Mercator (UTM) projection with WGS-84 spheroid and datum in order to make it usable further in the project.

7. After completion of the georeferencing work, the geo-referenced cadastral vector was superimposed on the orthoimage and the village boundary on the orthoimage was delineated referring to the boundary of the geo-referenced cadastral maps. While delineating the village boundary, utmost care was taken for the parcels of adjacent village, which should not be omitted by the new village boundary delineation. The boundary may be slightly adjusted to ensure seamless meeting of boundaries of adjacent villages subject to the non-displacement of cultural features already fixed. Now, all coordinates of each plot of the village were available within the drawn boundary village polygon.
8. The activities at para 2–7 were done by the agency in the laboratory and quality checked by ORSAC. Activity at para 1 being a separate pre-survey activity was assigned to other agency by the State Government as considered necessary.
9. After the notification under section 3, the Tahasildar held a Village-Level (Garamsabha) Meeting with all villagers and PRI members where he explained the purpose and the method of survey, demonstrated the method to the villagers and sought their cooperation during the entire process for preparation of their land records. Here, he explained the processes to be followed during (1) survey for boundary delineation, (2) obscure and Gharabari survey, and (3) verification during preparation of Draft Record of Right under Rule7(3) and Rule 9(4). The villagers were intimated specifically about their duty of filing Self-Declaration under Sub-Rule-1 of Rule-4 and the importance of remaining present during enquiry under Rule 9(4) positively. Besides, villagers explained regarding the provisions of grievance redressal mechanism under the Act for the people to raise objection to the recording for their appropriate disposal in course of the final preparation of Record of Right. The villagers were also informed that at every stage as stated above, the date and plot-wise calendar plan would be published in the village 7 days in advance of the starting of the process, so that the landowners/interested persons of the concerned plots might participate in the process on the scheduled date and time. The minutes of this village meeting recorded, signed by villager-participants, PRI members and agency, surveyors which remained as a part of the statutory file being maintained from the date of notification till date of final publication of Map.

10. Once the village boundary delineated over the orthoimage, the vendor submitted the soft copy of the same to ORSAC for quality checking. A cluster of villages was put to quality checking (QC) by ORSAC together with the two villages. During QC, proper georeferencing of the village over orthoimage as a whole was verified. While georeferencing of the villages boundary was made and it particularly verified the overlaps or gaps between boundaries of villages. At some places, there were obscured areas for which the delineation of the village boundary was not possible. The obscured areas/overlaps and gaps between the boundary line of villages, if any, were identified for field survey by the agency in the presence of revenue department field functionaries of the concerned Tehasil. ORSAC provided these plot numbers coming within obscured areas/overlaps/gap areas, for field demarcation by the agency.
11. In case there was road, canal, river along the village boundary, the same was delineated on the orthoimage as per ground reality taking the cue from the existing cadastral map.
12. Where the River/Stream/Nala was passing between two villages, the principle of "Dhar Dhur and Nispa Nispi" as provided under previous Survey procedure was adopted.
13. At the time of delineating village boundary, portion(s) of the cadastral village map was not identifiable on the orthoimage due to obscurity in the image or due to nonavailability of physical sign of plot bund in the field for different reasons. At times, there might appear a gap/overlap of boundary plot bunds while delineating the village boundaries of adjoining villages on the orthoimage as stated at para 11 above. In that event, coordinate data of the obscured portions of such boundary line were captured using DGPS/ETS by the Surveyor (Agency) with the assistance of Amin/ARI/R.I and preferably in the presence of concerned landowners. The surveyor of the agency went to field with a map showing the plot numbers to be surveyed for the purpose. Before initiation of boundary survey, the agency provided the date scheduled for survey of boundary and obscured area adjoining the boundary to Tahasildar in the following Format no 1 which was published in the village.

Format-1

Boundary plot Survey Schedule

Date of Submission:
Name of the Village____________
Name of the Tahasil____________

Date	Obscured plots over/along Village boundary to be surveyed.

Surveyor of the Agency

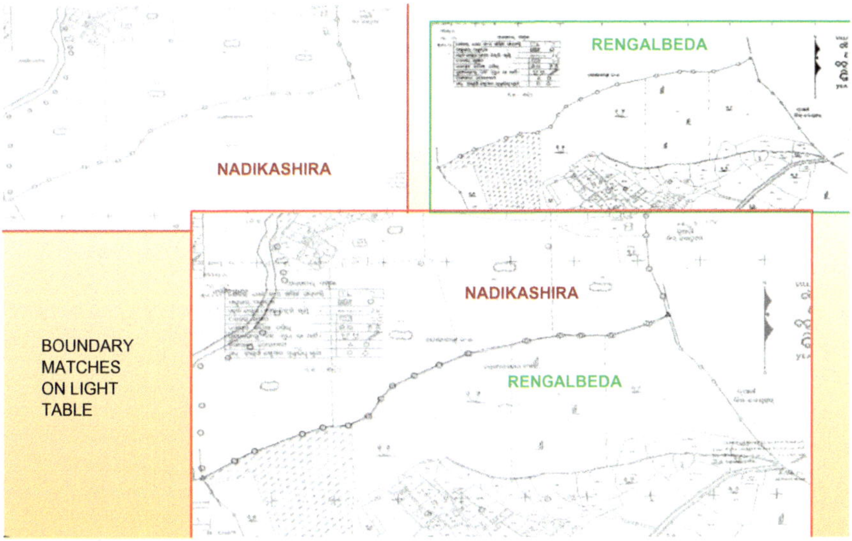

Fig. 1 Glass table matching

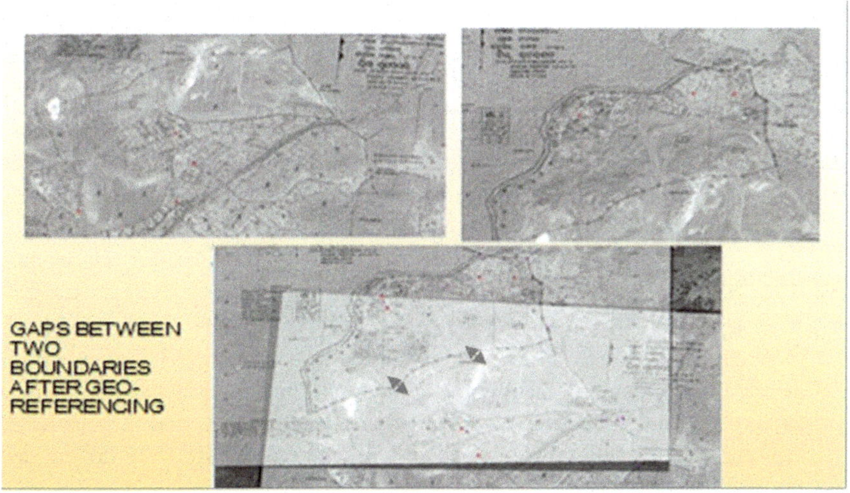

Fig. 2 Gaps between two adjacent geo-referenced village bounadries

While Fig. 1 shows there is no gap between two village boundaries of NADIKASHIRA and RENGALBEDA, when checked in glass table, the image of same two villages, when drawn over orthophoto as shown in Fig. 2, it comes to notice that there is a gap.

14. While demarcating the obscured portion(s) of the boundary line of a particular village, the obscured plots adjoining the boundary plots of that village were surveyed through ETS/DGPS simultaneously.
15. All necessary DGPS/ETS observations were taken from the existing GCP (Primary and Secondary) in the village. A report of survey in duplicate in the prescribed Format no 2 prepared, one copy of which retained at the Tahasil level for official record and other copy handed over to the surveyor of the agency for submission at ORSAC.

Format-2

Boundary plot Survey Report

Name of the Village_____________
Name of the Tahasil_____________

Date	Obscured / invisible plots over Village boundary surveyed.	Remarks

Surveyor of the Agency **Amin/ARI /RI**

16. The boundary line was drawn by the agency after incorporating the captured ETS/DGPS coordinate data on the orthoimage and the boundary was tentatively delineated.

Vectorisation of Plots from Orthoimage:

19. After tentative delineation of Village boundary, the agency captured the inner plot boundaries exactly as per the plot bunds on the orthophoto following mirror principle in GIS environment in the laboratory. The boundary line portion data captured by the agency was verified in GIS environment and the quality of drawing the inner plots were checked and cleared by ORSAC following the mirror principle.
20. During plot vectorisation, the agency identified the obscured areas and made a list of plots coming within such areas referring to the existing cadastral map as per the following Format no 3. Such list was certified by ORSAC.

Format-3

(Obscured plots for survey)

Name of the Tahasil	Name of the Village	Plot numbers to be verified in the field.	Remarks
			ORSAC

21. Similarly, for Gharabari areas, the vendor referred the old ROR and identified a Gharabari polygon taking the Gharabari plots and an additional buffer zone of clearly identifiable plots surrounding the Gharabari areas from the geo-referenced cadastral map and identified the same area over the orthoimage. The agency captured the clearly interpretable road and clearly visible Gharabari sections within the Gharabari polygon(s). Agency took A3 size print out of the drawn Gharabari polygon(s) with sections.
22. In addition, the vendor listed out the plots which appear in RoR as available in Bhulekh but could not be identified from the image due to the absence of field bunds, and provided the same to Tahasildar well in advance before start of survey to facilitate demarcation of those plots by the Tahasil Amin/RI relying on the information given by Tahasildar. Those plots needed to be demarcated during the survey of obscure and gharabari areas together with the newly mutated plots/plots where statutory was granted under any law in force like OGLS Act, FR Act, OPLE Act, etc., which had not been updated in Bhulekh. After demarcation of these plots, the mutation plot numbers were entered by the agency in the field book which reflected in the preliminary draft map. Plots identified in the field but not mutated were given temporary numbers against the original plot number as per Sabik map.
23. Before going for the survey for obscure and Gharabari areas and demarcation of the plots as mentioned above, the agency prepared date wise schedules in the following Format no 4 for publication in the concerned village by the Tahasildar to ensure adequate peoples' participation.

Format-4

Schedule of Obscured and Gharabari area survey/additional plot demarcation

Date of Submission:
Name of the Village____________
Name of the Tahasil____________

Date	Plots to be surveyed / demarcated

Surveyor of the Agency

Obscured Area Survey:

24. The agency surveyed the obscured areas as per the schedule. The agency with the help of Amin/ARI/RI reached the obscured area with the cadastral map in hand giving tick mark on the obscured plots and captured the coordinates using DGPS/ETS. Where the obscured area was inaccessible, like bushy/thorny/water logged/swampy/marshy etc. the survey was taken up at few points in both ends of the area and the line drawn in the map accordingly. The sketch map of the survey was provided by the agency to ORSAC.

Gharabari Survey:

25. The agency started the Gharabari survey section wise from the Ground Control Point (Auxiliary point/SCP/PCP) and reached the polygon point of the Gharabari portion. While capturing plot-wise coordinate data, the sketch map at hand was corrected accordingly in red ink. Finally, the agency had made sure that all Gharabari plots were brought to the map. After all the plots were captured, the complete sketch map of the Gharabari area was prepared by the agency in duplicate under the signature of both the surveyor of the agency and revenue field functionary of the Tahasil office. One copy of the sketch map was given to the Tahasil for office record and another copy was taken by the agency surveyor.

Additional Plots Survey:

26. For demarcating the plots which appeared in RoR of Bhulekh and not identifiable in the field due to absence of field bunds, the Amin/RI identified the demarcation points in the field to the agency. For the purpose, he referred the concerned mutation case records/available records with the concerned landowners, produced at the time of survey. The agency captured the coordinate values of the points of demarcation as shown by the Amin/RI in the field.
27. After the survey of obscure, Gharabari and additional plots were completed, a report was prepared in the following **Format no 5** mentioned below in duplicate. One copy was retained with the Tahasildar for official record and another copy was given to the agency.

Format-5

(Gharabari and obscured area/ additional plot survey report)

Name of the Village____________
Name of the Tahasil____________

Date	Gharabari / obscured /Additional plot numbers surveyed	Remarks

Signature, Name
Name and designation
Surveyor of the Agency

Signature,

Revenue authority

28. The agency after getting the coordinate values of points relating to obscure, Gharabari areas and additional plots, integrated them in the orthoimage to prepare a draft map of the village. The survey data was submitted to ORSAC by the agency in native format as available in ETS and DGPS in a magnetic media. The preliminary draft village map so prepared by the agency submitted for quality checking by ORSAC. After getting clearance from ORSAC, the hard copy of prepared village map was generated by agency and handed over the map to the concerned Tahasildars duly signed by the agency for their verification as per Rule 7(3) along with comparison table of areas in Form no 6(T) of the Technical Manual for Cadastral Survey using Modern Technology.
29. **Verification of Draft Map Under Rule 7(3)**:
After getting the draft maps from the agency, the same was verified by different revenue field functionaries as provided in the Rule 7(3) of the Odisha Special Survey and Settlement Rules, 2012. Here, the verification was for ensuring that all plots are available in the field, as well as the plots in the existing RoR and plots where new statutory right created under different acts were brought to the map and the map was a mirror of the existing plots.
Before going to the field for verification, the Amin made a plot-wise study. He did the following activities one by one.

    (i) The old plot area as provided by the agency in Form 6T was verified with Form no 6 and error if any in Form 6T was rectified in red ink.
    (ii) Then he read Table 6T. This gave him a picture as to which plots had merged/divided and how the new plot had been created. By reading Table-6T for analysis of the new plot, he compared the area of the new plot(s) vis-a-vis the area of the old plot(s).

30. This reading of Form 6T gave him a picture of how the plot as shown in the newly surveyed map had been carved; whether by division or merger from adjoining plot(s) or remained as before. If it was found that the area of the plot tallied with

the original plot and no new plot had been created in the meanwhile, the Amin proceeded for verification of the next plot. But where variation was noticed in the area of the newly drawn plot with that of the total area of corresponding old plot, the possible reason of variation was to be assessed. For this, he had to compare the area of adjoining plots from Form no 6T to have a firsthand knowledge of the possibility of increase or decrease in area of any old plot. Where any plot was found divided/merged, the sum total of the area of divided plots was compared with the area of original old plot(s). This homework was done first before the start of field verification as per the calendar plan.

It was needed to be remembered here that the mutated plots and additional plots demarcated by Amin during survey by the agency could only be verified on the map with the survey report jointly prepared by agency and Amin during field survey since field bunds were not available in such plots.

Here, the Tahasil staffs started field verification from one end and verified all the plots including all the government lands, Gochara, Rakshita, and Sarvasadharana common property resources like land belonging to public religious institutions and the like one by one accordingly to ensure that all the plots available in the field are in the map including the mutated plots and additional plots stated in para 26 above. The area of any Hal plot was found decreased and excess was found in its boundary plot(s), as noticed during comparison of Form 6T, they redemarcated and revised coordinate data captured using ETS/DGPS for correction of the map. If the plot area had increased, without affecting any of its boundary plot(s), there was no need of reducing the plot area by redemarcation at this stage.

31. While verifying a bigger plot in the sabik map which divided into many small plots, the team took the bigger plot area in the sabik map into consideration and compared its total area with the sum total area of the subdivided smaller plots. Here again if the total area increased/decreased the same was to be verified and marked in the map as stated above.
32. As far as verification of plot area of each smaller plot was concerned, the same was examined after hearing the interested persons and after verifying their records/title deeds, etc., efforts were made to dispose of all these smaller plot cases together on the same day. Similarly, where some plots were merged on the ground and a total new geometry of smaller plots had evolved, the total area of these plots was compared with their corresponding sabik plot(s).
33. If any change like division of any plot or merger of plot/part of plot was noticed during field verification, the said changes were marked in the map in red ink and coordinate data captured by the agency in presence of RI/Amin/ARI. The 6T Form, as revised were updated accordingly by correcting area and adding new plots to the list and new temporary numbers were assigned to each in the same manner as the agency had assigned at the time of generating the map after survey.
34. The map verified was cross checked by the statutory authorities as per rule who put their signatures at the bottom of the map after verification.

35. The agency regenerated three copies of the revised map, one in the gram panchayat, one in Tahasil office for display, and another copy along with the checked map signed copy was retained in the Tahasil.
36. Soon after 7(3) verification was completed, the Tahasildar submitted a completion certificate in Format no 6 to Director Land Records Surveys and Consolidation, Board of revenue, Odisha, Cuttack.

Format-6

Verification certificate under rule 7(3)

This is to certify that, the verification of map in respect of the following Village(s) under rule 7(3) of OSS&S Rule 2012 has/ have been completed.

Name of village(s)	Date of receipt of map from agency	Date of completion of the verification
-1-	-2-	-3-

Signature of the Tahasildar

Constitution of Team:

37. The team constituted as per rule without waiting for verification by the field functionaries so that by the time the verification was complete, the team must be in a position to carry forward the verification under Rule 9(4) of OSS&S Act and Rule 2012.

Verification Under Rule 9(4):

38. Soon after the display of the map in G.P and Tahasil office, the team updated the map prepared under Rule 7 and proceeded to prepare the Record of Rights in Form no 7 which was the Basic Record of Right. From this draft, Record of Right was prepared after hearing the objections by Tahasildar.
All pending mutation cases including Form 3 cases received and pending till 9(4) verification started were transferred to the team after the issuance of notice by Tahasildar by post to the purchasers. All these mutation cases were disposed and Land Record in the Village under OSS&S Act 2012 was created for the landowners.
The team prepared a date-wise calendar plan of the plots which should be published in the village at least three days in advance with a request to landowners to remain present during the stage.
The verifications for the earmarked plots as per the schedule was done in the forenoon and inquiry relating to ownership was done in the camp in the village/G.P same day in the afternoon session. The team did this ownership inquiry in the following field verification report **Format no 7** as given below referring to Form nos 3, 4, 5, 6, copy of sabik map, and the newly prepared revised map.

Format no-7

Field Enquiry Report

Sl no	Sabik Khata No	New Hal Plot	Report of the Team	Remarks
		Sabik Plot(Red ink)		
			Signature Member 1 Signature Member 2 Signature Revenue Supervisor Supervised by Tahasildar	

Map Verification:

39. Plot-to-plot verification was done for creation of revised map on the basis of ownership of landowners. If the Hal map showed a change in configuration (shape/geometry), the same was also inquired. Here, the team verified the configuration of the plot in the field with Sabik map and Hal Map and found out the reasons of changes in the plots and prepared the Land Record ownership wise and revised the map as per the captured coordinate data by the agency.
40. The verification work started from Govt. land, including land belonging to different State Govt. and Central Govt. Departments, their agencies, common property resources like Road, Rail line, Canal, Stream, Nala, River, land belonging to Public religious institutions, like Matha, temple, Mosque, Church, cremation, burial ground, Playground, Park, Gochara and the like. The Team first identified and prepared a list of such plots from Register in Form no 6 and compared the area of those plots with the area and configuration in sabik map and as recorded in Form no 6(T), as updated during Rule 7(3) verification.
41. During field verification the Team took note of all lands vested to State in Land acquisition proceedings subsequent to the date of final publication of Sabik Record of Rights on obtaining a copy of statement of such lands from the LAO/Special LAOs of the area to ensure that land vested to State in LA proceedings were recorded in name of State where the land still stands recorded in name of private individuals as per sabik Record of Right.
42. If area of the plot was found decreased, the area of its boundary plots within the same village was cross checked with their corresponding sabik area. Where it came to notice that the decreased area had been included within boundary plot(s), the team marked it in red ink in the map and captured the co-ordinate data of the redemarcated area.
43. After field verification of one plot was over, the team gave a new number to the plot as Hal plot number and prepared the field enquiry report accordingly. This

numbering was done in serial following the same principle as was being done for numbering during yadast of last settlement.

44. The statutory authorities cross checked the verification of the map as provided under rule 9(4) and put their signatures at the bottom of the map after verification.
45. While preparation of Field Verification Report and preparation of preliminary Record of Right, the ownership of the land recorded as per the "Rayatai Jami Record Kariba Pranali" and "Sarakari Jami Record Kariba Pranali" and the Govt orders issued from time to time.
46. The four types of Khata under the ownership of Govt. ROR of Govt owned lands recorded as "Abadajogya Anabadi, Abada ajogya anabadi, Rakshita, Sarbasadharana" under column 1 of the ROR below the khata number column. Accordingly, at the time of preliminary ROR and Draft ROR, the said information was recorded.
47. The status wise separate khata were prepared under private ownership, e.g., 1. Stitiban, 2. Pattadar, and 3. Bebandobasta. The Khata serial in the village was in that order. The Govt. khata was the last khata. The status recorded under the column khata number as provided in the preliminary ROR, Draft ROR, and the Final ROR.
48. The classification of the Govt. and private land was rationalized following "Kisam Niyamabali" published by Govt. of Odisha. All Govt. lease lands were recorded in the ROR in "Pattadar" status. The conversion of land leased for agricultural purpose by Tahasildar under section 8A of OLR Act without jurisdiction, the kisam of the land was recorded as per the Kisam Niyamabali but in the remarks column the following noting was recorded "Conversion to Gharabari is without jurisdiction". Land leased out for homestead purpose was recorded in "Gharabari Kisam" in Pattadar status.
    Where agricultural land was converted for nonagricultural purpose in violation of the provisions under 8 and 8A of OLR Act, the Kisam in the ROR was noted as per previous agricultural classification, with noting in remarks column as "Conversion under 8A not taken".
49. Where land was created/submerged due to alluvion or delluvion the provisions of Section 21 of the OLR Act applied. In case of Rivers, Nala, Stream shown in the boundary of the two villages, last settlement map was recorded following the old principle of "Dhar-Dhur Nispa-Nispi".
50. While physically verifying each plot, the team collected the information relating to additional layers as provided at **Annexure-3**, page 61 of the Technical Manual for Cadastral Survey using Modern Technology and indicated the same in the remarks column of the Field Enquiry report as noticed during field verification of the concerned plot which was indicated in the digitized GIS map in layers at a later stage.
51. All change of classification of land which had been done in contravention of statute like the change of classification of agricultural land for nonagricultural purpose where so ever was recorded as per the existing classification in the field, rent assessed accordingly but in the remark column of ROR the violation was noted.

52. The team prepared preliminary Record of Right in Form 7 and provided it to the agency for generation of Land Parcel Map (LPM) in the Format no 8 as given below which were distributed by the agency to the landowners with the preliminary ROR under seal and signature of the Tahasildar. Before distribution the LPM and ROR, the team verified the same and prepared as per the verification under Rule 9(4).

Format no -8

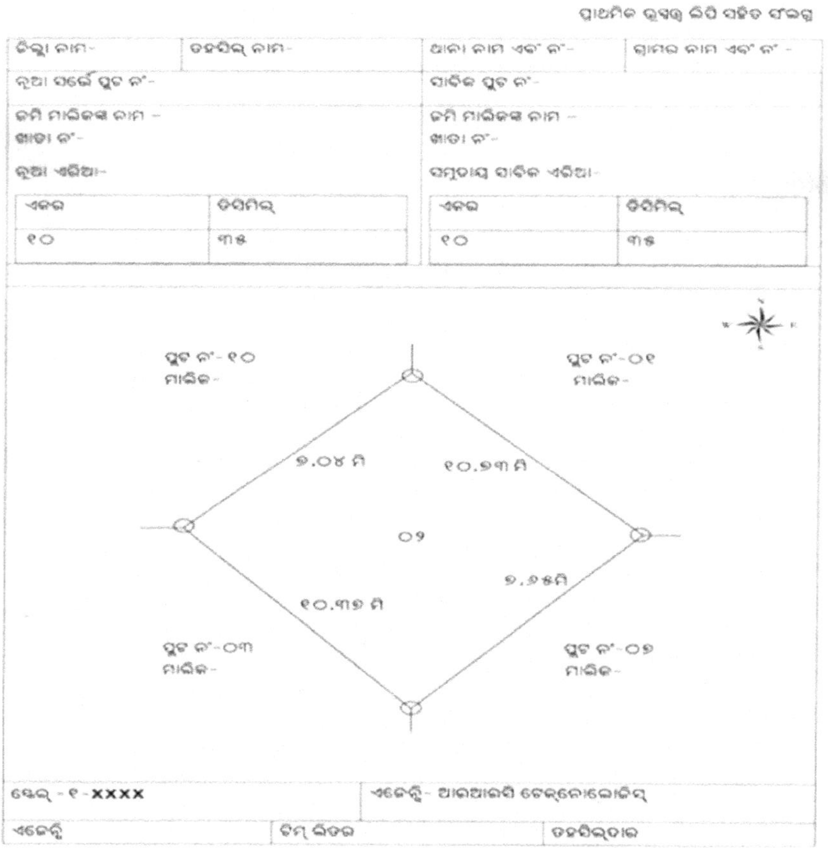

7 Results and Discussion

As per Government of India Centrally Sponsored Project, the Digital India Land Records Modernization Programme (DILRMP) which merged two existing Centrally Sponsored Schemes of Computerization of Land Records (CLR) and Strengthening of Revenue Administration and Updating of Land Records (SRA&ULR), as well as adding related system of Property Registration in the Department of Land Resources (DOLR), Ministry of Rural Development mandated revolutionary changes in creation of land records in digital format which includes entry of ownership records, continuous Mutations updating, digitization of maps and integration of textual and spatial data, survey/re-survey and updating of all survey and settlement records including creation of original cadastral records wherever necessary, computerization of registration and its integration with the land records, development of core Geographical Information System (GIS) and capacity building. The main objective of the DILRMP is to develop a modern, comprehensive and transparent land records management system in the country.

Thus, the Government of Odisha adopted DILRMP and initiated nine districts following two methodologies, i.e., HRSI and Aerial orthoimage with DGPS and ETS survey and completed two villages facing many technical challenges and those challenges are sorted out within the ambit of Acts and Laws provided under Odisha Special Survey and Settlement Act, 2012, where ever it is felt that new laws are required to deal with issues relating to maturation process, the same is referred to Revenue and Disaster Management Department, Government of Odisha to create the laws under the Act to sort out the issues.

The completion of High-Tech Survey for two villages now paves the way to go ahead with the creation of Digital Land Record and RoR for the current nine Districts and the same road map is to be rolled over to other 21 Districts of the Odisha State, once the 9 districts are completed successfully. The Digital Land Record in the form of a map and its RoR are given as Figs. 3 and 4.

Acknowledgements The authors would like to express their gratitude to the Principal Secretary to Govt. of Odisha, Revenue, Disaster Management and Director Land Records Odisha, Joint Secretary to Govt. of Odisha, Revenue and Disaster Management, Joint Director, Map Publication and Assistant Director, Computer Land Record, and Office of Director Land Records and Surveys for their support and encouragement during the project execution. The authors are also thankful to the engineering and technical personnel, ORSAC used for field survey as well as laboratory processing work during carrying out this study.

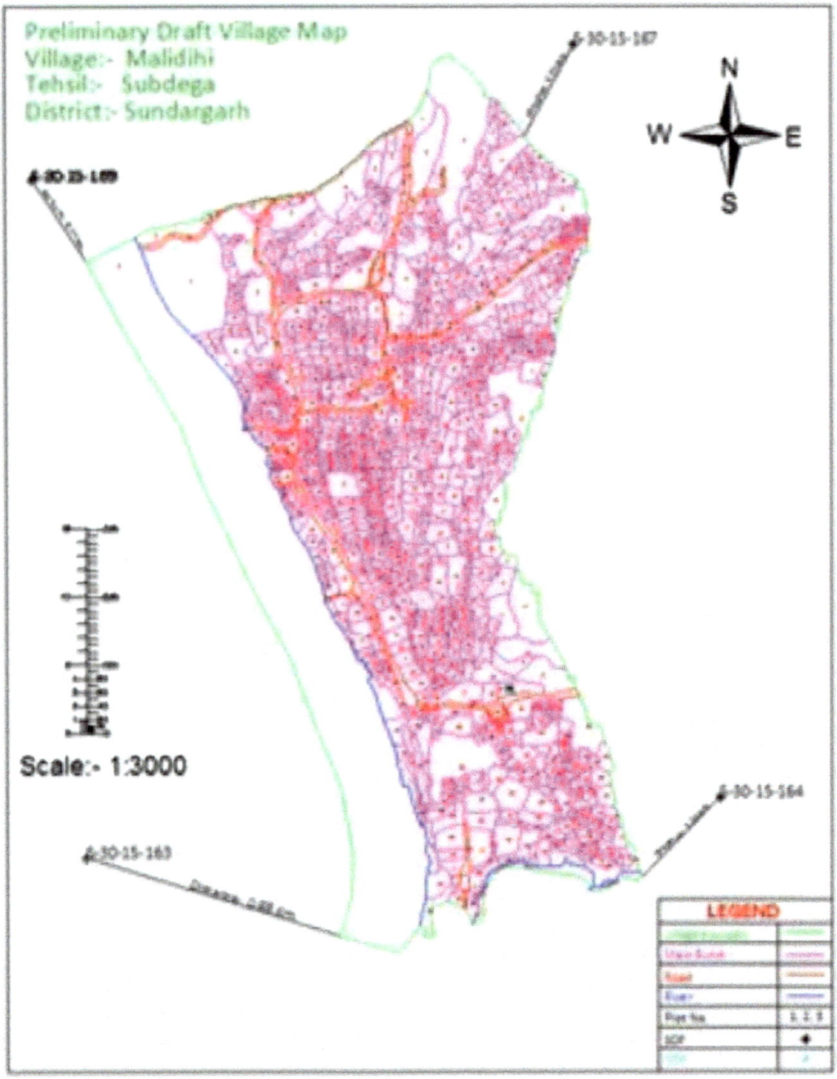

Fig. 3 Resurvey map of Malidhi village, Subdega Tehasil, Sundaragarh District

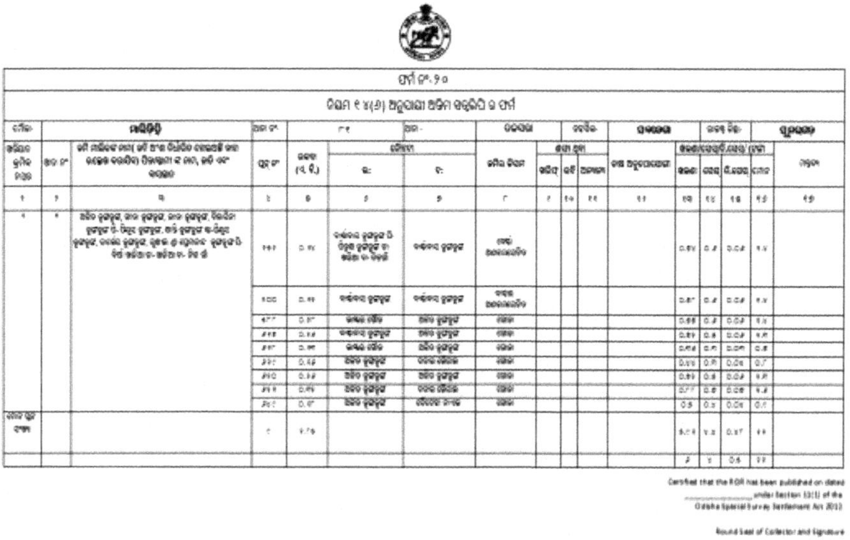

Fig. 4 Final ROR for Malidhi village, Subdega Tehasil, Sundaragarh District

References

The below references were went through by the Authors during the preparation of the Paper

1. Siva Subramanian KS, Amitabh S, Manda S (2003) Evaluation of digital elevation models created from different satellite images. In: Proceedings of map India conference 2003. New Delhi, India, Jan 2003
2. Jacobsen K (2003) Orthoimages and DEMs by QuickBird and Ikonos. In: Proceedings of EARSeL "remote sensing in transition". Ghent, pp 513–525
3. Jayaprasad P, Narender B, Pathan SK, Ajai Generation and validation of DEM using SAR interferometry and differential GPS supported by multispectral optical data. J Indian Soc Remote Sens 36(4):313–322
4. Satirapod C, Rizos C, Wang J (2001) GPS single point positioning with SA off: how accurate can we get? Surv Rev 36(282):380–386
5. Clavet D, Lasserre M, Pouliot J (1993) GPS Control for 1:50,000-scale topographic mapping from satellite images. Photogramm Eng Remote Sens 59(1):107–111
6. Wilkie DS (1990) GPS location data: an aid to satellite image analyses of poorly mapped regions. Int J Remote Sens 11(4):653–658
7. Barbarella M, Mancini F, Zanni M (2003) Rectification of high resolution satellite data: evaluating accuracy for map updating. In: Proceedings of the ASPRS/MAPPS conference
8. Barbarella M, Mancini F, Zanni M (2003) Processing of high resolution satellite data for map updating. In: Proceedings of 30th international symposium on remote sensing of environment. Honolulu, Hawaii, USA, Nov 2003

Accuracy Assessment of the Digital Elevation Model, Digital Terrain Model (DTM) from Aerial Stereo Pairs and Contour Maps for Hydrological Parameters

Odai Ibrahim Mohammed Al Balasmeh and Tapas Karmaker

Abstract In this study, some of the widely used data for elevation models were compared based on hydrological parameters such as slope, aspect, flow direction, and slope length and steepness factor (LS factor). The study considers the comparison among ASTER GDEM, SRTM DEM, topographic maps, and aerial photography at 430239 predetermined tests points. Bilinear interpolation technique was used to interpolate the data at these testing points. The result shows the digital elevation model (DEM) from topographic maps has relatively higher vertical accuracy (RMSE = 5.40 m), compared to ASTER GDEM (7.10 m) and SRTM DEM (15.07 m), while comparing with digital surface model (DSM) from stereo pairs. For validation, we used 47 ground control points (GCPs) using GPS. The results show the vertical accuracy are relatively higher for DSM (RMSE = 1.11 m), followed by DEMs from topographic maps (4.10 m), ASTER GDEM (7.36 m), and SRTM DEM (12.22 m). The DSM matches closely with the result of topographic DEM in case of slope, LS factor curves, aspect, and flow direction. The result with ASTER DEM matches better than results with SRTM data in all the parameters but both of them show poor match with the results from DSM data.

Keywords Digital elevation model · Hydrological parameters · Aerial photography · Accuracy assessment

1 Introduction

Nowadays, DEM has a major role in various field of study including geography, geology, and environment sciences. The quality or accuracy of DEM is considered as the main factor for application in a particular field depending on required parameters. There are various methods to produce DEM from satellite images, field survey (for

---

O. I. M. Al Balasmeh · T. Karmaker (✉)
Department of Civil Engineering, Thapar Institute of Engineering and Technology, Patiala 147004, Punjab, India
e-mail: tapas1976@gmail.com

J. K. Ghosh and I. da Silva (eds.), *Applications of Geomatics in Civil Engineering*, Lecture Notes in Civil Engineering 33, https://doi.org/10.1007/978-981-13-7067-0_35

small area), LiDAR technology, Aerial photos, and RADAR. Depending on the data used, the accuracy also varies significantly.

The scientific community is facing issues in determining the optimum spatial resolution of DEM as a part in decision making in hydrological problems and its analysis [1]. Hydrological parameters such as slope, aspect, flow direction, drainage network, and watershed can be derived by automated techniques from DEMs [1, 2]. Sometimes in hydrological studies, high accuracy in absolute elevations may not be required, instead the accuracies in slope, drainage network, aspect, etc., are required. Previous researchers suggested to use multiple DEMs for all hydrological problems based on their studies on the relation of DEMs spatial resolution (cell size) with the accuracy of hydrological analysis [1–3]. The errors in DEMs will affect the modeling of natural parameters and processes [4, 5]. Wu et al. [6] and Vaze et al. [7] reported that DEM has significant implications on hydrological parameters and error in watershed area and slope increases with coarser resolution [4–7].

At present, many tools and softwares are available to extract the topographic and hydrological parameters from DEM. A variety of freely available DEMs data with variable spatial resolution, e.g., Geographical Survey Institute (GSI) of Japan (GSI-DEM) with 10 m, Advanced Space-Borne Thermal Emission and Reflection Radiometer Global Digital Elevation Model (ASTER GDEM) with 30 m, Shuttle Radar Topographic Mission (SRTM) with 30 arc second (resampled to 3 arc second) are suitable to model the terrain of earth in worldwide very accurately [8]. However, the verification of these data is required by comparing between DEM at multiple reference points with measured in situ high precision data [8–14].

In this present study, comparison among SRTM, ASTER DEM and DEM derived from topographic map with reference DSM derived from stereo pair aerial photos, and validate the results with ground control points (GCPs) collected using differential global positioning system (DGPS).

2 Materials and Methods

2.1 Study Area

Humrat Es-Sahen is extended from the western part of As-salt city near Kufur Huda to King Abdullah Channel (KAC) in the Jordan valley (35° 36.8′–35° 42.2′ E and 32° 4.7′–32° 7.5′ N). Topographically, the basin is of concave shape and it is topographically known as the Humrat Es-Sahen Depression. The elevation ranges from 1300 m above mean sea level (AMSL) eastern of salt city, to a low of 200 m under mean sea level (UMSL), which constitutes the lowest point. The topography of the study area varies from nearly flat area to the mountainous regions with steep slopes. This variation of elevation made the area attract attention from various researchers from different scientific disciplines. The streams have flow directions mainly from east to west due to the topography (Fig. 1).

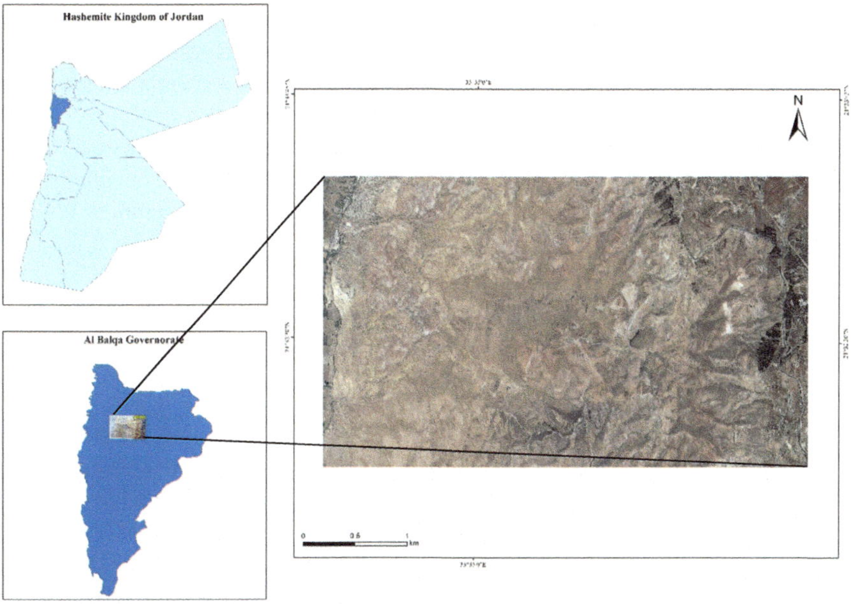

Fig. 1 Location of study area

2.2 *Digital Elevations*

The DSM from stereo pair aerial photos has compared with three different digital elevation models with different spatial resolutions. The available DEMs were collected from open sources such as SRTM DEM and ASTER GDEM, and paid sources for topographic maps. The topographic maps were digitized from contour lines with contour interval of 20 m with scale 1:50,000 by Royal Jordanian Geographical Center (RJGC). The contour lines were manually digitized and the DEM was generated using ArcGIS 10.2.

The SRTM data is available as 3 arc second (~90 m resolution) DEMs [1]. The area coverage of DEM data is available as ($5° \times 5°$). The ASTER GDEM (Ver. 2) shows significant improvements over the previous release. Though, most of users have advised that the data contains glitches and artifacts which that will be effectiveness for use in certain applications. NASA nor METI/Japan Space Systems (J-space systems) are provided DEM data, where it won't be responsible for any misuse of the data. The area coverage of DEM data is available as $1° \times 1°$ with 30-m ground resolution [2].

The digital surface model (DSM) has been developed using stereo pairs of color aerial photos. These were collected from the archives of aerial photos at the Jordanian Royal Geographical Jordanian Centre (JRGC). These photos were collected using a Leica RC 30 Aerial Camera Systems (ACS) with focal length of 153.28 mm and scale of 1:25,000 along two strips with an approximate 60% overlap and 20% side lap in flight. The photos had scanned with a photogrammetric precision scanner with 20 μm resolution, which approximately 0.50 m of ground pixel resolution. The basic processes for DEM generation from aerial photos using stereo correlation are based on the interior and exterior orientation [3].

The validation of these data was carried out using predetermined 47 GCPs collected using DGPS (with <5 mm accuracy). The analysis was carried out after reprojecting the DEMs using UTM projection system (zone 36 N), with datum as WGS84 [5].

3 Methodology

The point-to-point comparisons were conducted by extracting points (x, y) locations and (z) altitude from DSM and examining these points with the corresponding points from topographic DEM, SRTM, and ASTER GDEM data. The bilinear interpolation method was used for data extracting from DEMs files using DSM points [4].

The accuracy of each DEM has been examined depend on the reference DSM from aerial photos, the DEM has been calculated the differences with the same location in DSM using Eq. (1).

$$\mathrm{Elev}_{\mathrm{Error}} = \mathrm{Elev}_{\mathrm{examined}} - \mathrm{Elev}_{\mathrm{Reference}} \tag{1}$$

where $Elev_{examined}$ refers to the examined point from DEM, $Elev_{Reference}$ is the checked point from DSM, and $Elev_{Error}$ is the difference in elevation or error. The positive values represent the elevation in particular location of the DEM source is under the reference DSM at the same location, and the negative means the DEM source is above the reference DSM at the same location. The statistical parameters such as mean error, standard deviation error (*STD*), Sample Variance (σ^2), and root mean square error (*RMSE*) values were calculated for vertical accuracy.

3.1 Hydrological Parameters Analysis

DEM is a vital source of topographic data and its various derivatives provide substantial information regarding the terrain characteristics [6]. Several hydrological parameters were estimated such as slope, aspect, curvature, flow direction and slope length, and steepness (LS) factor using ArcGIS 10.2 software.

3.2 Slope, Aspect, and Flow Direction

The slope, aspect and flow direction are the primary derivatives of DEM. Slope (i.e., rate of change of elevation) influences the velocity of surface and subsurface flow, soil water content, erosion potential, soil formation. and several other earth surface processes [6], and thereby an important component in various environmental applications such as land capability classification, vegetation mapping, hazard zonation, etc. [7].

Aspect (or slope azimuth; the orientation of the line of steepest descent) is also an important parameter, controlling the availability of incident solar radiation and it is used to explain various environmental factors, such as potential evapotranspiration [8].

Flow direction is the method which calculates the flow for single cell depends on its steepest downslope neighbor and gives eight possibilities for directional flows. The flow direction as algorithm done using ArcGIS 10.2, Table 1 summarizes these directions.

3.3 Curvature

In the terrain parameters, the curvature one of the important parameters used with slope and aspect, the profile (rate of change of slope down a flow line), and plane (rate of change of aspect along a contour) curvature are most commonly used in different environment aspects.

Profile curvature gives an indication on the flow acceleration, zones of enhanced erosion, and deposition and sediment transport processes, whereas plan curvature has significant implications on flow convergence and divergence as well as soil water properties [7].

Table 1 Summarize the aspect and flow directions classes

Class	Flow direction	Aspect
1	–	Flat
2	1	North
3	2	Northeast
4	4	East
5	8	Southeast
6	16	South
7	32	Southwest
8	64	West
9	128	Northwest

3.4 LS Factor

In empirical models like Universal Soil Loss Equation (USLE) and its modified and revised forms, the topographic control on soil loss is elucidated by introducing LS factor (also known as topographic factor) indicating the length (L) and steepness (S) of terrain [8]. It calculates a spatially distributed sediment transport capacity that has significant implications in landscape assessment because it explicitly accounts for flow convergence and divergence [9].

4 Results and Discussion

4.1 Accuracy Assessment of DEMs

The vertical accuracy of DEM derived from topographic map, SRTM DEM and ASTER GDEM, were evaluated and compared with DSM derived from stereo pair aerial photos. According to the National Mapping Program Standards and Specifications [10] for the collection, processing and quality control of DEMs, minimum number of 28 test points per DEM is required (20 interior points and 8 edge points) to verify the accuracy.

The results of accuracy assessment for each DEMs with respect to the DSM as a reference elevation are shown in Table 2. The distribution of the residuals or errors for each DEMs (as point-to-point evaluation) shows linearly relation Fig. 2.

Among these DEMs sources, the errors of DEM which derived from topographic maps were very small and nearly zero. The ASTER GDEM has variation in elevation errors between 0 and 60 m, whereas SRTM DEM has the high error between −10 and 80 m. However, the DEM derived from topographic maps has relatively low RMSE (5.396 m), the ASTER GDEM very close RMSE (7.103 m) and SRTM DEM has higher RMSE (15.074 m).

Table 2 The residual among different DEMs with DSM

Residual	Sample points	Mean error	RMSE
Topo-DSM	430,239	0.36	5.40
ASTER-DSM	430,239	25.85	7.10
SRTM-DSM	430,239	20.84	15.07

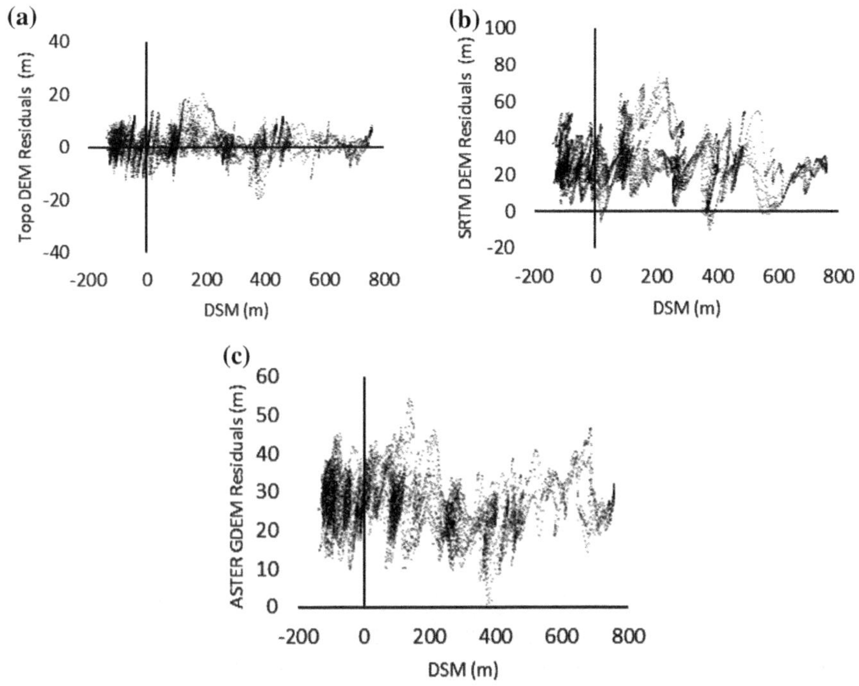

Fig. 2 The residuals of **a** Topo DEM. **b** SRTM DEM and **c** ASTER GDEM variation with DSM

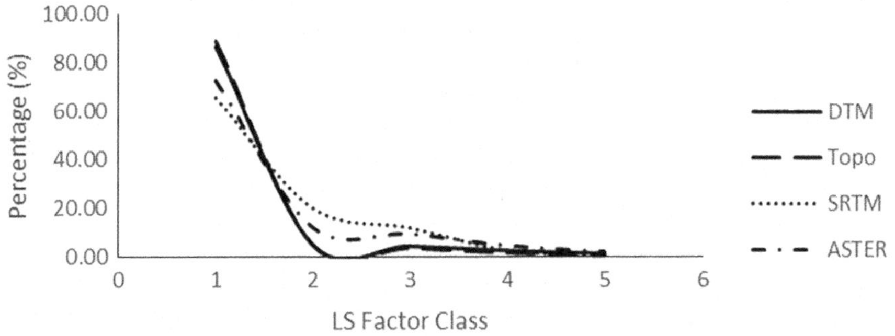

Fig. 3 The scatter plot of GCPs with **a** DSM. **b** Topo DEM. **c** SRTM DEM and **d** ASTER GDEM

4.2 *Comparison of Hydrological Parameters*

The estimated hydrological parameters were compared to evaluate the ability and the impact of spatial resolution of using different DEM sources in hydrological applications.

The analysis of different DEMs sources has been used in comparing according to the reference DSM. Figure 3 shows the difference among the slope data while

Table 3 Summarized of slope classes for DEMs

Slope	DSM	Topographic DEM	SRTM	ASTER
<5°	6.27	12.23	10.45	7.25
5°–15°	38.19	40.23	61.31	45.98
15°–25°	36.14	36.65	24.83	32.62
25°–35°	14.29	10.44	3.41	10.35
35°–45°	4.99	0.45	0.00	3.74
>45°	0.12	0.00	0.00	0.06

Table 3 summarizes the results of analysis the percentage cumulative and shows the relative relation among the DEMs used with respect to the DSM.

The main difference between aspect and flow direction is that the aspect identify the downslope direction of the change rate in each cell, and the flow direction identifies the flow path from each cell to its steepest downslope.

The analysis shows a significant variation in LS factor in between 2 and 3, however, these difference decreases after 4 and become almost negligible (Fig. 4).

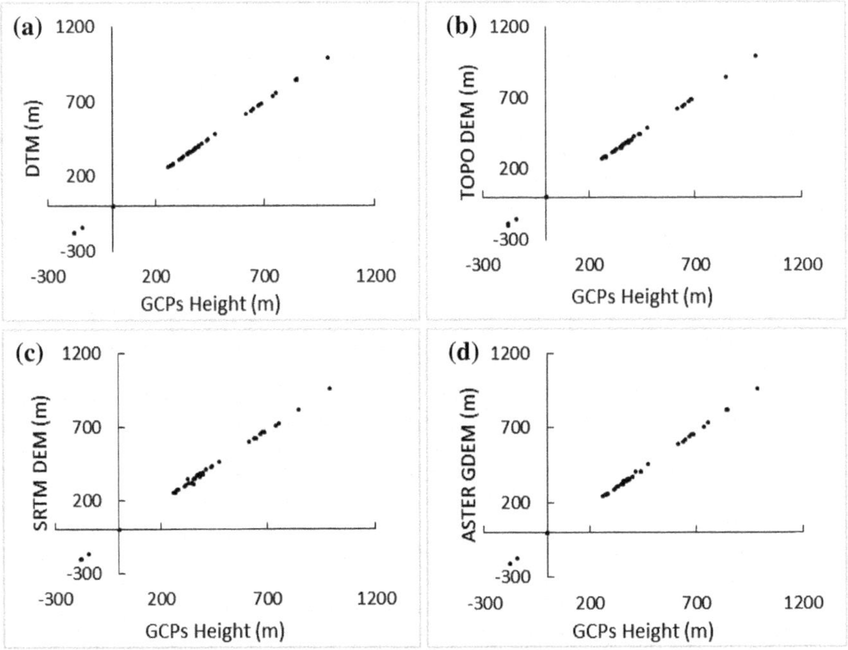

Fig. 4 The percentage of LS factor for different DEMs

Table 4 The basic statistical characteristics of 47 GCPs and DEMs

Parameters	GCPs	DTM	Topo	SRTM	ASTER
Max	989.48	989.48	842.60	962.25	959.18
Min	−179.12	−181.33	0.00	−203.45	−207.87
Mean	398.11	403.57	308.38	329.00	373.68
SD	282.83	273.68	226.56	268.84	280.59

4.3 Validation

The 47 GCPs used for validation was distributed randomly in the study area, and used to calculate the RMSE for each DEM. Results indicate that the DSM has the most accurate DEM and it can used as a reference for further studies. Topo DEM shows better elevation data, and ASTER GDEM is found to give the closest match with GCPs. Table 4 summarizes the statistical estimations with the characteristics of these DEMs and GCPs.

In Fig. 4 shows the relationship between the different DEMs with GCPs. A linear relation was appeared in all cases. In DSM and SRTM DEM, the linear relationship was better than Topo DEM and ASTER GDEM shows variation in few points.

The residual errors of the elevation are shown that, the residual errors in DSM are lying near zero (6 to −3 m), Topo DEM shows the errors are in a range between 15 and −10 m, ASTER GDEM has the errors between 0 and 35 m), and SRTM DEM has the errors between 50 and −20 m

5 Conclusion

This study compared four different DEMs, DSM from aerial photos, Topo DEM from topographic maps, SRTM DEM, and ASTER GDEM. The vertical accuracies were examined to find the most accurate DEM and these are validated with 47 GCPs collected using DGPS with vertical accuracies less than 10 cm. The statistical analysis of these errors shows that the DEM from topographic map is the best. However, as open source, ASTER GDEM gives better accuracy. On the other hand, hydrological parameters analysis show the Topo DEM from topographic maps has the closest match among other DEMs (SRTM and ASTER).

References

1. Deilami BR, Al-Saffar MRA, Sheikhi A, Bala MI, Arsal D (2013) Comparison of surface flows derived from different resolution DEM. Int J Eng & Technol IJET-IJENS 13(1)
2. Garbrecht J, Martz LW (1999) Digital elevation model issues in water resource modeling. In: Proceedings of the 19th ESRI international user conference presentation, San Diego, CA
3. Horning N et al (2010) Remote sensing for ecology and conservation. Oxford University Press, New York
4. Thomas J, Prasannakumar V, Vineetha P (2015) Suitability of spaceborne digital elevation models of different scales in topographic analysis: an example from Kerala, India. Environ Earth Sci 73:1245–1263. https://doi.org/10.1007/s12665-014-3478-0
5. Florinsky IV (1998) Accuracy of local topographic variables derived from digital elevation models. Int J Geogr Inf Sci 12(1):47–62. https://doi.org/10.1080/136588198242003
6. Wu S, Li J, Huang GH (2008) A study on DEM-derived primary topographic attributes for hydrologic applications: sensitivity to elevation data resolution. Applied Geogr 28(3):210–223. https://doi.org/10.1016/j.apgeog.2008.02.006
7. Vaze J, Teng J, Spencer G (2010) Impact of DEM accuracy and resolution on topographic indices. Environ Model Software 25(10):1086–1098. https://doi.org/10.1016/j.envsoft.2010.03.014
8. Pakoksung K, Takagi M (2016) Digital elevation models on accuracy validation and bias correction in vertical. Model Earth Syst Environ 2:11. https://doi.org/10.1007/s40808-015-0069-3
9. Jarvis A, Rubiano J, Nelson A, Farrow A, Mulligan M (2004) Practical use of SRTM data in the tropics: comparisons with digital elevation models generated from cartographic data. International Centre for Tropical, Agriculture (CIAT), Cali, p 32
10. Miliaresis GC, Paraschou CVE (2005) Vertical accuracy of the SRTM DTED level 1 of Crete. Int J Appl Earth Obs Geoinform 7:49–59
11. Hirt C, Filmer MS, Featherstone WE (2010) Comparison and validation of the recent freely available ASTER-GDEM ver1 SRTM ver4.1 and GEODATA DEM-9 s ver3 digital elevation models over Australia. Aust J Earth Sci 57:337–347
12. Kolecka N, Kozak J (2013) Assessment of the accuracy of SRTM C-and X-band high mountain elevation data: a case study of the Polish Tatra Mountains. Pure Appl Geophys. https://doi.org/10.1007/s00024-013-0695-5
13. Forkuor G, Maathuis B (2012) Comparison of SRTM and ASTER derived digital elevation models over two regions in Ghana: implications for hydrological and environmental modeling. In: Piacentini T (ed) Studies on environmental and applied geomorphology. InTech, Rijeka, pp 219–240
14. Nikolakopoulos KG, Kamaratakis EK, Chrysoulakis N (2006) SRTM vs. ASTER elevation products comparison for two regions in Crete, Greece. Int J Remote Sens 27:4819–4838

Study of Subsurface Roughness Impact on GPR Performance Using Modelling and Simulation

Narayana Rao Bhogapurapu, Dharmendra Kumar Pandey, Keesara Venkata Reddy and Deepak Putrevu

Abstract Numerical modelling and simulation for Ground-Penetrating Radar (GPR) were done to characterize and quantify the effect of subsurface roughness on GPR performance at 500 MHz operating frequency. An open-source electromagnetic solver was utilized for modelling multiple scenarios with various subsurface roughness and its numerical simulations. The detailed quantitative analysis for GPR signal losses was attempted in different subsurface roughness scenarios by using Normalized Energy (NE) and Signal-to-Clutter Ratio (SCR). This work also included the different fractal window sizes, i.e. variation of height of roughness with mean surface which define dynamic range of roughness. Quantitative results and analysis showed continuous decreasing trend in SCR with respect to increase in subsurface roughness. There was also a significant drop observed in SCR with increasing window size. The overall loss of signal has been quantified with respect to smooth interface. Hypothesis test has also been carried out to ensure and validate the subsurface roughness effect on GPR performance via signal-to-clutter ratio.

Keywords Fractal dimension · GPR · Normalized energy · Signal-to-clutter ratio

---

N. R. Bhogapurapu (✉)
NIT Warangal, Warangal 506004, India
e-mail: narayanarao.bhogapurapu@gmail.com

D. K. Pandey · D. Putrevu
Space Applications Centre (ISRO), Ahmedabad, India
e-mail: dkp@sac.isro.gov.in

D. Putrevu
e-mail: dputrevu@sac.isro.gov.in

K. Venkata Reddy
Department of Civil Engineering, NIT Warangal, Warangal 506004, India
e-mail: kvreddy@nitw.ac.in

J. K. Ghosh and I. da Silva (eds.), *Applications of Geomatics in Civil Engineering*, Lecture Notes in Civil Engineering 33, https://doi.org/10.1007/978-981-13-7067-0_36

1 Introduction

Ground-penetrating Radar (GPR) is a nondestructive geophysical technique primarily used for the detection of objects buried beneath the earth's surface or located within a visually opaque structure [1]. A GPR system uses the difference in the permittivity of both the target and the surrounding medium to detect a target. Due to advancements in research and studies, many accurate and efficient models are available to study the electromagnetic wave interactions with medium and also to model complex physical scenarios in time domain [2, 3] and frequency domain [4, 5].

Many studies have been done on the surface roughness effect on scattering of EM waves [6, 7]. It is well known that subsurface roughness induces high variability in GPR signal. This is most likely due to the distortion introduced by the twice traversed rough subsurface interface as the propagating wave encounters it on its way to and from the targets of interest [8]. Few studies have shown subsurface impact on GPR radargram but with limited quantitative analysis of signal losses across B-scan [9, 10].

In this work, various subsurface roughness conditions using fractal dimension which is a measure of roughness of the features were numerically simulated and quantitatively analysed for SCR degradation. Rough surfaces in this study were generated using fractal statistics [11]. A time domain solver, gprMax is an open-source software that simulates electromagnetic wave propagation, using the Finite-Difference Time-Domain (FDTD) method [12], for the numerical modelling of GPR [13]. The methodology followed in the present work is given in Sect. 2.

2 Methodology

Considering different roughness conditions using fractal dimension as a measure of intensity of roughness, various roughness conditions were modelled and simulated with by keeping all the other domain conditions constant. In order to compare quantitatively, a reference scenario with smooth subsurface interface was also simulated. Simulated output radargram of electric field strength at each cell for the whole scenario was examined for quantitative analysis by deriving the normalized energy for each scan and calculating the Signal-to-Clutter Ratio (SCR) with reference to smooth Interface scenario as in Fig. 1.

2.1 Numerical Modelling and Simulations

In order to study the impact of subsurface roughness at interfaces on GPR output response, different scenarios with varying subsurface roughness and window sizes were modelled and simulated using gprMax v3.1.1 [13] at 500 MHz GPR operating frequency. Fractal dimension from 0.25 to 3 with an interval of 0.25 was taken to

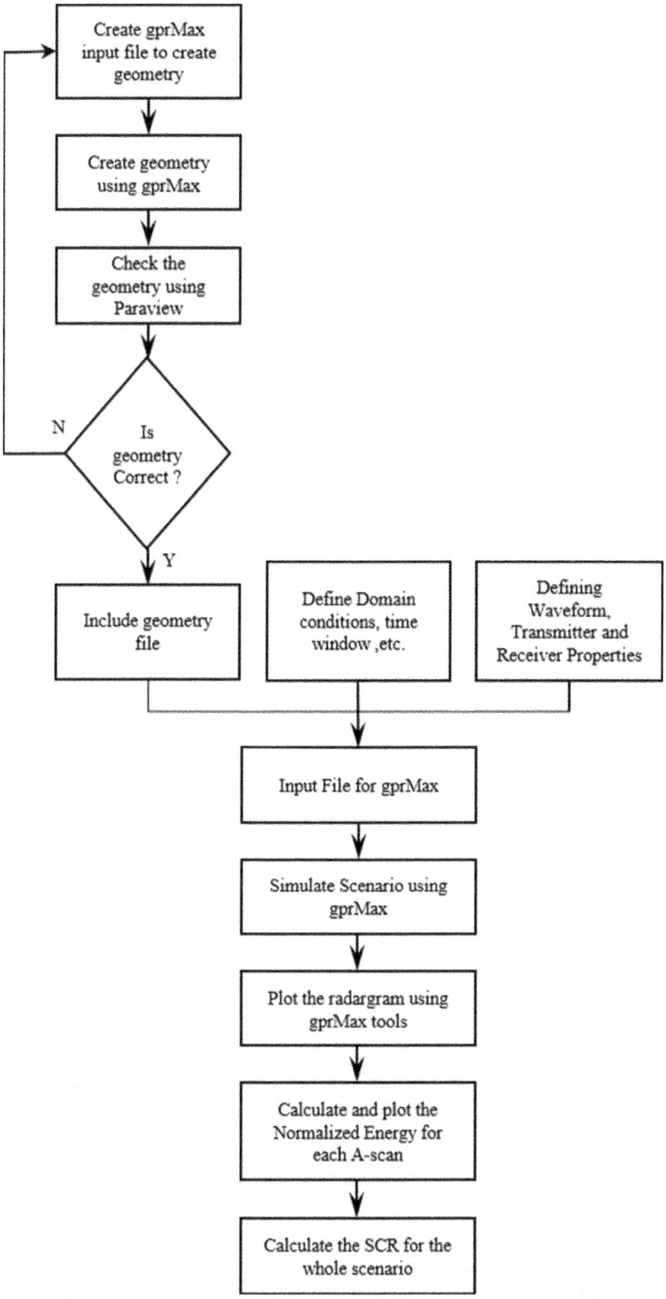

Fig. 1 Flowchart of methodology adopted in the study to numerically simulate and analyse a typical scenario

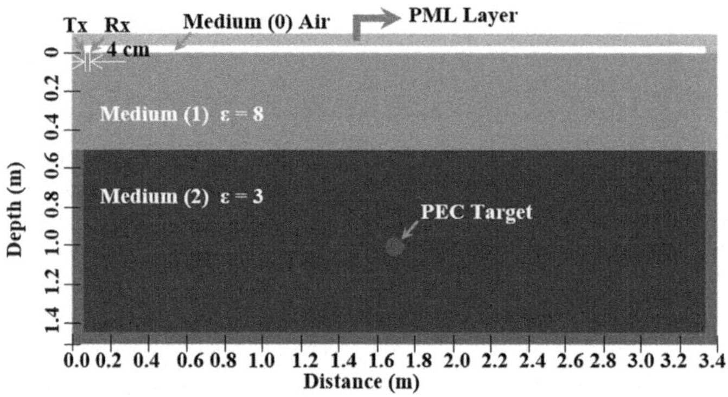

Fig. 2 A typical geometry model of smooth interface scenario

generate rough surface at the interface of medium-1 and 2 with varying window size from 2 to 11 cm with step of 3 cm. Window size of fractal dimension define the dynamic range of roughness variation with respect to reference surface.

A circular cylinder as per Fig. 2 with 5 cm radius, dielectric constant (ε_r) equal to 1 and infinite conductivity as a reference target in all numerical simulations with its centre located at 50 cm from the mean depth of the rough subsurface interface is considered for the present study. GPR scan length of 3.2 m was considered for all the scenarios bounded by Perfectly Matched Layers (PML) [13]. The generated rough interface is in between two media having ε_{r1} equals to 8 (top) and ε_{r2} equals to 3 (bottom). Fractal dimension varying from 0 to 3 was taken with a step interval of 0.25 to generate subsurface rough interface with different window sizes (2, 5, 8 and 11 cm).

The waveform used in transmitting antenna for all simulations was the first derivative of a Gaussian waveform of unit amplitude and central frequency of 500 MHz with a 4 cm constant/common offset between transmitter and receiver throughout the scan with a time window of 21 ns.

2.2 Quality Parameters

The parameters considered to quantitatively evaluate the performance of GPR radargram are Normalized Energy (NE) and Signal-to-Clutter Ratio (SCR). Calculating the energy from the field strength and normalizing it with the maximum energy xth scan are as follows.

$$Normalized\ Energy\ (x) = \sum_{i=1}^{n} \left(\frac{field\ strength\ (x, i)}{\max.wave\ amplitude} \times \Delta x \right)^2 \quad (1)$$

where $n = \frac{t}{\Delta t}$ no. of iterations for a single scan;

$$\Delta t \leq \frac{1}{c\sqrt{\frac{1}{\Delta x^2} + \frac{1}{\Delta y^2} + \frac{1}{\Delta z^2}}} \quad (2)$$

Courant–Friedrichs–Lewy (CFL) condition (Eq. 2); t = time window; Δx, Δy and Δz, are discretisation size in respective directions. To quantitatively derive the effect of subsurface roughness effect we used SCR as a measure of the quality of reflected signal. By considering the smooth interface as the reference signal, the calculation of SCR is as follows:

$$SCR = \frac{mean\,(NE_{Smooth\ interface})}{mean\,(NE_{Rough\ interface})} \quad (3)$$

3 Results and Discussions

For quantitative analysis of GPR radargram to calculate NE of each A-scan and SCR, the reflection of the direct and ground-bound bounce were excluded by mean removal method as these were common in all scenarios, and also suppresses the actual reflection from interface and target response due to high magnitude.

The resulting SCR showed a monotonically decrease in trend with increase in roughness reflecting subsurface roughness which impacted on GPR signal. From the Fig. 3, it was clear that the window size also shows an inverse impact on SCR and the trend of drop in SCR changing from linear to exponential with the increase in window size.

The comparative losses of GPR signal with respect to smooth interface were investigated to derive maximum and minimum loss of signal for each window size.

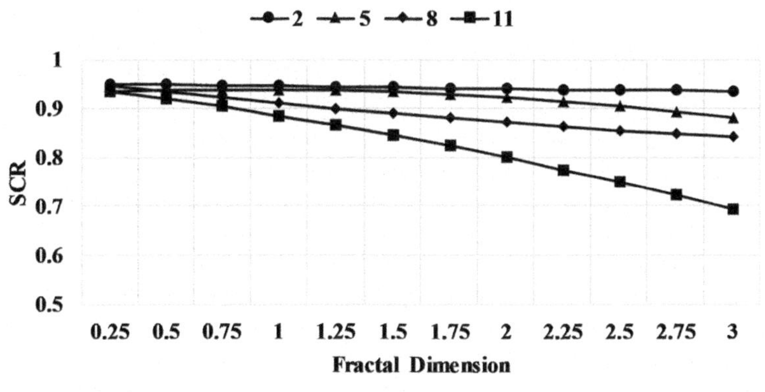

Fig. 3 SCR plot for different window sizes

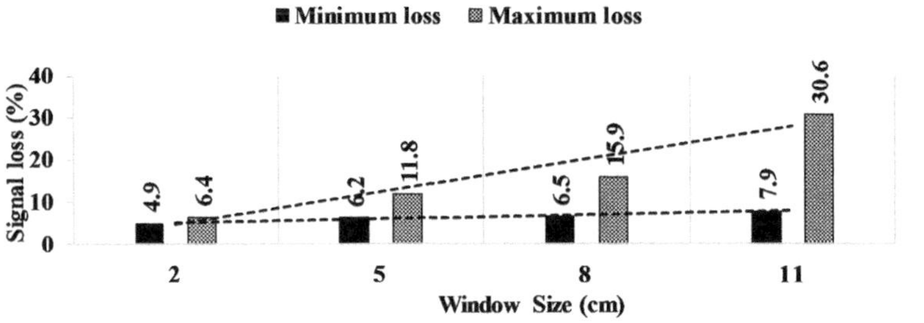

Fig. 4 Loss of GPR signal for different window sizes at an operating frequency of 500 MHz

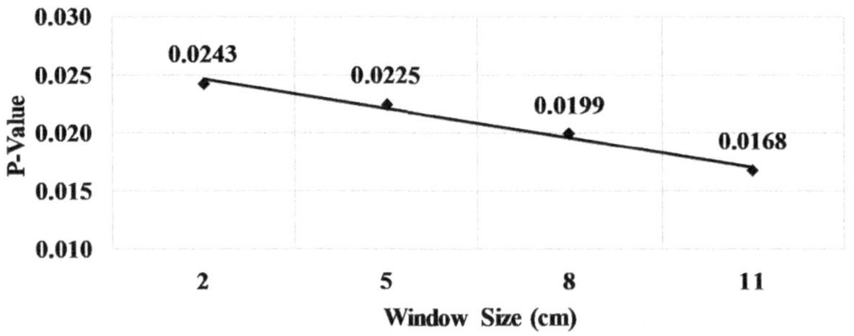

Fig. 5 P-values of t-test corresponding to different window sizes.

From Fig. 4, it is observed that there is an increase in range of losses with increase in window size as for 2.5 (2 cm)–22.7 (11 cm). For nearly smooth case, the effect of window size is not significant compared to very rough case as the loss increased from 4.9 to 7.9% (3%) but, for very rough case losses increased from 6.4 to 30.6% (24.2%).

Since the ideal smooth interface is impractical, the minimum losses associated with the different window sizes are difficult to avoid.

Hypothesis test has been carried out using t-test by taking normal distribution function as PDF to test the results against statistical accuracy. The results of t-test also revealed that there is more probability of impact of roughness with increase in window size as the decrease in p-value from 0.024 (2 cm) to 0.017 (11 cm) in Fig. 5.

4 Conclusions and Recommendations

In this study, the impact of subsurface roughness at interfaces on GPR SCR was analysed and quantified. Various rough subsurface scenarios were numerically modelled and simulated. The above quantitative analysis represents and quantifies the losses and impact of subsurface roughness on GPR output response. Numerical simulations show that there is a strong impact on GPR SNR for different rough interfaces. SNR losses in GPR signal is more at higher frequency. Based on this observed SNR analysis, it also recommends the requirement of increasing transmitted power of GPR to get rid of subsurface roughness effect for proper detection of buried interesting targets in a similar scenario where it is expected to have subsurface roughness.

Acknowledgements The authors are grateful to Shri Tapan Misra, Director (SAC), Dr. Raj Kumar, Deputy Director (EPSA), Dr. Arundhati Misra, Group Director (AMHTDG) of Space Applications Centre and Prof. Deva Pratap, Head, RS & GIS Division, Dept. of Civil Engineering, NIT, Warangal for their support and keen interest in this work.

References

1. Daniels DJ (2004) Ground penetrating radar, 2nd edn. Institute of Engineering and Technology, London
2. Giannopoulos A, Diamanti N (2003) Ground penetrating radar modeling of rough underground interfaces. In: 9th meeting of environmental and engineering geophysics, Prague, Czech Republic
3. Lampe B, Holliger K (2003) Effects of fractal fluctuations in topographic relief, permittivity and conductivity on ground-penetrating radar antenna radiation. Geophysics 68:1934–1944
4. Saillard M, Maystre D (1990) Scattering from metallic and dielectric rough surfaces. J Opt Soc Am A 7:982–990
5. Narayanan RM, Pardipuram R, Rudquist DC (1994) Statistical characteristics of simulated radar imagery from bare soil surfaces: effects of surface roughness and soil moisture variability. IEEE Trans Geosci Remote Sens 32:159–168
6. Van der Merwe A, Gupta IJ (2000) A novel signal processing technique for clutter reduction in GPR measurements of small, shallow land mines. IEEE Trans Geosci Remote Sens 38:2627–2637
7. Dogaru T, Carin L (2001) Time-domain sensing of targets buried under a Gaussian, expontential, or a fractal rough interface. IEEE Trans Geosci Remote Sens 39:1807–1819
8. Giannopoulos A, Diamanti N (2008) Numerical modelling of ground-penetrating radar response from rough subsurface interface. Near Surf Geophys 357–369
9. Fu L, Liu S, Liu L (2012) Numerical simulations and analysis for airborne ground penetrating radar. In: 14th international conference on ground penetrating radar, Shanghai, China, June 2012
10. Ciarletti V, Martinat B, Reineix A, Berthelier JJ, Ney R (2003) Numerical simulation of the operation of the GPR experiment on NETLANDER. J Geophys Res Am Geophys Union 108:E4 8028
11. Turcotte DL (1997) Fractals and chaos in geology and geophysics. Cambridge University Press, New York, USA
12. Taflove A, Hagness SC (2000) Computational electrodynamics, the finite-difference time-domain method, 2nd edn. Artech House, Norwood, MA, USA
13. Warren C, Giannopoulos A, Giannakis I (2016) gprMax: open source software to simulate electromagnetic wave propagation for ground penetrating radar. Comput Phys Commun

A Conceptual Framework of Public Health SDI

Ashutosh Kumar Tripathi, Sonam Agrawal and R. D. Gupta

Abstract From several years, public health is one of the major problems faced by several countries including India. This requires proper Public Health Information System (PHIS) that is safe, reliable, and efficient. At present, most of the public-health-based information and management system are lacking the support of real-time data uploading, data visualization, data analysis, and decision-making. To overcome these issues, Spatial Data Infrastructure (SDI) became the promising approach through which the improvements can be made. SDI provides an environment in which users can share and access the geospatial data with the common set of policies, standards, and network protocols. The main objective of the present work is to conceptualize an SDI framework for public health using open-source resources so that users can access the health information easily from the developed system. Moreover, the concept presented in the paper can be applied onto the larger scale.

Keywords GIS · Public health · SDI · Framework

1 Introduction

The rising level of population, climate change, urbanization, deforestation, and several other factors are adversely affecting the human health and increasing the risk of different kinds of diseases all over the world including India. These factors are affecting our society and public health. From several years, Geographic Information

A. K. Tripathi (✉) · S. Agrawal
GIS Cell, Motilal Nehru National Institute of Technology Allahabad, Prayagraj 211004, Uttar Pradesh, India
e-mail: rgi1551@mnnit.ac.in

S. Agrawal
e-mail: sonam@mnnit.ac.in

R. D. Gupta
Civil Engineering Department, Motilal Nehru National Institute of Technology Allahabad, Prayagraj 211004, Uttar Pradesh, India
e-mail: rdg@mnnit.ac.in

J. K. Ghosh and I. da Silva (eds.), *Applications of Geomatics in Civil Engineering*, Lecture Notes in Civil Engineering 33, https://doi.org/10.1007/978-981-13-7067-0_37

System (GIS) technology is playing an important role by providing several spatial tools that are very helpful for public health [1]. Research papers related to the use of GIS in public health, and causes of diseases and its outcomes have been studied to understand the need and importance of Public Health Information System (PHIS). These papers are related to the effects of air pollution [2], malaria disease distribution [3], socioeconomic factors in PHIS [4], etc. The classical approach to manage and access public health data is very tedious and time-consuming process. Therefore, it is necessary to combine the collection, visualization, management, and access of public health data with new technology. To develop the PHIS, information about spatial key entities like geospatial location, administrative boundary, etc. and nonspatial entities like sociotemporal information, environment study, disease information, etc. are necessary.

Many desktop-based public health solutions are provided by different organizations. Desktop-based solution requires professional who has in-depth knowledge of GIS tools and software. But, with the development of the Internet, the health services became easily available and accessible. Standards and policies were defined for easy access of web-based geospatial application through the Internet [5]. PHIS is very useful for the public. Most of the public health organizations collect health data locally when they required and make decision on them. However, this process is very unorganized with limited scope to access the public health information [6]. Due to such drawbacks, concept of Spatial Data Infrastructure (SDI) became the promising approach to enhance the overall capability of the system.

SDI provides the capability to discover, access, process, visualize, and manage the geospatial data [7]. The goal of SDI is not limited to the sharing of the spatial data, it also facilitates the documentation of the information and development of different data standards [8]. SDI is mainly a distributed network of computing nodes which exists in the different scales (e.g., local, state, national, and global) [9]. The geospatial datasets are deployed on these distributed nodes for geospatial analysis, processing, and decision-making. From several years, various developed and developing countries have started to develop the National Spatial Data Infrastructure (NSDI).

Interactive web-based PHIS can be developed by the collaboration of the public health with SDI. Several challenges faced by public health can be overcome through the public health SDI by providing different disease dispersion maps where user or decision-makers can make assessment. Furthermore, public health SDI also provides different services and sharing mechanisms that make improvement in the conventional approach to share the public health information [10].

The main objective of the present work is to develop a conceptual framework of SDI for public health using open-source resources. The proposed public health SDI framework is based on the three-tier client–server architecture.

2 SDI and Its Components

To understand the necessity of geospatial information within the different sectors, several countries all over the world have started to think to develop a geospatial infrastructure. Therefore, the concept of SDI comes into the real world that facilitates the discovery, publish, access, and sharing of geospatial information. The primary goal of SDI is to share and exchange the standardized geospatial data to the end user. Different standards and policies have been provided by different organizations through which sharing of data became easy. Several authors have defined SDI differently. Rajabifard et al. [11] stated that "SDI is fundamentally about facilitation and coordination of the exchange and sharing of spatial data between stakeholders from different jurisdictional levels in the spatial data community." The hierarchical relationship of different levels of organization is presented by Rajabifard et al. [11]. It includes the development of SDI from local, regional, and national to global level that requires the support of top to bottom level SDI participants. Figure 1 presents the components of SDI and their dynamic relationship.

As illustrated in Fig. 1, the core components of SDIs are people, data, policy, standard, and access networks. Rajabifard et al. [12] categorize these components in two parts, first part considers the fundamental role of people and data, while second part considering the standard, policy, and access networks as a technical component. People are here representing stakeholders, decision-makers, different agencies, and users. Data is the core component of SDI which includes different categories of datasets that share between producers and users. To recover technical barriers, technical components like access network works as a distributed network mechanism to exchange the spatial information. To define the policy, several factors are considered apart from exchanging of data like data security, data privacy, data recovery, authentication, and good governance. Moreover, standards can also perform the key role in SDI, which are defined for data, metadata, and services [13]. SDI provides service-oriented model for data discovery, data processing, and application for broad range of users [14]. SDI functionality involved service-oriented architecture to construct the dynamic, flexible, and configurable environment.

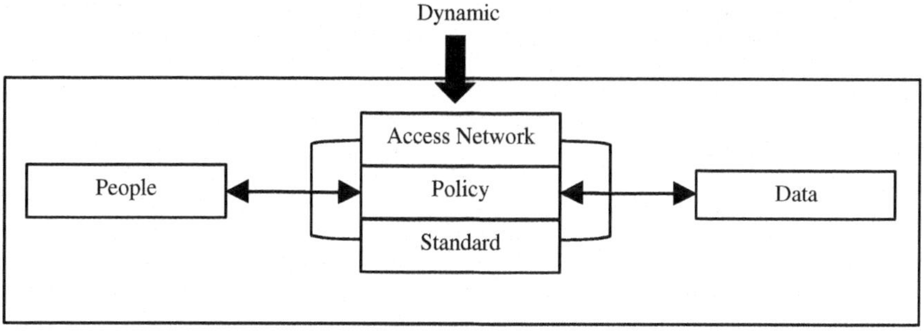

Fig. 1 Nature and relationship between SDI components. Adapted from Rajabifard et al. [12]

3 Collaboration of SDI with Public Health

The sharing of public-health-based information is crucial for preventing harmful disease, responding to emergency services and makes it understandable to the public and producers. Many health organizations do not have professional tools to map health-based information which is necessary to take immediate actions at emergency [6]. Over the last decade, GIS plays an important role within public health sector. With the availability of several desktop-based tools, it became easy to develop the desktop-based emergency system. But it requires GIS professionals having in-depth knowledge of GIS. Desktop-based solution requires lot of time, money, and skills to analyze the spatial data and render the results for interpretation.

With the growth of information technology and enhancement of the Internet, several web-based solutions become available to access the public health information. It is estimated that 80–90% of government data are related to the geographical data [15]. These datasets associated with a particular location like latitude and longitude, address, area code, census information, and administrative boundaries. The same is true for health data. These location-specific data translated into the information and integrated with the different datasets [16]. In year 2003, the term "Geographic Information Systems" is added into the controlled vocabulary thesaurus of US National Library of Medicine [17]. Such step spreads the use and importance of GIS in the field of health researches.

Several developing and developed countries all over the world are facing the problem of irregular spatial distribution of health information [18]. Health facilities in the rural areas or the remote locations are dealing with many difficulties to access the health-based information [19]. Most of the research till now considers the adoption of GIS tools for the analysis and management of public health information, while limited research has been found that includes integration of public health with the information infrastructure. Thompson et al. [20] advised that implementation of healthcare system demands geospatial data, but most importantly data harmonization and data sharing are the keys to improve decision-making process. Boulos [21] has illustrated the potential of different geospatial tools in several public health problems. It also includes the comprehensive analysis of national-level spatiotemporal health information infrastructure. Croner [5] identified that several key elements are necessary to develop a national public health information infrastructure such as involvement of highest level authority, current status of health resources, technical introduction, and geospatial readiness.

Several other factors are also necessary to develop the sustainable PHIS to discovery access and manage public health information. As the public health database involved heterogeneous data gathered from different sources such as different health agencies, census data, epidemiological data, etc, it requires geospatial and statistical analysis to identify the final outcomes. There are several barriers in effective communication between different participating agencies because different health agencies have different sets of communication protocols, policies, standards, and service interface [20]. Moreover, for the effective PHIS, local and national agencies work collectively to gather the public health data from distinct geographical locations, and it is necessary to adopt such system which eliminates the duplication of data [22]. It is identified that local-level agencies have better information about their geographical area and other factors than national-level agencies. So, the participation of local-level agencies is much more required for the sustainable public health SDI.

Collaboration of public health and SDI provides the effective environment for the public health organizations, decision-makers, stakeholders, and users. SDI represents information infrastructure by including institutional arrangements as well as people, standards, and protocols for the effective sharing of geospatial information. The same is considered in the case of public health SDI.

4 Generic Framework of Public Health SDI

The generic framework of public health SDI is presented in Fig. 2. The basic goal of this framework is to provide the interoperability of data and services by including standards and policies provided by SDI. Open Geospatial Consortium (OGC) and several other standard bodies like International Organization for Standardization (ISO) provide such standards and services for the interoperability. Interoperability ensures that the geospatial data can be searched, located, and retrieved from anywhere either from individual desktop or service provider.

OGC has introduced open web service that is based on the SOA and also provided different geospatial specifications such as Web Map Service (WMS), Web Feature Service (WFS), Web Coverage Service (WCS), Catalog Service for Web (CSW), and Web Processing Service (WPS) [6]. WMS has the capability to produce maps dynamically rather than to access specific georeferenced data. WFS defines the interfaces to access geospatial features on the web. WCS provides the interface to find geographical coverage area from the whole geospatial datasets, i.e., it returns actual raster details as per the user request. WPS offers standardized interfaces through which spatial processing functionalities like publish, discovery, and bind operations can be performed over the network. CSW defines interfaces for discovery and access of catalog data. In general, these service specifications integrated with the specific geospatial application develops the interactive web GIS application to perform common set of actions like search, discovery, and access.

The developed public health SDI framework includes three modules which are user-specific application, service interface, and distributed health datasets. Accessing public health information through standard service interface is important for data accessibility and interoperability. Through the standard service interface, improvement can be made in conventional PHIS. The creation and management of health database are present in distributed manner within public health SDI because different health organizations cooperate with each other. This framework provides several advantages to PHIS, for example, better accessibility of health data, easy interaction of public because of visualization of health information in maps, and most importantly cost reduction of healthcare application.

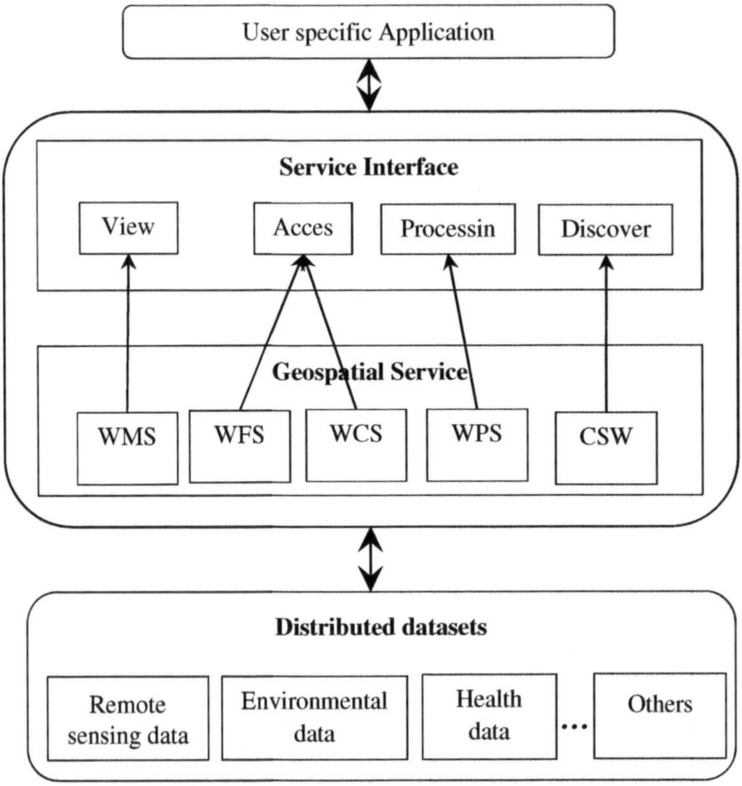

Fig. 2 Framework of public health SDI

5 Methodology Adopted

The proposed public health SDI framework is based on the three-tier client–server architecture. It requires free and open-source software and resources to manage the public health database, to publish the geospatial data on to the server and to authorize the users. All these responsibilities are managed by the administrator. Figure 3 presents the layered architecture view of public health SDI framework. It includes three layers which are data layer, application layer, and client layer.

The client layer consists of web browser, desktop application, or mobile client through which user can make request and get response from the application layer. This layer provides different geospatial functionalities like discovery, visualization, processing, and accessing of geospatial data. Further, client can also perform actions like zooming and querying on the interactive web maps that display on the client layer.

The application layer is used to develop web-based application through which user can request for different services like data, catalog, or processing services. The web-based application has been developed by using HTML, CSS, JavaScript, JSP,

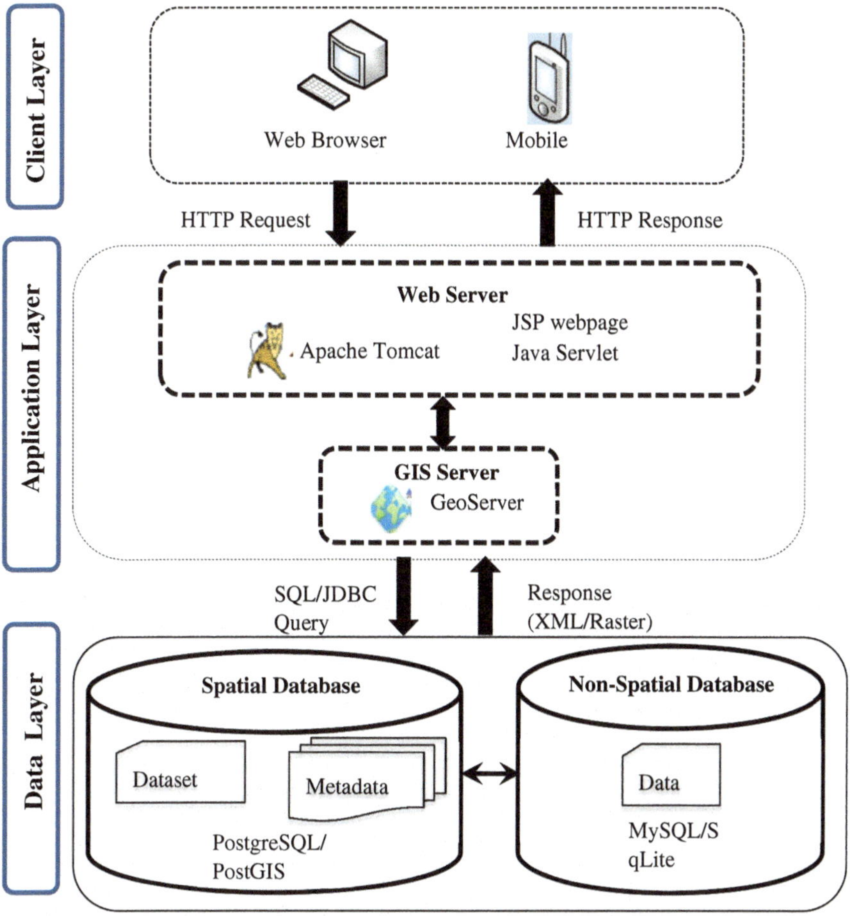

Fig. 3 An open system architecture for public health SDI

and Servlet. Apache Tomcat is used as a web server and GeoServer is used as GIS Server to publish the public health datasets onto the server. GeoServer is an open-source software which provides OGC compliant services such as WMS, WFS, WCS, and WPS for sharing geospatial data on to GIS server.

The data layer contains the public health datasets, for example, geospatial locational data in latitude and longitude values, census data, disease information, emergency information, etc. These public health datasets are uploaded onto the PostgreSQL/PostGIS database. PostgreSQL/PostGIS databases are used to store spatial and nonspatial data. The proposed public health database helps users to analyze, manipulate, and decide. The integration of public health data across the web provides users the facility to access detailed health information.

6 Conclusions

The paper presents the public health SDI framework which is interoperable and facilitates accessibility of data in a secure manner. This framework is used to overcome the issues of public health information and management with the support of SDI. The main advantage of public health SDI is that it provides efficient mechanism for discovery, access, and processing of public health data to the end user. The paper presents the three-tier client–server architecture on which public health SDI framework works.

The main focus is to use the free and open-source software like Apache Tomcat, GeoServer, PostgreSQL/PostGIS, and Quantum GIS to develop the web-based PHIS. The use of such open-source software in the specific geospatial application benefits many users to deliver information efficiently on global platform. This framework can be extended at national level with ease at affordable cost.

References

1. Mathys T, Boulos MNK (2011) Geospatial resources for supporting data standards, guidance and best practice in health informatics. BMC Res Notes 4:19
2. Buzzelli M, Jerrett M, Burnett R, Finklestein N (2003) Spatiotemporal perspectives on air pollution and environmental justice in Hamilton, Canada, 1985–1996. Ann Assoc Am Geogr 93:557–573
3. Martin C, Curtis B, Fraser C, Sharp B (2002) The use of a GIS-based malaria information system for malaria research and control in South Africa. Health Place 8:227–236
4. Mitchell R, Dorling D, Shaw M (2002) Population production and modelling mortality—an application of geographic information systems in health inequalities research. Health Place 8:15–24
5. Croner CM (2004) Public health GIS and the internet. J Map Geogr Libr 1:105–135. Adv Geosp Inf Collect Arch
6. Gao S, Mioc D, Yi X, Anton F, Oldfield E, Coleman DJ (2009) Towards web-based representation and processing of health information. Int J Health Geogr 8:3
7. Nebert DD (2004) Developing spatial data infrastructures: the SDI cookbook
8. Warnest M, Feeney MEF, Rajabifard A, Williamson IP (2002) Fundamental partnerships driving spatial data infrastructure development within Australia. Cartography 31:11–20
9. Rajabifard A, Williamson IP (2001) Spatial data infarstructures: concept, SDI hierarchies and future directions. In: Proceedings of GEOMATICS'80 conference, p 10
10. Granell C, Fernández ÓB, Díaz L (2014) Geospatial information infrastructures to address spatial needs in health: collaboration, challenges and opportunities. Futur Gener Comput Syst 31:213–222
11. Rajabifard A, Williamson IP, Holland P, Johnstone G (2000) From local to global SDI initiatives: a pyramid of building blocks. In: 4th global spatial data infrastructure conference, Cape Town, South Africa, pp 13–15
12. Rajabifard A, Feeney M-E, Williamson IP (2002) Directions for the future of SDI development. Int J Appl Earth Obs Geoinf 4:11–22
13. Crompvoets J, Bregt A, Rajabifard A, Williamson I (2004) Assessing the worldwide developments of national spatial data clearinghouses. Int J Geogr Inf Sci 18:665–689
14. Yang CP, Cao Y, Evans J, Kafatos M, Bambacus M (2006) Spatial web portal for building spatial data infrastructure. Geogr Inf Sci 12:38–43

15. FGDC. Homeland security and geographic information systems-how GIS and mapping technology can save lives and protect property in post-September 11th America. https://www.fgdc.gov/resources/whitepapers-reports/white-papers/homeland-security-gis
16. Foster RW, Ryttersgaard J (2001) The Nairobi statement on spatial information for sustainable development. In: FIG working week 2001, Nairobi
17. Richards T, Croner C, Rushton G, Brown C, Fowler L (1999) Geographic information systems and public health mapping the future. Public Health Rep 114:359–373
18. Yang DH, Goerge R, Mullner R (2006) Comparing GIS-based methods of measuring spatial accessibility to health services. J Med Syst 30:23–32
19. Hu R, Dong S, Zhao Y, Hu H, Li Z (2013) Assessing potential spatial accessibility of health services in rural China: a case study of Donghai county. Int J Equity Health 12:35
20. Thompson J, Eagleson S, Ghadirian P, Rajabifard A (2009) SDI for collaborative health services planning. In: Global spatial data infrastructures world conference, Rotterdam, The Netherlands
21. Boulos MNK (2004) Towards evidence-based, GIS-driven national spatial health information infrastructure and surveillance services in the United Kingdom. Int J Health Geogr 3:1
22. Goodchild MF, Fu P, Rich P (2007) Sharing geographic information: an assessment of the geospatial one-stop. Ann Assoc Am Geogr 97:250–266

Influence of Hyperparameter in Deep Convolution Neural Network Using High-Resolution Satellite Data

Ashish Soni, Radhakanta Koner and Vasanta Govind Kumar Villuri

Abstract In recent years, Deep Convolution Neural Network (DCNN) has evolved in the various fields of computer vision and performed state of the art in image classification task including feature extraction in high-resolution satellite imagery. During the training process, DCNN is required to achieve the highest accuracy but framework should be optimized with the help of hyperparameters (knobs) to reach that accuracy. In this study, we investigated the effect of various hyperparameters like batch size, learning rate, and other additional factors like test ratio w.r.t different variants of gradient optimizer algorithms, which can help to give the direction for the searching of better hyperparameter setting so that DCNN could achieve better classification accuracy with less computation time. Herein, we used SAT4 and SAT6 datasets to train the ALEXNET framework.

Keywords Machine learning · Remote sensing · Optimizers · SGD · AlexNet

List of Abbreviation

ADAM	Adaptive Moment Estimation
ANN	Artificial Neural Network
CNN	Convolution Neural Network
DCNN	Deep Convolution Neural network
FE	Feature Extraction
GPU	Graphics Processing Unit
LR	Learning Rate

A. Soni (✉) · R. Koner · V. G. K. Villuri
Department of Mining Engineering, Indian Institute of Technology (Indian School of Mines) Dhanbad, Dhanbad, India
e-mail: ashishsoni@me.ism.ac.in

R. Koner
e-mail: rkoner@iitism.ac.in

V. G. K. Villuri
e-mail: govindvilluri@gmail.com

J. K. Ghosh and I. da Silva (eds.), *Applications of Geomatics in Civil Engineering*, Lecture Notes in Civil Engineering 33, https://doi.org/10.1007/978-981-13-7067-0_38

NN	Neural Network
ReLU	Rectified Linear Unit
RS	Remote Sensing
SGD	Stochastic Gradient Descent
LR	Learning Rate

1 Introduction

In the last few years, computer vision and machine learning gave their admirable contribution in the area of pattern recognition. The most dominant application is pattern recognition in the real-world objects and remote sensing (RS) datasets are the most suitable for the real-world application [1, 2]. Nowadays, due to the rapid growth in the RS technology, the large amount of datasets is available. These datasets are having varying spectral and spatial resolutions due to the dynamic range of height like satellite or aerial imagery that covers the whole earth in less than a day. The image with high spectral resolution, the patches, or scene are comparatively clear for feature extraction [3] and semantic classification task [4].

In literature, there is a variety of algorithms available for classification of images that assign every pixel to a certain class. In addition to these approaches, more advanced techniques are also available which enhance the performance of classifier by combining the information from neighboring pixel values. These approaches are based on the high interclass separability depending on the different class and also depending upon the variability of single pixel value or of some neighboring pixels. On the large-scale setting, these approaches are not powerful enough [5] and the current sensor does not contain high spectral resolution, so it is difficult to distinguish object classes by their texture.

In last decades, excessive research has been done in the remote sensing image classification, focusing on classification algorithms and suitable feature descriptor to extract the feature. Conventional FE and image analysis techniques have been used such as Scale-Invariant feature transforms (SIFT) [6], Histogram of Oriented Gradient (HOG) [7], and Mathematical Morphology (MM). These methods were used in a variety of domains like medical imaging, remote sensing, handwritten data, etc. [8]. Spectral classification, decision tree, ANN, and Support Vector Machines (SVM) are some of the supervised approaches that have been developed for hyperspectral and multispectral image analysis.

Artificial Neural network (ANN) follows the pattern of the biological brain, trying to operate various functions on the massively interconnected node (called neurons). A typical NN framework consists of several hidden layers connected in feedforward fashion with each other, by which "Deep" feature comes. DCNN has dominated in the field of computer vision that gives the impressive result with relatively high accuracy at object recognition task. It has evolved in the field of remote sensing which includes land use images segmentation and image classification task [8, 9].

Moreover, DCNN requires a large amount of labeled data for training framework adequately for both to learn FE and classify the unlabeled images.

2 Deep Learning Architecture

The deep learning architecture is emerging fast in last few years because of bundle amount of dataset and impressive result in the field of classification and information retrieval [9–11]. DCNN mainly comprises feature extraction and classification process. Convolution operation stacks with the set of convolution filter followed by pooling operation and activation function for feature extraction. Whole aforementioned series of operation is applied on the dataset to extract feature hierarchically and classify the fully connected layers shown in Fig. 1. A model performed by various series of convolution operations (matrix operations), throughout the operation different numbers of the parameters need to be learned from data so that model is generalized enough to achieve maximum accuracy. However, there is also another kind of parameters involved in machine learning that cannot be learned directly during the training. These parameters define abstract properties of the model like how the model is optimized and converge fast. These parameters are called *Hyperparameters* or *Knobs*.

During the training process, hyperparameters play the prominent role in converging the network accurately and fast. Searching for the optimized hyperparameter for CNN is vigorous task known as *Fine-tuning*. There are several strategies like random, grid, and manual search widely used in optimizing the hyperparameter. It is empirically found that the random search is more efficient than grid search and manual search techniques [11–13]. In this study, the random search is performed to analyze the hyperparameter optimization using various optimizers [14] with respect to different hyperparameters to get the direction of search for optimized value. These knobs are usually fixed before training. Different knobs like learning rate, batch size, momentum activation function, and far more are explained in Sect. 3.

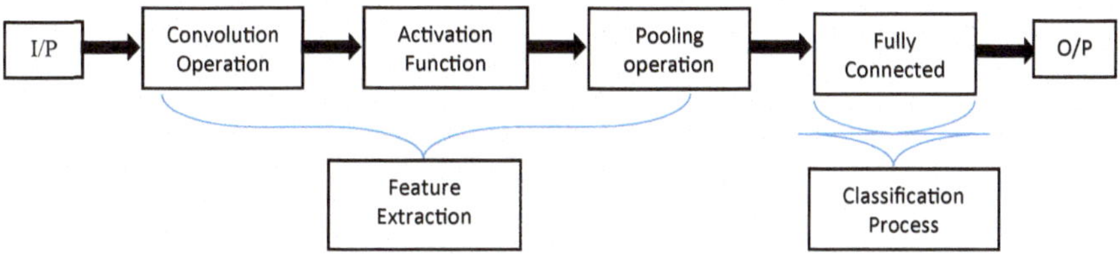

Fig. 1 Illustration of DCNN

3 Hyperparameters for DNN

3.1 Preprocessing of Input Data

In general, the pixel values of satellite imagery are integer and vary according to the number of bits. In order to avoid abnormal gradient and network instability, neural network parameter and activation function are randomly initialized between (0, 1). The normalization of the image means adjusting and distributing the pixel value of images under normal distribution. The normalization function is defined in Eq. 1 as follows:

$$X_{\text{out}} = \frac{\left(\frac{X}{Max}\right) - \mu}{\sigma} \tag{1}$$

where

Max maximum pixel value in the image,
μ, σ mean and standard deviation (SD),
X_{out} normalized image.

3.2 Activation Function and Dropout

In the DCNN architecture, feedforward network works as a linear model where neurons are used to pass the information to other neurons in the network. During training, the main objective is to find out the optimal value of all the parameters and activation function not only to evaluate the output but also to add some nonlinearity to the output. There are some commonly known activation functions like sigmoid function, hyperbolic function, Rectified Linear Units (RELU), and Parametric RELU (PRELU) [15]. The most commonly used activation function is RELU because it converges fast during the training process. RELU discard all values if the input is less than 0 otherwise output value is equal to input if the input is greater than or equal to zero. A general function can be expressed as the following equation:

$$f(x) = \max(0, x) \tag{2}$$

where x = input value.
Another important parameter to consider is dropout. It is a technique to prevent overfitting in the network and prevent neuron to co-adapting enough. The term "Dropout" is defined as the elimination of the neurons temporarily from hidden layer during training process [16]. Both parameters, i.e., activation function and dropout provide convolution kernel to acquire the feature from the images for object recognition.

3.3 Loss Function

In the image labeling problem, the feature set is derived from the input domain, and the aim is to find out the correct label associated with the image. The whole set of the process of optimizing the weight and biases helps to minimize the loss/cost function between the predicted value and target value. Consider L set of classes, encoded in the binary vector of L length, for example, three classes which are encoded in the form of 100,010,001 otherwise 0. At the top last layer, Softmax function is used to normalize the output value of probability distribution, i.e., in the range of [0, 1] and added up to 1. In this work, common cross-entropy loss function is defined as

$$L = -\frac{1}{n}\sum_{i=1}^{n}\sum_{k=1}^{l} y_k^i \log y_k^i \tag{3}$$

where L is defined as number of label classes,

Y^i is predicted class label for n training sample, i.e., i = 1, 2, 3, 4, 5, …, n.

3.4 Gradient Descent Algorithms

During the training process, the weight is adjusted and modified by the computing gradient vector, which defines the amount of weight required to increase or decrease according to the loss. The loss (or error) is calculated as the difference of input and output vectors, and computing the average gradient for that input while adjusting the weight accordingly. This process will be repeated until the function stops decreasing. These algorithms are generally known as backpropagation algorithms. The generalized equation of SGD are given below [17]:

$$W_{t+1} = W_t - V_{t+1},\ V_{t+1} = \lambda V_t + a\nabla W \tag{4}$$

$$B_{t+1} = b_t - U_{t+1},\ U_{t+1} = \lambda U_t + a\nabla b \tag{5}$$

where a is learning rate (sometimes called step size),

∇W and ∇b	are gradients of loss function w.r.t W and b,
λ	is momentum to accelerate the SGD, and
t	is number of epoch.

In real world, the deep learning practitioner uses the most frequent optimizer, i.e., SGD, but scientists use optimizers like Mini-Batch Gradient, Nesterov Accelerated Gradient (NAG), Momentum, ADAGRAD, Adaptive Moment Estimation (ADAM), and RMSProp [14]. These optimizers converge fast, whereas the ADAM optimizer works well in practice and outperform other adaptive techniques.

3.5 Learning Rate (LR)

It is the most important hyperparameter of machine learning techniques [18]. The NN is trained by gradient descent algorithms to modify the weight of neurons. After each iteration, the loss function is calculated w.r.t each weight with the help of backpropagation algorithm and then subtracted. So, during the learning process, each derivative is multiplied by the small value called as LR before subtracting it from its corresponding weight. In Eqs. 4 and 5, "a" is defined as LR. The value of LR with standardized input ranges less than 1 and greater than $10e^{-6}$ but it is not strictly bounded within the parameter and can depend on the model also.

3.6 Batch Size

Another important hyperparameter in machine learning architecture is batch size. Its value varies from 1 to 1000 while in CNN processing, the value is the multiple of 2, i.e., 2^4, 2^8, etc. Usually, the batch size depends on the training time, if the batch size is larger the computation time will be faster but that will increase the computation cost as well. As we select the batch size, generally it can be adjusted with other parameters while other hyperparameters can be further fine-tuned [18].

3.7 Number of Iterations

This hyperparameter is defined as the amount of time for any framework to train in order to avoid the effect of overfitting of the network. The process of *early ending* is considerably helpful in avoiding overfitting by keeping the track of error in the training progress. It is also used to analyze the effect of other hyperparameters that can cause overfitting and configure the hyperparameter to optimize the performance with too larger capacity within small training time. Due to this reason, it might be useful to stop the training process to analyze the effect of individual hyperparameter. In this experiment, we train the framework for "t" iteration and analyze if the framework is converging or not. During analysis, if framework shifts toward the optimal result with current configuration of hyperparameter, then process of early ending is not required.

3.8 Weight Decay

Weight decay (regularization coefficient λ) is a method to decrease the overfitting by the addition of regularization that improves the generalization of NN. During the training process, weight decay is common in use because after each update, weights are multiplied by the factor of less than 1 which can prevent the further growth of weight. It has advantages like suppressing the irrelevant data in weight vector by selecting smallest vector to solve the learning problem. Additionally, if the appropriate size is chosen, it can overcome the effect of static noise on target data, thus improving the generalization process [19].

3.9 Number of Hidden Units

The NN (say DCNN) consists of various layers of neuron, and thus all the other layers except input and output layers are known as hidden layer/unit. Each multilayer NN has a size that controls the capacity and plays a very prominent role in the NN optimization. It is found that large networks do not give significant results in generalization but need high computational cost, especially in the case of CNN. Empirically, it was analyzed that the pyramid structure gives less efficient results as compared to the layer of the same size, but obviously various supervised tasks are data dependent [7].

3.10 Test Ratio

Before performing the training on DCNN, we randomly split the data training dataset into two categories, i.e., training data and validation data. The ratio of these two is defined as the test ratio [20]. The test ratio is used to check the model generalization at certain steps during the process which can help to analyze whether model is performing well or going to *underfitting* or *overfitting*.

4 Experiment

4.1 Experimental Setup and Dataset

The amount of dataset is huge and the framework performs spatial convolution operation, and the whole operation needs to accelerate. The whole experiment is carried out in the FUJITSU Workstation CELSIUS M740 which is having NVIDIA Quadra K2200 GPU with the dedicated memory of 4096 MB. NVIDIA provides deep learning library, i.e., Compute Unified Device Architecture (CUDA). These tool kits integrate with Tensor flow-GPU version and include GPU-accelerated libraries, debugging, and optimization tools. The use of GPU makes calculation faster up to 140x [21] (Fig. 2).

Fig. 2 Sample of SAT 6 dataset (scene include road, building, grassland, tree, and barren land)

In this experiment, SAT 4 and SAT 6 are used to evaluate the framework. SAT images are airborne images, extracted from the National Agriculture Imagery Program (NAIP) dataset [17, 22]. SAT 4 dataset consists of image patches of 5,00,000 including the four classes, i.e., barren land, trees, grassland, and a class that consists of all land other than the above three classes. SAT 6 dataset consists of 405,000 image patches of the same size as of SAT 4 covering six land classes like barren land, trees, grassland, roads, buildings, and water bodies. Size of all the image patches is 28 * 28 pixels with 1 m spatial resolution, consisting four spectral resolution, i.e., R, G, B, and NIR (near infrared) [23, 24].

4.2 AlexNet Framework

In 2012, during ImageNet Large-Scale Visual Recognition Challenge (ILSVRC) competition, AlexNet (DCNN) provided extraordinary result in the pattern recognition task. It is scaled version of LeNet framework [11] with deep and larger network that could learn the intricate design pattern of the input images. The AlexNet used ReLU as nonlinear activation function instead of the conventional hyperbolic function, and therefore resulting in faster convergence. During the design of the process, Krizhevsky et al. [15] used dropout functionality to reduce the problem of overfitting and augmentation process to increase the volume of the dataset. Additionally, this model runs on GPU mode instead of CPU which allows the faster training time (see Fig. 3).

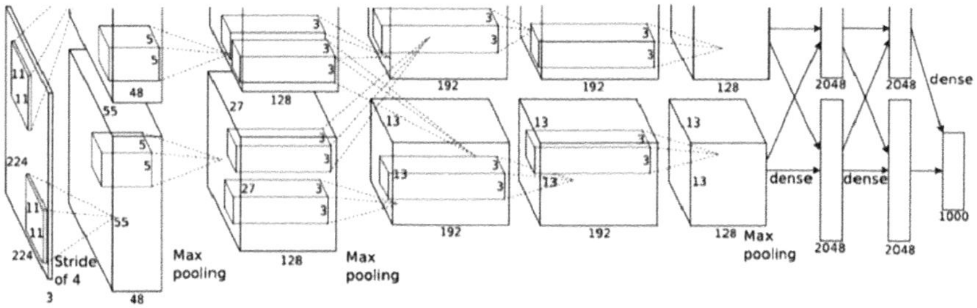

Fig. 3 AlexNet framework [1, 15]

4.3 Hyperparameter Analysis

In this section, we investigated the effect of various hyperparameters with respect to the different backpropagation optimizers in the AlexNet model. During training, batch size, learning rate, and test ratio have been analyzed w.r.t different optimizers like Momentum, SGD, Adam, etc. so that it can help to determine the direction for faster convergence of model.

Figure 4a reports the influence of increased learning rate, and the graph clearly explains that too low or too high learning rate directly affects the training time as well as the accuracy. When LR is increasing from 0.001 to 0.01, Momentum is showing almost constant accuracy of 99% but further that at 0.01 and 0.03, and it is decreased to 94% and 86%, respectively. SGD achieved accuracy of 99.04 at 0.001 and decreased with accuracy of 97.65 and 96.14% at 0.003 and 0.01, respectively. At LR of 0.03, SGD suddenly drops in the accuracy up to 86%. Adagrad is started with lower accuracy of 98.33 in comparison with Momentum and SGD and decreased up to 7% at 0.03 LR. Additionally, other optimizers like Adam and RMSProp work well up to 0.003 but as we increase the step size (LR), framework gives the poor performance and end at the accuracy of 40%. On another hand, the training time between all optimizers differs not much as LR increased.

Figure 4b illustrated the analysis of training accuracy and training time w.r.t the batch size. At any point in time, if batch size is increased to double, the training time will be reduced to half as it is independent of the type of backpropagation optimizer. In terms of performance, Momentum does not get affected much by increasing the batch size. On the other hand, RMSProp and Adagrad show the gradual change in the efficiency of the framework and gradual increment in the performance of the ADAM optimizer. The SGD optimizer works well in different batch sizes and shows an improved performance from 91 to 98%.

The relationship between the training accuracy and the time with respect to the test ratio is shown in Fig. 4c. At 10% of test ratio, the Momentum, Adam, and RMSProp have shown the efficiency of 98% and started decreasing with slow rate up to the test ratio of 80% and then a sudden fall in the accuracy was observed at 90%. RMSProp fails to converge at the 80% test ratio. On the other hand, SGD and Adagrad started

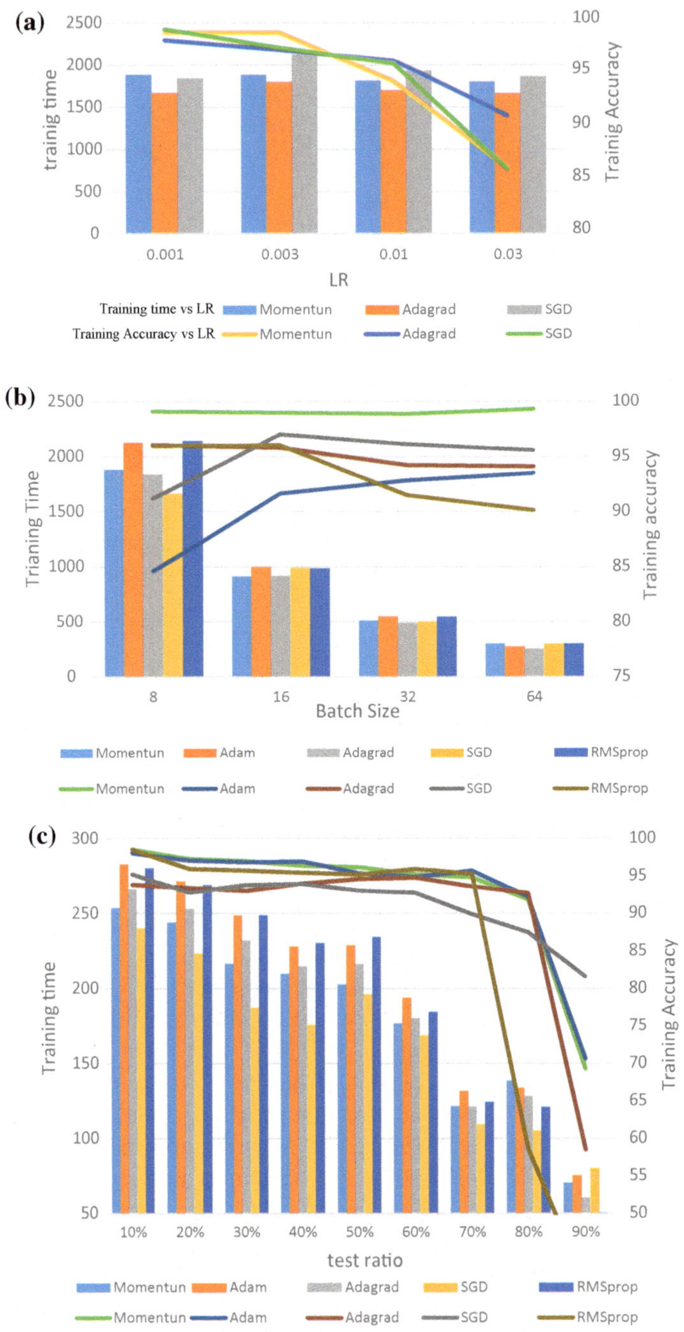

Fig. 4 The influence of hyperparameter with learning rate, batch size, and test ratio w.r.t training accuracy and training time (all the parameters are running in GPU mode only) **a** learning rate, **b** batch size, **c** test ratio

with low accuracy (approx. 95%) and as the test ratio approaches toward 90%, the SGD is able to manage the drop in efficiency up to 12%, whereas Adagrad fails to converge at the point of 90% test ratio. Additionally, SGD shows low training time among all the optimizers. Momentum also performed well in terms of training time and accuracy both.

5 Conclusions

Deep convolution neural network has various applications in the field of Image processing including Segmentation, classification, and object detection. During training time, DCNN framework needs fine-tuning which evolved various hyperparameter configurations. This study concludes that learning rate, batch size, and test ratio aids in converging the network efficiently and faster as follows:

- The optimizer, i.e., Momentum, Adagrad, and SGD provide accuracy of 99%, 98%, and 99%, respectively at LR of 0.001. As the learning rate increased to 0.03, 13% drop occurs in the Momentum and SGD whereas Adagrad having comparatively less drop of 7%. Additionally, other optimizers like Adam, RMSProp are not able to converge.
- Batch size directly affects the training time, i.e., "*if we double the batch size of the framework the training time is reduced approximately to half*". In terms of accuracy, Momentum attains the constant accuracy of 99% at every point. Adam and RMSProp attained the accuracy of 96% and reduced to 2% and 5%, respectively, at the batch size of 64. However, Adagrad and SGD started with the lower accuracy of 86 and 91.58% but its increment of 2 and 5% at the batch size increases to 64.
- Test ratio also plays an important role in the testing of overfitting. Almost all the optimizers work efficiently to the test ratio of 80% but later fail to converge and drop the accuracy up to 60% except SGD which gives the accuracy of 81% at 10% test ratio. On the other hand, time also decreases gradually as training size reduces.

References

1. Krizhevsky A, Sutskever I, Hinton GE (2012) ImageNet classification with deep convolutional neural networks. In: Proceedings of 25th international conference on neural information processing systems, vol 1, pp 1097–1105
2. Iglovikov V, Mushinskiy S, Osin V (2017) Satellite imagery feature detection using deep convolutional neural network: a Kaggle competition
3. Ruiz L, Fdez-Sarría A, Recio J (2004) Texture feature extraction for classification of remote sensing data using wavelet decomposition: a comparative study. In: International archives of the photogrammetry, remote sensing, XXXV, pp 1682–1750
4. Hare JS, Lewis PH, Enser PGB, Sandom CJ (2006) Mind the gap: another look at the problem of the semantic gap in image retrieval. Electron Imaging 607309–12

5. Maggiori E, Tarabalka Y, Charpiat G, Alliez P (2017) Convolutional neural networks for large-scale remote-sensing image classification. IEEE Trans Geosci Remote Sens 55:645–657
6. Kobayashi T (2014) Dirichlet-based histogram feature transform for image classification. In: Proceedings of the IEEE conference on computer vision and pattern recognition, pp 3278–3285
7. Larochelle H, Bengio Y, Louradour J, Lamblin P (2009) Exploring strategies for training deep neural networks. J Mach Learn Res 1:1–40
8. Scott GJ, England MR, Starms WA, Marcum RA, Davis CH (2017) Training deep convolutional neural networks for land-cover classification of high-resolution imagery. IEEE Geosci Remote Sens Lett 14:549–553
9. Castelluccio M, Poggi G, Sansone C, Verdoliva L (2015) Land use classification in remote sensing images by convolutional neural networks. 1–11
10. Lecun Y, Bengio Y, Hinton G (2015) Deep learning. Nature 521:436–444
11. LeCun Y, Bottou L, Bengio Y, Haffner P (2001) Gradient-based learning applied to document recognition. Intell Signal Process 86:306–351
12. Larochelle H, Erhan D, Courville A, Bergstra J, Bengio Y (2007) An empirical evaluation of deep architectures on problems with many factors of variation. In: Proceedings of the 24th international conference on machine learning—ICML '07, pp 473–480
13. Bergstra J, Bengio Y (2012) Random search for hyper-parameter optimization. J Mach Learn Res 13:281–305
14. Ruder S (2016) An overview of gradient descent optimization algorithms. 1–14
15. Krizhevsky A, Sutskever I, Hinton GE (2012) ImageNet classification with deep convolutional neural networks. Adv Neural Inf Process Syst 1–9
16. Srivastava N, Hinton G, Krizhevsky A, Sutskever I, Salakhutdinov R (2014) Dropout: a simple way to prevent neural networks from overfitting. J Mach Learn Res 15:1929–1958
17. Basu S, Ganguly S, Mukhopadhyay S, DiBiano R, Karki M, Nemani R (2015) DeepSat—a learning framework for satellite imagery
18. Bengio Y (2012) Practical recommendations for gradient-based training of deep architectures. BT—neural networks: tricks of the trade, 2nd edn, pp 437–478. arXiv:1206.5533
19. Krogh A, Hertz JA (1992) A simple weight decay can improve generalization. Adv Neural Inf Process Syst 4:950–957
20. Raschka S (2014) Python machine learning
21. Bergstra J, Frédric B (2010) Deep learning on GPUs with theano. Learn Work 1–2
22. Zhong Y, Fei F, Liu Y, Zhao B, Jiao H, Zhang L (2017) SatCNN: satellite image dataset classification using agile convolutional neural networks. Remote Sens Lett 8:136–145
23. Albert A, Kaur J, Gonzalez M (2017) Using convolutional networks and satellite imagery to identify patterns in urban environments at a large scale. 1357–1366
24. Ye D, Li Y, Tao C, Xie X, Wang X (2017) Multiple feature hashing learning for large-scale remote sensing image retrieval. ISPRS Int J Geo-Inf 6:364

Comparison of Various Indices to Differentiate Built-up and Bare Soil with Sentinel 2 Data

Prajjwal Singh Rahar and Mahesh Pal

Abstract Accurate mapping of built-up area and bare soil, having similar spectral characteristics is an important task for urban mapping using remote sensing data. In order to classify both built-up area and bare soil accurately, several indices with Landsat data are proposed in the literature. This study is planned to compare the performance of several indices proposed to accurately differentiate built-up area and bare soil using Sentinel 2 data acquired over Kurukshetra (Haryana) during April 2017. All together nine indices related to built-up area and bare soil identification and proposed in literature were derived with Sentinel 2 data. Different dataset combinations were used for classification in four land cover classes. Support vector machine classifier was used to classify different combination of images used in this study. Comparison of results in terms of area for both built-up area and bare soil using classified images and field visit suggest that a combination of six bands of Sentinel 2 data with six indices was found performing well in comparison to other combinations and was able to differentiate both built-up area and bare soil accurately.

Keywords Sentinel 2 · Built-up area extraction index · Bare soil index · New built-up index · Normalized difference built-up index · Urban index

1 Introduction

With the population growth rate of 2.3% and an estimated population of about 1.3 billion as on January 2016, Indian urban population has increased from 28% in 2001 to 32% in 2015 [1]. Presently India is facing the transformation of traditional agricultural society into a modern industrial and urban society. With the continuing urban expansion, Indian urban land expanded by about 1.5 million ha from 1955 to

P. S. Rahar
Department of Civil, NIT Kurukshetra, Kurukshetra, India
e-mail: rahar.sunny@gmail.com

M. Pal (✉)
Department of Civil, NIT Kurukshetra, Kurukshetra 136119, India
e-mail: mahesh.pal@nitkkr.ac.in

J. K. Ghosh and I. da Silva (eds.), *Applications of Geomatics in Civil Engineering*, Lecture Notes in Civil Engineering 33, https://doi.org/10.1007/978-981-13-7067-0_39

1985, primarily through encroachment on agricultural land, and its urban growth is continuing. Impervious surfaces such as roads, parking areas, rooftops of buildings, etc., in urban areas are manmade features and emerged as an indicator of the degree of urbanization, which itself is a major indicator of the environmental quality of an area [2]. To determine the extent of urban expansion and urban land use planning and management, multi-temporal remote sensing has effectively been used [3]. Land use mapping primarily employs the multispectral remote sensing data and image classification methods. With the availability of Landsat and Sentinel two data free of cost, they are effectively being utilized for various urban studies. The availability of remote sensing data greatly helped in mapping and managing earth resources, but urban built-up area extraction from moderate spatial resolution satellite data (e.g., Landsat, 30 m spatial resolution) is challenging because of significant intra-urban heterogeneity and spectral confusion between built-up area and the bare soil. As the spectral characteristics of urban built-up area and bare soil are similar, it may lead to confusion among these classes thus generating erroneous maps. To overcome intra-class confusion and to improve land cover classification accuracy for urban studies, different indices derived from the actual remote sensing data have been used. These indices help in separating various complex land covers, but still separating some land covers poses a major problem. Mapping the built-up and bare land and their proper differentiation during urban mapping is an important research area because proper differentiation between built-up area and bare land can be used as an urban development and environmental quality index.

Keeping in view the confusion, in accurate classification of built-up area and bare soil using moderate resolution satellite data, different indices enhancing the chances of better discrimination between built-up area and bare soil are proposed in the literature [4]. Indices are the most common form of spectral enhancement techniques used to improve the classification accuracies of built-up areas and bare soil. These are basically empirical relations, which are formed by combining spectral band ratios of satellite image data used for the study under consideration. Normalized difference built-up index (NDBI) [5], normalized difference soil index (NDSI) [6], normalized difference bareness index (NDBaI) [7], index-based built-up index (IBI) [8], normalized difference impervious surface index (NDISI) [9], enhanced built-up and bareness index (EBBI) [10], biophysical composition index (BCI) [11], bare soil index (BI) [12] and ratio normalized difference soil index (RNDSI) [13] are some of the indices used by various researchers to improve the predictive accuracy of built-up area and bare soil using medium resolution satellite datasets. In spite of the availability of several indices, it is not possible to justify the usefulness of one or other to correctly differentiate between built-up area and bare soil using Landsat data. Keeping in view this problem, this study aims in using a high-resolution (10 m) satellite data provided by Sentinel 2 to discriminate built-up area from bare soil. To judge the suitability of different indices in better discrimination of bare soil and built-up area, nine indices were derived from Sentinel 2 data and used in combination with the actual dataset to judge their influence on discrimination between built-up area and bare soil. Support vector machine classifier was used to classify different combination of images used in this study.

2 Dataset Used

The study area used for this work covers the city of Kurukshetra (Haryana) lying to the north of New Delhi, Capital if India. Sentinel-2 satellite imagery obtained from ESA websites for the year 2017 (10/04/2017) was used. Sentinel 2 acquires data in 13 wavebands (bands 2–4, 8: 10 m, bands 5–7, 8a, 11–12: 20 m, bands 1, 9, 10: 60 m spatial resolution). Several band combinations were used to classify images of the study area with both Landsat 8 and Sentinel 2 data. The aim of this study is to classify the used image into four classes: bare land (bare soil), vegetation cover, built-up area, and water bodies and calculating the area occupied by these classes so as to compare the results of the different land cover areas corresponding to the particular set of data Relative distribution of bare land and built-up area obtained by classifying the Sentinel 2 data was verified by doing a field survey of the area.

To derive various indices from the image, bands 11 and 12 of Sentinel 2 were under-sampled to a resolution of 10 m to use them for bare soil index/built-up area index calculation and further classification in combination with indices and bands 2–4 and 8. Indices are empirical relations that are derived from various spectral band ratios of satellite image data. These ratios help in spectral enhancement of selected classes hence making the results of supervised classification more accurate. The technique of rationing bands involves, at its most basic form, dividing the spectral response value of a pixel in one image (or band) with the spectral value of the corresponding pixel in another image (or band). This is done in order to suppress or normalize, varying effects such as viewing angles, sun shading, atmospheric effects, soil difference, and so on. It is also applied to maximize sensitivity to the feature of interest, such as the relative health of vegetation. To achieve this most indices go beyond simple band division to include differencing, weighting, and the introduction of other variables. Keeping in view the usefulness of several indices proposed in the literature to discriminate bare sol from the built-up area, nine indices as suggested in Table 1 are derived and used in combination with the actual image for further classification (Table 2).

3 Support Vector Machine (SVM)

The SVM is based on statistical learning theory [14] and works on the principle of optimal separation of classes. In the case of a two-class pattern recognition problem in which the classes are linearly separable, the SVM selects from among the infinite number of linear decision boundaries the one that minimizes the generalization error. Thus, the selected decision boundary (represented by a hyperplane in feature space) will be one that leaves the greatest margin between the two classes, where the margin is defined as the sum of the distances to the hyperplane from the closest cases of the two classes [14]. For linearly non-separable classes, the restriction that all training data of a given class lie on the same side of the optimal hyperplane can be relaxed by

Table 1 Various Indices used for differentiation of built-up land from bare soil

Index	Formula	Reference
BAEI	$\frac{B4+0.3}{B3+B11}$	[15]
BCI	$\frac{\frac{TC1+TC3}{2}-TC2}{\frac{TC1+TC3}{2}+TC2}$	[16]
BSI	$\frac{(B11+B4)-(B8+B2)}{(B11+B4)+(B8+B2)}$	[4]
IBI	$\frac{\frac{2*B11}{B11+B8}-(\frac{B8}{B8+B4}-\frac{B3}{B3+B11})}{\frac{2*B11}{B11+B8}+(\frac{B8}{B8+B4}-\frac{B3}{B3+B11})}$	[17]
NBAI	(B12 − (B8/B2))/(B12 + (B8/B2))	[15]
NBI	$(B3 * B8)/B4$	[15]
NDBI	$(B8 - B4)/(B8 + B4)$	[15]
NDSI	$(B12 - B2)/(B12 + B2)$	[13]
UI	(B12 − B8)/(B12 + B8)	[17]

Table 2 Various datasets used for classification

S. no.	Dataset combination
1	6 band of Sentinel 2 + baei
2	6 band of Sentinel 2 + bsi
3	6 band of Sentinel 2 + nbai
4	6 band of Sentinel 2 + nbi
5	6 band of Sentinel 2 + ndbi
6	6 band of Sentinel 2 + ui
7	6 band of Sentinel 2 + nbai + bsi
8	6 band of Sentinel 2 + baei + bsi + nbai + nbi + ndbi + ui
9	4 band of Sentinel 2 + nbai + bsi

introducing a slack variable. In this case, the SVM works by selecting a hyperplane that maximizes the margin and at the same time minimizes a quantity proportional to the number of misclassification errors. For nonlinear decision surfaces, the feature vector is mapped into a higher-dimensional Euclidean space (feature space) and using the concept of a kernel function to reduce the computational cost.

Several user-defined parameters are required to achieve optimal performance by the SVM classifier for land cover classification. While dealing with multiclass land cover classification problems, choice of a suitable multi-class approach, optimal value of regularization parameter (C), type of kernel and kernel specific parameters need to be selected for effective implementation of SVM. For the SVM based classification in this study the "one against one" multiclass approach, a radial basis kernel function, which has been found to work well with remote sensing datasets, with kernel specific parameter (γ), was used. In order to find the optimal value for each of the user-defined parameters with SVM classification algorithm, a trial and error method was used with data set combination 1 (Table 2; $C = 100$ and $\gamma = 1$) and the same parameters were used for all dataset as mentioned in Table 2 for initial classification.

4 Results and Analysis

Different band combinations (Table 2) were used to classify the used area in four classes using support vector machine classifier. Area for different classes from classified images obtained while using various band combinations were calculated in Km^2. Results from Table 3 and field visit to the study area using classified images suggests that area calculation for classes bare soil and built-up area as provided by dataset 8 (Table 2) are found to be closer to the ground data than other band combinations. Results also suggest that dataset combination 3 and 5 also perform equally well to the dataset 8 in differentiating bare soils from built-up area. This suggests that the indices NDBI and NBAI works equally well to differentiate bare soil and built-up area with the used study area. As the dataset 8 (Table 2) was considered providing the best estimate in differentiating bare soil and built-up area, classified images obtained from datasets 6–8 are provided in Fig. 1.

Keeping in view the improved performance of SVM classifier with dataset 8, several trials using a different combination of user-defined parameters (Table 4) were also carried out so to note the change in classified area for bare soil and built-up area with different values of parameters. A field study was conducted using classified images obtained using SVM classifier for various values of user-defined parameters.

Comparison of classified images and field visit to several selected sites suggest that SVM classifier with $C = 50$ and $\gamma = 0.1$ perform best with dataset 8. In order to compare the change in classified areas, results of two classified images obtained by using $C = 50$ and 130 and $\gamma = 0.1$ (Fig. 2) are reported using five areas during

Table 3 Percentage area using various datasets

Dataset no.	Classes (percentage area)			
	Bare soil	Water	Crop	Built up
1	19.9	1.42	55.81	22.87
2	19.39	1.33	56.52	22.76
3	19.71	1.4	56.23	22.66
4	20.32	1.38	55.41	22.89
5	19.72	1.37	56.2	22.72
6	19.41	1.36	56.54	22.7
7	19.17	1.37	56.79	22.67
8	19.81	1.28	56.18	22.73
9	19	1.23	56.76	23.01

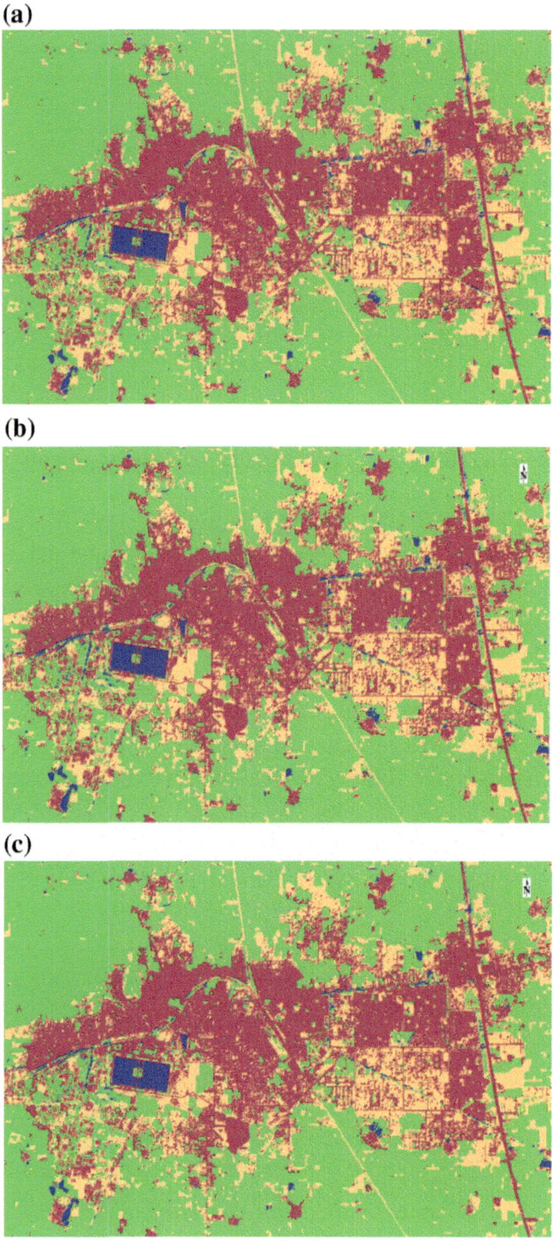

Fig. 1 Classified images using dataset eight (**a**), dataset seven (**b**) and dataset nine (**c**)

Table 4 Optimisation of SVM parameters for dataset 8

	6 + BAEI + BSI + NBAI + NBI + NDBI + UI		Classes (percentage area)			
	γ (RBF Kernel parameter)	C (regularization parameter)	Bare soil	Water	Crop	Built-up area
1	0.5	700	20.86	1.93	54.22	22.99
2	0.5	1000	20.85	2.09	53.95	23.11
3	0.6	1100	20.92	2.1	53.89	23.1
4	0.6	1500	21.07	2.1	53.8	23.02
5	0.7	1700	21.1	2.11	53.74	23.05
6	0.8	700	20.86	1.93	54.22	22.99
7	0.9	700	20.86	1.93	54.22	22.99
8	0.9	1000	20.85	2.09	53.95	23.11
9	1	1100	20.92	2.1	53.89	23.1
10	1	100	19.81	1.28	56.18	22.73
11	1.5	1100	20.92	2.1	53.89	23.1
12	0.1	50	20.91	1.52	55.41	22.16
13	0.1	70	20.76	1.54	55.29	22.41
14	0.1	90	20.81	1.58	55.16	22.45
15	0.1	110	20.71	1.59	55.14	22.56
15	0.1	130	20.68	1.61	55.12	22.59

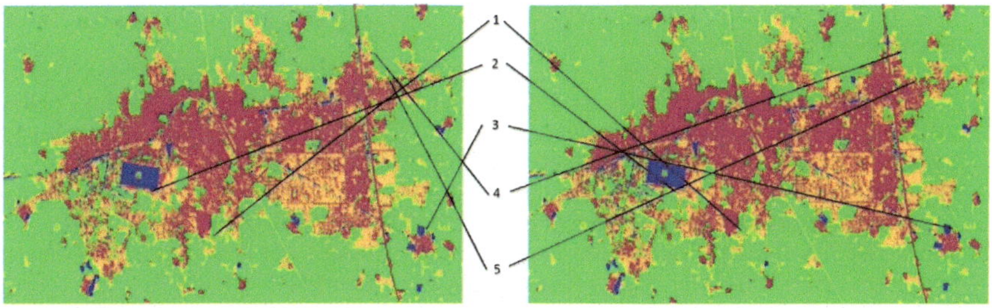

Fig. 2 Comparison of classified images of the study area for two different values of user-defined parameters with SVM classifier

the field visit. In first image area 1 is classified as vegetation, while in second image it is classified as bare land, for area marked as 2, first image classified it as built-up while second image classified it as bare land, for area 3, seconnd image misclassified some trees for water while first image provided better representation of the ground features. For area marked as 4, first image is suggesting it to be bare soil while second image shows it as water whereas for area 5, first image suggests it to be built-up area while second image showed it as bare land. A comparison of above five areas in classified image suggests better performance by SVM for user-defined parameters having C = 50 and $\gamma = 0.1$ values.

5 Conclusion

Sentinel 2 image acquired over Kurukshetra was used to differentiate bare soil and built-up area using several indices. A major conclusion of this study is that a combination of six indices with six bands of Sentinel 2 data performs better in differentiating both classes. Results also suggest comparable performance by several combinations, which may be because of the use of large training dataset during classification. Another conclusion from this study is that one needs to find an optimal value of user-defined parameters for different dataset as the performance of SVM classifier is found to vary with change in the number of bands in the input dataset. Further study is being planned to judge the effectiveness of different indices with limited training sample sizes.

References

1. World Bank Report (2015). http://data.worldbank.org/indicator/SP.URB.TOTL.IN.ZS. Accessed on 3 Apr 2016
2. Weng Q (ed) (2007) Remote sensing of impervious surfaces. CRC Press, Boca Raton, Florida
3. Donnay JP, Barnsley MJ, Longley PA (2001) Remote sensing and urban analysis. In: Donnay JP, Barnsley MJ, Longley PA (eds) Remote sensing and urban analysis. Taylor and Francis, London, United Kingdom and New York, N.Y., pp 3–18
4. Piyoosh AK, Ghosh SK (2017) Development of a modified bare soil and urban index for Landsat 8 satellite data. Geocarto Int. http://dx.doi.org/10.1080/10106049.2016.1273401
5. Zha Y, Gao J, Ni S (2003) Use of normalized difference built-up index in automatically mapping urban areas from TM imagery. Int J Remote Sens 24(3):583–594
6. Takeuchi W, Yasuoka, Y (2004) Development of normalized vegetation, soil and water indices derived from satellite remote sensing data. In: C-9.4 remote sensing applications, 25th ACRS 2004, Chiang Mai, Thailand
7. Zhao H, Chen X (2005) Use of normalized difference bareness index in quickly mapping bare areas from TM/ETM+. In: Proceedings of IEEE geoscience and remote sensing symposium, 2005. IGARSS'05, vol 3, pp 1666–1668
8. Xu H (2008) A new index for delineating built-up land features in satellite imagery. Int J Remote Sens 29914:4269–4276
9. Xu H (2010) Analysis of impervious surface and its impact on urban heat environment using the normalized difference impervious surface index (NDISI). Photogram Eng Remote Sens 76(5):557–565
10. As-Syakur AR, Adnyana I, Arthana IW, Nuarsa IW (2012) Enhanced built-up and bareness index (EBBI) for mapping built-up and bare land in an urban area. Remote Sens 4(10):2957–2970
11. Deng C, Wu C (2012) BCI: A biophysical composition index for remote sensing of urban environments. Remote Sens Environ 127(12):247–259

12. Li S, Chen X (2014) A new bare-soil index for rapid mapping developing areas using Landsat 8 data. Int Arch Photogram Remote Sens Spat Inf Sci 40(4):139–144
13. Deng Y, Wu C, Li M, Chen R (2015) RNDSI: A ratio normalized difference soil index for remote sensing of urban/suburban environments. Int J Appl Earth Obs Geoinf 39(7):40–48
14. Vapnik V (1998) Statistical learning theory. Wiley, New York
15. Bouzekri S, Lasbet AA, Lachehab A (2015) A new spectral index for extraction of built-up area using Landsat-8 data. J Indian Soc Remote Sens 43(4):867–873
16. Sun G, Chen X, Jia X, Yao Y, Wang Z (2016) Combinational build-up index (CBI) for effective impervious surface mapping in urban areas. IEEE J Sel Top Appl Earth Obs Remote Sens 9(5):2081–2092
17. Sekertekin A, Marangoz AM (2015) An erdas imagine model to extract urban indices using Landsat 8 satellite imagery. Int J Sci Technol Res 4(8):62–67

Assessment of SCATSAT-1 Backscattering by Using the State-of-the-Art Water Cloud Model

Ujjwal Singh, Prashant K. Srivastava, Dharmendra Kumar Pandey and Sasmita Chaurasia

Abstract The SCATSAT-1 satellite data can be used for various applications in the field of agriculture. The main aim of the study is to investigate the water cloud model (WCM) for backscattering simulation by using the field-measured soil moisture in order to validate the SCATSAT-1 measured backscattering. WCM requires various input datasets for simulation of backscattering such as vegetation parameters A and B and soil parameters C and D, which can be estimated by Non-linear least square fitting method by using with experimental dataset. The results showed that the simulated WCM values are well correlated with the backscattering of SCATSAT-1 satellite data. However, it can be further improved when each parameter of WCM is generated by using the ground-based measurements. In this study, some progress has been made toward backscattering simulations using the SCATSAT-1; however, it can be further refined with the advancement in the retrieval algorithms and sensor sensitivity.

Keywords SCATSAT-1 · Backscattering · Water cloud model (WCM) · Agriculture

1 Introduction

Microwave backscattering coefficients are important variable in many land surface applications such as soil moisture retrieval (SM) and leaf area index estimation (LAI)

U. Singh (✉) · P. K. Srivastava
Institute of Environment and Sustainable Development, Banaras Hindu University, Varanasi, India
e-mail: ujjwalrsmt@gmail.com

P. K. Srivastava
e-mail: prashant.just@gmail.com

D. K. Pandey · S. Chaurasia
Space Applications Centre, ISRO, Ahmedabad, India
e-mail: dkp@sac.isro.gov.in

S. Chaurasia
e-mail: sasmita@sac.isro.gov.in

J. K. Ghosh and I. da Silva (eds.), *Applications of Geomatics in Civil Engineering*, Lecture Notes in Civil Engineering 33, https://doi.org/10.1007/978-981-13-7067-0_40

[1–5]. Backscattering is directly affected with the change in soil moisture and leaf area index. In past, water cloud model (WCM) was used to analyze the microwave scattering [1, 5–8]. Several studies have been reported that the WCM was a robust method for simulation of backscattering with the help of leaf area index and soil moisture property [9]. Previously, WCM simulated the backscattering by using the soil moisture and vegetation geometry for L, C, X and Ku frequency bands. In order to improve the performance of WCM, researchers attempted and modified the input parameters. After all such efforts, it indicates that the WCM is a promising model for backscaterring simulation with satisfactory accuracy [9].

In order to evaluate the applicability of SCATSAT-1, WCM model is used in this study. SCATSAT-1 is the current mission of ISRO with Ku-band Scatterometer which provides operationally global ocean wind products at different grid resolutions. SCATSAT-1 has incident angle of inner beam (HH polarization) of 48.90° with 1400 km swath while incident angle of outer beam (VV polarization) corresponds to 57.60° with 1840 km Swath. SCATSAT-1 level 4 data has 0.02° spatial resolution [10]. The land products of SCATSAT-1 could be used for various applications in agricultural field such as soil moisture and vegetation water content. There are rare literature studies contributing to the knowledge of SCATSAT-1 evaluation for backscattering phenomenon related to water cloud model through ground-based soil moisture and leaf area index. Therefore, in this study, an attempt has been made to evaluate the SCATSAT-1 satellite backscattering (HH-Ascending) with the help of WCM.

2 Water Cloud Model

The water cloud model is the cross section of the canopy σ^0 backscattered in the term of sum of vegetation $\sigma_v{}^0$ and underlying soil $\sigma_s{}^0$ latter this attenuation through the vegetation parameter. Incident angle is (θ) for equations [11].

$$\text{Whole canopy:} \, \sigma^0 = \sigma_v^0 + \Gamma^2 \sigma_s^0 \; \left(\text{Unit m}^2/\text{m}^2\right) \tag{1}$$

$$\text{Vegetation:} \, \sigma_v{}^0 = AV1 \cos\theta (1 - \Gamma^2) \left(\text{Unit m}^2/\text{m}^2\right) \tag{2}$$

$$\text{Soil:} \, \sigma_s{}^0 = C + Dh_V \; (\text{Unit dB}) \tag{3}$$

$$\Gamma^2 = \exp(-2BV1/\cos\theta) \; (\text{Unitless}) \tag{4}$$

where Γ^2 the two-way attenuation through the canopy and h_v is volumetric soil moisture content (cm^3/cm^3) and V1 is the vegetation descriptor as leaf area index. The parameters C and D for soil and parameters A and B for the vegetation. The parameter A, B, C, and D can be generated by nonlinear least squares fitting method by using experimental field dataset. The parameters A and B depend on the canopy geometrical structure, vegetation water content and parameters C and D on bare soil moisture condition, respectively. Based on (1)–(4), final equation can be written as

$$\sigma^0 = (AV1\cos\theta\{1 - [\exp(-2BV1/\cos\theta)]\}) + (\exp(-2BV1/\cos\theta)\sigma_s{}^0\}\left(\text{Unit m}^2/\text{m}^2\right) \quad (5)$$

and $\sigma_s{}^0$ unit dB converted here in m^2/m^2.

where A, B, C, and D are in order to 0.005, 0.0915, −10.59, and 0.7187. The V1 = V2 is the leaf area index parameter. The incident angle is (θ) 48.90°.

3 Study Area and Datasets

In this study, simulated backscatter using water cloud model has been compared over 60 field sample points of Varanasi district measured during extensive field campaign. All the sampling points are located within 65 km of radius in Varanasi district. These points are taken in agricultural field (mostly wheat, mustard, peas, pigeon pea, sugarcane, and vegetables) and divided into three groups depending on field conditions. The range of soil moisture is from 14 to 40%. These field samples were collected for all sample sites during the daytime from 9 a.m. to 5 p.m. First group contains all the 60 samples. Second (Group-1) is the dense vegetation cover in comparison to Group-2, which comprises sparse vegetation. The reason behind dividing the three groups is to see the impact of vegetation on backscattering performance. The distances between the two sample points are kept more than 2 km as per grid resolution of SCATSAT-1 Level-4 data products to get adequate number of samples. This field-measured soil moisture at all sample points was taken by portable hydra probe meter. Soil moisture is measured up to the 10 cm of soil depth. Soil moisture collection data is synchronized with SCATSAT-1 overpass time. The WCM parameter leaf area index is taken from Copernicus land service of SPOT-VGT, PROBA-V sensor of 1 km resolution. The Proba-V has field of view 102 degree with ground resolution 350 m and swath 2250 km [12].

4 Results and Discussion

The results showed the performance of water cloud model simulated using the experimental field data for backscattering and compared against the SCATSAT-1 Ku band backscatter. This experimental data was used to generate the water cloud model coefficient parameters A, B, C, and D. In this way, this work analyzed the SCATSAT-1 Ku band data sensitivity for soil moisture and vegetation cover.

Statistical results for the second group showed a higher degree of accuracy (r = 0.7673, RMSE = 0.5025, and bias = −0.05042) which is the best among the groups as compared to the third one which showed the lowest degree of accuracy (r = 0.3550, RMSE = 1.1327, and bias = −0.4280) as shown in Figs. 2 and 3.

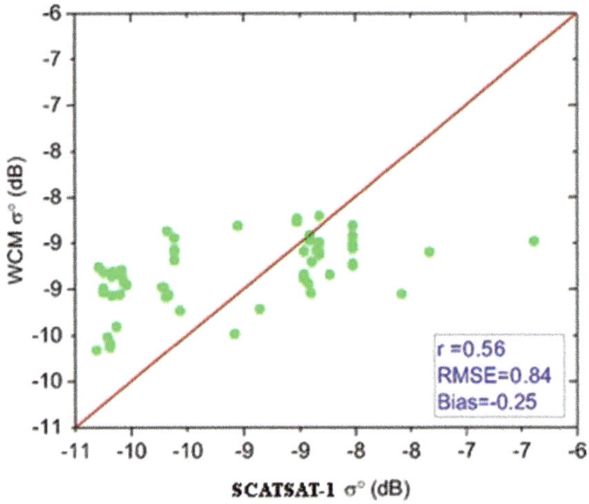

Fig. 1 Scatter plot between SCATSAT-1 and simulated WCM σ° for 60 samples

Whereas Group-1 showed intermediate results (r = 0.5562, RMSE = 0.8483, and Bias = −0.2472) as shown in Fig. 1. Group-2 SCATSAT-1 σ° is showing a close match with the WCM-simulated σ°. Group-2 statistical analysis has been shown in Fig. 2. With the help of statistical analysis, the best result was shown by Group-2. The Group-3 relation between SCATSAT-1 and simulated σ° is shown in Fig. 3 with its statistical result. Group-2-simulated σ° is the best matching with the SCATSAT-1 result in comparison to other two groups, which can be ascribed to the changes in vegetation density. There are some limitations found in this study (1) diverse agricultural practices with different irrigation scheduling including different types of crop canopy structure and (2) the water cloud model coefficients (A, B, C, and D) were based on single wheat growing period dataset. Therefore, increasing the sample size could provide more concrete results. Moreover, measurements of LAI and other ground-based parameters by using the field instruments may provide better results than the present case.

5 Conclusions

This initial study indicates that the SCATSAT-1 backscatter is having good match with simulated water cloud model backscattering over dense vegetation cover. This backscattering difference between SCATSAT-1 and simulated one can be attributed to the sensitivity to vegetation cover. Our study clearly indicates a number of challenges still required to be resolved by the scientific community, so that accurate products can be derived with the help of SCATSAT-1 satellite data for terrestrial applications. It is also obvious that to complement the requirements of SCATSAT-1 information for many practical applications, efforts should be made toward the investigation

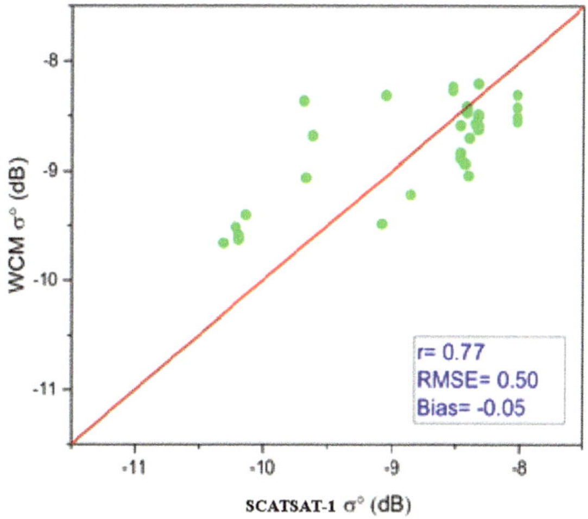

Fig. 2 Scatter plot between SCATSAT-1 and simulated WCM σ° over dense vegetation cover

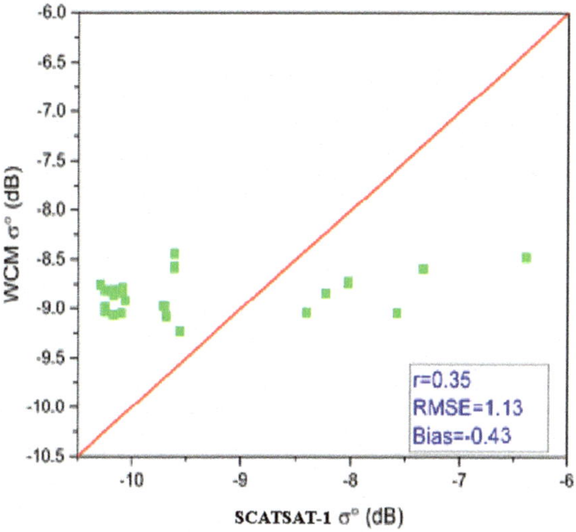

Fig. 3 Scatter plot between SCATSAT-1 and simulated WCM σ° with sparse vegetation cover

of the synergistic use of SCATSAT-1 with other co-orbiting satellites for different applications.

Acknowledgements We are thankful to the Meteorological and Oceanographic Satellite Data Archival Centre (MOSDAC) Space Applications Centre, ISRO for providing the SCATSAT-1 data. The authors are grateful to Space Applications Centre, ISRO, Ahmedabad for providing financial support under "SCATSAT-1 Utilization programme" of SAC-ISRO to conduct this research work.

References

1. Bouman B (1991) Crop parameter estimation from ground-based X-band (3-cm wave) radar backscattering data. Remote Sens Environ 37:193–205
2. Schmullius C, Furrer R (1992) Frequency dependence of radar backscattering under different moisture conditions of vegetation-covered soil. Int J Remote Sens 13:2233–2245
3. Moran MS, Vidal A, Troufleau D, Inoue Y, Mitchell TA (1998) Ku-and C-band SAR for discriminating agricultural crop and soil conditions. IEEE Trans Geosci Remote Sens 36:265–272
4. Dabrowska-Zielinska K, Gruszczynska M, Janowska M, Stankiewicz K (1993) Use of ERS-1 SAR data for soil moisture assessment. In: ESA/FAO/CEC/Telespazio workshop on use of ERS-1 SAR data for agricultural, forestry and environmental applications in central-eastern Europe, Frascati (Italy), 8–12 Nov 1993. ESA, (Year)
5. Lin D, Wood E, Beven K, Saatchi S (1994) Soil moisture estimation over grass-covered areas using AIRSAR. Int J Remote Sens 15:2323–2333
6. Attema E, Ulaby FT (1978) Vegetation modeled as a water cloud. Radio Sci 13:357–364
7. Mo T, Schmugge TJ, Jackson TJ (1984) Calculations of radar backscattering coefficient of vegetation-covered soils. Remote Sens Environ 15:119–133
8. Paris J (1986) The effect of leaf size on the microwave backscattering by corn. Remote Sens Environ 19:81–95
9. Ulaby FT, Moore RK, Fung AK (1986) Microwave remote sensing active and passive-volume III: from theory to applications
10. SCATSAT-1 Level 4 Data Products Format Document. https://www.mosdac.gov.in/
11. Ulaby F, Allen C, Eger Iii G, Kanemasu E (1984) Relating the microwave backscattering coefficient to leaf area index. Remote Sens Environ 14:113–133
12. Proba-V ESA. https://earth.esa.int/web/guest/missions/esa-operational-eo-missions/proba-V

Building Footprint Extraction from Very-High-Resolution Satellite Image Using Object-Based Image Analysis (OBIA) Technique

A. P. Prathiba, Kriti Rastogi, Gaurav V. Jain and V. V. Govind Kumar

Abstract Building footprints are an important input in several urban applications such as master plan preparation, development of 3D city building models, rooftop solar energy potential estimation, tax compliance evaluation, or study of population distribution in cities. The very high spatial resolution (VHR) image is invariably required for the extraction of building footprints. However, the conventional pixel-based approaches have limited success in building footprint extraction owing to inherent heterogeneity of the urban environment. In this research, we have therefore applied Object-Based Image Analysis (OBIA) for building footprints extraction from Cartosat-2 series data. The image was segmented into several objects on the basis of spectral and spatial homogeneity of pixels. The objects were thereafter classified using nearest-neighbor classification approach. Finally, these classified objects were further subjected to segmentation into smaller objects and classified using decision-rules. The combination of supervised nearest-neighbor classification with decision-rules resulted in an accuracy of over 82.5% in the extraction of building footprints. The results of the study will be used to develop a 3D city model of Ahmedabad city and assess the changes in the built-up volume in the city.

Keywords Building footprint extraction · OBIA · Nearest-neighbor classification

---

A. P. Prathiba (✉) · V. V. Govind Kumar
Indian Institute of Technology (Indian School of Mines) Dhanbad, Dhanbad, Jharkand, India
e-mail: prathi.gem@gmail.com

V. V. Govind Kumar
e-mail: vgkvilluri@iitism.ac.in

K. Rastogi · G. V. Jain
Space Application Centre (ISRO), Ahmedabad, Gujarat, India
e-mail: kritirastogi@sac.isro.gov.in

G. V. Jain
e-mail: gvj@sac.isro.gov.in

J. K. Ghosh and I. da Silva (eds.), *Applications of Geomatics in Civil Engineering*,
Lecture Notes in Civil Engineering 33, https://doi.org/10.1007/978-981-13-7067-0_41

1 Introduction

The urban settlements cover about 0.5% of the earth's land surface area, but it holds more than half of the world's population [1]. The transition from rural to urban indicates the shift from an agriculture-based economy to the manufacturing and service-oriented economy. As per the study by the United Nations, by 2050 85.9 and 64.1% (source: World Bank record) of the developed and developing world will be urbanized respectively. Figure 1 shows the proportion of urban population in the total population in India from the year 1961–2011.

Urbanization leads to numerous problems like overcrowding, traffic congestion, growth of slums, insufficient resources, etc. To satisfy the basic needs of people and for dynamic growth, advanced technology is needed to be associated with urbanization. Building footprint is one of the recent technology, which can be used as a source for several urban applications like the development of 3D city models, tax compliance evaluation, rooftop solar energy potential estimation, master plan preparation.

Traditionally the aerial photography is used for urban studies owing to its high spatial resolution. With the advancement of technology, commercial satellites capable of imaging at sub-meter spatial resolution are becoming popular and easily accessible. Satellite images give detailed visibility of terrestrial features by reducing the heterogeneity of each pixel, which leads to the shift from aerial to satellite images for urban applications [2]. For urban features classification and information extraction, spatial information is more important than the spectral information. High-resolution satellite images (high spatial information) provide detailed information on the objects.

The main challenge with the very-high-resolution satellite images is a higher level of details. If the resolution of satellite image increases, the information of the finer objects comes to the picture, but it may reduce the accuracy of the digital classification in urban areas. For the effective identification of the object, the pixel size should be

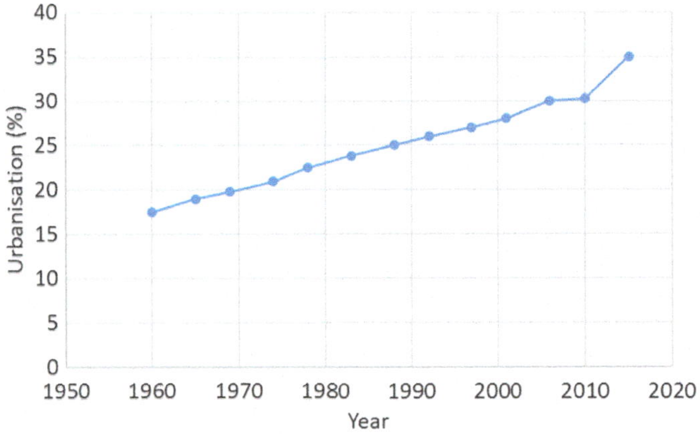

Fig. 1 Urbanization in India

half of the diameter of the smallest object of interest and a minimum four pixels are required for object identification [3]. The precision of the classification can be increased by adding the spectral information to the high-resolution image of the same spatial resolution. Visual interpretation of the urban objects using the multispectral image is advantageous. The advantages of both spectral and spatial resolution can be utilized effectively by fusion of multispectral image with the high-resolution panchromatic image [4].

Each pixel in the image carries the spectral and spatial information of the surface features such as color, texture, shape, etc., but by viewing individual pixels, it is difficult to identify. Object-based image analysis (OBIA) is the technique which is performed on the basis of an object, which human can interpret. OBIA groups the homogeneous pixels to form an object [5]. However, different objects have typical spectral reflectance, which helps the users to identify the object.

Image segmentation segments the image into several clusters by modifying the morphology of objects according to defined criteria, which serves as the framework for further analysis. eCognition provides several segmentation algorithms based on both top-down and bottom-up strategy. In which, multiresolution segmentation algorithm works on bottom-up strategy, based on relative homogeneity it merges the pixel with their neighbor. The main concept behind the multiresolution segmentation in eCognition is Fractal Net Evolution Approach (FNEA) [6]. Homogeneity criteria for multiresolution segmentation are the combination of scale, smoothness and compactness parameter [7]. The parameter for image segmentation is interpreted through the trial and error method. After several experiments, the most suitable parameter for multiresolution segmentation is selected.

The objects are classified into various classes based on hierarchy after completion of segmentation. The main type of classifiers is nearest-neighbor classifier or membership classifier. The standard nearest-neighbor classification is carried out by selecting the training samples of different location and varied reflectance for each classes using visual interpretation. The object-related features like layer value, texture, geometry, position, relation to neighbor objects and user-defined indices are used for the execution of fuzzy-based classification. The appropriate threshold value for each feature should be determined by the spectral reflectance curve of different classes, for accurate classification. After classification, the vector file for each class can be exported with object-related feature value for further analysis.

2 Study Area and Data Used

The study area used for the analysis is Ahmedabad, which is the sixth biggest city and seventh largest metropolitan city of India. The city in the first stage of the proposed smart city plan. As per census 2011, the population of the city was 6.3 million and it is estimated that the population will be 8.16 million in 2018. By 2021, there could be almost 9 million individuals living in the city [8]. The population density of the city is 9900 persons per square kilometer. For the study, three subsets area of 1 km^2 with

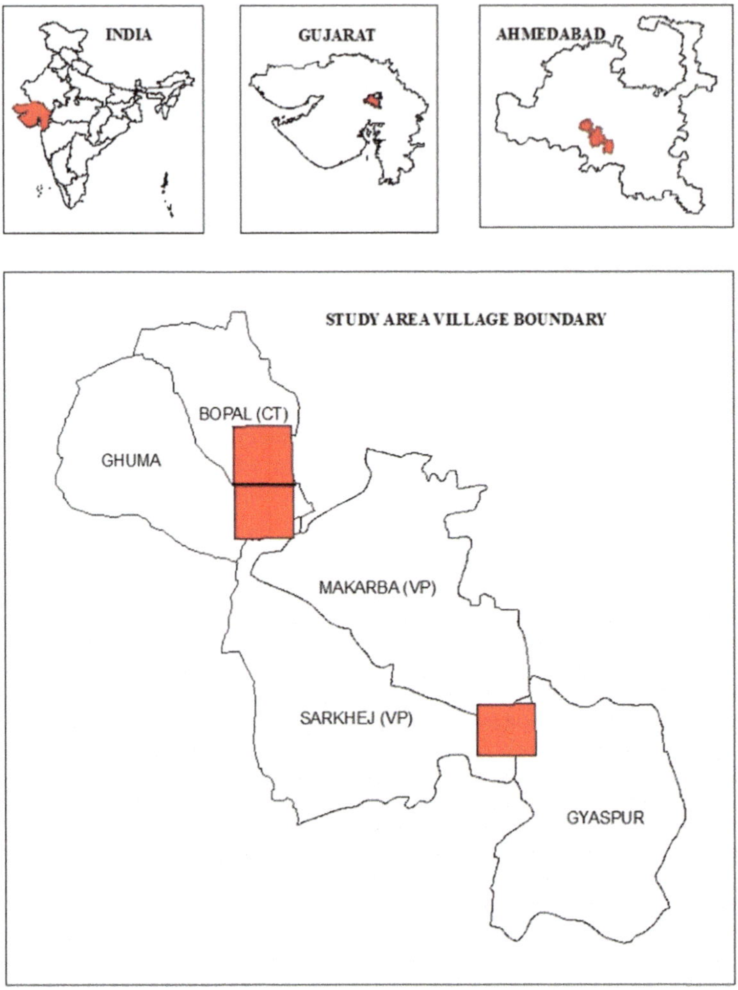

Fig. 2 Study area taken at few villages of Ahmedabad, Gujarat, India

different built-up type and density are selected to compare the accuracy of work. Figure 2 shows the location of the study area and areas selected for detailed accuracy assessment.

The very high spatial resolution image acquired by the Cartosat-2 series satellites were used in this study. The Cartosat-2 series satellites launched by Indian Space Research Organization (ISRO) offers sub-meter spatial resolution data in panchromatic mode. The captured images are useful for several cartographic, urban and land applications. The panchromatic sensor takes a single band image at less than 1 m spatial resolution in wider bandwidth but multiband sensor takes an image in the

Table 1 Specification of Cartosat-2 sensors

Satellite sensor	Period of capture	Spatial resolution	Spectral bands
Cartosat-2 series merged product	October 2016	<1 m	B1-Blue B2-Green B3-Red B4-NIR

accurate multiple bands at a moderate resolution [9]. The specifications of the data used for the study are given in Table 1.

3 Methodology

The overall flow of methodology is shown in Fig. 3. The details of each of the steps are discussed in the subsequent sections.

3.1 Image Preprocessing

The quality of the image taken from the sensor depends on its radiometric resolution, cloud cover, wavelength and bandwidth of the band, etc. Based on the application, either of these parameters specifications are considered. For urban studies, the image with high spatial resolution with multispectral bands is mostly preferred. In this study, the merged product of panchromatic high spatial resolution image with corresponding low-resolution multispectral image was used. The images were georeferenced with the corresponding geo-rectified reference images. The alignment of the image with respect to known coordinates is very important for the subsequent process of work.

3.2 Object-Based Image Analysis (OBIA)

Object-Based Image Analysis (OBIA) groups the pixels into meaningful clusters, on the basis of spectral, contextual and textural information. This approach helps to develop the ruleset based on statistics using spectral, spatial characteristics (such as shape, rectangularity fit, length, etc.), textural parameter, relation to neighbor objects for extraction of image objects [10]. It helps in discriminating the spectrally similar objects that improve the classification process, which cannot be obtained in pixel-based method. OBIA segments the image into several objects that are the building blocks for further classification. eCognition is the first commercial software for OBIA, previously it was known as "Definiens" [11].

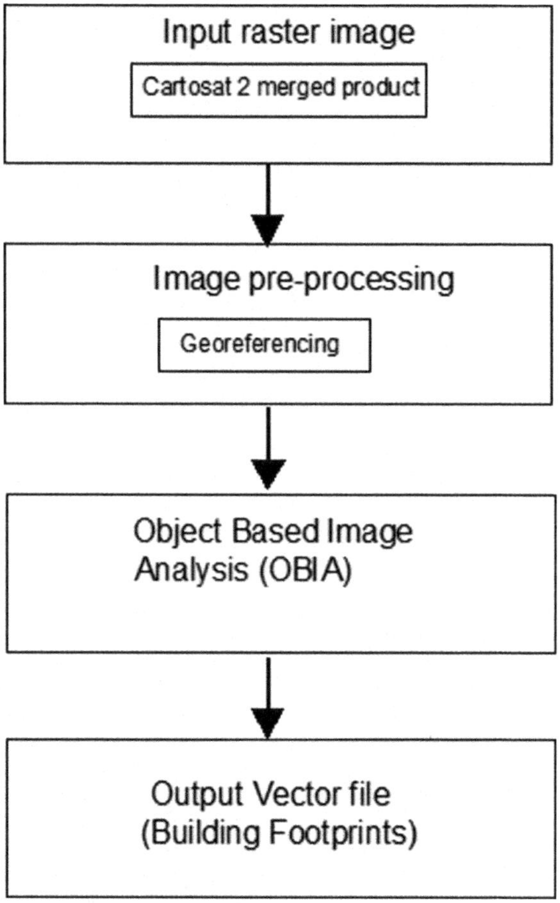

Fig. 3 Methodology used for building footprint extraction from very-high-resolution satellite image

The process is completely based on programmable workflow and in-built with Fractional Net Evolution approach [12]. Compared with single pixels, the segmented objects have additional spectral information like minimum, maximum values, mean of band, standard deviation, etc. [13]. Figure 4 shows the complete process of OBIA. Process in OBIA is carried out in two phases, segmentation followed by classification.

3.2.1 Image Segmentation

Image segmentation is a fundamental framework for object-based image analysis. In the object-based classification prototype, the region with a homogenous group of pixels is considered as objects. Multiresolution segmentation [12] in eCognition software is working on the basis of the Fractal Net Evolution Approach (FNEA),

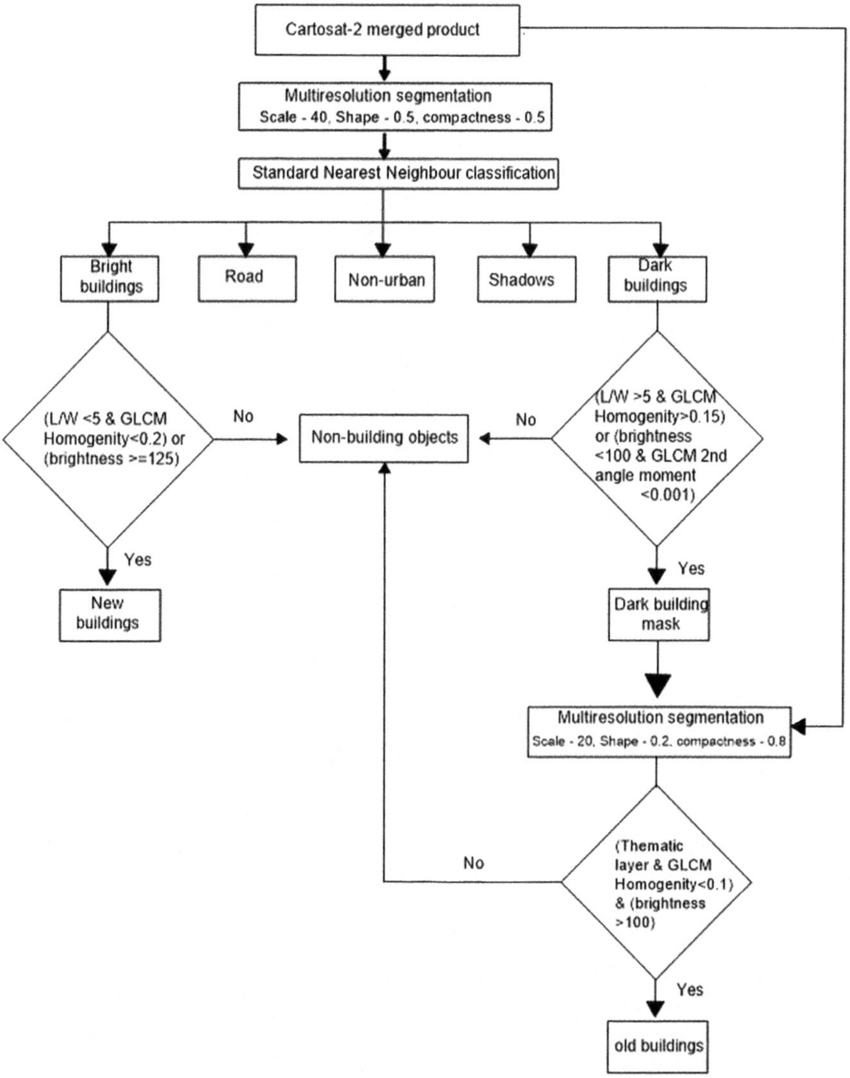

Fig. 4 Stages and algorithms used at various stages of OBIA. The output from each stage is used as input for the subsequent stage and building footprints are extracted

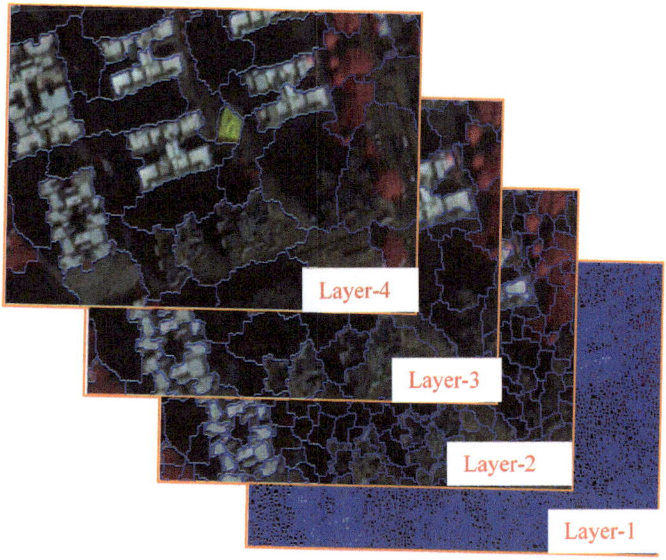

Fig. 5 The image is segmented at various scale parameters with same shape and compactness factor. Layer 1 is shown in the pixel size, layer-2 at the finer scale of 10, layer-3 at a moderate scale of 25 and layer-4 with the object scale of 50

which segments the image on the basis of pixel homogeneity [14]. Figure 5 shows the objects formed with different parameters of segmentation.

The main parameters considered for the multiresolution segmentation are scale, size and compactness factor. At different scales, the weighted values ranging from 0 to 1 are assigned to compactness and shape factor. To effectively control the homogeneity of the object, the summation of shape and compactness factor should be one. The scale parameter depends on the level of detail required for accurate classification, larger the scale (e.g. 100) larger the object size. Compactness factor maintains the smoothness of the object boundaries, whereas shape factor adjusts the spatial and spectral homogeneity of the objects. Generally, a larger optimal segmentation scale is used for lower spatial resolution and vice versa [15].

Multiresolution segmentation is tested at different scale, shape and compactness factor. We consider the scale levels range from 10 to 100, with the shape and compactness factor ranges from 0.2 to 0.8 for the analysis. The scale level of 10 gives a higher level of detail compared to scale 100. After trial and error process, the most suitable parameters for segmentation are selected.

In the whole OBIA process, image segmentation is executed in two stages. In the first stage, multiresolution segmentation is processed with scale level 40, shape and compactness factor 0.5 to execute the basic level of classification. During the second stage, a particular class of interest is further segmented with scale level of 20 to obtain a higher level of details and shape factor is maintained as 0.2 to give more attention to spectral homogeneity.

3.2.2 Image Classification

Object-based image classification on the segmented objects can be executed using nearest-neighbor classifier and membership function. To classify through standard nearest-neighbor (SNN) classifier, the feature space like bands, indices, geometry, and texture are defined for each class. A set of objects for each class are selected as a training sample, in order to assign the classes for the segmented object. In SNN classifier, each object in the image is marked either 0 or 1 [16]. The classifier will check each and every object based on the training samples whether it belongs to the particular class or not.

The image segmented with the scale of 40, the shape and compactness factor of 0.5 is classified through SNN classifier. The image objects are classified into 7 basic classes such as bright buildings (Buildings with higher spectral reflectance), dark buildings (Buildings with lower spectral reflectance), vegetation, soil, road, water bodies, and shadows. The common feature space like brightness, mean, texture is defined for each class. The additional feature class, geometry is used for delineation of the road using its linear characteristic. To increase the accuracy of SNN classification, buildings are classified into two classes because the newly constructed buildings reflectance are higher compared to old buildings.

Membership function classifies the image based on fuzzy logic, where each object in the image has a value between the ranges 0–1. Based on the knowledge and spectral reflectance curve, threshold values are assigned in each stage of classification and image objects are checked whether it satisfying the threshold value or not. The ruleset and constraints are defined after several sets of trial and error process. eCognition provides the several feature spaces to assign the threshold value for classification and it also permits the users to define their own indices. Grey Level Co-occurrence Matrix (GLCM) is used to acquire features like homogeneity, heterogeneity, dissimilarity, etc., by estimating the statistics of pixels. The features derived from texture information are used to differentiate the pixel having the same spectral reflectance.

In the first stage of classification, the object class bright building is used. There are some errors in SNN classification, because the reflectance of soil and concrete payment is similar to new building reflectance. The errors are eliminated by applying threshold values to geometry and texture. Length to width ratio is an important parameter for road classification. Homogeneity of soil and building are quite different, so those parameters are used for building delineation. Even though some of the commercial building object geometry is similar to the road object geometry, so the brightness value above 125 is classified as a new building. Similarly, the dark buildings (building 2) is taken as a class of interest and the condition is applied vice versa and classified as non-building objects. The objects which are not satisfying the conditions are retained in the original class (obtained from SNN classification). Further, object class dark building is exported as a vector layer.

In the second stage of classification, the vector layer is used as input along with a raster image for classification. Few of road and shadow pixel reflectance is similar to that of the dark building, some of them are wrongly classified in dark building (Fig. 6). To remove the unwanted pixel, the objects are subjected to further segmentation.

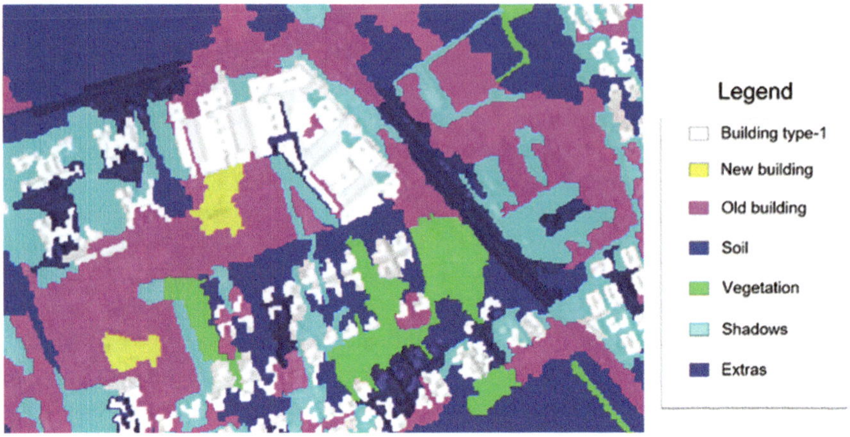

Fig. 6 Classified image; yellow color shows the omitted non-building objects from the building class

Segmentation is carried out only on the thematic layer with the shape factor of 0.2, scale 20, and compactness 0.8 [17]. Spectral and texture properties are used for further classification of old buildings. In the thematic layer, Brightness >100 and GLCM Homogeneity <0.1 is classified as Old buildings). The new and old building are exported and dissolved for further analysis.

4 Results and Discussion

The OBIA method defined above automatically delineate the building footprints from high-resolution satellite images. The created ruleset can be used for all sets of the high-resolution image with some minor modification in threshold value. Three subsets of the study area are taken, with discrete building type and density to check the accuracy of building footprint extraction (Fig. 7). In the subset region, the building footprints are also extracted manually to compare the accuracy of automatically extracted building footprint.

Buildings in a well-distributed built-up area are extracted very well with high accuracy. Old buildings in the study are not completely extracted by the developed ruleset. If ruleset is modified to extract the old buildings, it adversely increases the commission error. The accuracy of the automatic method with respect to the manual is 82.5% (Fig. 8a). Red color indicates buildings, which are common in both manual and automatic method. The green color shows the unidentified buildings in the automatic method and blue color shows non-building objects. The density of buildings is higher in Fig. 8b compared to Fig. 8a. An organized construction is done in Fig. 8b. So, building footprints are easily delineated. The accuracy of buildings from OBIA technique (Fig. 8b) compared with the manual method is 80%. In Fig. 8c, the build-

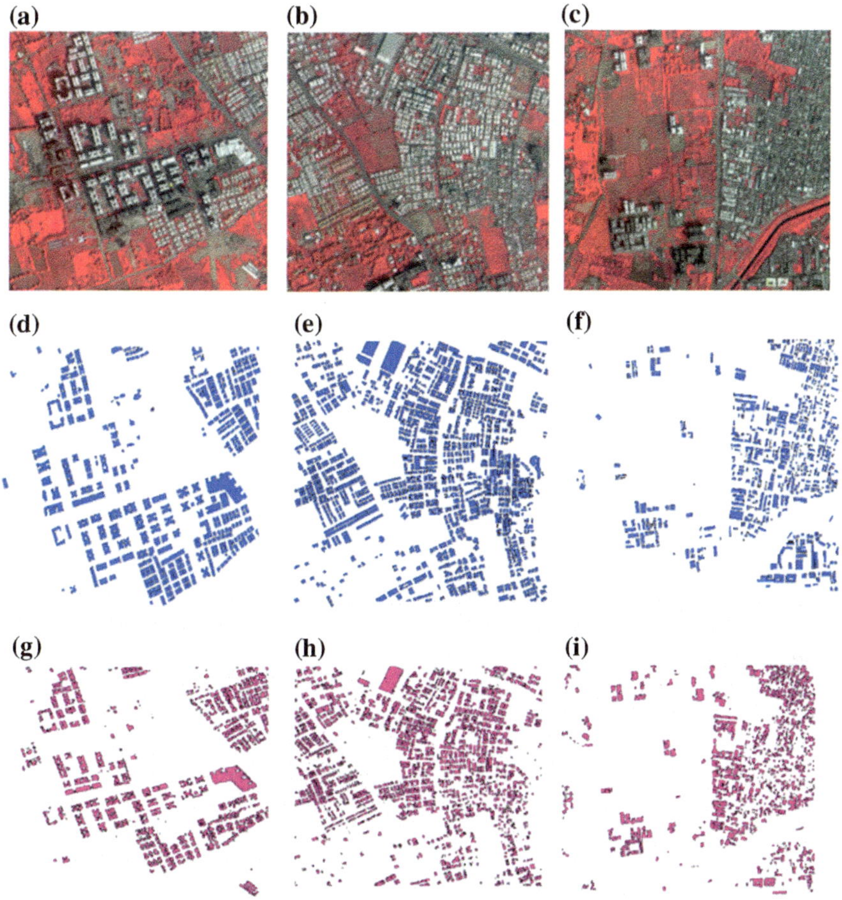

Fig. 7 **a–c** Subset of study area at various building type and density. **d–f** are manually extracted building footprint through visual interpretation. **g–i** are automatically extracted buildings through OBIA

ings (overcrowded slum areas) are very congested and constructed in an unplanned manner. Moreover, it is very difficult to point to individual building visually. However, the created ruleset is executed to validate the algorithm. The accuracy of the automatic method in overcrowded slum areas (Fig. 8c) compared to visual interpretation is around 69.2%. The accuracy of building footprint extracted automatically at a well-distributed built-up area, crowded built-up area and overcrowded built-up area are given in Table 2. The result shows that the accuracy of the automatic method for well-distributed built-up area is comparatively higher than obtained in crowded built-up and overcrowded informal settlements.

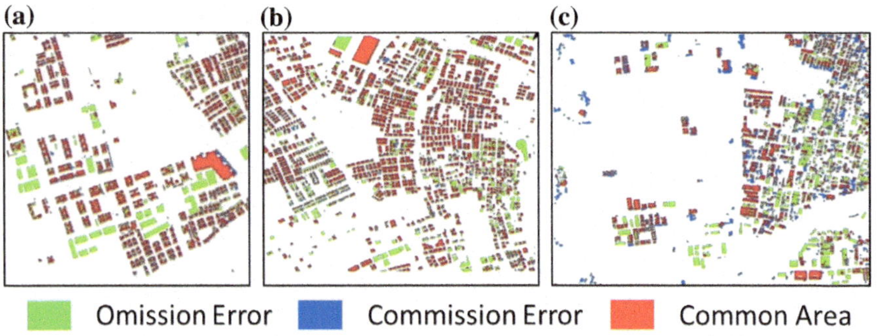

Fig. 8 Accuracy assessment of buildings extraction in (**a**) well-defined built-up area; **b** crowded built-up area; and **c** overcrowded areas

Table 2 Summary of OBIA accuracy

Urban areas	Reference area (m^2)	Correctly classified area (m^2)	Unidentified buildings (m^2)	Wrongly classified area (m^2)	Total classified built-up area (m^2)
Well-distributed built-up area	121,021.06	99,825.39 (82.5%)	21,195.67 (17.5%)	22,420.09 (18.5%)	122245.48
Crowded built-up area	200,202.02	160,161.67 (80%)	40,040.35 (20%)	57,769.89 (11.9%)	217931.56
Over-crowded slum area	72,170.76	49,942.16 (69.2%)	2,222.86 (30.8%)	1,726.36 (23.92%)	51668.52

5 Conclusions

The study proposed, a blend of object-based nearest-neighbor classification and rule-based classifier for building footprint extraction from very-high-resolution satellite images. The study developed a process-ruleset for object-based image analysis, which can be used for building footprint extraction from the very-high-resolution sensor with appropriate morphology-based modifications. The combination of supervised nearest-neighbor classification with decision-rules resulted in an accuracy of over 82.5% in the extraction of building footprints. The accuracy of feature extraction in densely populated areas is low due to the effect of spectral mixing of features and smaller size of objects. The results of the study will be useful in developing a 3D city model for Ahmedabad city and for assessment of changes in the built-up volume in the city.

References

1. Schneider A, Friedl MA, Potere D (2009) A new map of global urban extent from MODIS satellite data. Environ Res Lett 4:044003
2. Dahiya S, Garg PK, Jat MK (2013) A comparative study of various pixel-based image fusion techniques as applied to an urban environment. Int J Image Data Fusion 4:197–213
3. Jensen JR, Cowen DC (1999) Remote sensing of urban/suburban infrastructure and socio-economic attributes. Photogram Eng Remote Sens 65:611–622
4. Kpalma K, El-Mezouar MC, Taleb N (2014) Recent trends in satellite image pan sharpening techniques. In: 1st International conference on electrical, electronic and computing engineering. Vrniacka Banja, Serbia
5. Blaschke T, Burnett C, Pekkarinen A (2004) Image segmentation methods for object-based analysis and classification. In: Jong SMD, Meer FDV (eds) Remote sensing image analysis: including the spatial domain. remote sensing and digital image processing, vol 5. Springer, Dordrecht, pp 211—236
6. Lia, HT., Gua, HY., Hana, YS., Yanga, JH., Hanb, SS (2008) An efficient multi-scale segmentation for high-resolution remote sensing imagery based on statistical region merging and minimum heterogeneity rule. In: International archives of the photogrammetry, remote sensing and spatial information sciences, vol. XXXVII, Part B4. Beijing, pp 1257–1262
7. Im J, Jensen JR, Tullis J (2008) A object-based change detection using correlation image analysis and image segmentation. Int J Remote Sens 29:399–423
8. World cities population data. http://www.worldpopulationreview.com
9. Belfiore OR, Meneghini C, Parente C, Santamaria R (2016) Application of different pan-sharpening methods on worldview-3 IMAGES. J Eng Appl Sci 11:490–496
10. Myint SW, Gober P, Brazel A, Grossman-Clarke S, Weng Q (2011) Per-pixel versus object-based classification of urban land cover extraction using high spatial resolution imagery. Remote Sens Environ 115:1145–1161 (2011)
11. Lang S, Tiede D (2007) Definiens developer. GIS Bus 9(2007):34–37
12. Baatz M, Schape A (2000) Multiresolution segmentation: an optimization approach for high quality multi-scale image segmentation. In: Proceedings of the Angewandte Geographische Informationsverarbeitung XII. Beiträge zum AGIT Symposium. Salzburg, Austria
13. Hay GJ, Castilla G (2008) Geographic object-based image analysis (GEOBIA): a new name for a new discipline. In: Blaschke T, Lang S, Hay GJ (eds) Object-based image analysis. Lecture notes in geoinformation and cartography, pp 75–89. Springer, Berlin, Heidelberg
14. Willhauck G (2000) Comparison of object oriented classification techniques and standard image analysis for the use of change detection between SPOT multispectral satellite images and aerial photos. Int Arch Photogram Remote Sens, vol XXXIII, Part B3, pp 214–221
15. Ma L, Li M, Ma X, Cheng L, Du P, Liu Y (2017) A review of supervised object-based land-cover image classification. ISPRS J Photogram Remote Sens 130:277–293
16. Shackelford AK, Davis CH (2003) A combined fuzzy pixel-based and object-based approach for classification of high-resolution multispectral data over urban areas. Geosci Remote Sens IEEE Trans 41:2354–2363
17. Sharma SA, Agrawal R, Jayaprasad P (2016) Development of 3D city models using irs satellite data. Indian Soc Remote Sens 44(2):187–196

Role of Ground Control Points (GCPs) in Integration of Terrestrial Laser Scanner (TLS) and Close-range Photogrammetry (CRP)

Yogender, S. Raghavendra and S. K. P. Kushwaha

Abstract The need for GCPs is increasingly more important with the increase in higher accuracy requirements and increase in user expectations. GCPs (Ground control points) are necessary for orientation and placement of photographs and 3D models in the spatial coordinate system, and they play a key role in co-registration of two point clouds. This paper deals with the assessment of the role of Ground Control Points in Co-registration of CRP and TLS point clouds by point-pair selection methodology and Automatic co-registration algorithm. In this work, the point cloud is generated from multiple overlapping sequences of images using Close-range Photogrammetry (CRP) and Terrestrial Laser Scanning (TLS) for a building over the planar surface. GCPs were collected by the total station to register the TLS and CRP point cloud. Overlapping photographs are processed in Agisoft PhotoScan software. TLS point cloud was generated from Riegl VZ 400 and GCPs were used to geo-reference it in Cloud Compare software. Various subsets of both point clouds are co-registered by the point-pair co-registration method and by Automatic point detection fine co-registration. Two subsets for each of CRP and TLS point cloud are considered in such a way that one is having some common overlap and other is having no common feature. GCPs registered point clouds integrate precisely as compared to that of the non-registered point cloud. The RMS error achieved in case of geo-referenced point cloud co-registration was 0.0091645433 m and in non-geo-referenced co-registration was found to be 0.03327466 m. In this study, it was also found that error is significantly higher in Automatic point detection method as com-

Yogender (✉)
Department of Civil Engineering, National Institute of Technology Kurukshetra, Kurukshetra, India
e-mail: Yogender105@gmail.com

S. Raghavendra · S. K. P. Kushwaha
Photogrammetry and Remote Sensing Division, Indian Institute of Remote Sensing, ISRO, Dehradun, India
e-mail: raghav@iirs.gov.in

S. K. P. Kushwaha
e-mail: s.k.p.kushwaha92@gmail.com

J. K. Ghosh and I. da Silva (eds.), *Applications of Geomatics in Civil Engineering*, Lecture Notes in Civil Engineering 33, https://doi.org/10.1007/978-981-13-7067-0_42

pared to that of conventional point-pair selection co-registration. From this study, it is observed that higher accuracy of co-registration is achieved in the case of geo-referenced models by point-pair selection method. So, GCPs are the prerequisite for the effective and precise co-registration of 3D point clouds.

Keywords Close-range photogrammetry · Terrestrial laser scanning · Ground control points · Geo-referencing · Co-registration · Point cloud

1 Introduction

3D datasets are playing a vital role in analysing any modern targets virtually. Point cloud is a 3D dataset, which contains many numbers of points on the surface feature. In this research, point cloud has been generated using Terrestrial Laser Scanning (TLS) [1] and Close-range Photogrammetry (CRP) [2]. TLS is based on the principle of calculating the distance between the target and the laser by the time taken by the laser to the target and back multiplied by the speed of light divided by two [3]. CRP is based on the principle of epipolar geometry. Ground Control Points (GCPs) are the control points whose location is known and are helpful in geo-referencing the point clouds obtained from the TLS and CRP [4]. To cover the building, in this case, many scans were taken around the building and all the scans were from different locations with different origins, so to bring all the scans to a single coordinate system, co-registration of scans were done with the help of common tie points in the corresponding scans [5]. Automatic 3D point cloud registration is a primary issue in computer vision and remote sensing. One of the most commonly used arrangements is the outstanding Iterative Closest Point (ICP) algorithm [6]. Co-registered 3D models can be used for extracting fruitful information of the area under consideration. Therefore, the need of GCPs is more important with the increase in precision requirements and the rise in user expectations.

2 Study Site

For the present work, CSSTEAP building, IIRS (ISRO), Dehradun was chosen as a study area for this work. The location of the study site is shown in Fig. 1.

Fig. 1 CSSTEAP building, IIRS, ISRO, Dehradun

3 Data Acquisition

3.1 Close-range Photogrammetry

A total of 47 images were acquired all around the study area. The images were captured with Nikon D90 camera in which the focal length, lens aperture, and other parameters were kept fixed throughout the data collection process. During the entire data acquisition process, it was ensured that every two images must have some common overlap. It was tried to capture the images at a fixed distance all around the building (Figs. 2 and 3).

Fig. 2 Approach used for CRP data processing

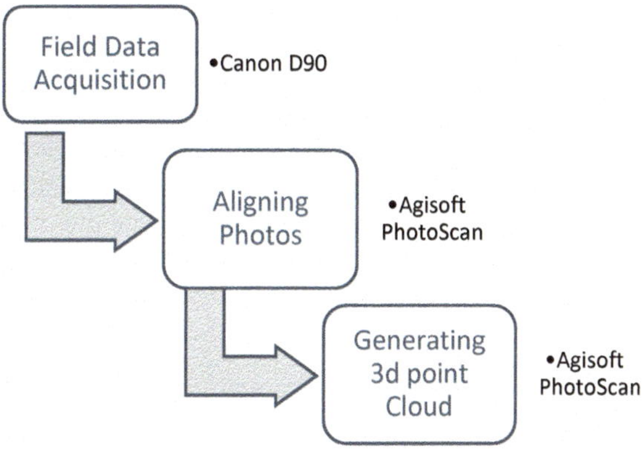

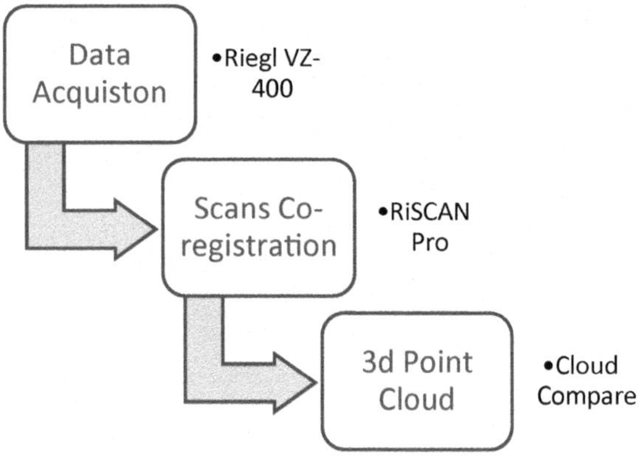

Fig. 3 Approach used for TLS data processing

3.2 *Terrestrial Laser Scanning*

For this study, Riegl VZ-400 laser scanner was used for the acquisition of laser data. Totally 12 scans were taken and were co-registered in RiSCAN Pro software. The distance of the instrument from the building was around 10–15 m. The scan resolution was set to be 0.02° for the detailed information of the study area and to minimise the complexity and scan time.

4 Data Processing

4.1 *CRP Data Processing*

The acquired overlapping images were aligned using image conjugate points using Structure from Motion (SfM) and then photogrammetric solutions were applied to generate the dense point cloud. CRP data processing was carried out with Agisoft Photoscan software.

CRP data processing has a very simple workflow as highlighted earlier. It is a four-step process involving the addition of photos, matching of points and image pairs, sparse cloud and dense cloud generation. In Photoscan, the accuracy was kept 'high' and pair preselection was kept 'generic'. For the generation of dense point cloud, the depth filtering was kept as 'moderate' and quality was kept 'high'. Generated dense point cloud might be altered and classified prior to export as.txt file or proceeding to 3D mesh model generation.

4.2 TLS Data Processing

First, all the TLS scans were co-registered in RiSCAN Pro software and then, poly data required was exported as *.txt* format file, which was further processed in Cloud Compare software.

4.3 Acquisition of Ground Control Points

Two base points were established and temporarily marked near the study area with the help of GPS in differential mode. Trimble R7 DGPS was used for this purpose. By orienting total station instrument Leica Pinpoint R1000 at these points, around 20–25 ground control points (GCPs) were taken over the planar surfaces which can be distinguished easily. It was tried to capture GCPs all over the study area. GCPs were identified by placing markers on the target points on the image of the study area.

4.4 Geo-referencing of Point Cloud

Both CRP and TLS models were in an arbitrary coordinate system. Point cloud generated through CRP and TLS were geo-referenced by manually aligning with the GCPs obtained earlier. Both the point clouds were geo-referenced in Cloud Compare software. Point-pair selection method was used to geo-reference both CRP and TLS models.

4.5 Subset Segmentation

Three similar subsets, each of CRP and TLS, were segmented in such a way that two of them have some common overlap and other subsets have no common feature. Similar point cloud subsets were segmented using the segment tool in Cloud Compare software. Different subsets of geo-referenced and non-geo-referenced models were segmented for further processing.

4.6 Co-registration

Subsets of geo-referenced and non-geo-referenced CRP and TLS models were co-registered using two algorithms in Cloud Compare software. The first method used was a conventional point-pair selection method in which point clouds were co-registered by manually picking four common align points of similar subsets of CRP and TLS. The second method used was automatic fine co-registration in which is based on a modified Iterative closest point (ICP) algorithm. The fine co-registration process was carried out for 50,000 points in the similar subsets. RMS error generated was considered as accuracy parameter for this work.

5 Results

5.1 Co-registration of Geo-referenced Point Cloud Subset

For the geo-referenced point cloud subset one in Fig. 4, the RMS error in co-registration was found to be 0.00929939 m and for section two, it was found to be 0.00415091 m. Subset one is registered using Point-pair selection method and the second one with ICP algorithm-based Automatic fine co-registration.

5.2 Co-registration of Non-geo-referenced Point Cloud Subset

For the non-registered point cloud subset one in Fig. 5, the RMS error in co-registration was found to be 0.0519184 m and for section two, it was found to be 0.0677572 m. Subset one is registered using Point-pair selection method and the second one with ICP algorithm-based automatic fine co-registration.

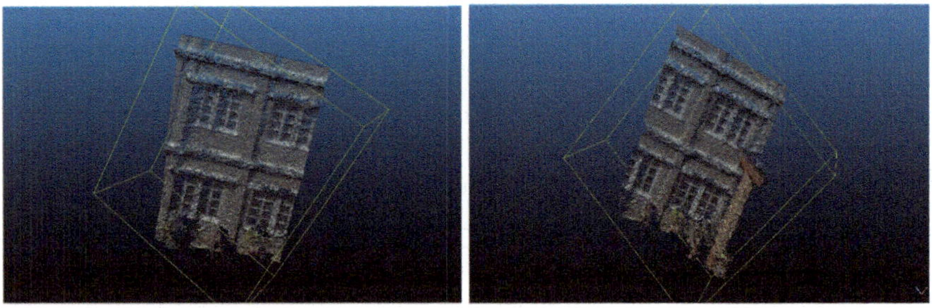

Fig. 4 Co-registered subsets of geo-referenced point cloud

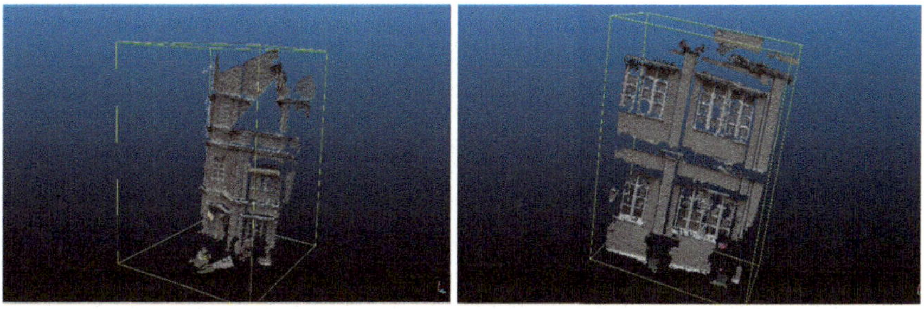

Fig. 5 Co-registered subsets of Non-geo-referenced 3D model

6 Conclusions

In this work, the assessment of the role of Ground Control Points in Co-registration of CRP and TLS point clouds by point-pair selection methodology and Automatic co-registration algorithm has been carried out for CSSTEAP building, IIRS, ISRO, Dehradun. GCPs registered point clouds integrate precisely as compared to that of the non-registered point cloud. The RMS error achieved in case of geo-referenced point cloud co-registration was 0.0091645433 m and in non-geo-referenced co-registration was found to be 0.03327466 m. In this study, it was also found that error is significantly higher in Automatic point detection method as compared to that of conventional point-pair selection co-registration. From this study, it is observed that higher accuracy of co-registration is achieved in the case of geo-referenced models by point-pair selection method. GCPs are the prerequisite for the effective and precise co-registration of 3D point clouds.

References

1. Burton G (2007) Terrestrial laser scanner, vol 4, pp 45–48
2. Luhmann T, Robson S, Kyle S, Harley I (2006) Close-range photogrammetry, vol 2
3. Pfeifer N, Briese C (2007) Norbert Pfeifer, Christian Briese. Scanning, pp 1–20
4. Tsakiri M, Lichti DD, Gordon SJ, Stawart MP, Franke J (2002) Comparison of digital photogrammetry and laser scanning, vol 0, January 2002, pp 39–44
5. Renaudin E, Habib AF, Kwak E, Chow J (2010) Optical images and terrestial laser scanning co-registration by the use of feature based methodology. Am Soc Photogramm Remote Sens Annu Conf 2010 Oppor Emerg Geospat Technol 2:2010
6. Gressin A et al (2013) Towards 3D lidar point cloud registration improvement using optimal neighborhood knowledge. ISPRS J Photogramm Remote Sens 79:240–251. https://doi.org/10.1016/j.isprsjprs.2013.02.019

Developing of Geoweb Application for Urban Planning

B. Sai Teja, K. Venugopal Rao, Y. Navatha, Reedhi Shukla and P. Sampath Kumar

Abstract The achievement of rapid growth that is both inclusive and sustainable presents formidable challenges for urban planning in India. Urban planning involves a lot of spatial implications. The planning process may have social, economic, and environmental effects. The emerging GIS techniques have various solutions for the urban planning. This paper presents the development of flexible and user-friendly geoweb application which has necessary tools that can be used as an alternative for the ease in urban planning. The main objective of this project is to develop a geoweb application for urban planning. And to develop a query builder so that it would be a useful tool for the urban planner and the users. In order to achieve these objectives, a detailed study was carried out to find the existing applications and technologies that are best suited for developing such a system and a methodology was prepared. After identifying software and technology stack required for the development, a spatial database containing all the data was prepared and published on GeoServer using Postgresql/PostGIS. A web interface was developed using HTML, CSS, JS, and PHP along with the OpenLayers to visualize and it was made responsive using the BOOTSTRAP. Hence, it can be used by various planners using the database and GIS tools in the applications for the town and country planning.

Keywords Geoweb · Urban planning · OpenLayers · GeoServer · PostGIS

B. Sai Teja · Y. Navatha
NIT Warangal, Warangal, India
e-mail: saiteja.ias@gmail.com

Y. Navatha
e-mail: navathayerram2002@gmail.com

K. Venugopal Rao (✉) · R. Shukla · P. Sampath Kumar
NRSC, Hyderabad, India
e-mail: venu_koppaka@nrsc.gov.in

R. Shukla
e-mail: reedhi_shukla@nrsc.gov.in

P. Sampath Kumar
e-mail: sampath_k@nrsc.gov.in

J. K. Ghosh and I. da Silva (eds.), *Applications of Geomatics in Civil Engineering*,
Lecture Notes in Civil Engineering 33, https://doi.org/10.1007/978-981-13-7067-0_43

1 Introduction

In the twenty-first century, the rapid growth of urban population is creating problems to the urban ecosystem. So, there is a need to solve these problems and make urban life comfortable. One remedy for this problem is developing the new urban areas. Hence, there is need for planning for these urban areas [1].

Nowadays, the advancement in technology and development provides an opportunity for the town and country planning authority and the city administration to plan and develop a fast-growing city [2]. Town and country planning has been introduced in India for a long time, mostly using the method of preparation and monitoring of development plans and town development of the 'blueprint' in which this method does the mapping in the drawing. This method is used to analyze potential and development problems, produce development plan, and review the town background [2]. However, this method makes it difficult to monitor the process of uncontrolled urban planning. Thus, the ability of Geographical Information System [GIS] is used to solve the problem of attribute data processing and spatial data simultaneously [3]. Then, GIS is a suitable technology tool to solve problems of urban planning [4].

Today, there is a lot of development in internet technology and web applications. The application of GIS has also been gradually upgraded to many forms like web GIS, mobile GIS. Thus, GIS is used for network applications. Extensive use of the Internet allows the spatial information resource on the network that is broadened continually thus making it accessible to a wide range of users. The main purpose of Web GIS is to realize the issue of spatial geographic information and data sharing on the Internet, to make the user be able to access GIS data through web browser, and to execute GIS functions like query, statistics, edit, filter, swapping, and many more.

Development of many technologies has given boost to the web GIS development. This emergence of the web GIS technique overcomes the shortcomings of traditional Distributed Object technique and provides the interoperable capability of cross-platform and cross-language in distributed environment [5]. The web services architecture establishes standard interconnection rules between services and information clients that nicely support the dynamic integration of data, which is the key to creating a spatial data infrastructure.

2 Data and Technology Used

The emerging new technologies and open source software's have made the geoweb applications to grow rapidly. The software development environment for geoweb application is also greatly simplified with easy-to-use development tools like Browser-side GIS APIs, such as the JavaScript Openlayer APIs, Geoext, Leaflet [6]. The OpenLayer API is emerging as one of the effective development environment for Geo Web 2.0 software applications. The GIS servers like Geoserver, Mapserver, and DEGREE are providing very simple to use tools for publishing spatial data into

Table 1 The tools and technology used are as follows

Activity	Software/tool used
GIS data creation, updating, editing etc.	Quantum GIS
Geospatial server Web server	Geoserver-2.12.1 Apache/Tomcat
Programming/Scripting	PHP, JavaScript
Styling of web application	CSS
Geospatial database	PostGreSQL-9.6.6, PostGIS
API	OpenLayers-2.13.1
Web responsive	BOOTSTRAP-4.0.0

various web services format in web browser environment. The geospatial database for importing the spatial data, PostgreSQL a powerful open source object relational database system is used with an extension of POSTGIS which provides spatial objects for PostgreSQL [7]. After a detailed study of various tools and technologies, the following tools are selected for development and deployment of web application for urban planning include various designing and programming languages along with mapping libraries frameworks, GIS software, and web servers in this study (Table 1).

3 Methodology

The development of geoweb application for urban planning for Haryana state. Spatial and nonspatial data of the study area is available and processed to obtain data that in a usable format. This involves conversion and correction of the data and the data is organized into PostgreSQL with spatial reference system EPSG 4326 with the help of extension POSTGIS to the database. This data are stored in the database according to the user's requirement and accessibility. The stored data is then published as service using the Geoserver enabling WMS (web mapping service) and WFS (web feature service) services.

The development of geoweb application is carried out using the open layers API using the programming scripts like PHP, JavaScript. A web interface is In order to develop the web interface an initial layout is to be prepared which helps in identifying where the different elements of the application have to be placed. An interface can be developed using HTML, CSS, JS, and PHP. Access to the data will then be enabled through the web interface which completes the development process.

The developed system will then be deployed through a webserver. The methodology adopted for the development process of the geoweb application for the urban planning is shown in the flow chart (Fig. 1).

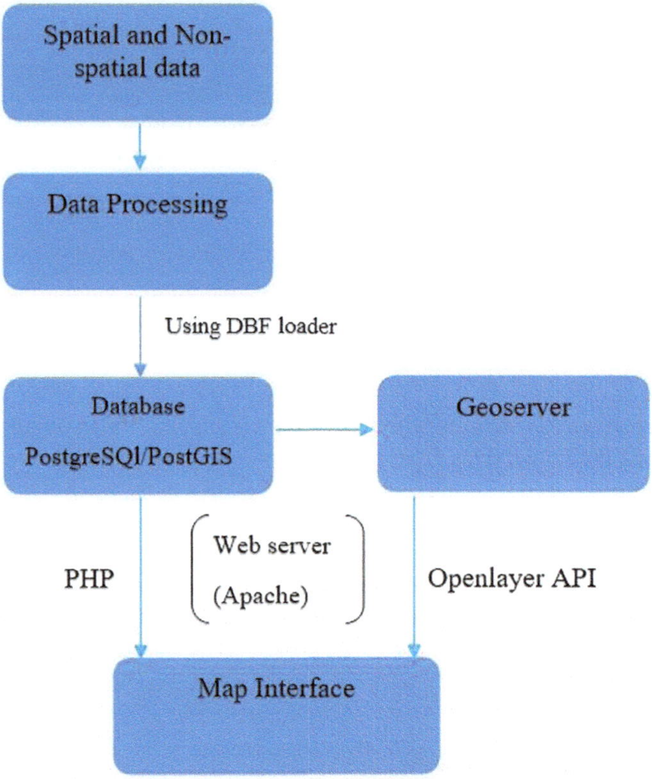

Fig. 1 Flow chart showing the methodology for developing of application

4 Results and Discussion

A geoweb application is developed considering various aspects of database, technology, architecture, and functionalities. Organizing and storing data at one place in a spatially enabled database such as PostgreSQL/POSTGIS. So that there would be easy access to the data from a single place. Figure 2 shows how the data is stored in PostgreSQL.

The data is maintained by modifying the data or by maintaining the versions of the data for updating the needs of the town planners, administrators and public. This is important to develop newer queries and analysis of the data.

Developing a geoweb application using open source tools and technologies. The emerging new trends in web technology and also the availability of various open source software have made to develop various applications provided at the user end without any dependency on commercial software. This has been a very cost-effective solution and also interoperable. A user-friendly interface has been developed (Fig. 3).

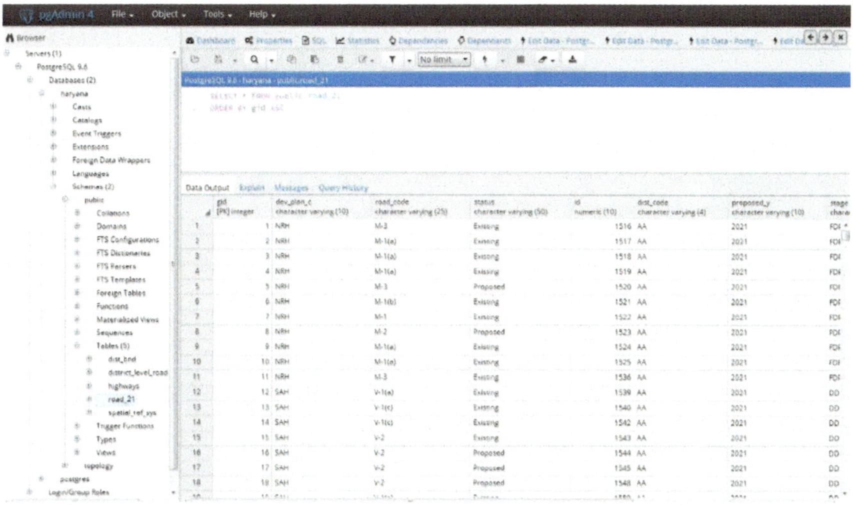

Fig. 2 Displaying the attribute data of a shapefile loaded in the database

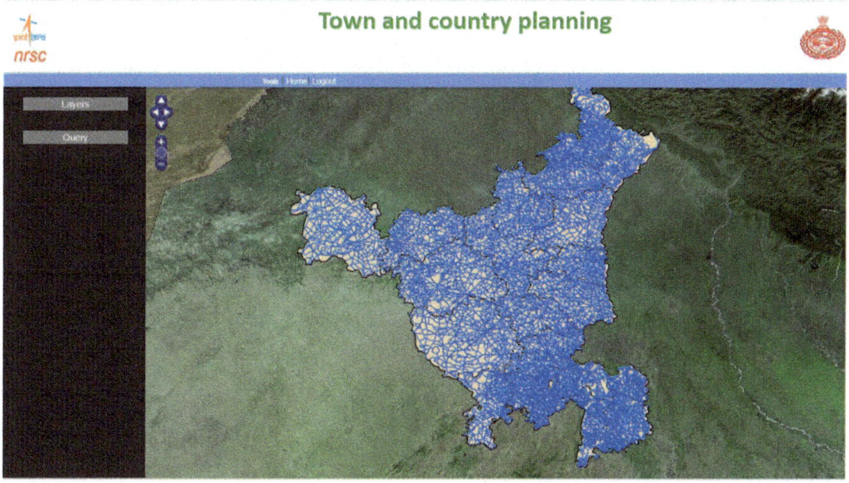

Fig. 3 Interface of the web application

Developing system specific and user specific queries. The user-specific queries such as identifying the infrastructure within a specified administrative boundary such as state or district.

Integrating database and web applications enables visualization, queries, analysis, and outputs. Geoweb applications are very useful for improving data management, visualization, and also the decision-making. As shown in Fig. 4, the data can be visualized after applying a query.

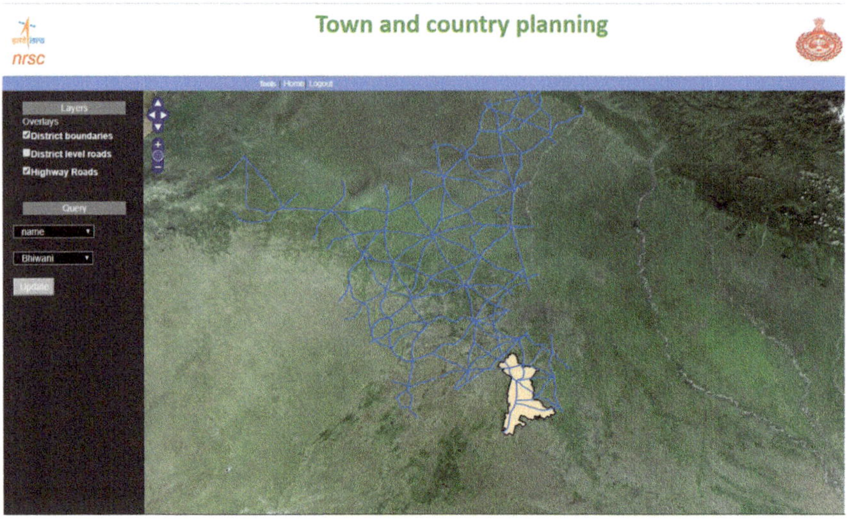

Fig. 4 Showing data after the query is applied

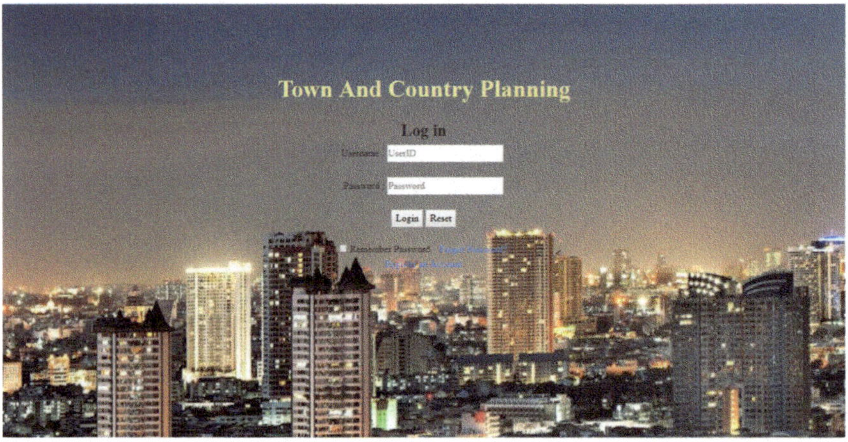

Fig. 5 Login page for the web application

Hierarchy-based login system for officials of the town and country planning and the public.

The official of the town and country planning can log in into the application and can create, alter, update, and query the data as the public users can only submit the reports (Fig. 5).

5 Conclusions

This application helps the officials of town and country planning for decision making and data management. Generally, data has to be submitted and approval of the data has to be done by the officials and this is a regular process. This application helps for communication among the town and country planning officials for day-to-day submission and approvals of data quickly. In future, this application becomes more robust as this maintains the transparency between the public and the government officials and quickens the decision-making process by the officials.

References

1. Chavare S, Pratapkumar UP (2013) Application of geoinformatics in urban planning and management. Indian Streams Res J 3(3)
2. Devaraju A, Herman NS, Sahib S (2017) Building of the enabled web-based GIS participation system: a tool to enhance community participation in city development plan. Faculty of information and communication technology, Technical University of Malaysia Melaka (UTeM)
3. Mondino EB, Barbara D (2016) GIS advanced tools for urban growth reading and management for best practices in town planning
4. Selamat MH, Selamat A, Othman MS, Shamsuddin NH, Zukepli NI (2012) A review on geographical information system (GIS) in town planning: Malaysia experience. Geoinf Int J (GIIJ)
5. Gupta A, Sharma MP, Snehmani MS (2014) Development of web GIS based digital avalanche Atlas. Int J Sci Eng Technol Res (IJSETR) 3(4)
6. Kumar S, Karnatak H, Shukla R, Jha CS (2012) Geo-web application for greening India: an open source technology development, OSGEO-India: FOSS4G 2012. In: First national conference, open source geospatial resources to spearhead development & growth
7. Karnatak HC, Shukla R, Sharma VK, Murthy YVS, Bhanumurthy V (2012) Spatial mashup technology and real time data integration in geo-web application using open source GIS—a case study for disaster management, Geocarto international
8. Karnatak H, Shukla R, Sampath Kumar P, Raghavaswamy V, Krishnamurthy YVN (2012) Online updation of city master plan using ISRO bhuvan services and open source software in web browser environment. In: OSGEO-India: FOSS4G 2012, first national conference, open source geospatial resources to spearhead development and growth
9. Karnatak H, Shukla R, Sampath Kumar P, Raghavaswamy V, Krishnmurthy YVN (2012) Online updation of city master plan using ISRO Bhuvan services and open source software in web browser environment OSGEO-India: FOSS4G 2012. In: first national conference, open source geospatial resources to spearhead development & growth
10. Karnatak H, Singh H, Shukla R, Saran S (2012) Development of GeoWeb application using open source technology: an innovative approach for disaster mitigation & management, OSGEO-India: FOSS4G 2012. In: First national conference, open source geospatial resources to spearhead development & growth

A Proposed Framework Approach for Mapping Glacier Hazard Zones

Rahul Nijhawan and Josodhir Das

Abstract The main aim of this study is to propose a new ensemble framework model for the prediction of snow avalanche prone regions. The proposed approach is based on the following pillars: Support vector machine classifier (SVM) and recent state-of-the-art ensemble learning techniques, Bagging and MultiBoost. On the basis of current literature, it has been observed that such techniques have been rarely used in this field. The study was conducted on the surrounding region of Gomukh, Uttarakhand, India. Several parameters were computed on which the base classifier was trained, viz., slope, aspect, surface curvature, precipitation, surface temperature, and snowfall. It was observed that the proposed ensemble frameworks outperformed with the current state-of-the-art SVM classifier. The highest classification accuracy was observed by the MultiBoost ensemble framework (92.61%), followed by Bagging (88.93%), while the lowest classification accuracy of (78.98%) was produced by artificial neural network (ANN) classifier. Accuracy assessment was performed and the proposed models were evaluated using receiver operating characteristics (ROC) curves and statistical measures.

Keywords SVM · Ensemble · MultiBoost · Bagging

1 Introduction

Sudden flow of a massive volume of ice or snow down a cliff is called an avalanche. Several parameters play a major role in the occurrence of an avalanche such as surface temperature, slope direction, slope steepness, climate conditions, and vegetation of the area, terrain conditions and snow pack conditions. Avalanche basically occurs when the amount of snow being deposited as a result of snowfall is such that it fractures the most fragile portion of snow on the slope [1]. Stability in snowpack structures could be easily studied from the snowpack parameters [1, 2]. Many studies are performed to estimate the mass balance and energy-related processes occurring

R. Nijhawan (✉) · J. Das
Department of Earthquake Engineering, Indian Institute of Technology Roorkee, Roorkee, India
e-mail: rahulnijhawan2010@gmail.com

J. K. Ghosh and I. da Silva (eds.), *Applications of Geomatics in Civil Engineering*, Lecture Notes in Civil Engineering 33, https://doi.org/10.1007/978-981-13-7067-0_44

in snow cover using the meteorological variables [3–5]. Analyses for the release of wet snow avalanche, temporally was done by designing an expert system by [6]. Knowledge base for the expert system was created by analyzing the changing behavior of wet snow avalanche at Italian Ortles Alps and was illustrated by fuzzy logic rule base. Amount of deposition of wet snow avalanche served as the output for the inference method. Parameters of the release area, aspect, slope, elevation, and surface curvature were analyzed by [7], and fuzzy set theory algorithms were used for analysis. Potential release areas were marked on the basis of terrain features obtained by knowledge as given by [8]. Terrain characteristics, forest cover, and amount of snow fall impact snow avalanche as was observed by [9]. Snow depth was measured by a combination of in situ and remotely sensed data by [10]. Assessment of avalanche hazard was also done by spatially linking Bayesian networks to GIS by [11]. As snow cover is highly dynamic both spatially and temporally so short observation cycles are preferred. Due to inaccessibility to certain snow-covered areas, accurate data collection becomes tough by field visit. While with the help of remote sensing large snow-covered area could be easily accessed at different spatial resolutions as per the requirement [12]. Remote sensing and GIS have been widely used to study terrain parameters based hazard zonation [13, 14].

Recently, ensemble techniques have become very popular among researcher as of their capability to improve the classification accuracy. These methods are data mining approaches. They utilize the predictive machine learning models in order to merge several base classifiers in order to improve the classification accuracy compared to single classifier. They have been observed to give very promising results in near research as confirmed by [15]. Some of the most widely used ensemble algorithms include AdaBoost, Bagging, Dagging for classifying the real world problems. More recent ensemble algorithms include MultiBoost, Random Subspace, and Rotation Forest. Use of such techniques has been hardly made in the field of mapping avalanche-prone regions. While these techniques have been widely used in the field of computer science [16], medical [17], and banking [18].

The main aim of this study is to analyze and compare the performance of different ensemble frameworks using support vector machine and ensemble techniques (MultiBoost and Bagging) for prediction of avalanche-prone zones in the surrounding region of Gomukh, Uttarakhand. The models used are evaluated by means of receiver operating characteristic curve and statistic measures. Analysis is done and training parameters are computed using Arc Gis 10.2 and MATLAB 2013 software.

2 Study Area

The study area covers surrounding region of Gomukh, Uttarakhand, Himalayas in India as shown in Figs. 1 and 2.

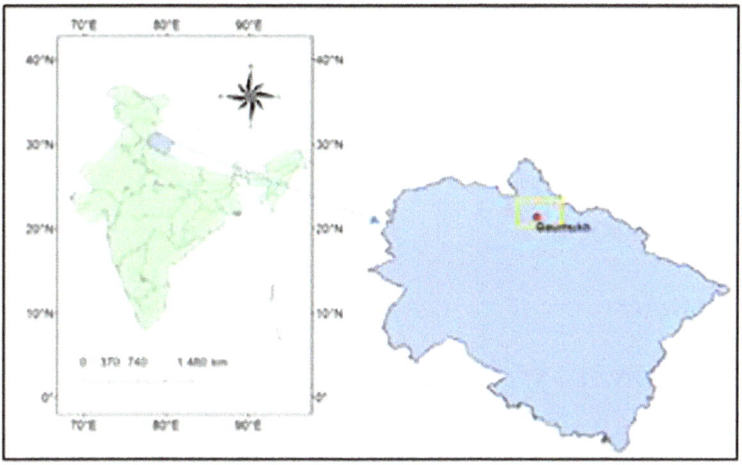

Fig. 1 Location of the study area (Surrounding region of Gomukh, Uttarakhand, Himalayas)

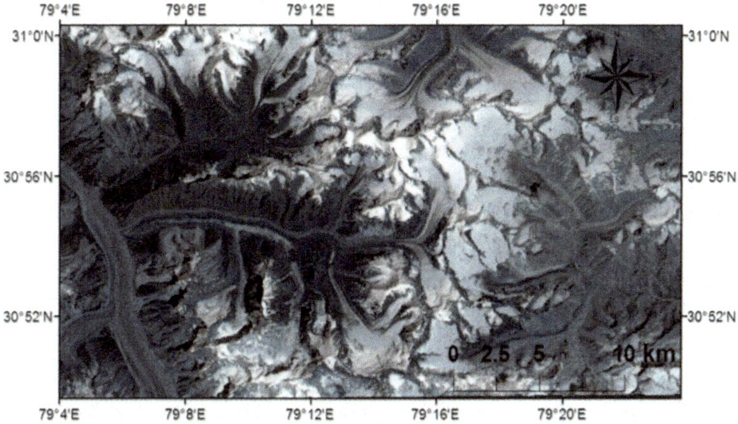

Fig. 2 Landsat ETM + multispectral scene of the study area

3 Data Sources

The LANDSAT satellite data (cloud free) used for the study was collected from the USGS website. SRTM digital elevation models (spatial resolution 30 m) were downloaded for the study area from the USGS/Bhuvan website. Details of the data used in the study are given in Table 1.

Table 1 Details of the data sources

Source	Scene ID	Date	Spatial resolution
LAND SAT 8	"LC81450392015148LGN00"	28-05-15	Visible, NIR, SWIR—30 m
SRTM	SRTM1N30E079V3	23-09-14	30 m

4 Methodology

4.1 Classifiers

4.1.1 Support Vector Machine (SVM)

The recent study has indicated that SVM-based approaches on remote sensing studies have high supervised classification potential. Studies have shown that SVM has much high classification capability compared to the other popular approaches including artificial neural networks and decision trees, also compared to the traditionally used classifiers such as maximum likelihood classification [19]. They were originally structured for performing binary classification but several approaches were developed to extend it for multiclass classification [20].

SVM aims at fitting most optimal hyperplane separating the classes by means of training samples present at boundaries of class distributions known as support vectors. While remaining samples in the training data are ignored as do not yield any information for hyperplane location [21]. So, high accuracy is achieved with minimum set of training data and high degree of generalizability.

SVM is generally applied in situations where the two classes could be separated linearly in a q dimensional space by means of training data given as $\{x_i, y_i\}$, where i = 1, …, r, and $y_i \in \{1, -1\}$. The aim is to develop a classifier that has accurate generalization capability. Several hyperplanes can be inserted for separating the classes in the feature space, but one among them has the capability to generalize accurately. This hyperplane lies separating the two classes such that all the samples of the class lie on one side for the second class on the other side. Structure is such that the distance of the closest training data for both the classes is maximum.

Hyperplane is given by the equation $wx + b = 0$, where w is perpendicular to the hyperplane, x is simply a point present on the hyperplane, and b represents the bias. For two classes the hyperplane is defined as $wx_i + b \geq +1$, for $y_i = +1$ and as $wx_i + b \leq -1$, for $y_i = -1$. Both equations together could be represented as

$$y_i(wx_i + b) - 1 \geq 0$$

On these two hyperplanes, the training samples are known as support vectors.

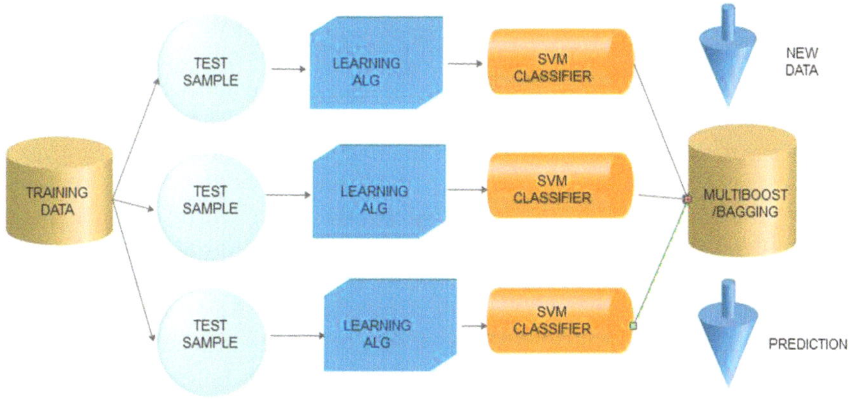

Fig. 3 Flow diagram of proposed ensemble framework

4.2 Ensemble Learning Frameworks

4.2.1 Bagging

This technique was first brought into picture by Breiman [22], one of the initial ensemble framework. Each classifier is driven by means of bootstrap samples. The first stage is forming sub-training subsets from the available learning sets by means of random sampling with replacement approach. Base classifiers are then trained using these sub-training sets. Then, the results of these base classifiers are merged using weighted majority voting.

4.2.2 MultiBoost

This approach was introduced by which originates from the combination of Wagging and AdaBoost. Its main purpose is to reduce both bias and variance and avoid overfitting problem. First, AdaBoost technique is used to allot weight values to instances, and thereafter Wagging technique is worked through to change the weight values based on the past classifier performance [23] (Fig. 3).

5 Results and Discussion

Validation dataset that was not used in the training phase is used for evaluating the performance of the three models. It could be seen from the table that AUC of 0.921 was observed for SVM with MultiBoost, with classification accuracy of 92.61% which was near ahead of SVM with Bagging model with AUC 0.886 and classification accuracy 88.93%, followed by SVM with AUC 0.865 and classification accuracy 82.83% and RF with AUC 0.842 and accuracy 80.11%. While the minimum classification accuracy (78.98%) was observed from current state-of-art model with AUC value of 0.831. ROC curves for first three models are shown in Fig. 4. The complete details of the statistics are shown in Table 2 for training phase and Table 3 for validation phase. It could be seen that the highest value of sensitivity (80.91%) was observed by SVM model indicating the probability to rightly classify the pixels belonging to avalanche zone to the avalanche-prone class, followed by SVM with MultiBoost with sensitivity of (80.88%). The lowest sensitivity (71.45%) was observed from SVM with Bagging model. Whereas the highest specificity (90.75%) was observed by SVM with MultiBoost model indicating 90.75% of pixels which are not affected by avalanche, belongs to non-avalanche prone class, which was followed by SVM with Bagging model (89.76%). Whereas the lowest specificity (79.54%) was observed

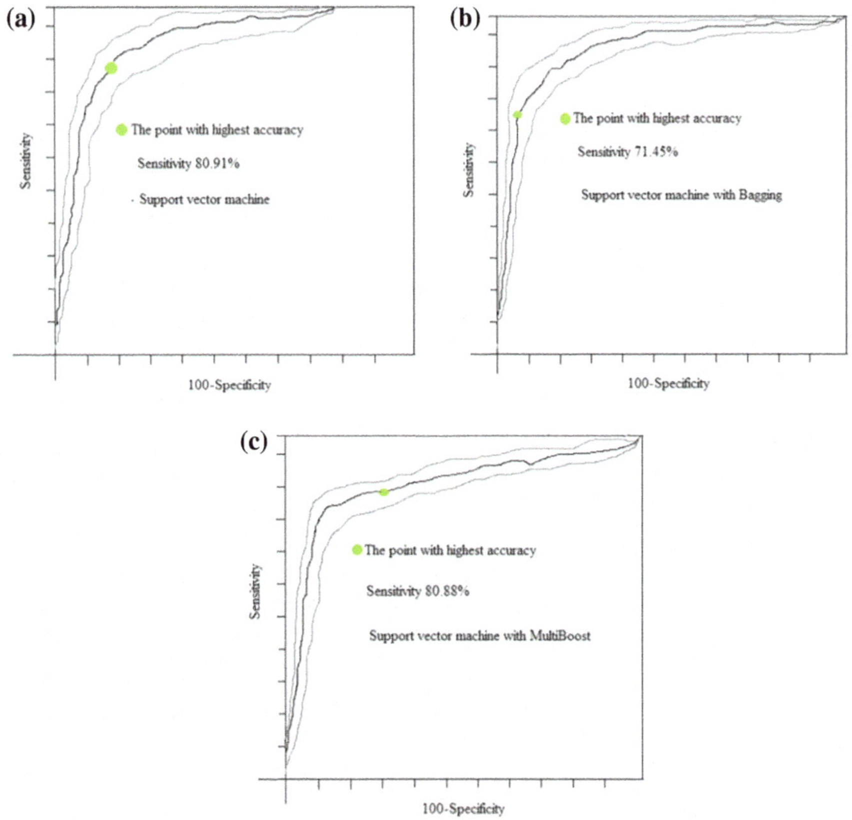

Fig. 4 Represent ROC curves for validation of three set of models. **a** SVM, **b** SVM with bagging, and **c** SVM with MultiBoost

Table 2 Results obtained for training for the SVM-Bagging-MultiBoost models

Parameters	ANN	RF	SVM	SVM with bagging	SVM with MultiBoost
TP (True positive)	2032	2212	2312	2323	2413
TN (True negative)	1988	2098	2109	2208	2245
FP (False positive)	276	289	306	243	234
FN (False negative)	160	180	189	190	95
Specificity (%)	80.214	81.234	82.432	86.987	87.965
Sensitivity (%)	85.543	86.321	89.23	88.54	92.012
Accuracy (%)	81.54	83.12	85.98	87.12	90.45
Kappa index	0.687	0.701	0.712	0.723	0.845
AUC	0.867	0.891	0.902	0.928	0.974

Table 3 Model validation

Parameters	ANN	RF	SVM	SVM with bagging	SVM with MultiBoost
TP (True positive)	667	678	735	587	754
TN (True negative)	654	698	756	865	824
FP (False positive)	78	87	98	57	67
FN (False negative)	73	79	100	205	76
Specificity (%)	79.54	81.32	83.77	89.76	90.75
Sensitivity (%)	76.98	78.88	80.91	71.45	80.88
Accuracy (%)	78.98	80.11	82.832	88.93	92.61
Kappa index	0.667	0.689	0.697	0.635	0.742
AUC	0.831	0.842	0.865	0.886	0.921

for ANN model. Based on the mentioned results, it was concluded that SVM with MultiBoost was the best proposed model for prediction of avalanche-prone regions.

6 Conclusion

In this paper, we have proposed a new ensemble framework model for prediction of snow avalanche prone regions. The framework is based on SVM classifier and state-of-the-art ensemble algorithms, i.e., Bagging and MultiBoost. In this paper, we employed Landsat 8 satellite imagery for surrounding regions of Gomukh, Uttarakhand, India. The base classifier was trained on several morphometric (slope, aspect, surface curvature) and other relevant parameters. The proposed framework outperformed the state-of-the-art (ANN, RF, SVM) classifiers. In general, employing the proposed ensemble architecture produced an infrastructure for future practical use of remote sensing resources for mapping of snow avalanche prone areas in the Himalayan region.

References

1. McClung D, Schaerer PA (2006) The avalanche handbook. The Mountaineers Books
2. Colbeck ILEJRM et al (1990) The optical properties and morphology of cloud-processed carbonaceous smoke. J Aerosol Sci 21(4):527–538
3. Anderson EA (1976) A point of energy and mass balance model of snow cover. NOAA technical report NWS 19, pp 1–150
4. Dozier J, Outcalt SI (1979) An approach toward energy balance simulation over rugged terrain. Geogr Anal 11(1):65–85
5. Marks D, Winstral A (2001) Comparison of snow deposition, the snow cover energy balance, and snowmelt at two sites in a semiarid mountain basin. J Hydrometeorol 2(3):213–227
6. Zischg A et al (2005) Modelling the system behaviour of wet snow avalanches using an expert system approach for risk management on high alpine traffic roads. Natural Hazards Earth Syst Sci 5(6):821–832
7. Ghinoi A, Chung C-J (2005) STARTER: a statistical GIS-based model for the prediction of snow avalanche susceptibility using terrain features—application to Alta Val Badia, Italian Dolomites. Geomorphology 66(1):305–325
8. Maggioni M, Gruber U (2003) The influence of topographic parameters on avalanche release dimension and frequency. Cold Reg Sci Technol 37(3):407–419
9. McClung DM (2001) Characteristics of terrain, snow supply and forest cover for avalanche initiation caused by logging. Ann Glaciol 32(1):223–229
10. Stoffel M, Bollschweiler M, Hassler G-R (2006) Differentiating past events on a cone influenced by debris-flow and snow avalanche activity—a dendrogeomorphological approach. Earth Surf Proc Land 31(11):1424–1437
11. Grêt-Regamey A, Straub D (2006) Spatially explicit avalanche risk assessment linking Bayesian networks to a GIS. Natural Hazards Earth Syst Sci 6(6):911–926
12. Singh P (2001) Snow and glacier hydrology, vol 37. Springer Science & Business Media
13. Barbolini M, Keylock CJ (2002) A new method for avalanche hazard mapping using a combination of statistical and deterministic models. Natural Hazards Earth Syst Sci 2(3/4):239–245
14. Gruber U, Margreth S (2001) Winter 1999: a valuable test of the avalanche-hazard mapping procedure in Switzerland. Ann Glaciol 32(1):328–332
15. Pham BT et al (2015) Landslide susceptibility assessment at a part of Uttarakhand Himalaya, India using GIS-based statistical approach of frequency ratio method. Int J Eng Res Technol 4:338–344
16. Abawajy JH, Kelarev A, Chowdhury M (2014) Large iterative multitier ensemble classifiers for security of big data. IEEE Trans Emerg Topics Comput 2(3):352–363
17. Kelarev AV et al (2012) Empirical study of decision trees and ensemble classifiers for monitoring of diabetes patients in pervasive healthcare. In: 2012 15th international conference on network-based information systems (NBiS), IEEE
18. Chen N, Ribeiro B, Chen A (2015) Comparative study of classifier ensembles for cost-sensitive credit risk assessment. Intel Data Anal 19(1):127–144
19. Foody GM, Mathur A (2004) A relative evaluation of multiclass image classification by support vector machines. IEEE Trans Geosci Remote Sens 42(6):1335–1343

20. Hsu C-W, Lin C-J (2002) A comparison of methods for multiclass support vector machines. IEEE Trans Neural Netw 13(2):415–425
21. Belousov AI, Verzakov SA, Von Frese J (2002) A flexible classification approach with optimal generalisation performance: support vector machines. Chemometr Intell Lab Syst 64(1):15–25
22. Breiman L (1996) Bagging predictors. Mach Learn 24(2):123–140
23. Webb GI (2000) Multiboosting: a technique for combining boosting and wagging. Mach Learn 40(2):159–196

Snow Cover Analysis in Chandra Basin of Western Himalaya from 2001 to 2016

Rakesh Sahu and R. D. Gupta

Abstract Snow cover and glaciers play an important and significant role in high-altitude areas. Chandra Basin, a subbasin of Chenab Basin, is situated in Lahaul–Spiti district, Himachal Pradesh, Western Himalaya and lying on the monsoon-arid transition region. To study the snow cover variation in Chandra basin from 2001 to 2016 for winter period (November–April), MODIS 8-day Terra (MOD10A2) data product is used. A Python script (Arcpy and Numpy package) is used to estimate the snow cover variation in Chandra Basin during the study period. This analysis shows that maximum average snow cover during 2001–2016 was in March (96.94%) and minimum in November (76.93%). Linear-trend-based analysis of snow cover area indicates that it is shrinking at the rate of 0.12% per year. The total reduction in snow cover area in the study area during 2001–2016 is 2.00%.

Keywords Snow cover · MODIS · Chandra Basin

1 Introduction

Snow cover and glaciers play an important and significant role in the Himalayan region to provide a natural source of water reservoirs as well as a large amount of freshwater over the year. Snow-covered area based study is important for better understanding the response of climate change and its variability, as the contribution of glaciers melt ice and melted snow cover in source of freshwater availability on earth's surface is 70% [1]. In Himalayan region along with streamline generation, snow-covered region and melted glacier ice play a significant and important role in generation of hydropower, to understand the local climate variability, strate-

R. Sahu (✉)
GIS Cell, Motilal Nehru National Institute of Technology Allahabad, Prayagraj 211004, India
e-mail: rgi1503@mnnit.ac.in

R. D. Gupta
Civil Engineering Department and Member in GIS Cell, Motilal Nehru National Institute of Technology Allahabad, Prayagraj 211004, Uttar Pradesh, India
e-mail: rdg@mnnit.ac.in

J. K. Ghosh and I. da Silva (eds.), *Applications of Geomatics in Civil Engineering*,
Lecture Notes in Civil Engineering 33, https://doi.org/10.1007/978-981-13-7067-0_45

gic future planning, and other developmental activities [2]. Study of snow cover is important as it is a component of climatological and surface balance system of area region [3, 4]. Various studies have been done which shows rising and falling trend of snow cover mapping in all over world based on climate change [5–11]. Information regarding snow cover extent in a particular basin is used to estimate the amount of water that it will produce during snow melting process [12]. In Himalayan region, seasonal- and monsoonal-based precipitations heavily affect the climate variability of the Himalayan river flow as well as the availability and distribution of water [13, 14]. Therefore, information regarding snow cover extent is significant and important component of earth system analysis.

Estimation of accurate Snow Cover Area (SCA) in Himalaya region is a time-consuming and tedious task due to its rugged topography. Remote sensing satellite data and related techniques provide a promising idea to analysis the spatial- and temporal-based changes of snow cover. Using space-borne sensors, effective snow cover mapping started since the early 1980s [15]. Various medium resolution sensors, i.e., GEOS, SPOT, AWiFS, Landsat Multispectral Scanner System (MSS), and Thematic Mapper (TM) have been used for analysis of snow cover variability in the past decade [16, 17]. These sensors have some limitations like small swath, band saturation, and inappropriate temporal and spectral resolutions [17–19]. Since the initiation of Moderate Resolution Imaging Spectroradiometer (MODIS) in year 1999 [20], this data has been frequently and globally used for measuring the spatial and temporal change analyses of SCA [10, 11, 21, 22]. The different products of MODIS data, i.e., land surface temperature and snow cover extent have been used in many studies and show good accuracy as compared to other sensors [23–25]. Due to rugged topography in different portions of Himalayan region, MODIS snow product may have variation in accuracy [26]. MODIS snow product, i.e., MOD10 is cautiously used in hydrological applications due to its less accuracy in transition snow cover region where melting of snow is fast, and nevertheless it plays a significant role in climatological studies [27].

The main aim of this research paper is to derive SCA variation in Chandra Basin during 2001−2016 using satellite data and remote sensing techniques. The present study also analyzes the temporal variation of snow cover in Chandra Basin.

2 Study Area and Data Used

2.1 Study Area

Chandra Basin is a subbasin of Chenab Basin situated in the northern ridge of Pir Panjal range of the Himalaya, in Lahaul–Spiti district, Himachal Pradesh, Western Himalaya. Latitudinal and longitudinal extensions of the study area range from 32°03′35″N to 32°45′50″N and 77°15′35″E to 77°50′50″E, respectively. The total area of Chandra Basin is ~$2.44*10^2$ km^2 and elevation range lies from 2800 to 6600 m

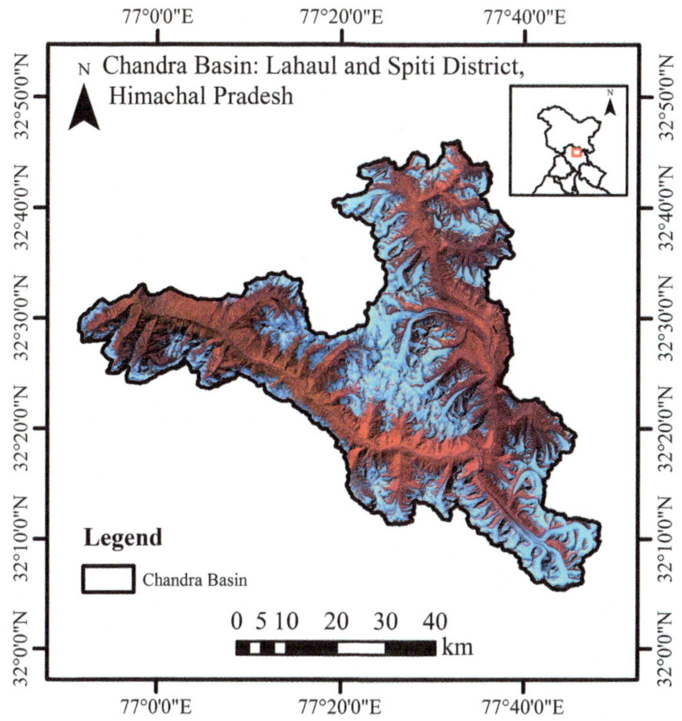

Fig. 1 Locational map of Chandra Basin with Landsat ETM+ as background image

amsl [28]. Total 201 glaciers lie in this basin and cover 29.54% of the basin area. Largest glaciers in terms of area in this basin are Bara Shigri and Samudra Tapu.

The Lahaul–Spiti valley region, falling in monsoon-arid transition zone, is heavily covered with snow and glaciated ice and receives precipitation due to Indian monsoons in summer and western disturbance in the winter [29]. The climate of this region is very cold, alpine, and glacial due to its high altitude and the region is not reachable during the winters [28]. In Chandra Basin, less metrological data is available which is mostly at Chhota Shigri Glacier showing the temperature and precipitation information [30, 31]. Dobhal et al. [30] reported the meteorological observations during 1987–1989 in Chhota Shigri Glacier for summer months (July–September) and indicated temperatures variation from −5.2 to +10.5 °C at 4600 m amsl. Figure 1 represents the locational map of Chandra Basin (the study area).

Azam et al. [31] discussed the climate setup based on automatic weather station installed in Chhota Shigri Glacier located at 4863 m amsl from October 1, 2009 to September 30, 2013, which shows January as a coldest month (−15.8 °C) and August as a warmest month (+4.3 °C). Geologically, the region is mainly determined by metamorphic rocks mostly of meso- to keta-zonal metamorphites, migmatites, and gneisses [32].

2.2 Satellite Data Used

MODIS 8-day Terra (MOD10A2) data product has been used for snow cover mapping. The Terra (MOD10A2) satellites used to generate snow cover product are launched by National Administrative Space Application (NASA). Terra (MOD10A2) data product was acquired during the study period from 2001 to 2016 from NASA website http://reverb.echo.nasa.gov/reverb. 8-day Terra (MOD10A2) data product is available since February 24, 2000 and satellite is still working. MODIS 8-day Terra (MOD10A2) data product available at 500 m resolution and comprises data of maximum snow cover extend over an eight-day period [24]. Acquired data format is Hierarchical Data Format-Earth Observing System (HDF-EOS), which also includes equivalent metadata. One tile of MOD10A2 data covers 1200 × 1200 km area and the acquired data is available in a sinusoidal map projection system [20].

Snow cover MODIS products were generated using Normalized Difference Snow Index (NDSI)-based algorithm and other criteria test [16]. In NDSI-based algorithm, generally four bands (band 1, band 2, band 4, and band 6) are used from given MODIS bands to create the land surface image (Table 1).

Reflectance value of snow is high in the visible region and absorption value is high in the short-wave region of electromagnetic spectrum. For NDSI value, MODIS Terra product reflection in band 4 (0.545–0.565 μm) and absorption in band 6 (1.628–1.652 μm) are used [16, 24, 33] (Eq. 1).

$$\text{NDSI} = (\text{band 4 - band 6})/(\text{band 4 + band 6}). \quad (1)$$

The generated NDSI value using satellite image for cloudy pixel will be very small, whereas snow-based pixel will be high. Hereafter, the NDSI algorithm also works as snow cloud pixel identification. Various studies suggested a threshold value (NDSI > 0.4) [16–18, 33].

Table 1 MOD10A2 (MODIS) bands used for snow cover product generation

Band No.	Bandwidth range (μm)
1	0.62–0.67
2	0.841–0.876
3	459–479
4	0.545–0.565
5	1230–1250
6	1.628–1.672
31	10.780–11.280
32	11.770–12.270

3 Methodology

The methodology adopted for creation of snow cover map is shown in Fig. 2 with the help of flowchart. Earth Observing System (EOS) data gateway has been used to download the MOD10A2 (Terra) data. In this study, the study area lies in MODIS product tile h24v05 (V005). MOD10A2 (Terra) snow data has not been directly used in sinusoidal map projection system.

For processing purpose, the downloaded data file is converted into UTM projection (UTM 43 N zone) with datum being WGS84. The study area is then clipped. These clipped images have different coded pixel values which divide the area into different classes. The different values of different classes are defined in Table 2.

Fig. 2 Methodology for creating snow cover map

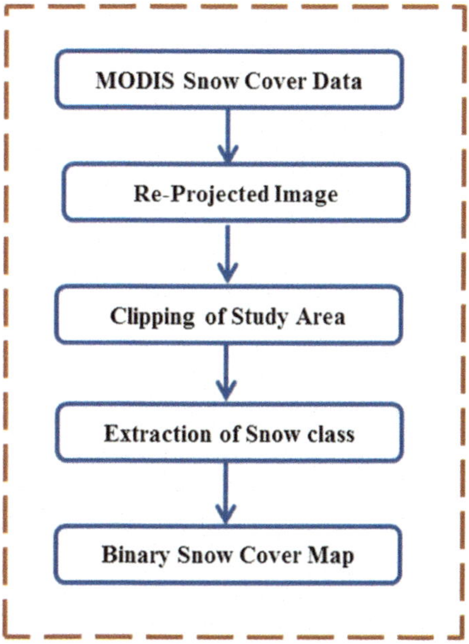

Table 2 MODIS Terra bands used to generate the MODIS snow cover products

Coded pixel values	Definition
0	Missing information
1	Without decision
11	Night time
25	No snow cover
37	Lake area
39	Ocean region

(continued)

Table 2 (continued)

Coded pixel values	Definition
50	Cloud cover
100	Lake ice covered by snow
200	Snow region
254	Detector is saturated
255	Fill value

These classes are generated by snow detection algorithm and pixel data are classified as snow cover, lack area, snow-covered lack ice, ocean region, cloud cover, and other classes. So, to separate snow cover pixel, we filter the images by giving pixel value as 200 and generate a snow-covered binary map [10]. Binary map contains two classes: snow and no snow which are shown in Fig. 3a, b for years 2001 and 2016, respectively. Total 362 binary maps at an interval of 8 days for 6 months in a year (November to April) for 16 years from 2001 to 2016 were generated for SCA determination. The whole process is done using Python script (Arcpy and Numpy). Finally, results are analyzed to draw meaningful conclusions, followed by the creation of different graphs and tables.

4 Results and Discussion

In this study, intra-annual SCA variation during the period 2001−20016 is carried out, and is shown in Fig. 4. It can be observed that SCA gradually increases from November to March, and then it decreases in April. The maximum average snow cover occurrence during 2001−2016 was in March (96.94%) and minimum in November (76.93%). The mean seasonal SCA during 2001−2016 of the study area was also derived. It is found that SCA variation was maximum in 2001−2002 (95.80%) and minimum in 2012−2013 (78.13%). Total SCA in winter was observed as 87.97% of basin area during 2001−2016.

Linear trend analysis of SCA is also carried out and is given in Fig. 5. It is analyzed that SCA is shrinking at the rate of 0.12% per year and an overall reduction in SCA is 2.00% in the study area during 2001−2016.

Snehmani et al. (2016) also derived mean seasonal SCA during 2001−2012 in Bhaga Basin using 8-day MOD10A2 and MYD10A2 snow cover products and found maximum in 2001–2002 (97.8%) and the minimum in 2005–2006 (86.1%). Maximum mean seasonal SCA variation is the same in Chandra and Bhaga Basin in the year 2001. Intra-annual SCA variation pattern is the same in Chandra Basin as in Bhaga Basin [10], the upper part of Sutlej Basin [11], and upper Bhagirathi Basin [34] in Winter.

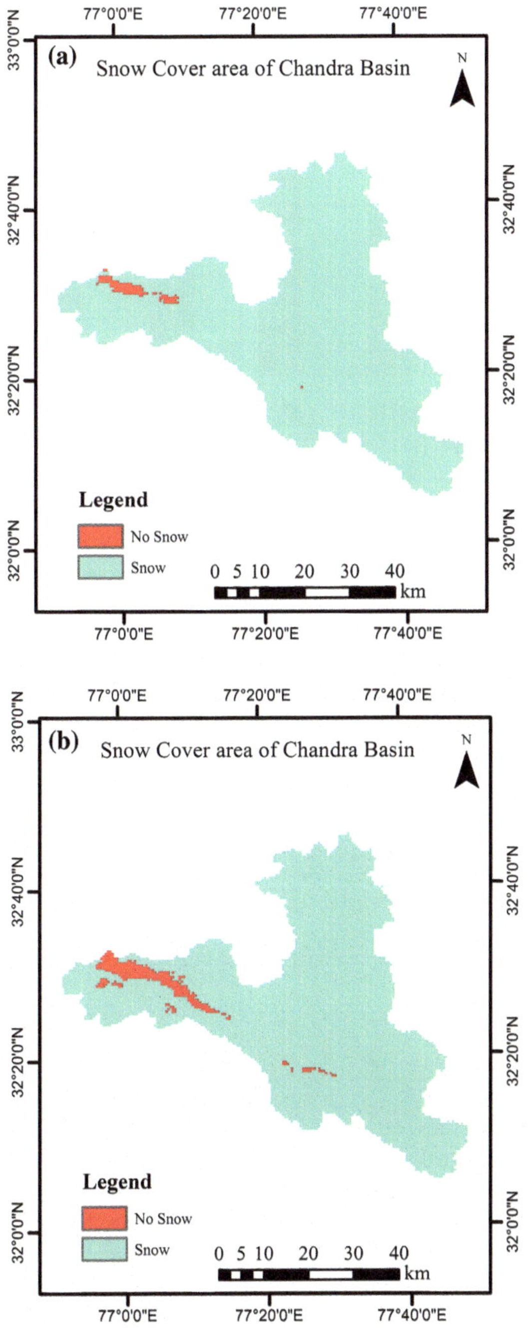

Fig. 3 Binary snow cover map of Chandra Basin **a** 2001 and **b** 2016

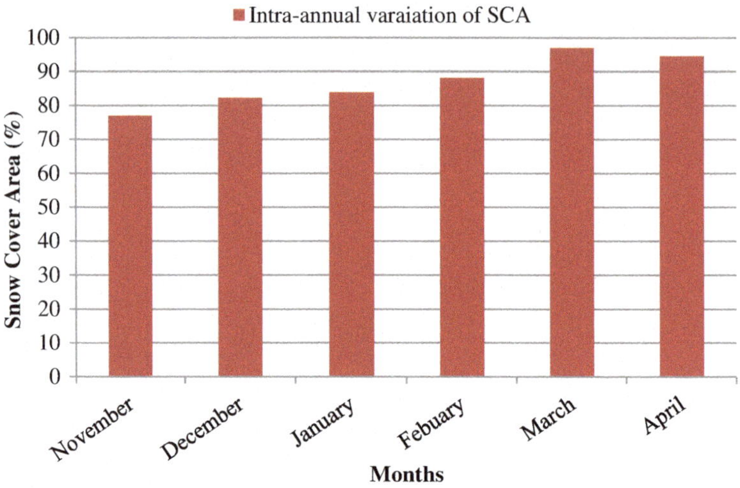

Fig. 4 Intra-annual variation of SCA during 2001–2016 in winter season (November–April)

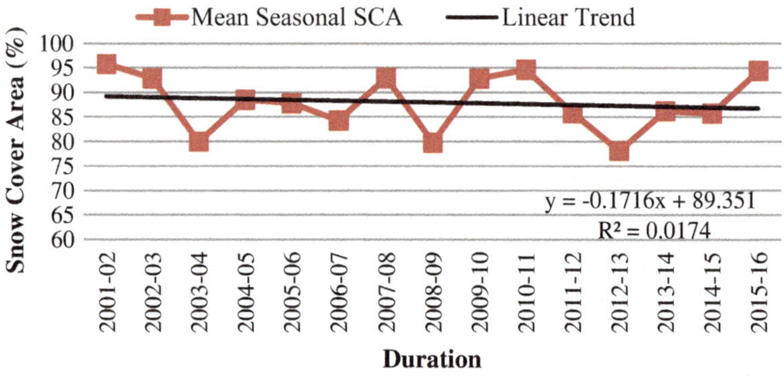

Fig. 5 Time series analysis of mean seasonal SCA from 2001 to 2016 for winter period

5 Conclusion

Estimation of accurate snow cover area in mountainous terrain like Himalaya is time-consuming and tedious task due to its highly rugged topography and complex climate condition. However, it is necessary to continuously monitor the SCA which can be used to estimate the mass balance of glacier and study of hydrological system. Present study computes and discusses the temporal characteristics of the snow cover variation using MODIS data products for Chandra Basin, situated in western Himalayas.

It is observed that shrinking of snow cover area in Chandra Basin is continuing from 2001 to 2016. SCA is shrinking at the rate of 0.12% per year and an overall reduction in snow cover area is 2.00% in the study area during the study period. There is a need to correlate this trend with climate parameters for better understanding in a future study.

References

1. Shiklomanov IA, Rodda JC (2003) World Water Resources at the Beginning of the Twenty-First Century Edited by. The press Syndicate of the University of Cambridge, Cambridge
2. Negi HS, Saravana G, Rout R (2013) Snehmani: monitoring of great Himalayan glaciers in Patsio region, India using remote sensing and climatic observations. Curr Sci 105:1383–1392
3. Foster J, Liston G, Koster R, Essaery R, Behr H, Dumenil L, Verseghy D, Thompson S, Pollard D, Cohen J (1996) Snow cover and snow mass intercomparisons of general circulation models and remotely sensed datasets. Am Metrol Soc 9:409–426
4. Salomonson VV, Appel I (2004) Estimating fractional snow cover from MODIS using the normalized difference snow index. Remote Sens Environ 89:351–360
5. Klein AG, Hall DK, Nolin AW (2000) Development of a prototype snow albedo algorithm for the NASA MODIS Instrument. Presented at the 2000
6. Vikhamar D, Solberg R (2003) Snow-cover mapping in forests by constrained linear spectral unmixing of MODIS data. Remote Sens Environ 88:309–323
7. Cherry J, Cullen H, Visbeck M, Small A, Uvo C (2005) Impacts of the North Atlantic oscillation on Scandinavian hydropower production and energy markets. Water Resour Manag 19:673–691
8. Jain SK, Goswami A, Saraf AK (2008) Accuracy assessment of MODIS, NOAA and IRS data in snow cover mapping under Himalayan conditions. Int J Remote Sens 29:5863–5878
9. Zhao Q, Liu Z, Ye B, Qin Y, Wei Z, Fang S, Environmental AR, Science E, Lands S (2009) Sciences a snowmelt runoff forecasting model coupling WRF and DHSVM. Hydrol Earth Syst Sci 13:1897–1906
10. Snehmani, Dharpure JK, Kochhar I, Hari Ram RP, Ganju A (2015) Analysis of snow cover and climatic variability in Bhaga basin located in western Himalaya. GEOCARTO Int 1–13
11. Shukla S, Kansal ML, Jain SK (2016) Snow cover area variability assessment in the upper part of the Satluj river basin in India. Geocarto Int 1–22
12. Maurer EP, Rhoads JD, Dubayah RO, Lettenmaier DP (2003) Evaluation of the snow-covered area data product from MODIS. Hydrol Process 71:59–71
13. Immerzeel WW (2010) Climate change will affect the asian water towers. Science 328:1382–1385
14. Dimri AP, Dash SK (2012) Wintertime climatic trends in the western Himalayas. Clim Chang 111:775–800
15. Dozier J, Marks D (1987) Snow mapping and classification from landsat thematic mapper data. Ann Glaciol 81:97–103
16. Hall DK, Riggs GA, Salomonson VV (1995) Development of methods for mapping global snow cover using moderate resolution imaging spectroradiometer data. Remote Sens Environ 54:127–140
17. Kulkarni AV, Singh SK, Mathur P, Mishra VD (2006) Algorithm to monitor snow cover using AWiFS data of RESOURCESAT - 1 for the Himalayan region. Int J Remote Sens 27:2449–2457
18. Xiao X, Moore B, Qin X, Shen Z, Stephen B (2002) Large-scale observations of alpine snow and ice cover in Asia: Using multi-temporal Vegetation sensor data. Int J Remote Sens 23:2213–2218
19. Dankers R, Jong SM De (2004) ing snow-cover dynamics in Northern Fennoscandia with SPOT Vegetation images. Int J Remote Sens 25:37–41
20. Riggs GA, Hall DK, Salomonson VV (2006) MODIS snow products user guide to collection 5

21. Kripalani RH, Kulkarni A, Sabade SS (2003) Western Himalayan snow cover and Indian monsoon rainfall: a re-examination with INSAT and NCEP/NCAR data. Theor Appl Climatol 78:1–18
22. Dietz AJ, Wohner C, Kuenzer C (2012) European snow cover characteristics between 2000 and 2011 derived from improved MODIS daily snow cover products. Remote Sens 4:2432–2454
23. Parajka J, Bloschi G (2006) Validation of MODIS snow cover images over Austria. Hydrol Earth Syst Sci 3:1569–1601
24. Hall DK, Riggs GA (2007) Accuracy assessment of the MODIS snow products. Hydrol Process 21:1534–1547
25. Tang B, Shrestha B, Li Z, Liu G, Ouyang H, Raj D (2013) Determination of snow cover from MODIS data for the Tibetan Plateau region. Int J Appl Earth Obs Geoinf 21:356–365
26. Pu Z, Xu L, Salomonson VV (2007) MODIS/Terra observed seasonal variations of snow cover over the Tibetan Plateau. Geophys Res Lett 34
27. Gascoin S, Hagolle O, Huc M, Jarlan L, Dejoux J-F, Szczypta C, Marti R, Sánchez R (2015) A snow cover climatology for the Pyrenees from MODIS snow products. Hydrol Earth Syst Sci 19:2337–2351
28. Sangewar CV, Shukla SP (2009) Inventory of the Himalayan glaciers: a contribution to the international hydrological programme. Special Publication, Geological Survey of India, 34: An Updated Edition (594 pp)
29. Bookhagen B, Burbank DW (2010) Toward a complete Himalayan hydrological budget: Spatiotemporal distribution of snowmelt and rainfall and their impact on river discharge. J Geophys Res 115:1–25
30. Dobhal DP, Kumar S, Mundepi AK (1995) Morphology and glacier dynamics studies in monsoon-arid transition zone: an example from Chhota Shigri glacier, Himachal-Himalaya, India. Curr Sci 68:936–944
31. Azam MF, Wagnon P, Patrick C, Ramanathan A, Linda A, Singh VB, Vincent C, Ramanathan A, Linda A, Singh VB (2014) Reconstruction of the annual mass balance of Chhota Shigri glacier, Western Himalaya, India, since 1969. Ann Glaciol 55:69–80
32. Kumar S, Dobhal DP (1997) Clitnatic effects and bedrock control on rapid fluctuations of Chhota Shigri glacier, northwest Himalaya, India. J. Glaciol. 43:467–472
33. Hall DK, Riggs GA, Salomonson VV, Digirolamo NE, Bayr KJ (2002) MODIS snow-cover products. Remote Sens Environ 83:181–194
34. Joshi R, Kumar K, Pandit J, Palni LMS (2015) Variations in the seasonal snow cover area (SCA) for upper Bhagirathi Basin, India. Dyn Clim Chang Water Resour Northwest. Himalaya Soc Earth Sci Ser 9–22

Long-Term Dynamics of Mangrove Forest in Andaman, India

Subha Chakraborty, Swati Saha, Debaleena Majumdar and Debajit Datta

Abstract Mangrove forests in Andaman district, India are chosen for investigation based on its high diversity and rich species composition. The long-term understanding of distribution and dynamics of mangrove forest in Andaman region is still inadequate. This study aims to generate the detail changes of mangroves during past decades. It also focuses on the key factors responsible for spatial changes during the period. Historical database are generated from past literatures and coupled with satellite image information to prepare the mangrove dynamics. This study finding revealed that the mangrove deforestation is high due to natural as well as anthropogenic stresses specially the impact of seismic and climate change and reclamation of land, agriculture pressure, and pollution. The outcomes of the study have been checked during field but due to the convenience of accessibility conditions only eastern part of Andaman island is considered for validation.

Keywords Change detection · Climate change · Coastal erosion image classification · Statistical correlation

1 Introduction

Mangrove forests, located in inter-tidal zones in tropical and subtropical region, [1] are very important for shoreline protection (Twilley et al. 1996; [2]). It helps to reduce the intensity of natural disasters and provide an important ecological and societal role including breeding and nursing ground of coastal habitats, food, medicines, and building materials for local communities. However, this valuable forest community is declining rapidly over the decades due to the intense impact of natural and anthropogenic pressure (Wilkie and Fortune 2003). Almost one-third of mangrove forest has been lost with the past 50 years [3].

S. Chakraborty (✉) · S. Saha
Department of Architecture, Town and Regional Planning, IIEST, Shibpur 711103, India
e-mail: subha.gis@gmail.com

D. Majumdar · D. Datta
Department of Geography, Jadavpur University, Kolkata 700032, India

J. K. Ghosh and I. da Silva (eds.), *Applications of Geomatics in Civil Engineering*,
Lecture Notes in Civil Engineering 33, https://doi.org/10.1007/978-981-13-7067-0_46

Drastic changes of mangrove have been also recorded in Indian subcontinent over the decades. Andaman faced the mangrove loss scenario due to population pressure and high natural impact. Earthquake, tsunamis, and tropical cyclone are main threats of destruction of mangrove and associated ecological communities over few decades. Global sea level rise with impact parameters plays dominant role in ecological changes. Increasing trends of sea temperature and salinity, changing nature of wind velocity and direction and unnatural pattern of rainfall impacts on coral and sea grass communities which plays a major role in ecological interconnectivity in mangrove and associated community. Seismological information shows a greater impact on mangrove dynamics as the region lies in seismic zone V outside the Himalayan belt and has experienced several earthquakes of moderate-to-large magnitude during the historic and recent past. Totally, 33 numbers of earthquakes are recorded during 1940–2017 which are above 6.5 magnitudes (Mw) [4]. The impact of the earthquake on mangroves has been identified as uprooting as well as inundation of landmass by seawater and resulting stagnation.

Apart from natural impacts, anthropogenic factors also play dominant role in mangrove change. In the north eastern part of the study area, Diglipur region, mangrove losses due to the increasing pattern of population pressure and tourism activities. Few artificial mangrove plantation projects are taken in last decades but the regeneration trend are very low comparing to loss. Land upliftment due to earthquake also loses mangrove area in northern and western part of the area. On the other hand, land subsidence in southern part provides natural mangrove nursing ground. In Port Blair, surrounding region agricultural land subsides after the tsunami and converted into young mangrove area which gained mangrove population.

Thereafter, the complete scenario of mangrove change in Andaman region is very complex in nature [5]. This work tries to illustrate the spatial locations of mangrove and the changing scenario of mangrove population in Andaman District, India which will help in further detail work on mangrove conservation.

2 Material and Method

2.1 Delineation of the Study Site

Andaman and Nicobar islands are a group of islands located in the juncture of Bay of Bengal and Andaman Sea (Fig. 1). The latitudinal and longitudinal extension of entire islands is from 6°N to14°N latitude and 92–94°E longitude, respectively. It is the largest archipelago system and the hotspot of biodiversity in Bay of Bengal. The total area covered by these islands is about 7950 km^2 in which 20% of land area is covered by mangrove forest. The topography of the region is a combination of low range hilly mountains to narrow valleys which results in an undulation terrain with steep slope to coastal plains. The climate of the region is an equatorial humid tropical with average temperature from 29 to 32 °C (maximum) and 22 to 24 °C (Minimum)

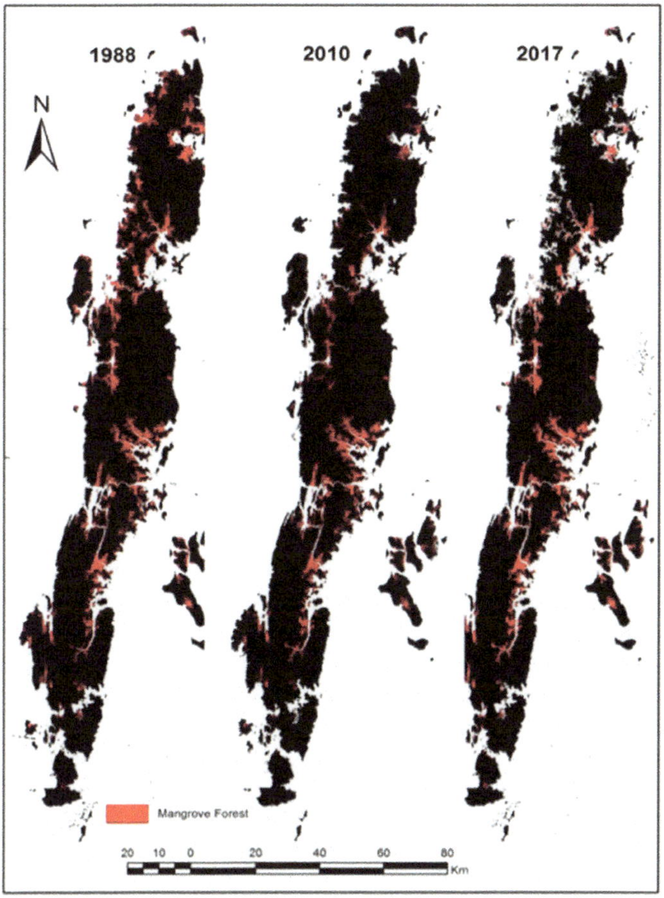

Fig. 1 Location of Mangrove area in Andaman district, India

and average rainfall is 3100 mm in a year. The maximum rainfall approximately 95% occurred in May to December. The average relative humidity ranges from 68 to 86% with mean wind speed at 5–15 km/h although the highest wind speed at 15 km/h occurred during June to August. The total population of the island was about 0.36 million in 2001 census and increased to 0.38 in 2011 census with growth rate of 7% in 2001–2011.

Mangrove areas are extracted using Landsat 5 TM scenes acquired on 1988 and 2009 and Sentinel 2 scenes acquired on 2017. Identifiable mangrove patches are delineated depends on their distribution along the coastal lines and around the river mouth regions based on classification techniques. The mangrove regions also verified through literature review information and high tide line data. We also use past literatures to illustrate the historical footprints of mangrove in the study location.

2.2 Data Sources

Multi-temporal satellite images [6] of Landsat Thematic Mapper (TM) of 1988 and 2010 and Sentinel 2 of 2017 covering entire study area were downloaded from freely available United States Geological Service (USGS) Glovis (http://glovis.usgs.gov), Global Land Cover Facility (GLCF) (http://glcf.umiacs.umd.edu/), and Earth Explorer (https://earthexplorer.usgs.gov/) web site for the present study. Open series maps are downloaded from Survey of India (http://soinakshe.uk.gov.in) at a scale of 1:50000 and ground control point were used for geo-referencing purposes. Files are processed in Universal Transverse Mercator (UTM) map coordinates at zone no 46 with WGS 84 datum. A GPS based field operation performed during December 2016 over the selected study locations. The field work supported in image interpretation and delineation of land cover types specially mangrove areas. It also generates the independent reference sites for accuracy assessment. The locations also used in geometric correction of the images acquire from USGS for the study. The field information generates detail of land use and land cover of the regions. It helps to delineate the mangrove patches, other vegetations, and converted land uses. We used Garmin eTrex 10 GPS unit to record the observed locations and capture every land use based transact and points. A detailed data directory was prepared after the field which describes site and transect identification information, date, time, description of landscape, characterization of land use, description of understory vegetation, the presence or absence of urban land use, and other site-specific comments. We also use Nikon 1 AW1 camera to capture the verity of land use and vegetation types of the region. The site location based field directory with site-specific photographs helps in satellite images interpretation.

2.3 Image Processing

All images were geometrically corrected using 30 Ground Control Points (GCP) collected during the field in December 2016 and root mean squared error (RMSE) of less than 0.5 pixel was attained using nearest neighbor resampling. Apart from geometric corrections, atmospheric corrections are performed on each image as sunlight and other atmospheric noise can interrupt image classification accuracy using Envi FLAASH module. The mathematical background of atmospheric corrections can be described as

$$L = \left(\frac{A_\rho}{1-\rho_e S}\right) + \left(\frac{B_{\rho_e}}{1-\rho_e S}\right) + L_\alpha$$

where:
ρ is the pixel surface reflectance
ρ_e is an average surface reflectance for the pixel and a surrounding region

S is the spherical albedo of the atmosphere
L_α is the radiance back scattered by the atmosphere
$A\ and\ B$ are coefficients that depend on atmospheric and geometric conditions but not on the surface.

2.4 Image Classification

Supervised classification is performed after atmospheric correction by identifying spectral classes and assigned them into training areas. For mangrove mapping, training areas were demarcated from false color composite images based on Landsat TM band combination of 5, 4, 3 and Sentinel 2 band combination of 6, 5 and 4 (Fig. 1). On the other hand, site-specific land use classifications are performed using a parametric decision rule, i.e., maximum likelihood scheme [7]. The accurate selection of training sites are represented real earth cover which were gathered through the information generated from reference data and these data were gained through maximum likelihood which used a complete multidimensional probability to prepare the likelihood of which pixel belongs to given classes. A confusion matrix was used to give a variety of estimates of the classification accuracy in a thematic map. The overall proportion of area correctly classified is the sum of diagonal of the confusion matrixes. The producer's accuracy refers to the probability that a certain land cover of an area on the ground is classified as such, while the user's accuracy refers to the probability that a pixel labeled as a certain land cover class in the map is really this class.

3 Result and Discussion

The major outputs of this present study are the compilation of thematic datasets, assessment of land use/land cover distribution in two specific locations in 1988, 2010, and 2017 as well as the dynamic or changes of mangrove cover over the study region during 1988–2017

3.1 Mangrove Change

Forest of the Andaman and Nicobar Island is rich in biodiversity with almost 2000 plant species belonging to 682 genera and 137 families, of which 14% are indigenous to the islands, 54% occur in other parts of the country while 32% are of extra-Indian origin. The marine environment with high precipitation over 8–9 months and equatorial condition enlighten various ecosystems in the region like corals, mangroves, and terrestrial forest. In this work, we focus extensively on mangroves. Mangroves of the study region are one of the most beautiful with high biomass contain. A total number of 38 mangrove species from 12 families and 19 genera were recorded from Andaman and Nicobar Island which are approximately 50% of the total numbers of global mangrove species (Ragavan et al. 2015). Maximum species richness found in Mayabunder, Diglipur, Middle Andaman and south Andaman according to the previous work Ragavan et al. However, mangroves are threatened due to high influences of natural and anthropogenic impacts during last few decades. The present analysis quantifies that from 1988 to 2017, mangrove forest in the Andaman district decreased by 30.65%. The change rate was not uniform between 1988–2010 and 2010–2017. From 1988 to 2010, mangrove areas are decreased by 33.29% and from 2010 to 2017 the area increased by 2.63%.

3.2 Site-Specific Land Use/Land Cover Assessment

Detail land use/land cover study is performed in two different sites over Andaman district to assess the land use dynamic in mangrove dominated regions. These are site-1 Diglipur Mayabunder area (Fig. 2) and site-2 Port Blair area (Fig. 4). The sites were chosen based on the high temporal changes of mangrove area. The categories of land use/land cover in 1988, 2010 and 2017 were extracted by knowledge-based interpretation of satellite images at 1:50000 scale. The categories are (1) Mangrove, (2) Tropical Evergreen Forest, (3) Tropical Mixed Forest and (4) Villages with Agriculture Land (Figs. 2 and 4). Due to limited open sources, cloud-free satellite images Port Blair region are classified based on two images on 1988 and 2017 (Fig. 5). However, in site-1 net, mangrove losses occurred during 1988–2017 are 25.52% whereas villages with agriculture land gained 22.88% during the period (Fig. 3). Apart from mangrove and villages site 1 loosed 0.3% tropical mixed forest and gained 0.13% tropical evergreen forests during the entire analysis periods (Fig. 4).

On the other hand, Port Blair region gained 2% mangrove, 0.33% tropical evergreen forest and 15% tropical mixed forest whereas loose 4% settlement and agriculture land during last 29 years (Fig. 5). These lose and gain statistics shows the unpredicted changes occurred due to the impact of natural deserter like tsunami, coastal flooding, earthquake, and growth of populations. Among various natural deserters tsunamis had greater impact on the region. Land subsidence, loss of land use, and land cover were high, on the other hand, new land addition, destruction of underwater ecology, and increase of new mangrove are the other impacts happened due to the giant tsunami of 2004. The major triggering factors of the high amount of mangrove loss in site 1 are increasing population pressure due to population and tourism growth whereas site 2 experienced increment of new mangrove region generated from agriculture and settlement domain due to the land submergence occurred in time of 2004 tsunami.

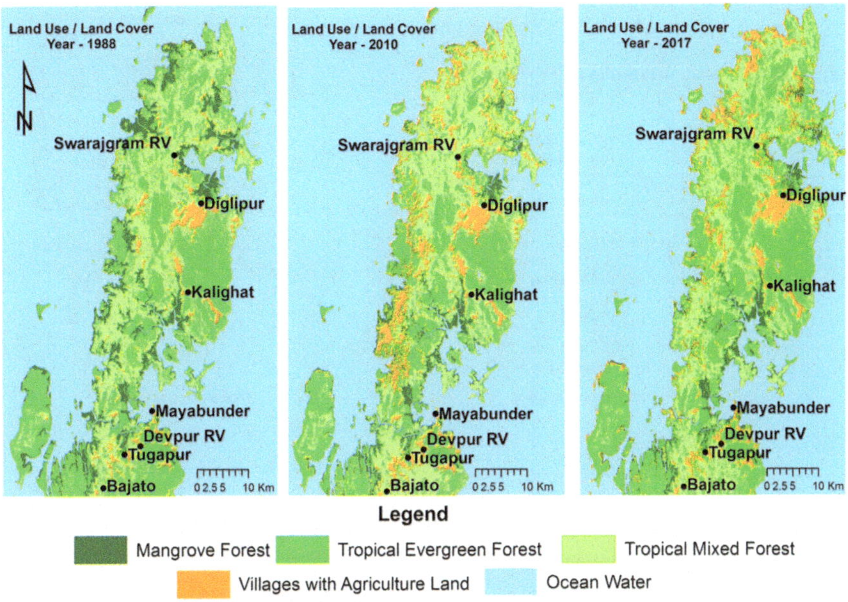

Fig. 2 Land use–land cover of Diglipur–Mayabunder (site-1)

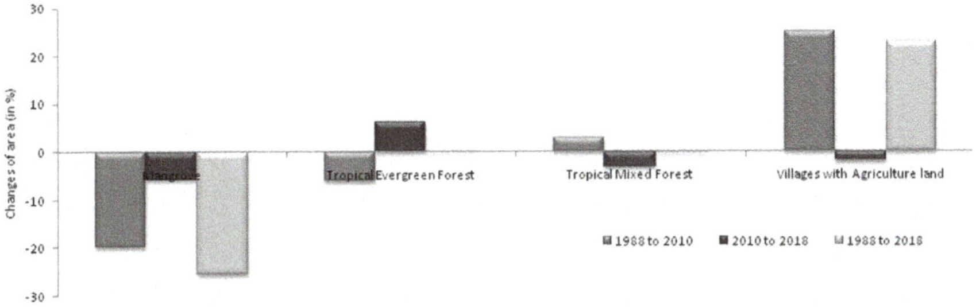

Fig. 3 Areal change statistics in percentage during 1988–2017 in site-1

4 Conclusion

The present work focuses on the overall footprint of mangrove zones and associated changes during the last four decades in the study region. Andaman and Nicobar archipelago is very complex in nature, thus it is very hard to analyze the entire area in one or two methods. Simultaneously, very limited locations of this archipelago are accessible in nature. In this context, remote sensing based analysis and indexes are playing an important role in monitoring and assessment. Furthermore, the entire region is highly vulnerable in term of seismological and climate change impacts along with anthropogenic activities. Ecological interconnectivity is also in high changing state due to natural and anthropogenic impacts. Detail study of each biotope is very

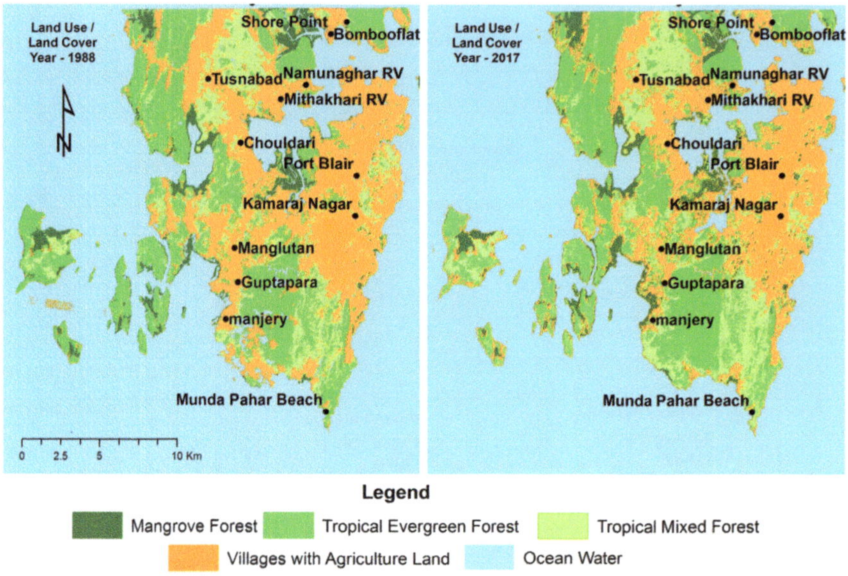

Fig. 4 LULC of Port Blair in 1988 and 2017 (site-2)

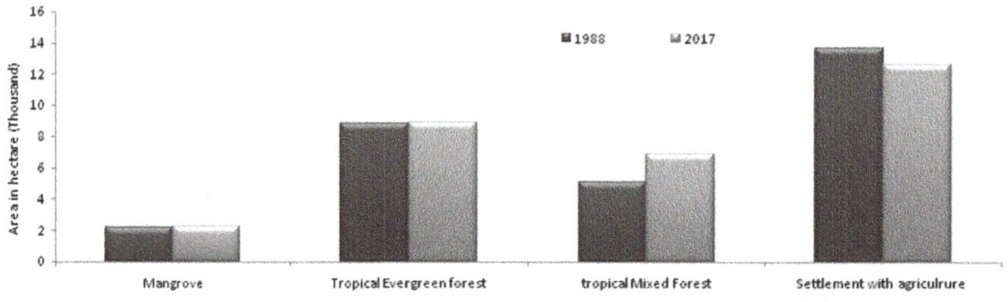

Fig. 5 Area statistics of different Land-use–land-cover during 1988–2017 in site-2

important to understand the detail relationship. Thus, further works need to focus on the interconnectivity and conservation practices of different ecological units in this region.

References

1. Bird E (2010) Encyclopedia of the world's coastal landforms, vol 1. Springer Science and Business Media. Coastal management sourcebooks 3, part 2 the acquisition, correction and calibration of remotely sensed data. http://www.unesco.org/csi/pub/source/rs10.htm
2. Sheridan P, Hays C (2003) Are mangroves nursery habitat for transient fishes and decapods? Wetlands 23:449–458

3. Alongi DM (2002) Present state and future of the world's mangrove forests. Environ Conserv 29:331–349
4. Rajendran K, Rajendran CP, Earnest A, Ravi Prasad GV, Dutta K, Ray DK, Anu R (2008) Age estimates of coastal terraces in the Andaman and Nicobar Islands and their tectonic implications. Tectonophysics 455(1–4):53–60
5. Ramesh DA, Vel AS (2011) Methodology of integrated coastal zone management plan preparation–case study of Andaman Islands. India J Environ Prot 2(6):750
6. Kamal M, Stuart P, Kasper J (2016) Assessment of multi-resolution image data for mangrove leaf area index mapping. Remote Sens Environ 176:242–254
7. Chakraborty S, Majumdar D, Sahoo S (2013, July) Comparative analysis of different supervised classification techniques using linear regression model. Zenith Int J Multidiscip Res 3(7):317–327 ISSN 2231–5780

Evaluation of CORDEX Multi-RCM for Indian Subcontinent Using NASA's RCMES

Saket Dubey and Sandeep Kumar Chouksey

Abstract Mean, minimum, and maximum surface air temperature, precipitation and cloudiness are evaluated for their strengths and limitation in the period (1991–2001) from the CORDEX regional climate model hindcast experiment. South Asia region has limited CORDEX data availability, 3 RCMs namely RCA4 (SMHI), REMO (MPI), and REGCM4 (ICT) were available for assessing their climate feature simulation capabilities. Regional climate model evaluation system (RCMES) is a powerful tool for characterizing strength, weakness, and uncertainty involved in the model. In this study, Indian region has been divided into 17 subregions in such a way that they represent spatial climatic variability of the country. These regions are assessed individually as well as collectively with CORDEX data which is available at a resolution of 0.5° for their biases and correlation against the gridded surface station data of climate research unit (CRU TS v. 4.01). Comparative study of simulated precipitation against CRU dataset revealed that systematic model biases do occur across all the models whereas basic climatological features are simulated reasonably well except for few regions including Himalayas as well as regions of Gujarat and Arunanchal Pradesh. The study also revealed that surface air temperature is simulated better in comparison to precipitation and cloudiness, multi-model ensemble mean outperforms individual models. Taylor diagrams are used to enhance studies effectiveness by incorporating correlation coefficient, RMSE and standard deviation among various models.

Keywords CORDEX · RCMES · India · Downscaling · RCM evaluation

1 Introduction

Simulation of precipitation, temperature, and cloudiness from General Climate Models (GCMs) is quite close and accurate for the study of the impact of climate change

S. Dubey (✉) · S. K. Chouksey
Department of Civil Engineering, NIT Raipur, Raipur, India
e-mail: saketdubey4@gmail.com

S. K. Chouksey
e-mail: chouksey.sandeep@gmail.com

J. K. Ghosh and I. da Silva (eds.), *Applications of Geomatics in Civil Engineering*, Lecture Notes in Civil Engineering 33, https://doi.org/10.1007/978-981-13-7067-0_47

at a global scale. These models describe the complex dynamic system of the Earth using numerical equations. However, they run at relatively coarse resolution grids that hinder their ability to capture fine scale details in climate pattern that are often desired, hence they are deemed unreliable to provide necessary information. The need for a better understanding of the climate change impact on the geophysical science of important sectors at a local scale is necessary. This shortcoming was however met by the use of Regional Climate Models (RCMs). "They are formed by nesting a physically based, small spatial resolution regional climate model within the grid of a GCM output. The GCM outputs are used as boundary conditions to drive the regional models" [1]. This process is termed as dynamical downscaling. Downscaling resolves the scale disagreement between the coarse resolution GCMs and the fine observation resolutions required for impact assessment. These high resolution models provide projections that can be used to evaluate changes brought about by climate that are sensitive to fine scale gradients. They account for the detail information which is useful in understanding the climate of regions having complex topography.

These simulations at a finer resolution with improved accuracy could potentially be a good source of input for hydrological models useful for impact studies. Evaluating fidelity and finesse of these limited area models, RCMs along with their ensembles for various climatic parameters are of utmost importance to gain insight of the numerous uncertainties associated in the assessment of their climate feature simulation capabilities. In this study, an ensemble of three models is selected for the climate projection because of the unreliability in using only one model. A single model cannot solely represent and depict the present, intermediate and future impacts of climate change over an area and may result in imprecise modeling, which can be subdued by incorporating multi-model ensemble approach. This approach can help reduce any epistemic uncertainty. Previous studies that have been carried out in evaluation include evaluation of CORDEX for South Africa [2], a study of surface climatology [3], cloud-based evaluation of regional climatic model [4], detailed description of working of RCMES [5], strength and weaknesses about RCMs for India [6], detailed description of RegCM4 [7–11], CORDEX scenarios for Europe [12], air temperature relation by HadRM3p [13], and performance assessment of CORDEX for Himalayas [14]. Data from these climate models have been used in a range of impact assessment studies [15–18] where it was observed that the use of these data yield effective results to assess the impact of anthropogenic activities [19].

2 Evaluation of CORDEX

The majority of the work in this study was performed using a comprehensive suite of software resources called Regional Climatic Model Evaluation System (RCMES). It is a toolkit which provides a suitable platform for the evaluation of regional and local scale climate data and earth system models by providing convenient access to large scale observed datasets (satellite, local gauging stations, and reanalysis data) as well as modeled data (via ESGF). RCMES also provides another advantage for performing visualization task as well as basic analysis of data.

Datasets and Data Processing

Data from Climate Research Unit (CRU) are obtained as observed data for mean precipitation, cloudiness, and mean, maximum and minimum surface air temperature for a baseline period 1991–2000 which are then plotted against model data of various RCMs from CORDEX evaluation experiment for south Asia. To take into account the uncertainties involved with reference data, tropical rain measuring mission (TRMM) [20, 21] satellite data at a higher resolution for precipitation is used to carry out uncertainty analysis.

For preprocessing the data, climate data operator (CDO) was used whereas shell scripts were used in combination with CDO operator to perform operations on multiple files.

In this study, the Indian subcontinent was divided into 17 subregions (Fig. 1) in such a way that they describe the spatial and temporal variability of climate over the country. Among the potential RCMs accessible worldwide, only three RCM models are available for South Asian Region (Table 1) with the boundary conditions from the ERA-interim reanalysis for baseline period (1991–2000) that were evaluated for their fidelity in describing various climatological features over the subcontinent.

Taylor diagrams were also used in addition to provide a graphical summary of how well the model and observed pattern complement each other in terms of correlation (Pearson correlation coefficient), their RMSE differences and ratio of variances [22].

Taylor diagrams help depict how well model and observed data comply with each other.

Table 1 Various RCMs and their institutions

Model (labels)	Institutions
REGCM4 (ICT)	Abdus Salam International Center for Theoretical Physics
RCA4 (SMH)	Sveriges Meteorologiska och Hydrologiska Institute
HadRM3P (UKO)	United Kingdom meteorological office. Hadley center

Precipitation, mean, minimum and maximum surface air temperature from all RCMs. Cloudiness from all RCMs except REGCM4

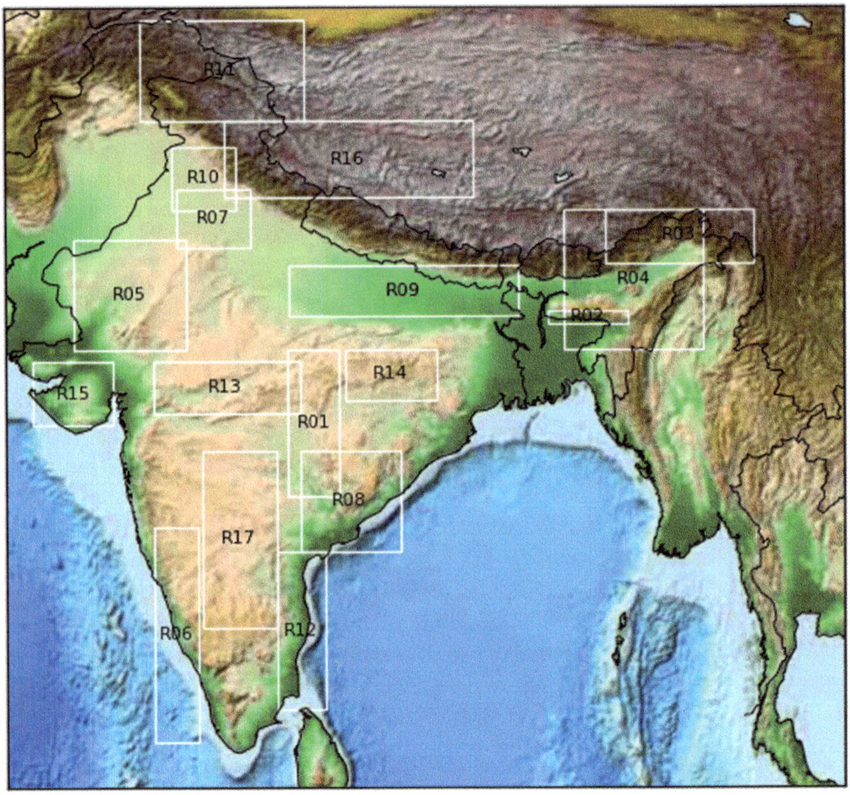

Fig. 1 Subregions used for the study

3 Results

3.1 Precipitation

As shown in Fig. 2 described by CRU that climatology over India is described by tropical monsoon which is characterized by seasonal reversal of wind, the tropic of cancer passes midway between the country which separates it into southern torrid and northern temperate zone. Position of the mountain ranges in the Himalayan region and the direction of rain-bearing winds greatly affect the precipitation over the country [23]. This region is characterized by immense spatial and temporal variability with a heavy rainfall of about 10.5 mm/day in Cherrapunji (Northeast India) to less than 1 mm/day in Jaisalmer (west).

All the RCMs represent basic climatological features reasonably well but biases in these models vary strongly. It is also seen that there is systematic variation of RCMs over regional scale for majority of areas. Most of the RCMs show wet bias for

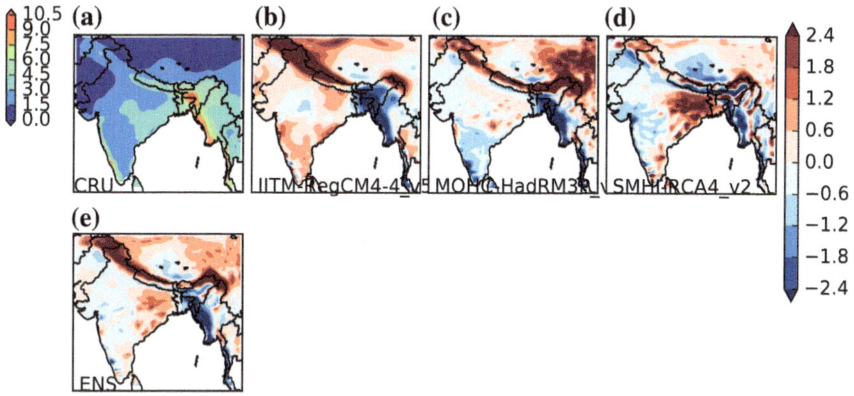

Fig. 2 Annual mean precipitation (mm/day) from CRU (**a**) and biases in the model RegCM4 (**b**), HadRM3P (**c**), RCA4 (**d**) and their ENS (**e**)

northern region of Himalayas, Himachal Pradesh and Nepal whereas strong dry bias is observed in the northeastern states. It is also seen that the constancy of these RCMs is more for the central region as compared to the peripheral region of the country which are mostly characterized by either mountain ranges or sea. For the western part of the country which basically includes Rajasthan and Gujarat, RegCM4 shows wet bias whereas all the other RCMs show dry bias. The southern region of the country is bounded by western and eastern ghats and no systematic regional correlation is seen in various RCMs in representing the precipitation. From Fig. 3, it is observed that seasonality of precipitation is simulated well by almost all the RCMs and despite having large inter-RCM variability ENS mean reasonably represents the reference data for most of the subregions.

Observation also shows that the regions which include high mountain ranges, affected by snowfall such as region 11 and region 16 are not represented effectively by RCMs which may lead to deleterious effect in representing regional atmospheric circulation while using these data for impact assessment studies. Portrait diagrams were used for evaluating the skill of various RCMs in simulating precipitation for all the subregions. Fig. 4 also shows that for most of the regions, value of correlation coefficient lies above 80% of CRUs precipitation for almost all RCMs in all the subregions except Himalayas, part of Himachal Pradesh and Arunachal Pradesh (region 11, 16, 3).

The strength of various RCMs in simulating phase of annual cycle can be assessed using correlation coefficient. All RCMs simulate phase of annual cycle reasonably well except for Himalayas, Himachal Pradesh, and parts of Eastern Ghats where RMSE values were also observed to be large. Standard deviation for various models as well bias are represented in Fig. 4 for various regions.

"The spatial variation of the annual-mean precipitation was evaluated for pattern correlation and standard deviation (Fig. 5) over the land area" which were observed using Taylor diagram in a similar way as done in [2]. Since typical biases for all

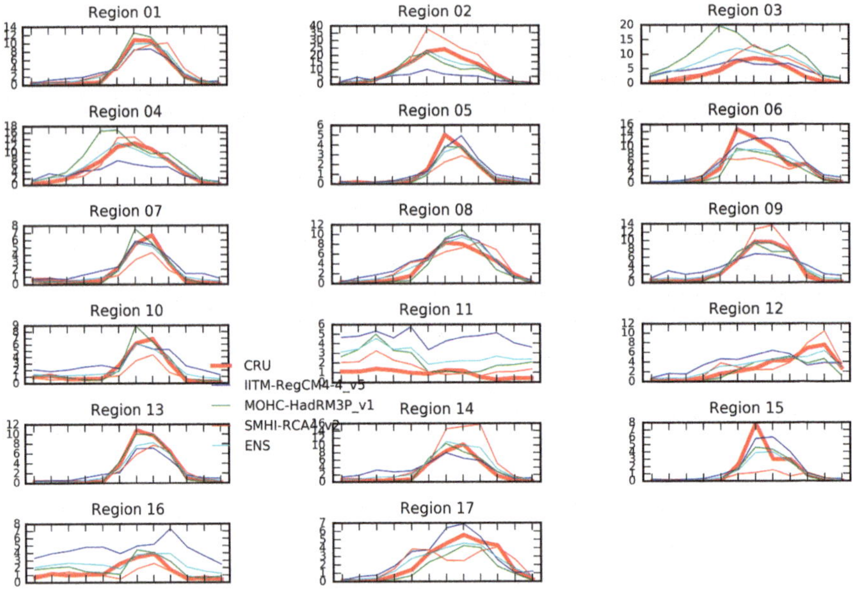

Fig. 3 Annual cycle of simulated precipitation (mm/day) for selected 17 regions. Bold red line represents CRU data, whereas blue, green, red and cyan represents RegCM4, HadRM3P, RCA4 and ENS mean respectively

the RCMs lies within 20% of CRU they simulate the mean precipitation reasonably well. RegCM4 shows a similar standard deviation as CRU whereas RCA4 shows a smaller value of RMSE and a better correlation. The correlation coefficient for all the RCMs shows a value greater than 0.6. ENS mean shows the smallest value of RMSE and value of correlation coefficient closest to 1.

RCMs also yield higher skill in simulating precipitation for summer as compared to winter (Fig. 6). In fact, the ability of these models in simulating winter rainfall is so low that for RegCM4 and HadRM3p the observation lies beyond representable range of Taylor diagram.

3.2 Surface Air Temperature

The evaluation of mean, maximum and minimum daily surface air temperature was also carried out for baseline period (1991–2000) against CRU data. The strength of RCMs in simulating temperature is much better as compared to precipitation. All RCMs simulate precipitation with high reliability but they overestimate the observed spatial variability (i.e., standard deviation value >1) for all the 3 temperatures. The pattern correlation for all the RCMs for all the three surface air temperatures lies above 0.95 (Fig. 7).

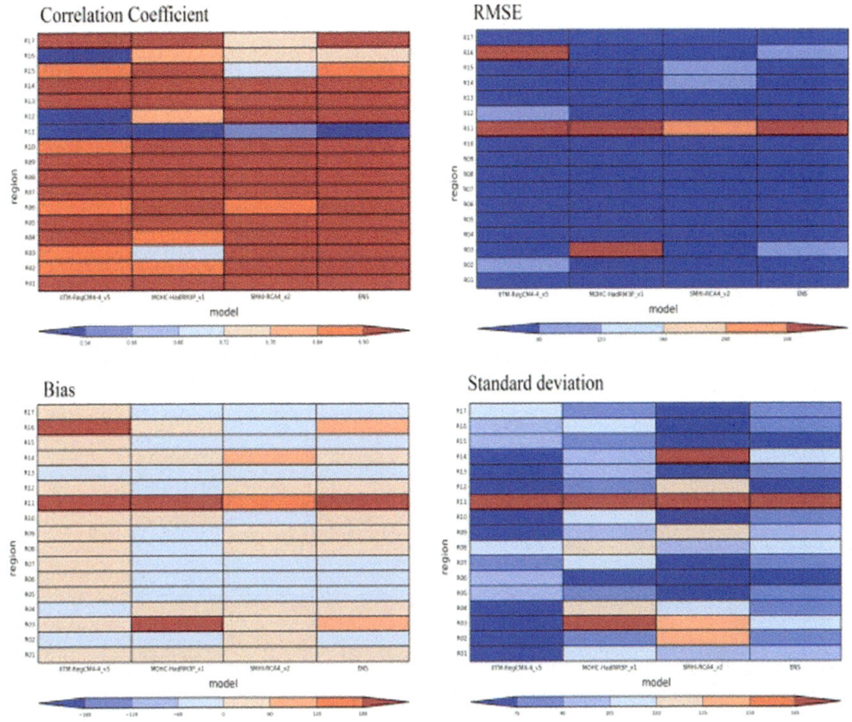

Fig. 4 Portrait diagrams showing correlation coefficient, RMSE, bias and standard deviation between simulated precipitation and CRU data

Value of RMSE was found minimum for ENS mean as compared to individual models in a similar manner as in the case for precipitation. From Fig. 8, it is observed that all the RCMs underestimate the value of temperature in the northern region especially for Himalayas and regions of Himachal Pradesh whereas for the region of Gujarat and Rajasthan all the RCMs slightly overestimates the value of temperature. Simulation of central part of the country is carried out with higher fidelity by all the RCMs (Fig. 9).

The phase of the annual cycle was also well simulated with correlation coefficient lying above 0.9 for almost all the regions. All RCMs perform better for the Torrid Zone as compared to the temperate zone. Figure 7 also shows that RCMs that perform better for Mean surface air temperature also performs well for other temperature variables.

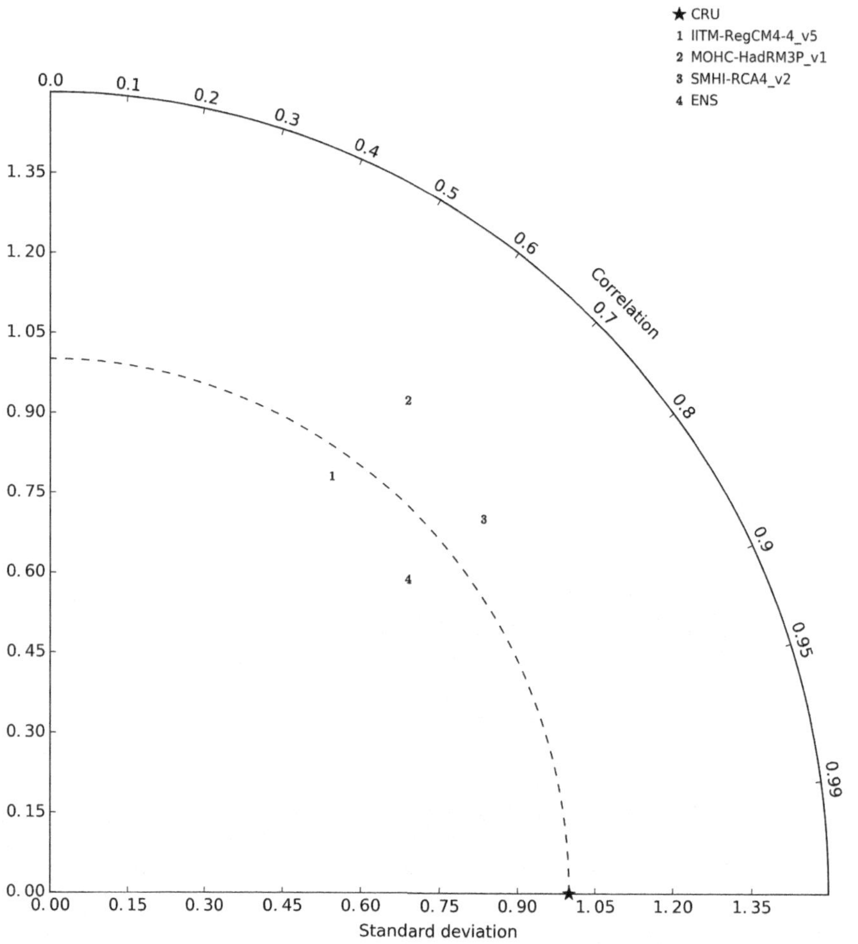

Fig. 5 Taylor diagram plotted for various models against CRU data (plotted as star mark) representing spatial pattern correlation and standard deviation for precipitation

3.3 Cloudiness

Most of the evaluation work that has been done in the past usually includes precipitation and surface air temperature and there is very less work that has been carried out in evaluating the fidelity of cloud fraction which is basically the fraction of the sky covered with the clouds which plays very important role in most of the climatological phenomenons. Due to limited availability of RCMs data for South Asian region only 2 RCMs, i.e., HadRM3p and RCA4 were evaluated for their strength and weaknesses in simulating cloud fraction. It was observed that HadRM3P slightly underestimates the spatial variability (i.e., standard deviation <1) whereas RCA4

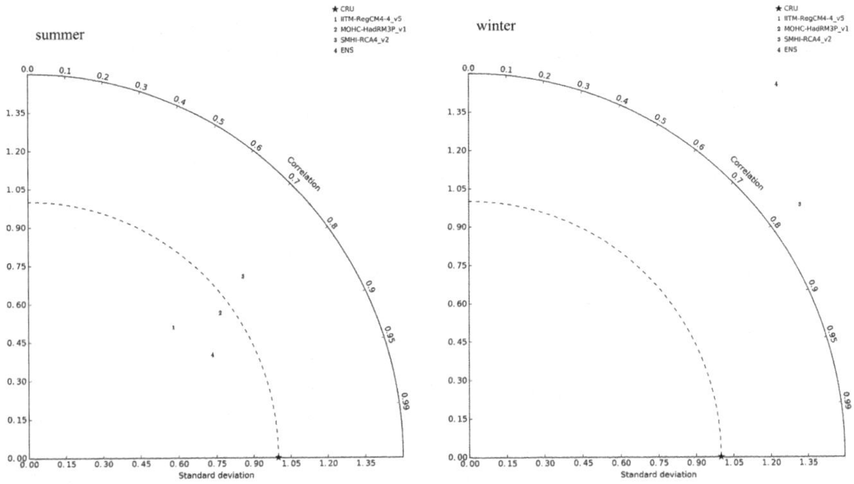

Fig. 6 Taylor diagram showing spatial pattern correlation and standard deviation for summer in the left and winter in the right

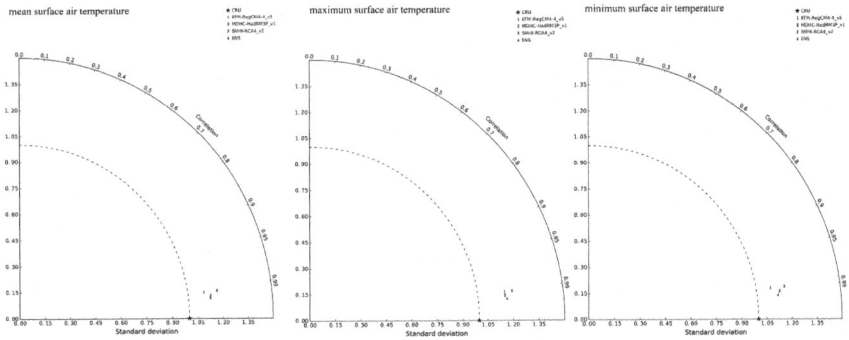

Fig. 7 Taylor diagram showing standard deviation and spatial pattern correlation for mean maximum and minimum surface air temperature

overestimate (standard deviation >1) (Fig. 10). HadRM3p shows a smaller RMSE value as well as correlation coefficient closer 1. In case of cloudiness, HadRM3p outperforms even the ENS mean as the values of RCA4 deviates by a reasonable margin from the CRU dataset.

From Fig. 11, it can be seen that both the RCMs as well as their ENS shows a negative bias (i.e., smaller value of cloud fraction) for the northern region and a positive bias for the southern region.

From (Fig. 11), it can also be seen that unlike precipitation or temperature, the bias of RCMs is outside ± standard deviation range and bias of ENS >20% for most of the region hence fidelity of RCMs in simulation of cloud fraction is comparatively less in comparison to precipitation and surface air temperature.

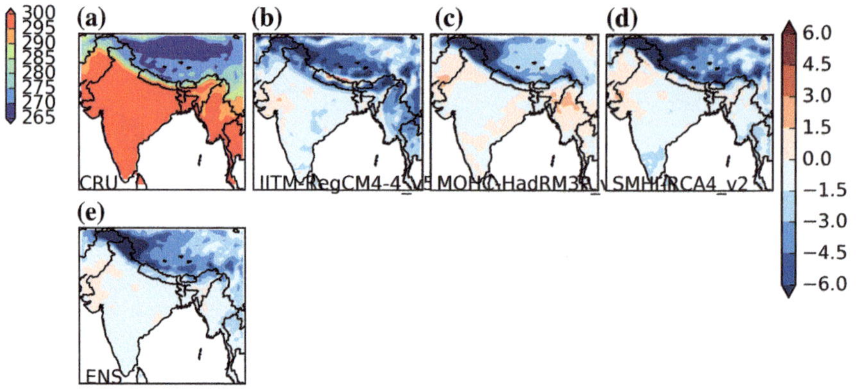

Fig. 8 Annual mean temperature (kelvin) from CRU (**a**) and biases in the model RegCM4 (**b**), HadRM3P (**c**), RCA4 (**d**) and their ENS mean (**e**). From Fig. 9, it can be seen that RMSE for most the RCMs lies within 40% of CRU and they show similar skill in simulating the annual cycle of all the 3 temperature for all 17 subregions

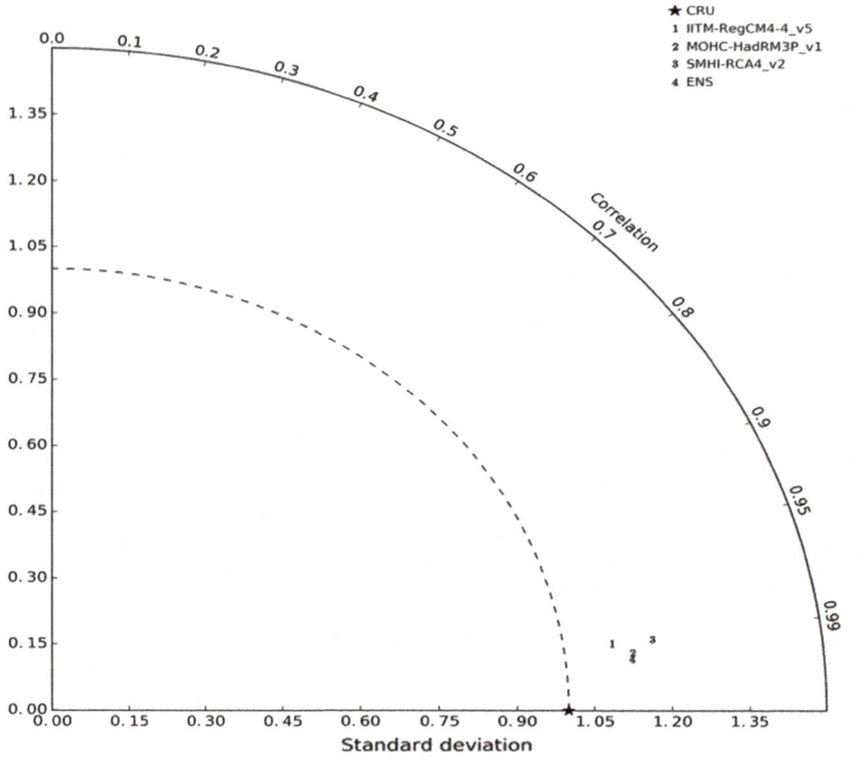

Fig. 9 Taylor diagram plotted for various models against CRU data (plotted as star mark) representing spatial pattern correlation and standard deviation for temperature (1991–2001)

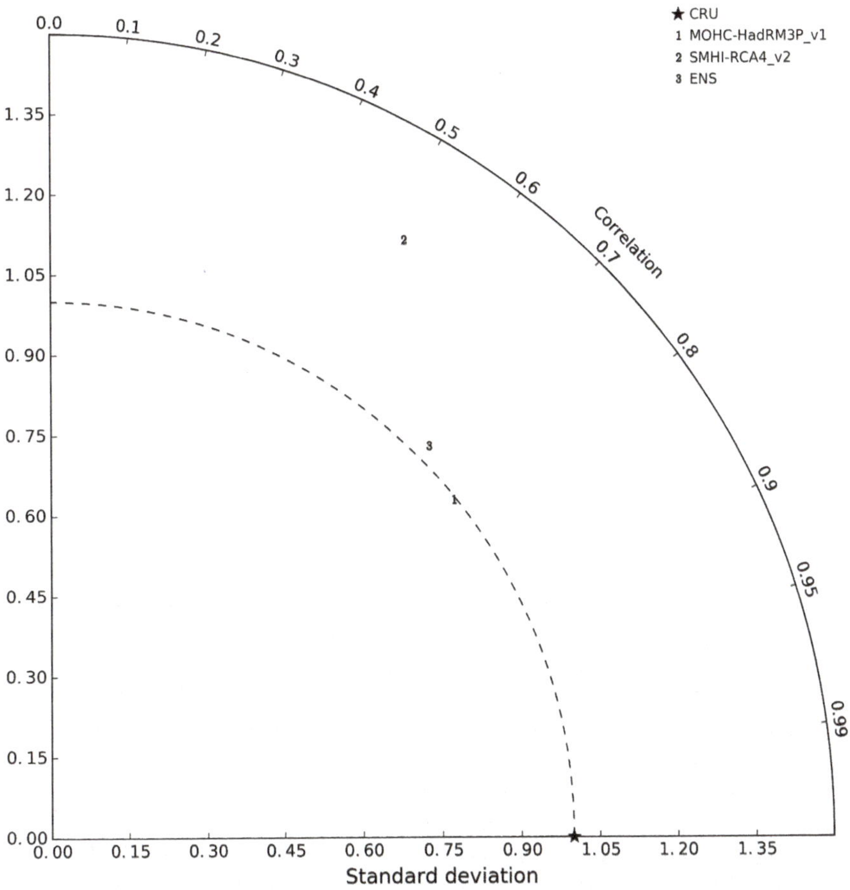

Fig. 10 Taylor diagram plotted for various models against CRU data (plotted as star mark) representing spatial pattern correlation and standard deviation for cloud fraction (1991–2001)

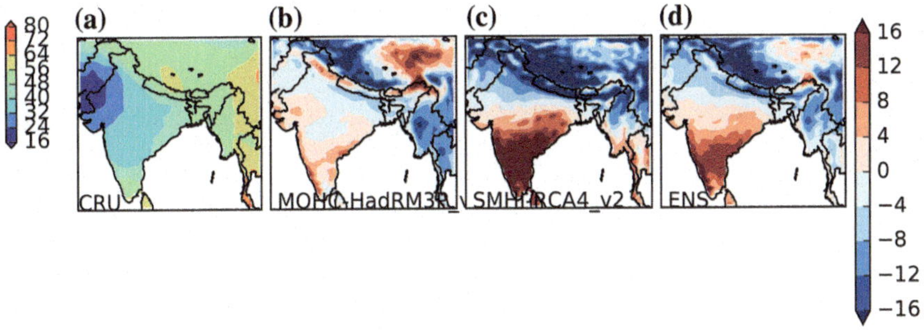

Fig. 11 Annual mean cloudiness (percentage) from CRU (**a**) and biases in the model HadRM3P (**b**), RCA4 (**c**) and their ENS (**d**)

3.4 Uncertainties Associated with Reference Data

Before carrying out any evaluation experiment, the knowledge of preciseness of the reference data and the uncertainties associated within them are of prior importance as every set of observed or reanalysis data obtained, there are uncertainties associated within them that maybe due to lack of local station density or malfunctioning of measuring instrument [24]. Uncertainty analysis in the reference data for precipitation in evaluation experiment is carried out by using 2 set of reference data, i.e., CRU and TRMM of precipitation for the baseline period (2001–2008).

The results obtained from CRU and TRMM were found to be similar in evaluating the spatial pattern of monthly precipitation, however, from Fig. 12, it can also be seen that HadRM3P and RCA4 yields higher correlation coefficient value when evaluating against TRMM. The difference between the precipitation evaluation based on TRMM and CRU may have resulted due to different observational platform, i.e., subjective surface observation for CRU and satellite based observation for TRMM. This shows that cross-examination of reference dataset is important for quality control of model evaluation experiment.

4 Summary and Conclusions

The evaluation of CORDEX-SOUTH ASIA for Indian subcontinent was carried out for hindcast period (1991–2000) using a set of CRU reference data for precipita-

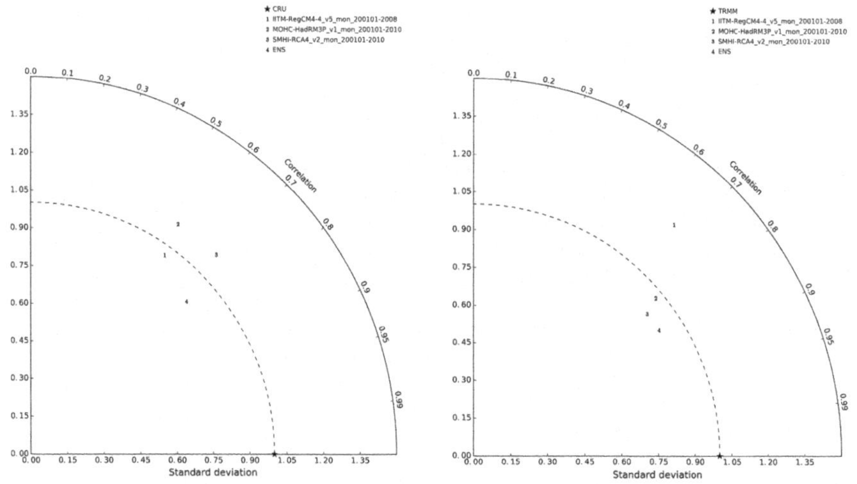

Fig. 12 Taylor diagram for spatial pattern (2001–2008) between simulated precipitation and CRU data on the left and TRMM satellite data for precipitation on the right

tion, mean, maximum, and minimum surface air temperature as well as cloudiness. A check for uncertainties involved in reference data was carried out using a different set of satellite-based TRMM data of precipitation for the baseline period (2001–2008). In the study, 3 RCMs for precipitation, surface air temperature, and 2 for cloudiness were incorporated.

Simulation of precipitation is carried out reasonably well by almost all the RCMs but systematic model biases do occur across all. Biases for these models are within 20% of CRU data. All RCMs show wet bias for northern part of country and a dry biases for eastern part. The central region which is characterized by plain terrain is simulated well in comparison to other parts of the country. Despite having inter-RCM variability ENS mean reasonably represent the reference data for most of the subregions. It was also seen that simulation of precipitation for summer was carried out in a better way in comparison to winter. The spatial pattern of precipitation is simulated well with correlation coefficients >60 for almost all the subregions except for Himalayan region and adjoining regions where large ENS precipitation bias is observed. All selected RCMs are incapable of simulating precipitation for such areas and thus extra caution should be taken during impact assessment study of these regions.

In simulation of surface air temperature for mean, maximum, and minimum, it was observed that simulation of temperature is carried out with much higher reliability in comparison to other variables such as precipitation and cloudiness. It was also found that the RCM which performed better for mean surface air temperature also performed better for minimum and maximum surface air temperature. Large biases in simulating temperature of the northern part of the country were seen and seasonal cycle and spatial distribution were simulated with very high fidelity having correlation coefficient for most of the subregions above 0.9. ENS mean bias is well within ± standard deviation range for almost all the domains. RCMs generally perform better for Torrid Zone in comparison to temperate zone.

All the selected RCMs have low ability in the simulation of cloud fraction which is the percentage of sky covered by clouds. ENS shows positive bias for the southern region and negative bias for the northern region. Spatial pattern is simulated well but spatial standard deviation varies widely. HadRM3p outperforms RCA4 as well as their ENS in projection simulations. For most of the subregions, the ENS bias is well outside ± standard deviation range thus use of these RCM data of cloudiness for impact assessment studies is not as reliable as it is for precipitation and surface air temperature.

Evaluation of precipitation using two reference data one local station based and other satellite station based showed that there can be substantial variation in evaluation results based on the preferred reference data. These results emphasize for the cross examination of the reference data for the uncertainties involved in them. The discrepancy of the model data and the occurrence of systematic bias of different RCMs show the lack of competence in these models in parameterization modeling of various atmospheric phenomena.

Acknowledgements The authors acknowledge the support provided by Climate Research Unit, School of Environmental Science, University of East Anglia for providing necessary data to complete this work and also to Department of Civil Engineering, National Institute of Technology, Raipur for providing guidance and necessary institutional support.

References

1. Wood AW, Leung LR, Sridhar V, Lettenmaier DP (2004) Hydrologic implications of dynamical and statistical approaches to downscaling climate model outputs. Clim Chang 62(1–3):189–216
2. Kim J, Waliser DE, Mattmann CA, Goodale CE, Hart AF, Zimdars PA, Jack C (2014) Evaluation of the CORDEX-Africa multi-RCM hindcast: systematic model errors. Clim Dyn 42(5–6):1189–1202
3. Kim J, Waliser DE, Mattmann CA, Mearns LO, Goodale CE, Hart AF, Boustani M (2013) Evaluation of the surface climatology over the conterminous United States in the North American regional climate change assessment program hindcast experiment using a regional climate model evaluation system. J Clim 26(15):5698–5715
4. Hart AF, Goodale CE, Mattmann CA, Zimdars P, Crichton D, Lean P, Waliser D (2011, May). A cloud-enabled regional climate model evaluation system. In: Proceedings of the 2nd international workshop on software engineering for cloud computing, pp 43–49. ACM
5. Mattmann CA, Waliser, D, Kim J, Ramirez P, Goodale C, Hart AF, Khudikyan S Model evaluation using the NASA regional climate model evaluation system (RCMES)
6. Singh S, Ghosh S, Sahana AS, Vittal H, Karmakar S (2017) Do dynamic regional models add value to the global model projections of Indian monsoon? Clim Dyn 48(3–4):1375–1397
7. Giorgi F, Coppola E, Solmon F, Mariotti L, Sylla MB, Bi X, Turuncoglu UU (2012) RegCM4: model description and preliminary tests over multiple CORDEX domains. Clim Res 52:7–29
8. Ozturk T, Altinsoy H, Türkeş M, Kurnaz ML (2012) Simulation of temperature and precipitation climatology for the central Asia CORDEX domain using RegCM 4.0. Clim Res 52:63–76
9. Raju PVS, Bhatla R, Almazroui M, Assiri M (2015) Performance of convection schemes on the simulation of summer monsoon features over the South Asia CORDEX domain using RegCM-4.3. Int J Climatol 35(15):4695–4706
10. Park JH, Oh SG, Suh MS (2013) Impacts of boundary conditions on the precipitation simulation of RegCM4 in the CORDEX East Asia domain. J Geophys Res Atmos 118(4):1652–1667
11. Oh SG, Suh MS, Cha DH (2013) Impact of lateral boundary conditions on precipitation and temperature extremes over South Korea in the CORDEX regional climate simulation using RegCM4. Asia-Pac. J Atmos Sci 49(4):497–509
12. Strandberg G, Bärring L, Hansson U, Jansson C, Jones C, Kjellström E, Ullerstig A (2015) CORDEX scenarios for Europe from the Rossby centre regional climate model RCA4. SMHI
13. Morabito M, Crisci A, Moriondo M, Profili F, Francesconi P, Trombi G, Orlandini S (2012) Air temperature-related human health outcomes: current impact and estimations of future risks in Central Italy. Sci Total Environ 441:28–40
14. Ghimire S, Choudhary A, Dimri AP (2015) Assessment of the performance of CORDEX-South Asia experiments for monsoonal precipitation over the Himalayan region during present climate: part I. Clim Dyn 1–24
15. Ghosh S, Mujumdar PP (2008) Statistical downscaling of GCM simulations to streamflow using relevance vector machine. Adv Water Resour 31(1):132–146
16. Ghosh S, Mujumdar PP (2006) Future rainfall scenario over Orissa with GCM projections by statistical downscaling. Curr Sci 90:396–404
17. Laprise R, Hernández-Díaz L, Tete K, Sushama L, Šeparović L, Martynov A, Valin M (2013) Climate projections over CORDEX Africa domain using the fifth-generation Canadian regional climate model (CRCM5). Clim Dyn 41(11–12):3219–3246

18. Panitz HJ, Dosio A, Büchner M, Lüthi D, Keuler K (2014) COSMO-CLM (CCLM) climate simulations over CORDEX-Africa domain: analysis of the ERA-interim driven simulations at 0.44 and 0.22 resolution. Clim Dyn 42(11–12):3015–3038
19. Sinha J, Sharma A, Khan M, Goyal MK (2018) Assessment of the impacts of climatic variability and anthropogenic stress on hydrologic resilience to warming shifts in Peninsular India. Sci Rep 8(1)
20. Huffman GJ, Bolvin DT, Nelkin EJ, Wolff DB, Adler RF, Gu G, Stocker EF (2007) The TRMM multisatellite precipitation analysis (TMPA): Quasi-global, multiyear, combined-sensor precipitation estimates at fine scales. J Hydrometeorol 8(1):38–55
21. Kummerow C, Barnes W, Kozu T, Shiue J, Simpson J (1998) The tropical rainfall measuring mission (TRMM) sensor package. J Atmos Oceanic Technol 15(3):809–817
22. Taylor KE (2001) Summarizing multiple aspects of model performance in a single diagram. J Geophys Res Atmos 106(D7):7183–7192
23. Ghosh S, Luniya V, Gupta A (2009) Trend analysis of Indian summer monsoon rainfall at different spatial scales. Atmos Sci Lett 10(4):285–290
24. Mastrandrea MD, Mach KJ, Plattner GK, Edenhofer O, Stocker TF, Field CB, Matschoss PR (2011) The IPCC AR5 guidance note on consistent treatment of uncertainties: a common approach across the working groups. Clim Chang 108(4):675

Biomass Estimation Using Synergy of ALOS-PALSAR and Landsat Data in Tropical Forests of Brazil

Vinayak Huggannavar and Amba Shetty

Abstract Satellite remote sensing technologies are currently tested and suggested as a tool in REDD+ (MRV, Measurement Reporting, and Verification). SAR (Synthetic Aperture Radar) has got an extensive application in the estimation of biomass due to its all-weather capabilities. L band radar signals penetrate the canopy more efficiently when compared to C band. Scientific biomass study using SAR has not been conducted in Para in spite of extensive field datasets being freely available under CMS (Carbon Monitoring System) project. This study aims in using various polarization combinations like HH + HV, HH − HV, HH + HV/HH − HV and vegetation index such as NDVI from the optical data. ALOS-PALSAR and Landsat 7 data acquired over Paragominas in Brazil, where field samples were collected in the form of transects. Regression analysis was performed using backscatter coefficients and field collected Above Ground Biomass (AGB). Semi-empirical model was developed to model AGB using various polarization combinations and NDVI as predictor variables. Combination gave higher R^2 value of 0.657 for biomass prediction. Multiple linear regression using NDVI and HH + HV as variables yielded R^2 of 0.73 during calibration and 0.363 during validation. There is future scope to use other vegetation indices such as RVI, EVI, etc., along with increased number of samples, which may yield more robust models with acceptable level of accuracy for practical application.

Keywords NDVI · SAR (Synthetic Aperture Radar) · AGB (Above Ground Biomass) · REDD+ · ALOS-PALSAR

V. Huggannavar (✉) · A. Shetty
NITK, Surathkal, Mangalore 575025, India
e-mail: vhuggannavar@gmail.com

A. Shetty
e-mail: amba_shetty@yahoo.co.in

J. K. Ghosh and I. da Silva (eds.), *Applications of Geomatics in Civil Engineering*, Lecture Notes in Civil Engineering 33, https://doi.org/10.1007/978-981-13-7067-0_48

1 Introduction

Earlier, collection of biophysical parameters for the forest inventory was tiring and time-consuming task. Comprehension of global C cycle is important in estimation of global terrestrial biomass to eliminate uncertainty [6]. Advent of remote sensing has made it possible to estimate above ground biomass by retrieving information from images acquired by satellites. Synthetic Aperture Radar (SAR) acquisitions are independent of cloud cover, weather and light conditions. SAR images are acquired using both airborne and space-borne platforms in various wavelength bands such as X, C, L, and P. Another Important advantage of using radar image is its penetration capability. L band radar signals penetrate the canopy more efficiently when compared to C band. And L band has performed considerably well in retrieving biomass when compared to C Band [3, 7, 9]. Moreover, SAR plays an important role in forest observation [15]. It was demonstrated that the sensitivity of SAR polarimetry is depending on the structure, density, and tree elements (i.e., trunk/stem, branches, and leaves) of the forests. Although there have been many biomass estimation studies conducted all over the world, there are limited studies done in estimating biomass using synergy of Optical and SAR images. Deforestation has cleared about 15% of the extensive forest on the Brazilian Amazon frontier. In response to the potential climatic effects of deforestation, policy makers have suggested reductions in emissions through deforestation and forest degradation and enhanced forest carbon stocks (REDD+) [14]. Application of different SAR and optical techniques in estimation of biomass will help in understanding the forest management ecosystem [10]. Combination of optical and SAR data has exhibited increase in accuracy level of biomass estimation [5]. This study was done using freely available datasets, as ALOS-2 datasets are expensive to procure.

The sensitivity of backscatter and saturation depends on site conditions and structure of forest [8, 11]. The frequency of SAR is directly proportional to the depth of wave penetration, which means that shorter wavelength can only penetrate the forest for a few centimeters, while longer wavelength can penetrate deeper and sometimes can interact with forest floor [8]. In addition to this, weather conditions may also influence dielectric properties of vegetation and soil surface [6]. The scattering and attenuation of signal depend on stand features, frequency, incidence angle, etc. InSAR is one more emerging technique to estimate biomass [13]. The present study aims to establish a relationship between ABG using SAR backscatter and NDVI as input variables and generate biomass maps of higher accuracy. In addition, freely available field and satellite data are used in this study.

2 Study Area

The study area is located in Paragominas, Brazil (Fig. 1). The study area is bound by extent coordinates of 48°54′15.607″ W, 3°35′15.339″ S to 48°19′43.994″ W, 3°43′24.833″ S. The dominant vegetation in this region is humid forest with predom-

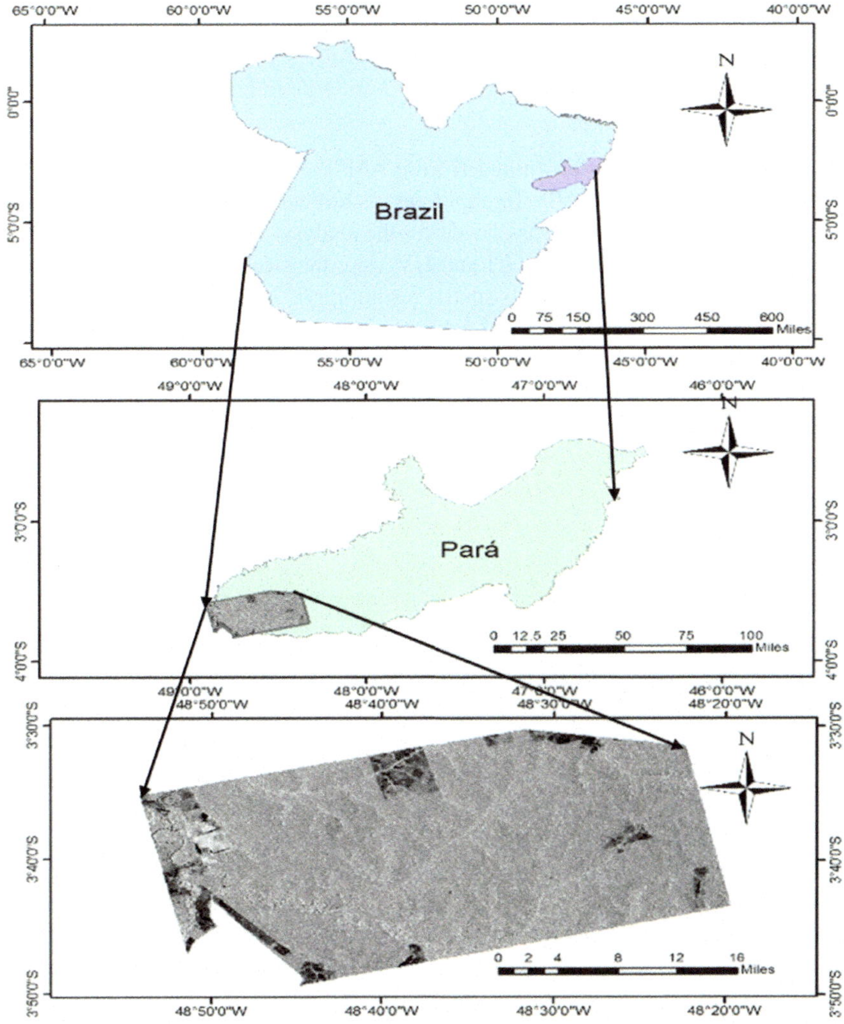

Fig. 1 Location map of the study area in State of Paragominas, Brazil

inantly oxisols perenefólia and ultisols soils. Paragominas was a large agricultural and timber center of the country, with its exploitation started in the 70s and expansion in the 80s. During the period 1989–1990, it is estimated that 67,845 ha were explored for wood intended to supply the 238 sawmills operating in the region. Paragominas was once considered the largest timber source of Brazil [4, 14].

3 Data Collection

3.1 Satellite Data

ALOS PALSAR data was downloaded from ASF's (Alaska Satellite Facility) Data portal for June month in 2010. To avoid any disturbance in backscatter due to soil moisture, dry season of year was chosen for the analysis. Data was acquired in Fine Beam Dual Polarization mode. HH and HV were the two polarization channels in which data was acquired. Data has spatial resolution of 20 m and was acquired at an incidence angle of 34.3°. ALOS PALSAR mission was decommissioned in the year 2011. As there was no suitable data available in 2011, 2010 data was used. Landsat 7 LaSRC (Landsat Surface Reflectance Corrected) Level 2 products were procured by ordering through USGS earth explorer website. Landsat 7 has spatial resolution of 30 m × 30 m.

3.2 Field Data

This data set provides measurements for diameter at breast height (DBH), commercial tree height, and total tree height for forest inventories taken at the Fazenda Cauaxi and the Fazenda Nova Neonita, Paragominas municipality, Para, Brazil. Also included for each tree are the common, family, and scientific name, coordinates, canopy position, and for dead trees the decomposition status. These biophysical measurements were made at Fazenda Cauaxi during 2012 and 2014 and at the Fazenda Nova Neonita during 2013. The data were collected under the project Sustainable Landscapes, a project supported by the United States Agency for International Development (USAID) and US Department of State [4]. Forest inventory surveys were conducted at the Fazenda Nova Neonita and Fazenda Cauaxi in the Paragominas municipality, Para, Brazil (). Total area inventoried at the Fazenda Cauaxi was 22 ha: 22 plots of 10,000 m^2. Plot sizes were 20 × 500 m with a 2 × 500-m subplot within a plot. Measurements were made at the Fazenda Cauaxi from 2012/01/27 to 2012/03/26 and also from 2014/02/18 to 2014/04/25. Commercial and total height measurements were only made in 2012. Trees with diameter at breast height (DBH) equal to or greater than 35 cm were accounted for and measured within the plot area. Total Height (m) was measured on live trees and standing dead trees using a clinometer and tape as the height to the highest point of the tree crown (Fig. 2).

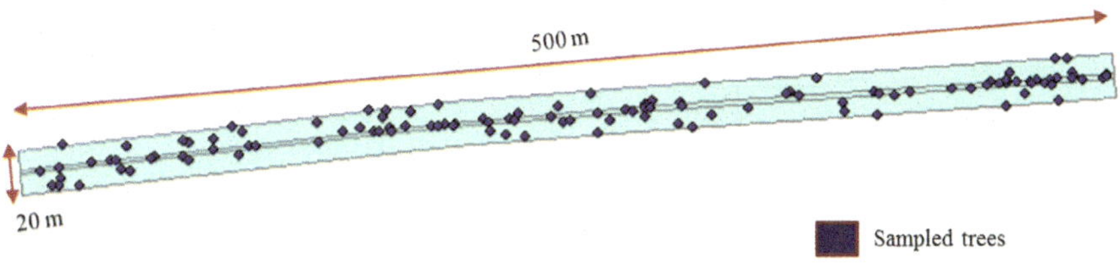

Fig. 2 Transects having dimensions of 500 m × 20 m with sampled trees inside

4 Methodology

4.1 Image Preprocessing

Level 1.5 product of ALOS was downloaded through ASF's data portal. Image is a 16 bit datatype with DN's (Digital numbers) varying from 0 to 65,535. SNAP software was made use of to process the image and the software automatically calibrates the ALOS image for Normalized Radar Cross Section (NRCS) using the following equation:

$$\sigma^{\circ}_{\mathrm{dB}} = 10 \cdot \log_{10}\left(\mathrm{DN}^2\right) - 83 \quad (1)$$

Fine beam dual pol data has spatial resolution of 20 m. Speckle filtering was done using refined lee filter to eliminate speckles in the image. Image was upscaled to 100 m × 100 m or 1 ha resolution as it is standard practice to measure the biomass in tons/ha units [2]. Layover and shadow effects on the image were reduced using SRTM 1 arc second Digital Elevation Model (DEM). This is done because if terrain slope is facing the sensor then returning signals get stronger relatively to the other side of the slope. Landsat 7 LaSRC products were used to generate NDVI maps using ArcMap 10.1 software. Landsat 7 products had gaps due to error in the detectors and were gap filled using IDL code in ENVI software. NDVI maps generated were also upscaled to a pixel area of 1 ha (Fig. 3).

4.2 Field AGB Calculation

Field AGB was calculated using generalized allometric equations for the study area [1]. AGB depends upon species of tree and can be estimated using allometric equations which are either species specific or generalized within the given area. As there are no species-specific allometric equations, generalized equations have been utilized to calculate the field AGB. Allometric equations take DBH (cm) and Height (m) as input variables and give AGB in tons/ha. The equation used for this study is as follows:

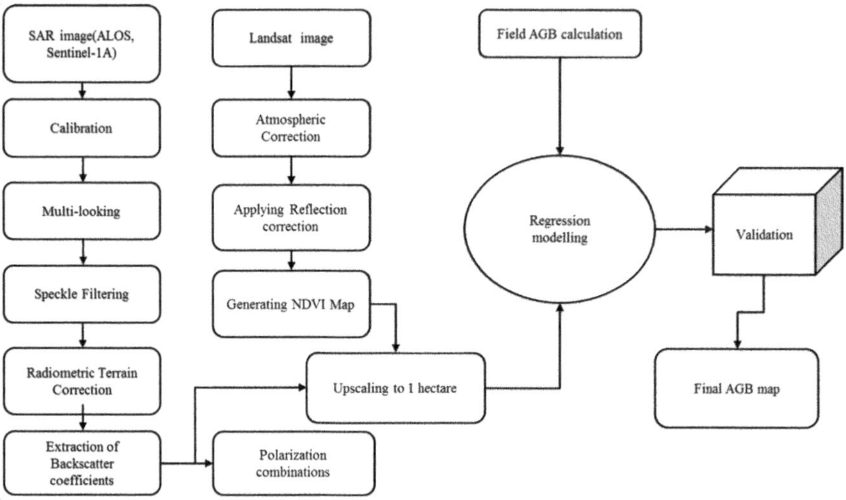

Fig. 3 Flow chart of overall methodology

$$AGB = 0.026 \cdot DBH^{1.529} \cdot H^{1.47} \tag{2}$$

4.3 *Plot Planning to Extract Pixel Values*

As we can see in Fig. 2, sampling of trees has been done in the form of transect which makes it difficult to extract pixel values for a 500 m long transect. To avoid the same, each transect was manually divided into 3 sample plots of size 90 m × 20 m, 30 m × 20 m and 60 m × 20 m resulting in plots of area 1800 m^2, 600 m^2, 1200 m^2, respectively. Generally, random sampling techniques are adopted to fix the plot size which are generally square in shape. As such square plots were unavailable, in this case randomly sampled plots were created to avoid bias in results. Finally, this resulted in 64 plots for the regression modeling of backscatter versus field AGB. In which 44 plots were chosen for modeling and 20 for validation. Plots covering the upscaled imagery were noted and values were extracted for NDVI, HH and HV images.

5 Results and Discussion

5.1 AGB Modeling Using Optical Data

Nonlinear regression was performed between field calculated AGB and NDVI. Logarithmic fit gave most appropriate results while also giving suitable R^2 values. NDVI was used as a predictor variable for modeling. Equation of the following form was used. Coefficient of determination and RMSE were used as performance indicators for the model

$$AGB = a * \ln(NDVI) + c \quad (3)$$

R^2 of 0.298 was obtained which is considerably low and is exhibiting poor relation with field calculated AGB.

5.2 AGB Modeling Using SAR Data

Generally, the regression function between field AGB and SAR backscatter ($\sigma°$) in dB is known to be nonlinear. It can be represented by logarithmic, exponential, and power functions until it reaches saturation. As the curve starts to reach this threshold value of AGB backscatter starts to saturate and becomes nearly constant beyond this AGB point [11, 12]. Coefficient of determination was used as performance indicator for the model. R^2 value of 0.604 and 0.345 was obtained for HH and HV polarization channels, respectively. HH backscatter is outperforming HV in estimation of biomass. Apart from this, other polarization combinations were also tested for biomass estimation. In this study, the following polarization combinations were tested (1) HH + HV (2) HH − HV (3) (HH − HV)/(HH + HV) are used (Figs. 4 and 5).

HV backscatter showcased better R^2 when compared to HH backscatter with value of 0.5592. Also, HH + HV polarization combination exhibited reasonably good (R^2 = 0.6574) relation. (HH − HV) and (HH + HV/HH − HV) combinations had lower relation with AGB.

5.3 AGB Modeling by Combining Optical and SAR Data

Multiple linear regression was carried out in order to generate higher accuracy biomass maps using polarization combinations and NDVI as predictor variables with AGB as a response variable. HH backscatter was considered because it was

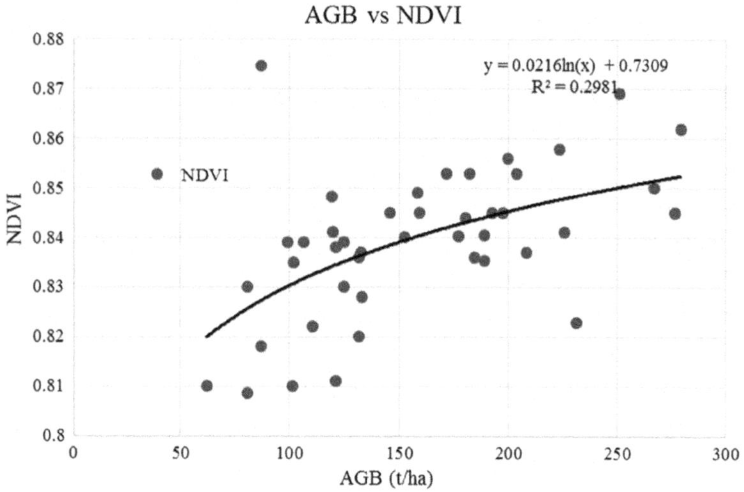

Fig. 4 Regression model for field calculated AGB and NDVI

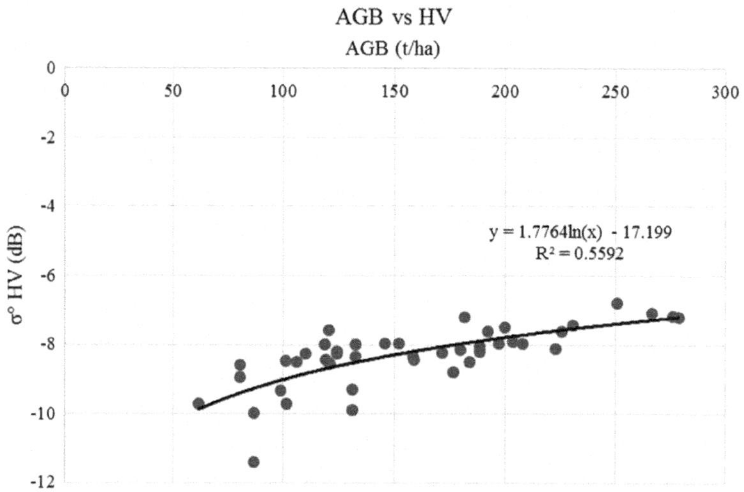

Fig. 5 Regression model for field calculated AGB versus HV

showing better relation with field AGB. HH backscatter along with NDVI showed increase in R^2 to 0.663 with RMSE of 31.01 t/ha.

$$\mathrm{AGB} = 26.21 * (\mathrm{HH} + \mathrm{HV}) + 1329 * \mathrm{NDVI} - 385.29 \tag{4}$$

Equation 4 was used to calculate predicted biomass. Model validation was done using linear regression between predicted and field AGB which resulted in R^2 of 0.3625 (Figs. 6 and 7).

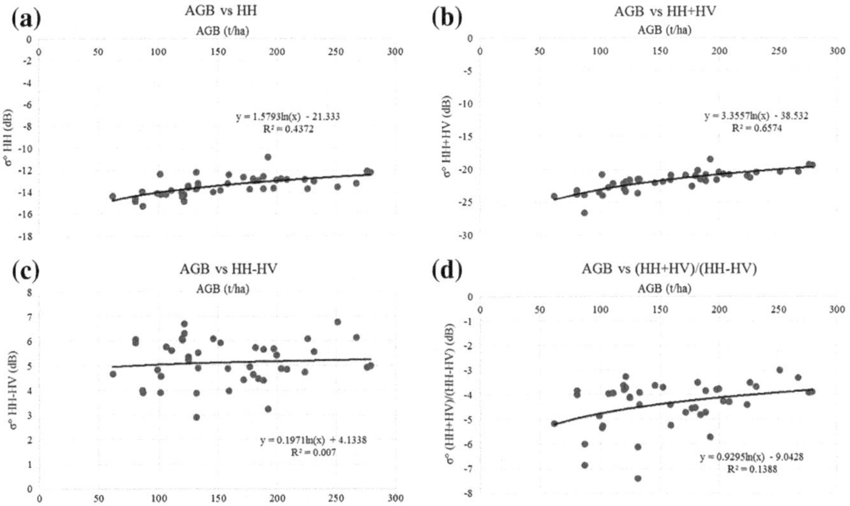

Fig. 6 **a** Regression model for AGB versus HV. **b** Regression model for AGB versus HH + HV. **c** Regression model for AGB versus HH – HV. **d** Regression model for AGB versus (HH + HV)/(HH – HV)

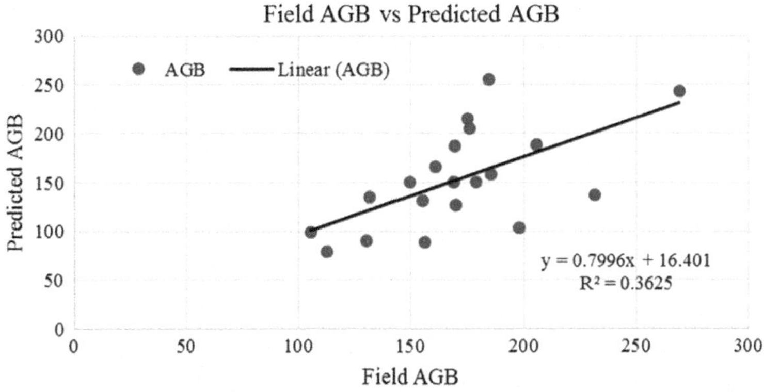

Fig. 7 Correlation between predicted and field calculated AGB

6 Conclusion

Synergy of optical and SAR data with different polarization combinations in estimation of biomass was assessed. HV backscatter has better sensitivity (R2 = 0.559) for AGB than HH backscatter (R2 = 0.437) which agrees with earlier research. HH + HV yielded R2 of 0.6574 which was outperforming all other backscatter combinations. Empirical model was developed to generate biomass maps and R2 of 0.732 was obtained during calibration and 0.363 during validation (RMSE = 43.18 t/ha). However, this result was generated without excluding any outliers so as to avoid bias

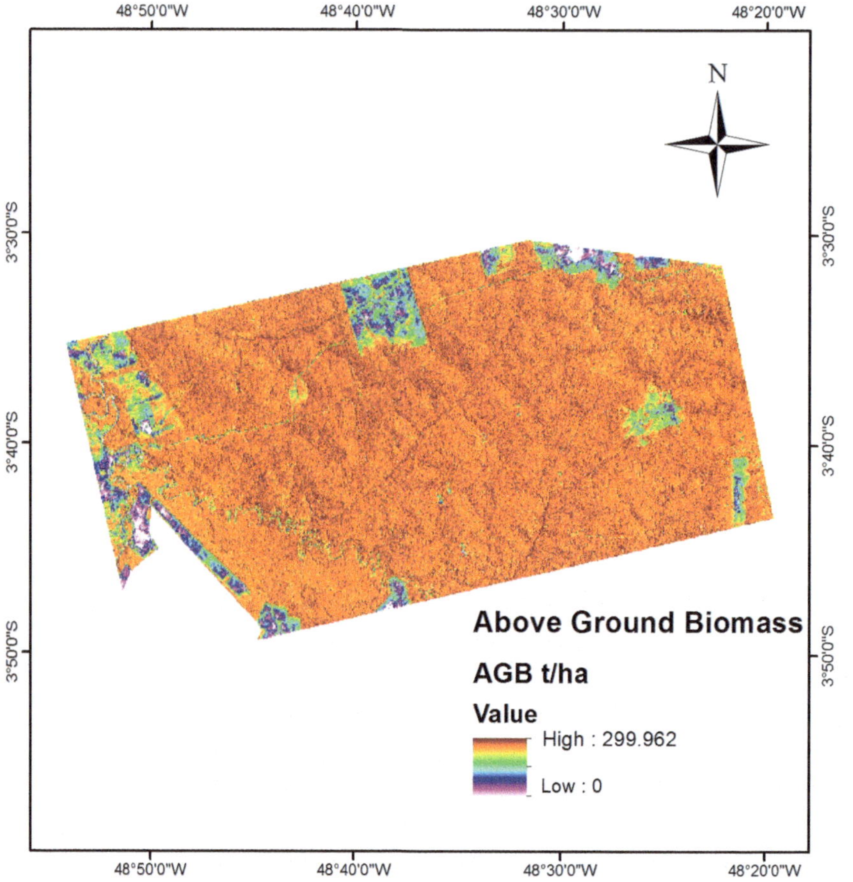

Fig. 8 Biomass map generated using combination of optical and SAR data

due to small number of samples. Also, Model generated is site specific and depends on structure of tree and weather of surrounding area. Tropical trees have higher density when compared to other species, resulting in attenuation of SAR signals. There is future scope to try other vegetation indices like RVI, EVI, etc., along with various polarization combinations. It is expected that with larger number of sample plots, modeling, and validation statistics become more robust [2]. Figure 8 depicts final biomass map generated using MLR equation.

Acknowledgements The author would like to thank ASF and ORNL DAAC for providing ALOS PALSAR dataset, Forest inventory datasets (CMS project), respectively.

References

1. Araujo TM, Higuchi N, de Carvalho Junior JA (1999) Comparison of formulae for biomass content determination in a tropical rain forest site in the state of para, Brazil. Forest Ecol Manag 117(1–3):43–52
2. Baig S, Qazi WA, Akhtar AM, Waqar MM, Ammar A, Gilani H, Mehmood SA (2017) Above ground biomass estimation of Dalbergia sissoo forest plantation from dual-polarized ALOS-2 PALSAR data. Can J Remote Sens 43(3):297–308
3. Castel T, Beaudoin A, Stach N, Stussi N, Le Toan T, Durand P (2001) Sensitivity of space-borne SAR data to forest parameters over sloping terrain. Theory experiment. Int J Remote Sens 22(12):2351–2376
4. dos Santos M, Pinage E, Longo M, Spinelli-Araujo L, Keller M (2015) Characterized edge effect with the use of lidar data in a degraded forest land-scape in the municipality of paragominas (pa). In: Proceedings of the XVII Brazilian symposium on remote sensing-SBSR, Joao Pessoa, Brazil, vol 25
5. Goh J, Miettinen J, Chia AS, Chew PT, Liew SC (2014) Biomass estimation in humid tropical forest using a combination of ALOS PALSAR and spot 5 satellite imagery. Asian J Geoinform 13:4
6. Hamdan O, Aziz HK, Rahman KA (2011) Remotely sensed l-band SAR data for tropical forest biomass estimation. J Trop Forest Sci 318–327
7. Harrell PA, Kasischke ES, Bourgeau-Chavez LL, Haney EM, Christensen NL Jr (1997) Evaluation of approaches to estimating aboveground biomass in southern pine forests using SIR-C data. Remote Sens Environ 59(2):223–233
8. Imhoff ML (1995) A theoretical analysis of the effect of forest structure on synthetic aperture radar backscatter and the remote sensing of biomass. IEEE Trans Geosci Remote Sens 33(2):341–352
9. Kellndorfer JM, Dobson MC, Vona JD, Clutter M (2003) Toward precision forestry: plot-level parameter retrieval for slash pine plantations with JPL airsar. IEEE Trans Geosci Remote Sens 41(7):1571–1582
10. Kumar KK, Nagai M, Witayangkurn A, Kritiyutanant K, Nakamura S et al (2016) Above ground biomass assessment from combined optical and SAR remote sensing data in Surat Thani Province, Thailand. J. Geogr. Inf. Syst. 8:506
11. Lucas R, Armston J, Fairfax R, Fensham R, Accad A, Carreiras J, Kelley J, Bunting P, Clewley D, Bray S et al (2010) An evaluation of the ALOS PALSAR L-band backscatter above ground biomass relationship Queensland, Australia: impacts of surface moisture condition and vegetation structure. IEEE J Sel Top Appl Earth Obs Remote Sens 3(4):576–593
12. Mitchard ET, Saatchi SS, Woodhouse IH, Nangendo G, Ribeiro N, Williams M, Ryan CM, Lewis SL, Feldpausch T, Meir P (2009) Using satellite radar backscatter to predict above-ground woody biomass: a consistent relationship across four different African landscapes. Geophys Res Lett 36:23
13. Tanase MA, Panciera R, Lowell K, Tian S, Hacker JM, Walker JP (2014) Airborne multi-temporal l-band polarimetric SAR data for biomass estimation in semi-arid forests. Remote Sens Environ 145:93–104
14. Verissimo A, Barreto P, Mattos M, Tarifa R, Uhl C (1992) Logging impacts and prospects for sustainable forest management in an old Amazonian frontier: the case of Paragominas. Forest Ecol Manag 55(1–4):169–199
15. Wang C, Niu Z, Gu X, Guo Z, Cong P (2005) Tropical forest plantation biomass estimation using RADARSAT-SAR and TM data of South China. In: MIPPR 2005: SAR and multispectral image processing, vol 6043. International Society for Optics and Photonics, p 60432Q

Geospatial Data Availability Through Map and Server—A Case Study

Bipin Chand Pandey

Abstract Now days free mapping tools are available to prepare a map, but they are not so much reliable and authentic, because the new mapmakers are not so much trend about principle, concept and methodology. So improperly designed map with lack of modern concept and technique do not effectively use for analysis and mapping. The study and analysis of area of this paper is lying between latitude of 17:19:0.5–17:30:07 (approx.) and longitude of 78:27:30 to 78:38:29 (approx). It covers an area of 400 km^2 (around 10 km from IISM, SOI, Uppal, Hyderabad). The area falls in Survey of India OSM Sheet No. E44M7 & E44M11. The study and its analysis using MS-V8, Erdas and GIS, and creation of public utility map in terms of spatial and non-spatial data is a challenging task. The heavy data volume can't be handled and analysis by traditional system of record keeping. This paper is to develop integrating concept of GIS for future smart city projects. It is very helpful for future city/urban planning and healthy city development. Finally it is Upload and displaying the Map of selected area of city on Internet browser with attribute using open source software Geo-server, Google Earth & Bhuvan Portal (for upload data). Those map which have spatial and non spatial information along with geometrical accuracy will have a massive demand not only in user community but also in infrastructure development in a systematic manner. Forthcoming smart city project, interactive revolution of Geospatial mapping tool will have a very powerful optimization.

Keywords Analysis and mapping · Smart city · Open source software · Infrastructure development · Geospatial mapping

---

B. C. Pandey (✉)
Indian Institute of Surveying and Mapping, Survey of India, Hyderabad, India
e-mail: bcppandey@gmail.com

J. K. Ghosh and I. da Silva (eds.), *Applications of Geomatics in Civil Engineering*,
Lecture Notes in Civil Engineering 33, https://doi.org/10.1007/978-981-13-7067-0_49

1 Introduction

Preparing and making good maps, which have a truly modern concept for developing city and urban area, is a very time taking, challenging and expensive task. There are different projection and datum are using for mapping purpose as per their basic requirement.

Now days free mapping tools are available to prepare a map, but they are not so much reliable and authentic, because the new mapmakers are not so much trend about principle, concept and methodology. So improperly designed map with lack of modern concept and technique do not effectively use for analysis and mapping.

In Present scenario urbanization and its requirement is a big issue and challenge for any country. Since day by day Population is rapidly increasing, due to this many problems comes in the city like Pollution, Population, Heavy traffic, Drinking water (Quality) problem, Collecting and treating wastewater, Drainage problem etc.

In Modern technique GIS have a power full concept and tool for addressing and analyzing geographic data and environmental issues. GIS is capable to handle large and diverse geographic dataset.

Improved hardware, software and networking technology have created opportunities for the build and benefit from GIS.

GIS is using for Utility/Guide mapping along with best of the breed software application capabilities. We can prepare accurate utility map and:

- Develop more efficient urban planning and for modern smart city concept.
- Make updating efficiently, simple and fast.
- Make better decision based on comparative study of metadata.
- Logically store spatial and non-spatial information.
- Work with entire digital viewing like Village, Tahsil, District, State, Country or World.
- Produce accurate map in more effectively using internet with WMS/WFS services.

2 Objective

The Study of this case is to develop integrating concept of GIS for future smart city projects. It is very helpful for future urban planning of healthy city development and also to rejuvenation the metro cities and improved the basic infrastructure. Finally it is displaying the study area of city on Internet browser with attribute using open source software Geo-server & Google Earth and uploaded on Bhuvan Server.

3 Data

- **Documents/Data Used**:
  - Survey of India Hard Copy of Hyderabad Guide Map of edition 2014.
  - Soft Copy of Hyderabad Guide Map of edition 2000.
  - Survey of India OSM Sheet No. E44M7, 11 of Scale 1:50,000.
  - Survey of India Topo Sheet No. 56 K/7, 11 (NE, NW, SE, SW) of Scale 1:25,000.
  - Imageries downloaded from Google Earth Pro for digitization/updation.
- **Hardware/Software Used:**
  - A high configuration computer system to carry out the project. (Processor—i3, Hard Disk—500 GB, RAM—2 GB, Latest Monitor Screen with keyboard and mouse).
  - HP Design jet 500PS Plotter to take hard copy of imageries/Map/necessary plots.
  - Arc GIS software available/new version, Erdas Imagine 10.1, Microstation V8.
  - Internet connection for downloading imageries (if latest imagery not available for updation).
  - All software tool of Microsoft office available/new version.
  - Java jdk (8u60-windows-i586) for background support.
  - Geo Server 2.7.2 from open source software.

4 Study Area

The study and analysis area of this paper is lying between latitude of 17:19:0.5–17:30:07 (approx.) and longitude of 78:27:30–78:38:29 (approx). It covers an area of 400 km^2. The area falls in Survey of India OSM Sheet No. E44M7 & E44M11 (Figs. 1 and 2).

5 Methodology

The Preparation Study and Analysis of Geospatial data using MS-V8 and GIS of specified area and creation of public utility map in terms of spatial and non-spatial data is a challenging task. The heavy data volume cannot be handled and analysis by traditional system of record keeping. Since the map are easily destroyed and displaced and analysis with map is a very tedious job and not fulfill user requirement in quick way. So GIS is a very useful and for different type of analysis of GIS data for creation of utility map, which is helpful to prepare future urban planning of healthy city development and its requirement.

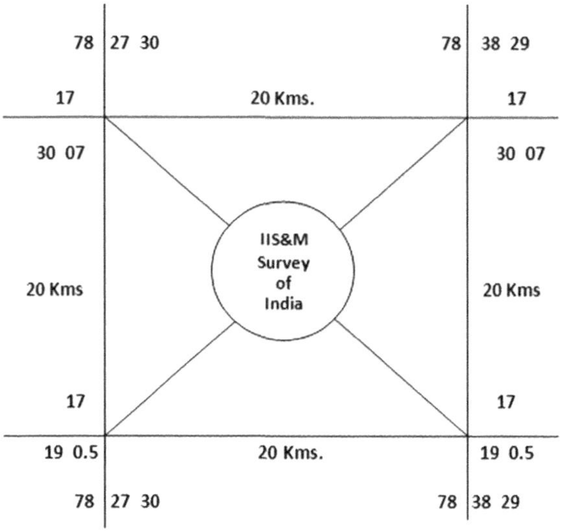

Fig. 1 Area layout

Proper planning was done before the initialization of project work. In the modern world, day by day population is increasing rapidly which required increasing the related facilities like communication network, educational institutions, religious places, industries, commercial and residential buildings etc. to lead the normal life for the human being.

To locate the details different types of maps are being used. But in conventional hard copy maps we were only able to locate the spot, whereas we were not able to get all related information of the details. Accordingly, it was decided to describe the spatial and non-spatial data of this area in GIS with proper entity, attribute and relationship.

To explore and enhancing Public utility capabilities with the help of this study. Route Network Analysis, Buffer analysis has done of different zone, and other closing facility available in the city for various Medical/Fire/Police emergencies.

Since the received vector data is not suitable for GIS analysis purpose. So prepared some important layer like Road, Railway, Residential Zone (Block), Religious feature, Boundary, Water feature in Micro-station V8 using Geo-referenced Hyderabad SOI Guide Map and update with Mosaic and Geo-referenced Google imagery, and prepare some public utility feature in GIS. The extent of imagery on Google Earth Pro imagery was downloaded in 36 tiles.

The study was planned and executed in a number of systematic stages as shown in the work flow diagram

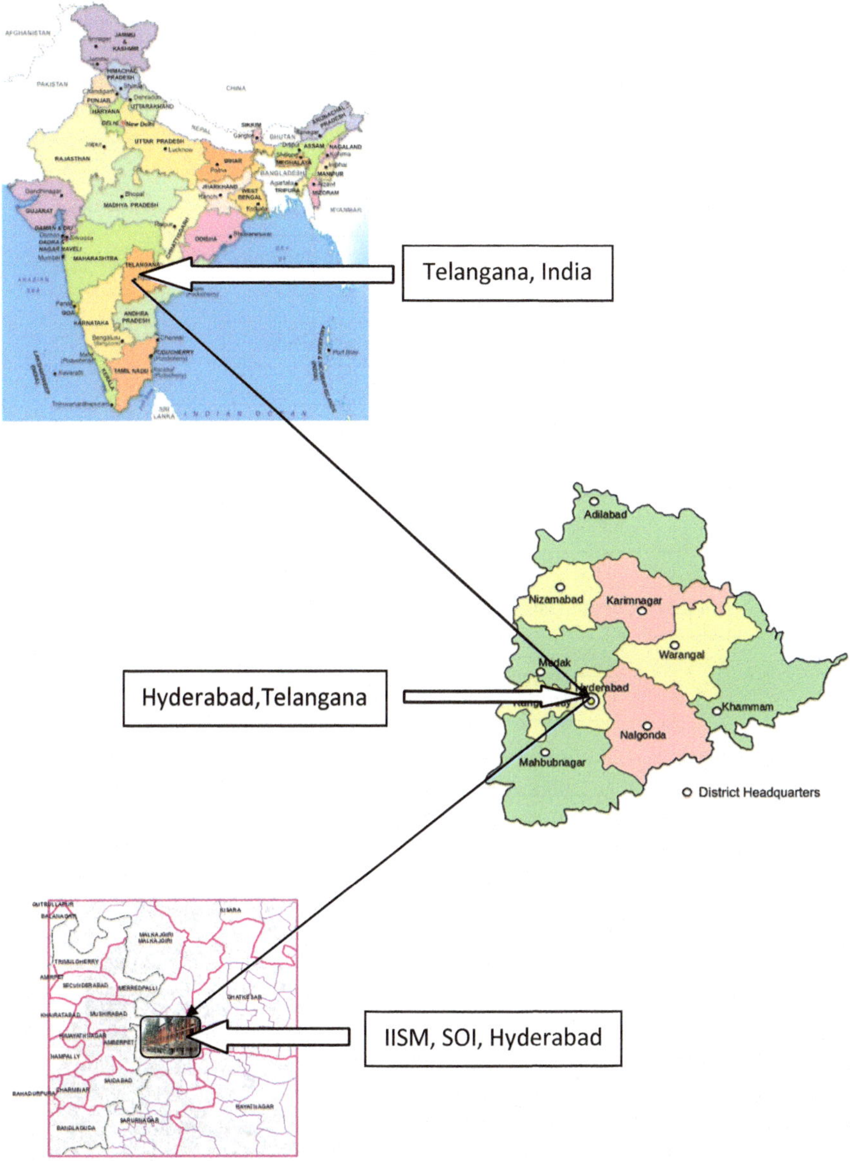

Fig. 2 Layout map of study area

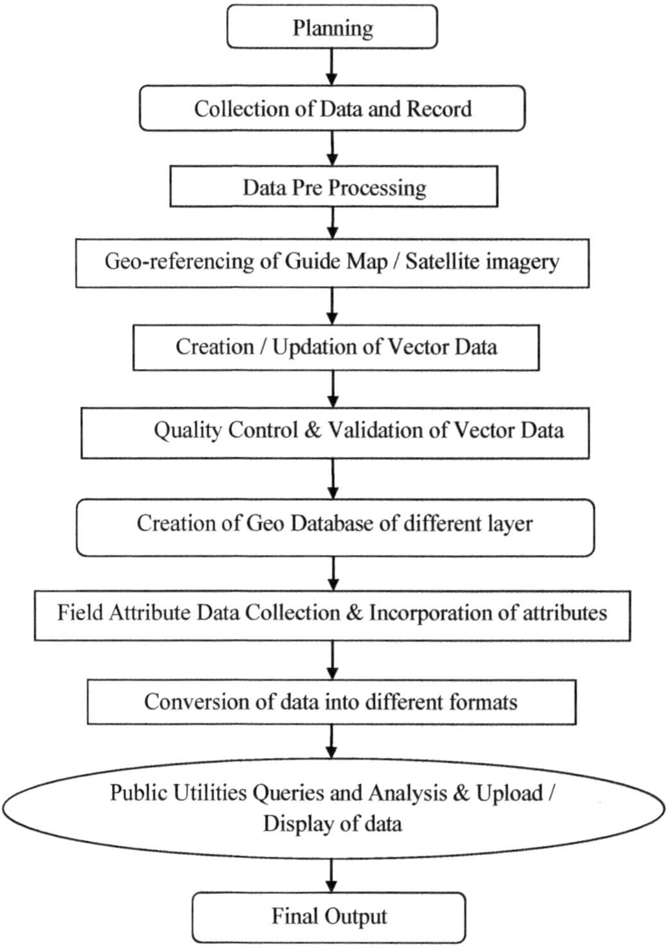

6 Observation

The study and analysis this Geo database is based on the utility planning for public and well distribution of resources and development. In present status public or user is very conscious about their time and money. It is reality that people are greatly enthusiastic in adapting to new technology and new way of working together and take lot of interest in knowing new ways of doing things. Secondly the people raise their status with political, economic and social affiliations, and express their solidarity in accordance with new tools and technology in their day to day work. The new generation feels social upheaval and enhancement in their social status along with rapidly growing technology. The people's representatives give all encouragement and support and stand by the people in introduction of new technology and in new ways of doing things, irrespective of their political affiliations. The Central and State planning department plays an important role to consider advancement of new mapping technology and make some solid steps about revaluation of mapping concept.

7 Scope of Paper

- Map having spatial and non spatial information will be massive useful for all Govt. Planning Department.
- Network analysis will be very useful for Crime, Route analysis, and also use to solve vehicle routing problem.
- Geo coding can also be done. For example we can find out details of every houses by telephone no's/mobile no. It is easy for tax collection and development funds.
- If this map uploaded with latest imagery on server and also have insertion facility then it will be have real time data for user community along with spatial and non spatial information.
- We can also add another layer into this geo database like Water Pipe line, Sever Line, Gas Pipe Line etc.
- Here we can observe how the land parcel data are converted into residential/commercial/industrial field.
- Govt. can also suggest to public domain about importance of Land. As day by day construction of building is going to beyond of limit and urbanization is also capturing a huge mass amount of land. Due to this the ecosystem is poorly affected. Therefore Land use for Residential/Commercial building construction should be as per zonal requirement and insure about healthy and proper utilization of land.

8 Final Output

If database prepared systematically and properly, then this map can be delivered in a very short time to user, whether it is hard copy or soft copy on demand (Figs. 3, 4, 5, 6, 7, 8 and 9).

9 Conclusion

During analysis Observed the Following realistic concept and realize to develop mapping feature along with user friendly interactive GIS for a healthy city development. This analysis says

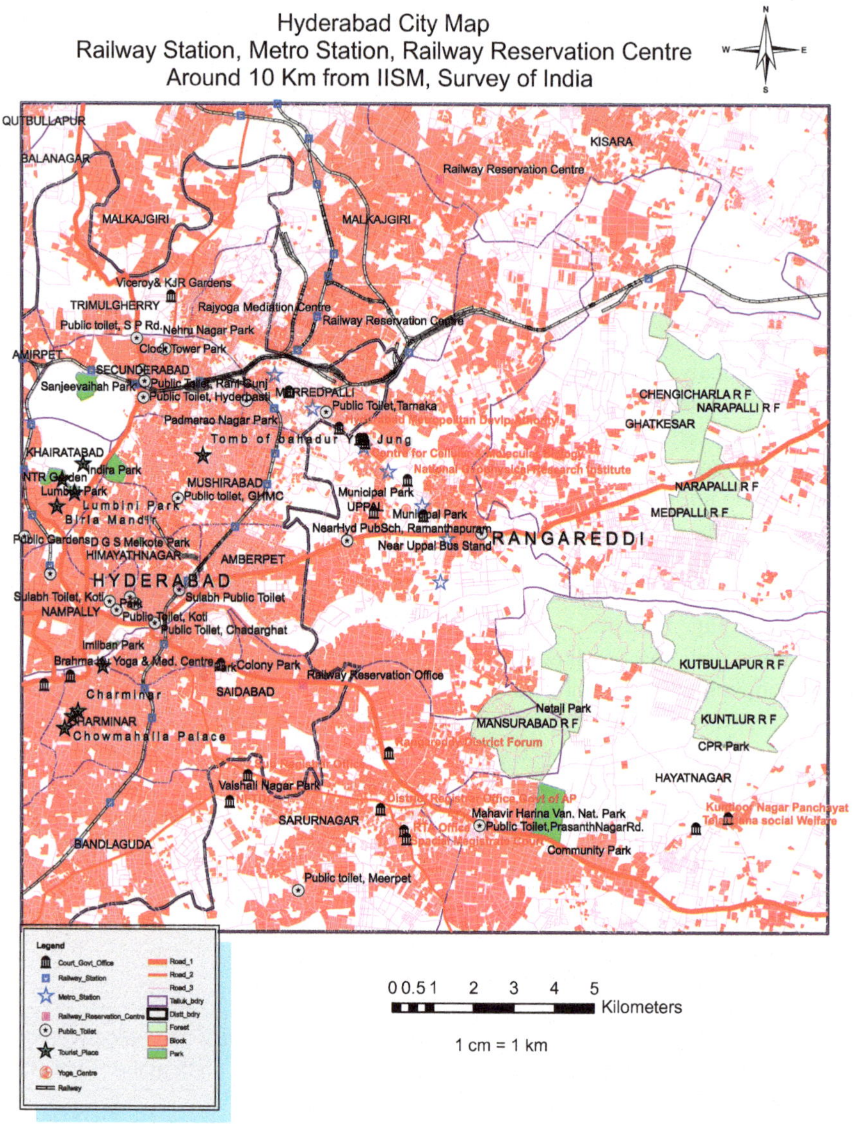

Fig. 3 GIS map of study area

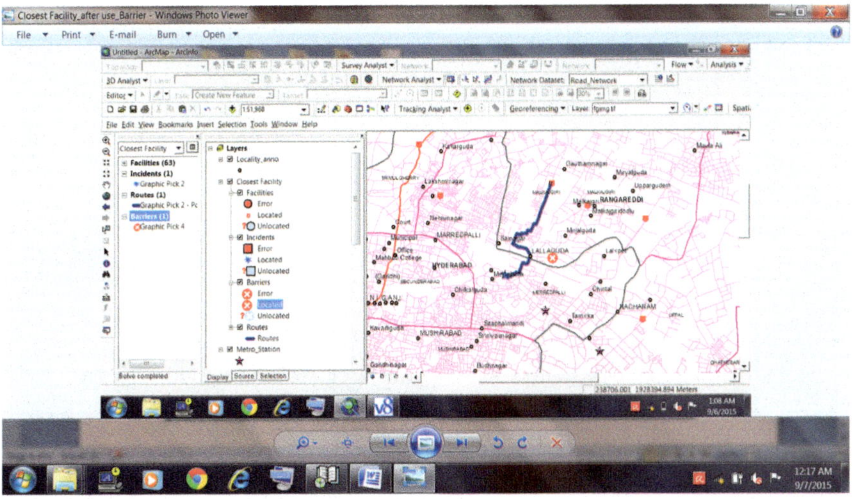

Fig. 4 Closest facility analysis

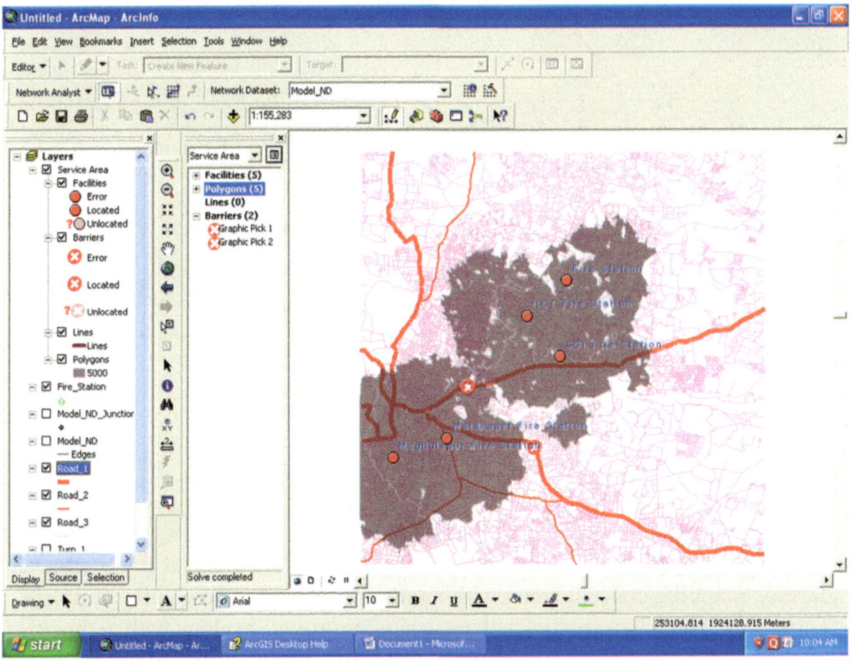

Fig. 5 Fire service area within 5 km

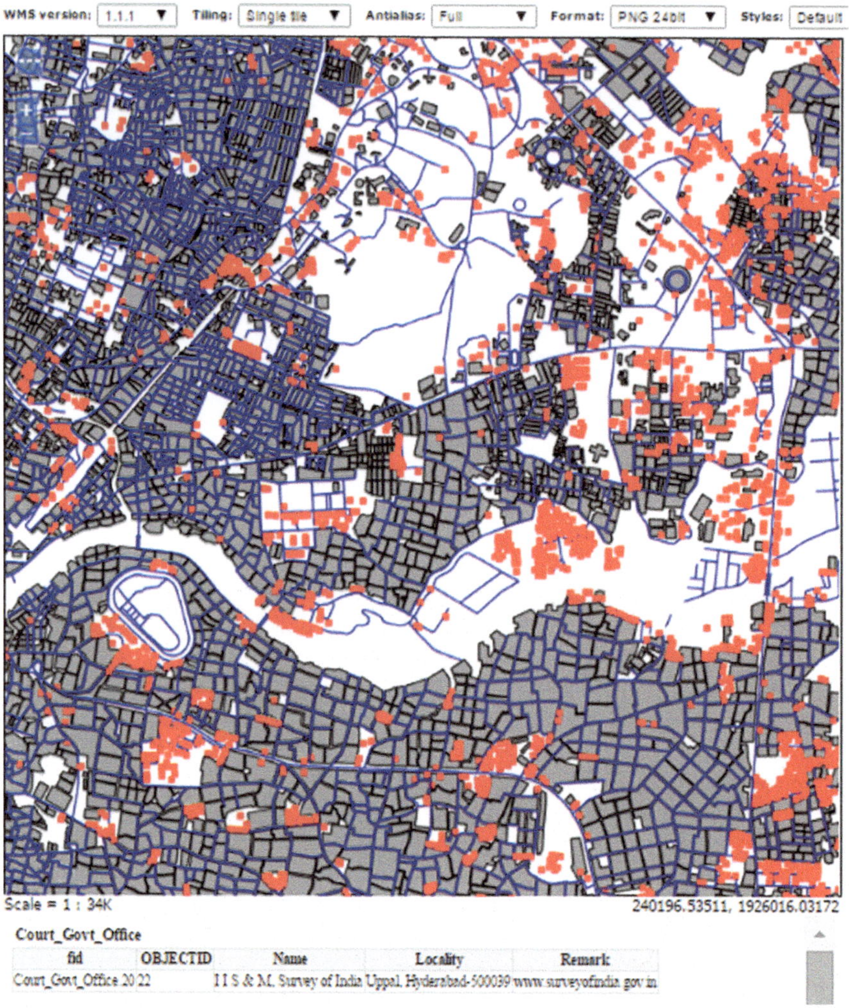

Fig. 6 Display of study area through on Geoserver

- In modern mapping scenario and quick result with accuracy, it is necessary to develop mapping concept along with modern technology and coming software.
- Those map which have spatial and non spatial information along with geometrical accuracy will have a massive demand not only in user community but also in infrastructure development in a systematic manner.
- Forthcoming smart city project, interactive revolution of Geospatial mapping tool will have a very powerful optimization.

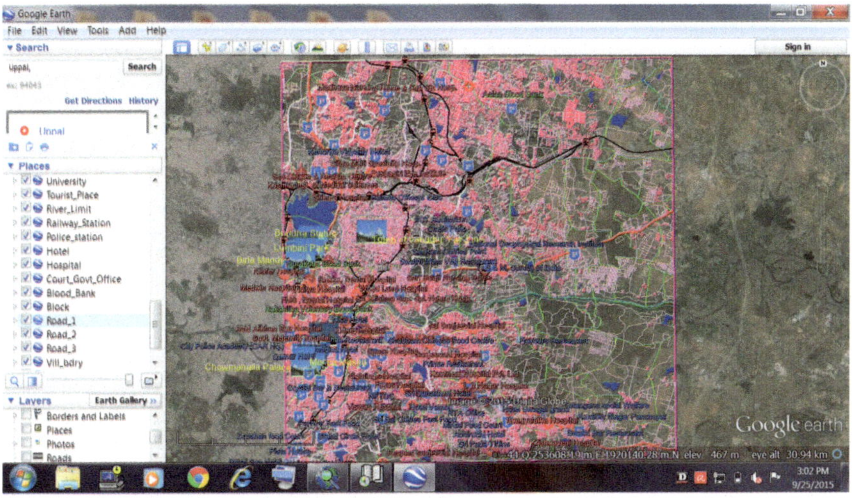

Fig. 7 Upload all layer of study area on Google Earth

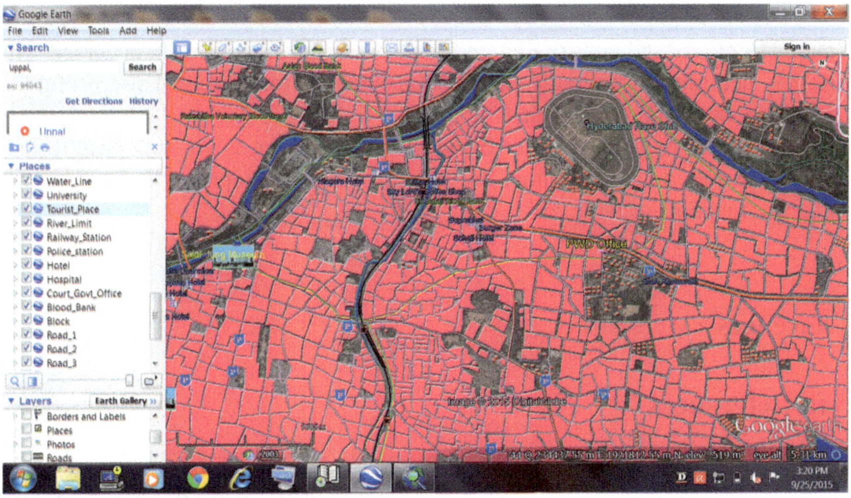

Fig. 8 Study area (a zoom portion) on Google Earth

- Observed that city is growing rapidly and extended also, have fulfill so many public real and modern necessities, but have a very less open area like Park/Garden, Foot over/under bridge, which is a very common and important factor for a healthy and smart city.
- During field attribute collection and verification of spatial data, Observed that many places, which are facing regular problem due to lack of drinking water. The

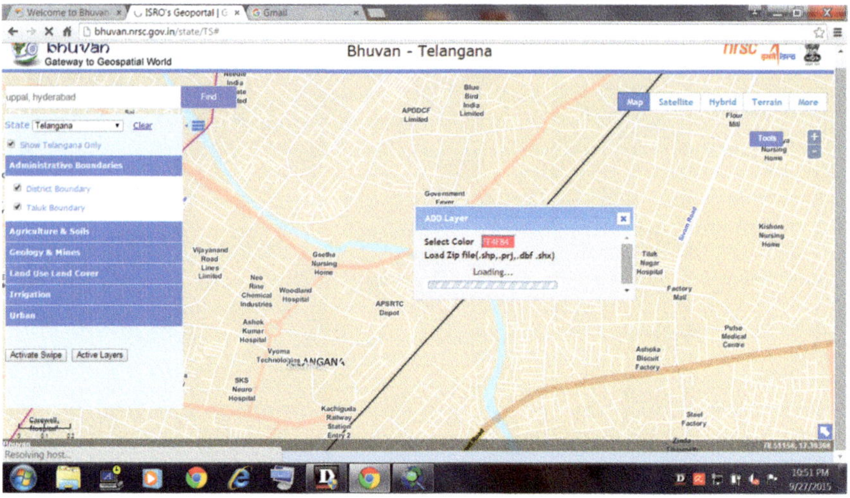

Fig. 9 Upload layers on Bhuvan Portal

mapping of city showing spatial and non spatial information with accuracy will be very helpful for future planning and management for Govt. and also useful for public to aware future problems.

- Now a day's Population is increasing and urbanization is growing, due to this reason city becomes unhealthy. So for systematic development, it is necessary upgrade mapping criteria.

References

Web References

1. google.earth.com
2. bhuvan.nrsc.gov.in
3. geoserver.org
4. surveyofindia.gov.in

Web-GIS-Based Interface for a UBA Selected Village

B. Pavan Kumar and K. Venkata Reddy

Abstract The Unnat Bharat Abhiyan (UBA) mission is to enable people of higher educational institutions, work along with people of rural areas by identifying their development challenges and evolving solutions for accelerating sustainable growth. Planners and policy makers have to depend on spatial infrastructure and non-spatial data for effective planning and decision-making. The Web-GIS with integration of both spatial and non-spatial data, i.e. household data which helps in extending the information accessibility for rural mass and the administrators in planning development activities as well as monitoring different rural programs by the government. This aim of this work is to provide interactive geospatial database, which helps in accessing the household information including spatial data for Ramulapalli village of Kannuru gram panchayat selected under UBA program. Household survey data as non-spatial data, and google earth image as Spatial data integrated into Geospatial environment and provide through web interface. Web-based interface, containing information that is helpful planning in development activities has been developed using MapGuide open source platform.

Keywords Non-spatial · Spatial · Unnat Bharat Abhiyan (UBA) · Village · Web-GIS

1 Introduction

1.1 General

Rural development is still facing key issues like less diversified occupation, lack of infrastructure in terms of education, health, road transportation, etc. Applications of GIS had a great relevance in the planning of rural areas development [1–3], land

B. Pavan Kumar (✉)
Civil Engineering Department, NIT Warangal, Warangal 506004, India
e-mail: bpavankumar@student.nitw.ac.in

K. Venkata Reddy
Department of Civil Engineering, NIT Warangal, Warangal 506004, India
e-mail: kvreddy@nitw.ac.in

J. K. Ghosh and I. da Silva (eds.), *Applications of Geomatics in Civil Engineering*, Lecture Notes in Civil Engineering 33, https://doi.org/10.1007/978-981-13-7067-0_50

management, drinking water assessment and planning rural facilities apart from those it can be aiding rural governance and panchayat administration activities to take a big leap in providing a single geospatial database [4]. Spatial information technology in recent years advanced at a remarkable pace. Spatial data is evolved as multidisciplinary with various platforms of data like ground survey results, Maps and charts, aerial surveys give aerial photographs and space-based as satellite images. GIS can be defined as system of computer, which includes hardware and software that is designed to allow to collect, manage, analyse and retrieve large volumes of spatial referenced data and associated attributes collected from a variety of sources [5]. In GIS, both spatial and non-spatial data, i.e. collected by surveys can be integrated and a set of spatially registered layers formed as shapefiles can be analysed independently or in combination.

1.2 Web-GIS

The Internet, which is a major carrier around the world for information, offering more meaningful direction in terms of planning better development. Necessary support from the advanced spatial information technologies with GIS availability at every walk of life as Web-GIS. Web-GIS, which is supporting multiple clients via internet protocols. Web-GIS components are a Map server, client-server and a Web browser. Web-GIS functioning is like the following:

- The Web interface, where clients can perform query on web.
- The queries or updates from clients are received using HTML.
- Server that performs the GIS operations and sends response to client via HTML.

Web-GIS is more advantageous than simple desktop GIS which operates on a personal computer, limiting access. The Internet provides an alternative solution to a Desktop GIS, with flexibility to work remotely, collaboratively and be reactive to macro influences. Web-GIS offers with tools to quickly import geo data and present it on a map in multiple formats.

Open source software's for Web-GIS as used by [6–9] are like GeoMajas, GeoServer (supports the WMS, WFS, WCS standards of Open Geospatial Consortium (OGC)), MapFish, MapServer, OpenLayers, OpenWebGIS, TileMill.

1.3 MapGuide Mastero

MapGuide Maestro is a free application that manages the spatial data in MapGuide open source. MapGuide mastero has various editor interfaces that come handy in feature loading, layer definitions, map definitions, web layouts and predeveloped fusion layouts. It uses Apache as the client server. Various GIS tools can be developed in this platform using XML formats.

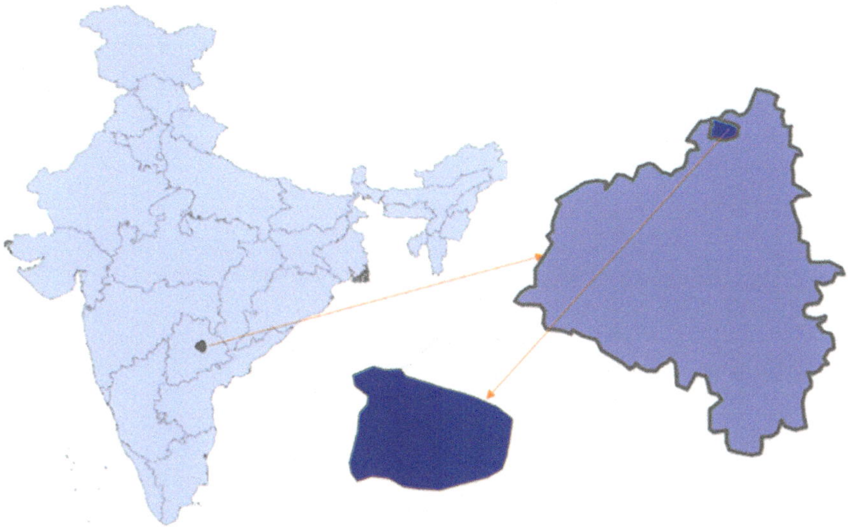

Fig. 1 Study area

2 Study Area

The study area is from India's 29th state Telangana, Warangal Urban district, kannuru gram panchayat. It is located between Latitude and Longitude of 18.205886°N–18.204216°N and 79.505174°E–79.508580°E. it as shown in Fig. 1.

3 Methodology

General Flowchart of Methodology used in developing a Web-GIS-based interface with MapGuide maestro was like data side various shapefiles preparation. Loading them into MapGuide server software and their handling in server side. In Web Interface viewing data in AJAX format. Figure 2 shows this general methodology with these three main parts of any Web-GIS interface.

3.1 Data Collection

Non-spatial data, which is an attribute data collected using a detailed household survey based on the questionnaire of UBA. Spatial data is developed using clipped high resolution google earth image.

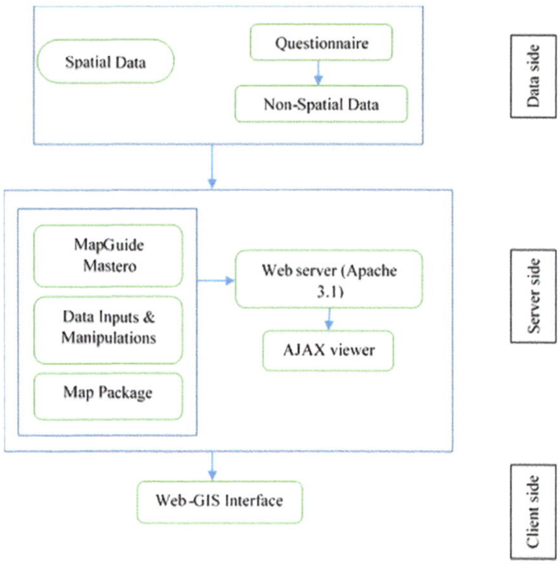

Fig. 2 Flowchart of methodology

3.2 *Shape File Creation*

The Spatial data is prepared from digitising houses and linking them with non-spatial data using house number as a common attribute and preparing shape file. Various other shape files like Road network and village boundary also prepared by digitising.

3.3 *MapGuide Mastero*

MapGuide maestro is an application to be used to manage spatial data of MapGuide open source. Map building process, in MapGuide maestro started by a root folder Ramulapalli and other five subfolders like data, layer definitions, map, web layout containing all spatial data. Figure 3 shows the view of the MapGuide mastero with loaded shapefiles.

In the left column of MapGuide maestro shows folder structure including the loaded data. With shapefile load procedure, feature sources were loaded to data subfolder and layer definitions to layer folder. Projection system for the uploaded file mentioned at the time of loading shapefile. Stylization of these shape file like attribute data labels and symbols, for example, theme editor for road network based on the type of roads, selecting type of variable to be visible on the map etc.. Now map definition was done by stacking of all the three shape files in a preplanned order.

Web layout was created with map as resource utilising layer definitions containing DMSG fusion templates nothing but a HTML coded webpage. The widget set like Menus (File Menu, Context Menu, Task Menu), Toolbars collection of vari-

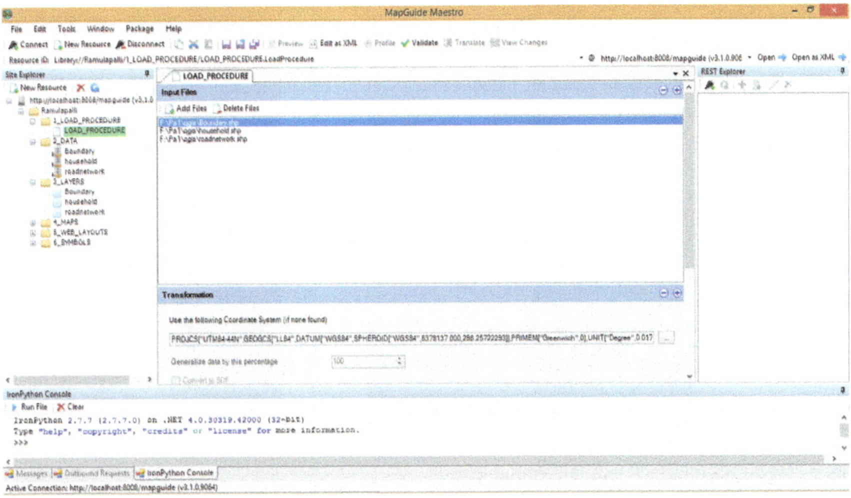

Fig. 3 Shapefile loaded with respective projections

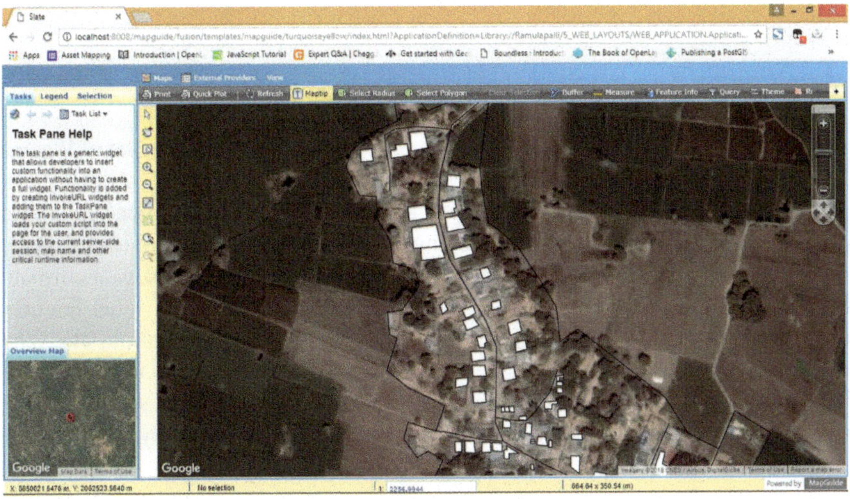

Fig. 4 Web-GIS interface with google satellite viewer in background

ous tools like print tools, navigation tools and analytical tools are included in the webpage. Panes set in the web interface like Map pane, Task pane, Information pane are included with certain ready available tools like buffer tools, search tools, etc., are included. External map sources like google street view and google satellite are included in the web interface. Web-GIS-based Interface for Ramulapalli village developed with google satellite as background shown in Fig. 4, whereas Fig. 5 shows all available tools in the web application.

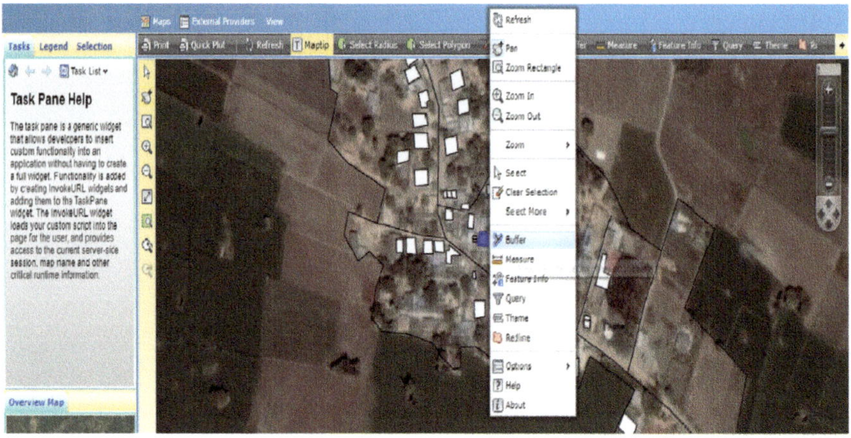

Fig. 5 Web-GIS interface with all available tools

4 Summary and Conclusions

4.1 Summary

Web-GIS application for Ramulapalli village of Kannuru gram panchayat of Warangal urban district has been developed. The interface of each tool possessing interactive capabilities with their resemblance to desktop GIS. Web application allows to perform every GIS analysis in planning like buffer analysis for roads, distance and area measurement as length of road, area of houses and query based on attribute data. There is a red line tool that helps in adding new shapefiles for other features like tanks and fields. Theme tool serves analysis like poverty education, etc., The number of beneficiaries for each government scheme, income status, land and irrigation availability, poverty status, education levels of each member and social security pension, etc., which are helpful for administration to plan accordingly.

4.2 Conclusions

The Web interface designed with the purpose of providing a usable and within reach geospatial environment. Web-GIS is a cost-effective GIS solution for database, which contains every rural related information spatially. GUI, i.e. developed uses open source GIS that helps users to perform spatial and non-spatial information retrieval, querying easily.

References

1. Alajangi S, Rao Pyla K, Eadara A, Prasad NSR (2013) Web GIS based information system for rural development. Int J Sci Res. ISSN (Online Index Copernicus Value Impact Factor 14, 2319–7064
2. Arora PK, Bhatia R, Parkash S, Sekhon BJS (2015) Web based rural geographic information system. In: Proceedings of 2015 IEEE international conference on electrical, computer and communication technologies, ICECCT 2015. https://doi.org/10.1109/icecct.2015.7226061
3. Kodge BG, Hiremath P (2012) Web based geo-spatial and village level information extraction system using FOSS. J Adv Inf Technol 3:222–227. https://doi.org/10.4304/jait.3.4.222-227
4. Adinarayana J, Raj FJ, Sharma V (2004) Village level information system—a tool for decentralized planning at district level in India. J Environ Inform 4:56–64
5. Garg PK (2008) Spatial planning of infrastructural facilities in rural areas around Roorkee, Uttarakhand, India. Gener J Am Soc Aging 2–15
6. Navatha Y, Venkat Reddy K (2015) Analysis and planning of infrastructural facilities in rural areas using facility index methods. J Geomat 9:237–244
7. Pispidikis I, Dimopoulou E (2015) Web development of spatial content management system through the use of free and open-source technologies. Case study in rural areas. J Geogr Inf Syst 7:527–540. https://doi.org/10.4236/jgis.2015.75042
8. Richard S, Vijayan PK (2012) Sustainable rural development with the application of Spatial Information Technology and Mahatma Gandhi national rural employment guaranteed scheme. Int J Geomat Geosci 2:1048–1061
9. Sitender, Satish Kumar, Reena (2012) Village information system—a case study of Muklan Village, Hisar, Haryana (India). Int J Res Soc Sci 2(2). ISSN: 2249-2496

Downscaling of Coarse Resolution Land Surface Temperature Through Vegetation Indices Based Regression Models

Kul Vaibhav Sharma, Sumit Khandelwal and Nivedita Kaul

Abstract In geoscience and remote sensing necessitate thermal imagery having high-resolution for various applications like estimation of the Land surface temperature (LST) analysis, thermal comfort, urban energy resources, forest fire, assessment of evapotranspiration, drought prediction, etc. We need accurate and sharp thermal images to explore surface temperature related phenomenon on frequent basis. The present physical and technological constraints have not allowed us to dig up remote sensing thermal data at high temporal and spatial resolution simultaneously. Hence, it is obligatory to construct a dynamic relation between low- and high-resolution satellite data to acquire enhanced thermal images. The present study evaluates three downscaling algorithms in our study area, namely, disaggregation of radiometric surface temperature (DisTrad), sharpening thermal imagery (TsHARP), and local model using seasonal Landsat 8 and MODIS data thermal imagery. The aggregated Landsat 8 LST of 1000 m resolution has been downscaled to 400, 300, 200, and 100 m using DisTrad, TsHARP, and the local model and compared with original Landsat 8 and resampled LST of matching level. The results have shown that LST downscaling technique performance varies over climate, surface feature and earth surface moisture conditions. The models have not performed well in surface having highest and lowest water content i.e. water bodies and arid sandy areas. Alternatively, regression-based downscaling accuracy is higher for NDVI > 0.3. For example, the accuracy of all algorithms is higher for the growing seasons (February and October) unlike the harvesting season (April). The root means square error of the downscaled LST increases from 400 to 100 m spatial resolution in all seasons. The downscaling algorithms gave realistic results of MODIS satellite thermal band to a spatial resolution of 200 m. The present study is an attempt to rationalize coarse resolution thermal image by using the association between earth facade vegetation indices and land surface temperature. The study aims to develop a robust LST downscal-

K. V. Sharma (✉) · S. Khandelwal · N. Kaul
Department of Civil Engineering, MNIT Jaipur, Jaipur, India
e-mail: kulvaibhavsharma@gmail.com

J. K. Ghosh and I. da Silva (eds.), *Applications of Geomatics in Civil Engineering*,
Lecture Notes in Civil Engineering 33, https://doi.org/10.1007/978-981-13-7067-0_51

ing algorithm for MODIS data at LANDSAT resolution. The downscaling methods successfully operate over a heterogeneous landscape and reduced thermal mixture effect to monitor the daily basis long-term environmental phenomena.

Keywords LANDSAT · MODIS · Downscaling · Regression · NDVI

1 Introduction

The earth surface thermal property and vegetation percentage have been found out directly related from the past literature. It has been observed that as an increase in vegetation has reduced the LST in both urban and rural areas. So we can say that the LST and NDVI physical associated negatively [1]. The LST-NDVI relationship has been examined in different climatic region and results shown a strong inverse connection between LST and NDVI [2]. The previously studied LST-NDVI relationship over various LULC classes and the relationship found out be linear and negative to each other [3]. The LST-NDVI correlation holds quite good in most of the earth covers apart from wet surfaces [4]. The downscaling thermal data study observed that NDVI and LST has been inversely associated with the study area [5]. The research in the agricultural fields of the USA during the crop growing season has been carried out and the results point out the unidirectional LST-NDVI linking [6]. So most of the past research in agricultural areas has mentioned a negative correlation between both surface temperature and earth surface vegetation fraction. In colder region like North–South pole ecosystem, high elevated areas from sea level and regions having higher vegetation temperature than soil, a positive correlation also exist between LST and NDVI [7, 8].

The enhancing of thermal data and increasing spatial resolution has been referred to as Downscaling LST data, i.e., disaggregating coarse resolution single pixel to new values using various numerical models [9]. In this research, LST downscaling models has been used to downscale coarser LST using regression analysis between Landsat 8 and MODIS thermal and vegetation values. We have utilized the property of earth surface vegetation indices using visual bands and surface thermal property from TIR band. The accessibility of vegetation indices (VI) at finer resolution make possible to conduct direct regression analysis between the input data to simulate LST at vegetation imagery scale.

LST-NDVI relationship depends based on various climatic and surface-related parameters, for example, incident radiation, surface reflectance, satellite path, atmospheric condition, water pixels, and surface emissivity, i.e., changes with geographic locations [10]. Earlier researches have highlighted the effect of geographical location on input parameter relationship. The regression equations between the input parameters of different degree changes its slope and intercept with location, vegetation percentage, and rainfall [11]. The spatial and temporal variability of earth thermal nature with surface vegetation cover has been studied by researchers [12]. The wetness index and moisture conditions have been one of the main factors contributing to variation in correlation of input parameters [13]. The type of crop being harvested and their cropping patterns changes over different climatic zones of earth [14]. The thermal properties of the land surface, evapotranspiration and net radiation, soil and vegetation water content observed that the slope and intercept have been site and scene-specific [15]. We have implemented an inverse association between LST-NDVI in our study area due to semi-arid climatic conditions but it has to first recognized before mounting LST downscaling model further.

The MODIS satellite operates daily basis benevolent thermal data of 1000 m and NDVI of 250 m, a downscaling of the thermal dataset at 100 m spatial resolution would indeed be a leap forward for urban thermal comfort, surface, and environment related studies. Therefore spatiotemporal analysis based on the factors mentioned above should be performed. It is experimental that the downscaling algorithm operates fine for agricultural and vegetated landscapes but not correctly validate the water body, river bed, and sandy areas. Regression-based models have been statistically contributed to perform on input data variation and various environmental factors. The models applied to MODIS thermal data to downscale to 100 m resolution with less than 1 °C accuracy. Three regression models are employed to downscale the LST of MODIS data using Landsat 8 data and results have been compared. The wide scope of research is present for developing thermal image/LST downscaling model over mixed landscapes, especially in India, where the agricultural plot size is smaller. In a country like India, various climatic regions with their inherent characteristics are present. Consequently, application downscaling model over different zones of India would help us to analyze regarding its reliability. However, the downscaling techniques have applied so far, often failed to address the problem of landscape heterogeneity and nonlinearity trends within the data during disaggregating LST. The visual quality of the downscaled LST is also affected due to boxy artifacts. As per available literature, a small number of studies practiced the downscaling model on actual coarse resolution MODIS data LST. To approach this predicament, the present research explores different parametric and nonparametric modeling techniques to downscale coarser resolution LST image and try to improve the visual quality of the image using residual modeling. The objective of the present study is to generate enhanced LST imagery on a daily basis using Landsat 8 and MODIS data input parameters to downscale MODIS thermal data. The test results for different seasons in study area having both urban, rural, and mixed landuse areas situated has been verified by original thermal data for reliability examination.

2 Study Area

The study area selected is Jaipur in the eastern part of Rajasthan State of India. The Jaipur urban agglomeration is geographically situated in subtropical arid climate having a dense settlement in old walled city. The study area selected is the Jaipur urban and its surrounding rural area in Rajasthan state of India. In perspective of the area covered, Rajasthan state is the largest state of India in which its capital city Jaipur

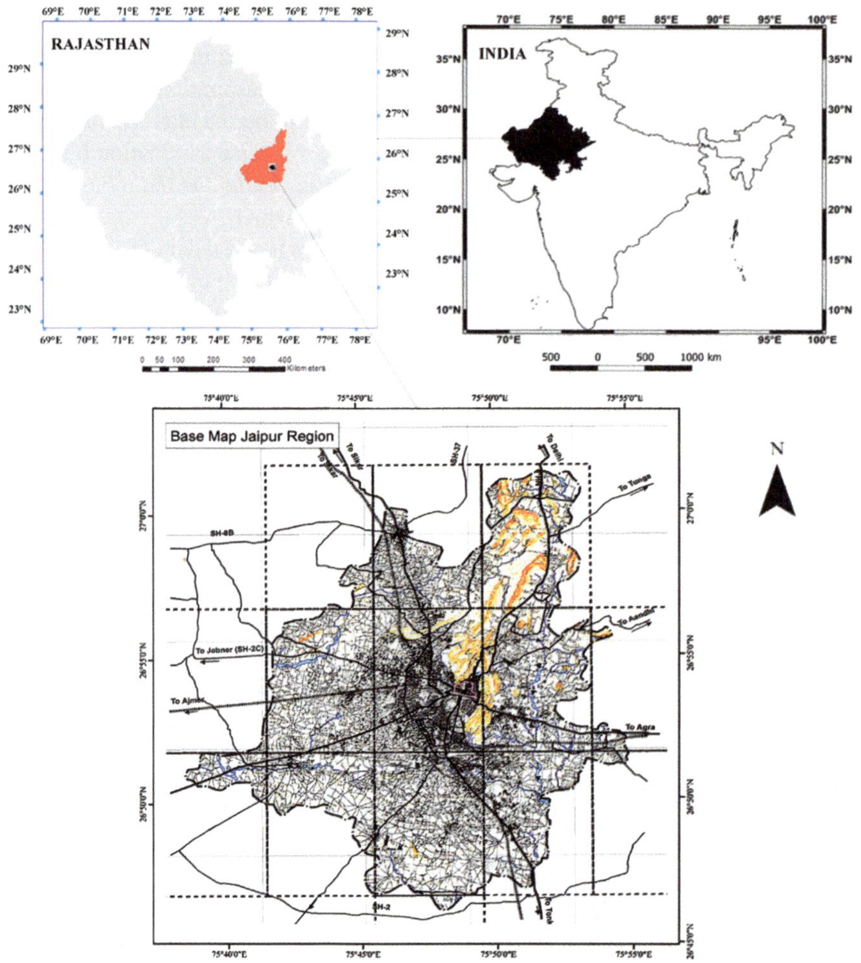

Fig. 1 Location of study area

is the tenth biggest city of India. The study area is located geographically between 26°30′18″–27°00′54″ North latitudes and 75°20′16″–76°10′40″ East longitudes. The day number selected for fusion algorithm is based on the three seasons in a year, i.e., summer (Mid of March to June), Monsoon (July to Mid of October), and winter (November to February). Landsat 8 satellite covers study area by 147/041 (path/row) in topographical positioning given in Fig. 1.

Landsat 8 images are utilized in this study. The Landsat 8 has two push broom (alongside scanner) instrument on board name as OLI and TIRS rather than whiskbroom sensors (across-track scanner) [3]. Landsat 8 satellite has three new groups of spectral bands later than added Landsat TM/ETM + sensor. The India Meteorological Department (IMD) of India designates four climatologically seasons for the study area, due to monsoon rain and cloudy weather it is difficult to get the cloud-free satellite data during monsoon seasons. The remote sensing images of February 15th, 2016, June 26th, 2016, and September 26th, 2016, on behalf of three seasons, have been in use for the study. The selected months for this study are representative of climatologically seasons of India. Entire dataset has been downloaded from Landsat 8 TIRS dataset (path/row: 147/41), MODIS satellite data band MOD11A1 and MOD09GQ of 1000 m and 250 m resolution have been adopted (Fig. 2).

The MODIS dataset has been further used in this study acquires radiometric (12 bit) sensitivity in various spectral (36) bands having 0.4–14.4 pm wavelength range. MODIS imageries nearby to Landsat 8 data have been clubbed before and after equitation days. The MODIS data thermal band radiometrically calibrated and atmospherically corrected with has been projected from Sinusoidal to UTM at 1000 m spatial resolution. We have utilized both aqua and terra platform of MODIS satellite to get surface reflectance products. The estimate of the ground level surface spectral reflectance without absorption or atmospheric scattering has been determined. The Landsat 8 TIR images have been resampled at MODIS band downscaling resolutions before operating at regression domain. The land surface temperature has been derived at original 100 m spatial resolution of TIR band. The spectral signatures between the classes of both satellite bands can be very much different from each other. The decision function is based on minimum values of (inverse cosine) spectral signature vectors of co-registered TIR and PAN bands. Sensor data processing of Landsat 8 data has been done by using ENVI and ERDAS Imagine software.

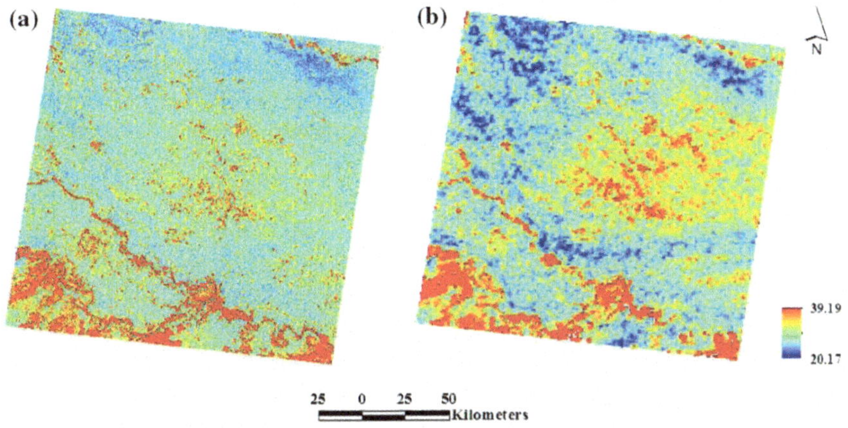

Fig. 2 LST images of **a** Landsat 8 (100 m) and **b** MODIS (1000 m) of study area

3 Data Aggregation and Downscaling Models

The NDVI of Landsat 8 study area has been derived from atmospherically corrected visible and near-infrared data of 30 m resampled to 100, 200, 300, and 1000 m spatial resolution over different forward kernel sizes (4 × 4, 8 × 8, and 32 × 32 pixels). A radiance-based aggregation procedure has been adopted in the testing algorithms to generate thermal images at diverse resolutions. The corrected radiances LST using Plank's function model have been resampled to 200, 300, and 1000 m spatial resolutions. To begin with, we checked results of downscaling from radiative transfer models with the resampled Landsat 8 thermal image. The LST has been downscaled from 1000 m to 200 and 100 m and validated with resampled original LST of comparable resolution. Results of the image enhancing modules may vary from one satellite to another due to change in wavelength, satellite position, azimuth angle, signal noise ratio, and surface spectral function. Therefore, thermal image sharpening models to generate replicated LST image may be adequate observation about the applicability and presentation of the algorithm.

3.1 DisTrad

This model tested over the agricultural fields of Oklahoma, the USA has been performed for enhancing remotely access earth surface temperature. The algorithm based on negative relationship between temperature and NDVI between for both high- and low-resolution satellite data. The downscaling model has operated over second-order polynomial regression. The second-order polynomial model works superior in higher and lower regression allocation of LST-NDVI. The DisTrad model for enhancing and model coefficients has been calculated based on the second-order polynomial least-square regression (SL_{Sr}).

$$\hat{LST} = a + b \times NDVI + c \times NDVI^2 \tag{1}$$

3.2 TsHARP

This model is an alteration of the DisTrad algorithm. This model operates a simple least-square (LS) regression function in preference to $SL_{Sr.}$ The fractional vegetation cover (fc) has been used in place of conventional NDVI approach. The polynomial function has been found more responsive to outliers compared to the simple LS. The present study uses both NDVI and fc as a covariate function with the LST. The TsHARP model Eqs. (2)–(4) given.

$$\text{Polynomial fit} \quad LST = a + b \times NDVI + c \times NDVI^2 \tag{2}$$

$$\text{fc transformation} \quad LST = a + b\left(1 - \left(\frac{NDVI\max - NDVI}{NDVI\max - NDVI\min}\right)^{0.625}\right) \tag{3}$$

$$\text{Simplified fc transformation} \quad LST = a{-}b(1{-}NDVI)^{0.625} \tag{4}$$

3.3 Local Model

The local model is an alteration of the previously attempted TsHARP algorithm for downscaling of LST values of coarser remote sensing data. This model operates moving window of different mask size regression technique in view of the restricted neighborhood for LST sharpening over an urban area has been applied. The thermal and vegetation association varies with change earth cover, moisture condition, vegetation type, and atmospheric conditions. We have implemented a local pixel informatics regression so that it targets a better accuracy in sharpening LST. LST and NDVI variation over both input satellite band data has been used for model building. We have used different kernel window sizes both overlapping or nonoverlapping of 3×3, 7×7, and 9×9 in which 7×7 nonoverlapping window size has given the higher accuracy so as recommended for the present study area.

4 Result and Discussion

The performance of the LST downscaling model due to the seasonal variation of the LST-NDVI relationship is examined in the present study. The remote sensing images of February 15th, 2016, June 26th, 2016 and September 26th, 2016 have been taken. The temperature and rainfall characteristics are different. The results have been comparatively far better in the agricultural land-use then in the water body and sand areas. Therefore, downscaling models are applied to actual MODIS thermal imagery and validated. In the tropical regions, temperature, rainfall, and vegetation density/cover vary in different seasons. It affects the LST-NDVI relationship. The Landsat-8 band 10 and 11 thermal images pixel values in digital numbers (DN) have been converted into surface reflectance, thereafter surface reflectance converted into the top of atmosphere radiance values. Surface leaving radiances for TIR bands has been calculated using atmospheric correction parameters and emissivity values. The surface emissivity has been calculated using vegetation proportion (Pv). The sun angle provided and band Multiplicative factor is available in the product metadata file. The proposed algorithms data set pair has been co-registered at nearest date available having cloud cover less than 1%. Initially, Landsat 8 satellite TIR band LST values at 100 m resolution have been resampled and co-registered to the MODIS band data pixel by pixel (Fig. 3).

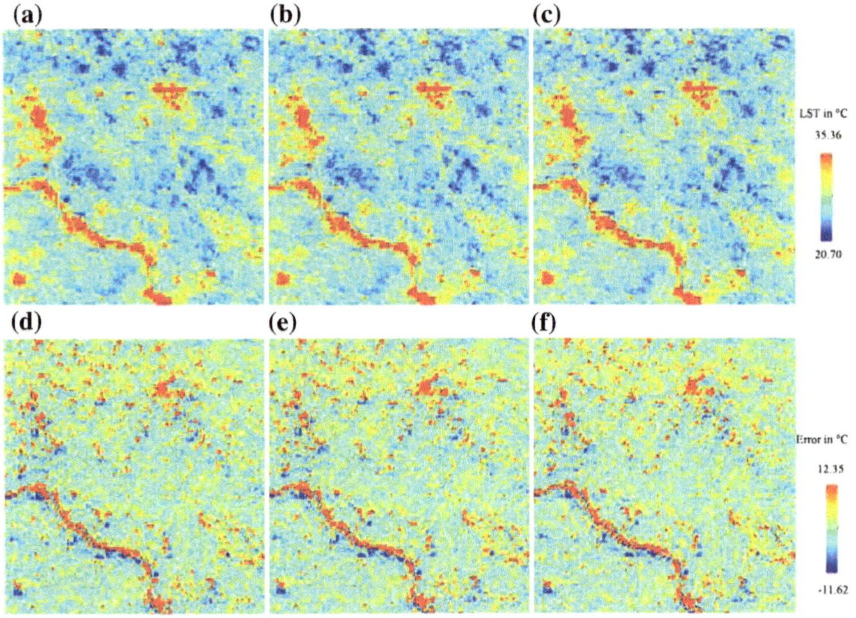

Fig. 3 LST downscaled using **a** Distrad, **b** TsHARP and **c** Local model and LST retrieval error are shown (**d**–**f**)

Table 1 Error classes of method used for LST downscaling

Error	0–0.5	0.5–1	1–1.5	1.5–2.0	2.0–2.5	2.5–3.0	>3.0
Method 1	18.75	19.24	8.37	8.66	13.17	15.8	16.78
Method 2	36.12	30.24	19.62	2.36	3.17	1.8	6.78
Method 3	45.45	38.21	8.12	3.71	2.16	1.7	0.65

The surface vegetation proportion is based on the cropping season and results have given less error for high vegetation time period. The monsoon seasons favors regression analysis between so NDVI > 0.35 has been significant to get precise results for urban and rural areas (Table 1).

Owing to less vegetated areas, most of the satellite capture area has transformed into current fallow with NDVI < 0.35 and LST > 40 °C (Table 2).

The present data point toward the LST-NDVI allocation is exaggerated by seasonal influences and vegetation cover conditions. The average NDVI of September is 0.56, indicating vegetation vigor. After rainy days, the initial phase of crop growth takes places and the mean NDVI is >0.50. In April, all croplands are converted to the current fallow due to harvesting and the mean NDVI is reduced to 0.26.

Table 2 RMSE (°C) of the disaggregate LST

	DisTrad			TsHARP			Local model		
Target	5	16	26	5	16	26	5	16	26
Resolution (m)	February	April	October	February	April	October	February	April	October
400	0.71	1.08	1.27	0.51	1.08	0.88	0.46	0.93	0.57
300	0.91	1.34	1.56	0.65	1.34	1.08	0.57	1.14	0.70
200	1.32	1.59	1.81	0.78	1.59	1.24	0.70	1.36	0.84
100	1.46	2.01	2.13	1.01	2.01	1.48	0.94	1.74	1.13

5 Conclusion

This study aims to downscale the coarse resolution MODIS data LST using its seasonal relationship with NDVI and to evaluate the downscaling models in this context with Landsat 8 data. The study has been undertaken over a heterogeneous landscape of humid subtropical regions, situated in Uttar Pradesh, India. Three satellite images representing the characteristics of different seasons have been analyzed. The results have also shown that LST is inversely connected with NDVI. The negative correlation holds good in the high vegetation areas and post-rainy days. Three different LST sharpening algorithms have been practical—DisTrad, TsHARP, and local model—on seasonal LST. The result signifies that the MODIS LST at 1000 m resolution can be enhanced to 100 m with an RMSE of 0.94 °C via aggregated Landsat 8 LST images. It has been observed that the seasonal variation directly affects the precision of the model used. During the high vegetation period, the error in temperature of the downscaling approaches has been lower compared to those in the low vegetation season. It indicates the superior working of LST sharpening models in high NDVI values. The NDVI has been used as the main supporting data for regression analysis for all the downscaling algorithms. The downscaling results has improved the visual quality of MODIS LST and better earth surface object recognition. The local model algorithms has outperformed others due to the application of rearrangement residual and removal of box-like feature that has been used in other models. The algorithms have downscaled MODIS 1000 m LST data to 200 m and 100 m resolution with RMSE of 0.41 °C and of 0.56 °C, respectively. The relationship has been analyzed through R^2 values which signifying the LST-NDVI relationship is higher in February (0.63) and October (0.77) compared to April (0.54).

Figure 4 shows that the results of the local model have been better than DisTrad and TsHARP in our study area. The local model results RMSE of 0.41 °C whereas DisTrad of 0.55 °C and TsHARP of 0.48 °C when compared with for original Landsat 8 LST values. The results have show LST-NDVI relationship as nonstationary with seasonal variation. The scatter plots shapes have been keep on changing throughout the LST-NDVI variation with seasons change. The coefficients (slope and intercept) of the regression equation vary in different seasons. The regression coefficients along with the confidence interval (95%) level have been used as prime significance values.

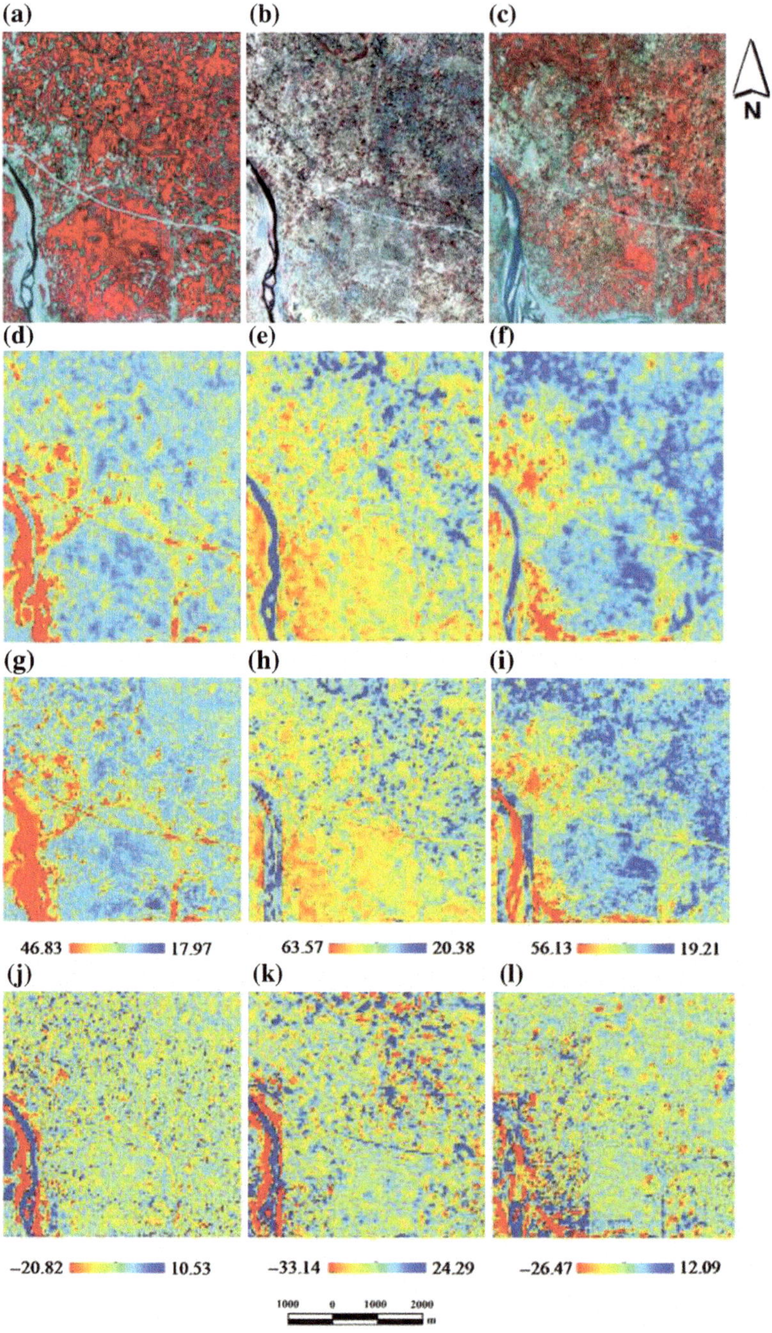

Fig. 4 Landsat 8 (February 5th, 2016, April 16th, 2016, and October 26th, 2016) (**a**–**c**), Actual LST 100 m resolution (**d**–**f**), predicted LST using the local model (**g**–**i**) and the corresponding errors are also shown (**j**–**l**)

The slope and intercept are not constant over various seasons. A significant decrease in slope and an increase of intercept are seen in April compared to September. The input data of February and October over upper and lower teals of NDVI shoes R^2 value of the polynomial regression model for October (0.78) is slightly higher to the linear regression model.

We have applied the TsHARP algorithm in our study area and can say that NDVI should be used as a proxy variable instead of fc. The performance of the model is not so accurate for NDVI < 0.35 and after that RMSE reduces rapidly (to an RMSE of less than 1 °C). The study has not drawn any firm conclusion to use fc and increase of model accuracy. This indicates the applicability of this approach in vegetated areas. Hence, in regions where the vegetation cover is very low, additional covariates such as elevation or soil moisture are necessary to improve the model's accuracy, which can be explored in further study. The local model provides the highest accuracy in LST downscaling compared to the other downscaling method. The DisTrad model has given the least correct values of LST due to mixing of information in urban pixels. It was found that the efficiency of the model decreases towards fine resolution due to the higher sub-pixel fickleness in fine resolution pixels having greater mix built-up nearby.

The analysis and estimation of the downscaling model results in other climatic regions along with actual MODIS data also needs to be tested further. More data sets within a season can be taken, and the average seasonal accuracy in LST downscaling can be estimated further. In view of the fact that the MODIS NDVI (250 m) and LST images (1000 m) are accessible on a daily basis globally. So downscaled LST image at 100 m spatial resolution would certainly extend the potential uses of LST for thermal comfort, vegetation, surface, and environment-related applications. In this study, new approaches have been presented for the enhancing MODIS TIR data using Landsat 8 data as supporting input to enhance its spatial resolution and details in thermal imagery. In this manner, the proposed algorithms consent (i) to enhance the spatial resolution of the TIR image while saving spatial points of interest of the PAN images and (ii) to improve the thermal data properties and its spread along the earth's surface with at finer resolution.

References

1. Howell TA (2013) Downscaling surface temperature image with TsHARP downscaling surface temperature image with TsHARP. https://doi.org/10.13031/2013.35876
2. Wang Q, Shi W, Atkinson PM, Zhao Y (2015) Downscaling MODIS images with area-to-point regression kriging. Remote Sens Environ 166:191–204. https://doi.org/10.1016/j.rse.2015.06.003
3. Chen X, Member S, Yamaguchi Y, et al (2012) Scale effect of vegetation-index-based spatial sharpening for thermal imagery: a simulation study by ASTER data. IEEE Geosci Remote Sens Lett 9:549–553
4. Agam N, Kustas WP, Anderson MC, et al (2007) Utility of thermal sharpening over Texas high plains irrigated agricultural fields. J Geophys Res Atmos 112:1–10. https://doi.org/10.1029/2007jd008407

5. Bindhu VM, Narasimhan B, Sudheer KP (2013) Development and verification of a non-linear disaggregation method (NL-DisTrad) to downscale MODIS land surface. Remote Sens Environ 135:118–129. https://doi.org/10.1016/j.rse.2013.03.023
6. Essa W, Van Der Kwast J, Verbeiren B, Batelaan O (2013) Downscaling of thermal images over urban areas using the land surface temperature—impervious percentage relationship. Int J Appl Earth Obs Geoinf 23:95–108. https://doi.org/10.1016/j.jag.2012.12.007
7. Kustas WP, Norman JM, Anderson MC, French AN (2003) Estimating subpixel surface temperatures and energy fluxes from the vegetation index—radiometric temperature relationship. Remote Sens Environ 85:429–440. https://doi.org/10.1016/s0034-4257(03)00036-1
8. Liu D, Zhu X (2012) An enhanced physical method for downscaling thermal infrared radiance. IEEE Geosci Remote Sens Lett 9:690–694
9. Rodriguez-galiano V, Pardo-iguzquiza E, Sanchez-castillo M et al (2012) Downscaling Landsat 7 ETM+ thermal imagery using land surface temperature and NDVI images. Int J Appl Earth Obs Geoinf 18:515–527. https://doi.org/10.1016/j.jag.2011.10.002
10. Kim J, Hogue TS (2012) Evaluation and sensitivity testing of a coupled Landsat-MODIS downscaling method for land surface temperature and vegetation indices in semi-arid regions. J Appl Remote Sens 6:63569-1. https://doi.org/10.1117/1.jrs.6.063569
11. Ha W, Gowda PH, Howell TA (2011) Downscaling of land surface temperature maps in the texas high plains with the TsHARP method. GIScience Remote Sens 48:583–599. https://doi.org/10.2747/1548-1603.48.4.583
12. Chen J, Ding F, Li Q et al (2016) Using a modified HUTS algorithm to downscale land surface temperature. Landsat-8 imagery: a case study of Xiamen City, China
13. Bonafoni S, Anniballe R, Gioli B, Toscano P (2016) Downscaling landsat land surface temperature over the urban area of Florence. Eur J Remote Sens 49:553–569. https://doi.org/10.5721/eujrs20164929
14. Gu Y, Wylie BK (2015) Downscaling 250-m MODIS growing season NDVI Based on multiple-date landsat images and data mining approaches. Remote Sens 7:3489–3506. https://doi.org/10.3390/rs70403489
15. Trishchenko AP, Luo Y, Khlopenkov KV (2006) A method for downscaling MODIS land channels to 250-m spatial resolution using adaptive regression and normalization. Remote Sens Environ Monit GIS Appl Geol VI:636607. https://doi.org/10.1117/12.689157

Optimization Models for Selecting Base Station Sites for Cellular Network Planning

Shikha Tayal, P. K. Garg and Sandip Vijay

Abstract Increasing number of base station sites with continuously growing customers not only lifted up the total cost of the cellular network but it also has radiation hazard issues affecting health. So, it is vital to select most favorable sites in the planning of cellular networks. For this, various site optimization models based on Meta-heuristic approaches (like Genetic Algorithm GA) and Geographical Information system GIS have been presented in this paper. Outcomes of this study will help us in developing a new model for placement of optimal number of base stations in Uttarakhand (study area). Paper concludes with the pros and cons of different models and also outlines all the necessary variables and/or parameters required for the study area.

Keywords Cellular network planning · Base station sites · Optimization · GIS · Genetic algorithm

1 Introduction

With increased competition between various service providers in the market, task of cellular network planning and resource optimization is becoming much more challenging. A good network planning will be the one which results in less infrastructural expenses by utilizing optimal resources while ensuring maximum customers' satisfaction. Planning a cellular network is an optimization task where appropriate values for a given set of system parameters are determined. To optimize the performance of

S. Tayal (✉)
Uttarakhand Technical University, Dehradun, Uttarakhand, India
e-mail: shikha.tayal@gmail.com

P. K. Garg
Department of Civil Engineering, IIT Roorkee, Roorkee, Uttarakhand, India
e-mail: pkgiitr@gmail.com

S. Vijay
Shivalik College of Engineering, Dehradun, Uttarakhand, India
e-mail: vijaysandip@gmail.com

J. K. Ghosh and I. da Silva (eds.), *Applications of Geomatics in Civil Engineering*,
Lecture Notes in Civil Engineering 33, https://doi.org/10.1007/978-981-13-7067-0_52

the system, the selected parameters are coverage, QoS (quality of service), capacity and cost. Coverage and capacity can be enhanced by placing a number of cellular towers at each required area. But in order to maximize the QoS and minimize the cost, selection of these sites must be scientifically planned. Apart from this, base station sites also have a great impact on radiation hazard issues on surroundings. So, it is vital to select potential base station sites while maintaining a balance between all the above attributes.

For real-world problems, selecting tower sites would depend upon geography, topography, population, physical constraints, and easy installation of radio towers in those locations. Selection of favorable base station sites in Uttarakhand State with varying topographic (demographic Map shown in Fig. 1) is a typical optimization problem. In order to deal with this problem, we need to understand the basic methods of solving an optimization problem.

In an optimization problem, first of all, an objective function is to be set in order to satisfy various conditions. In the beginning, a specified system configuration is selected in order to calculate error or cost function. Error function is simply the difference of previously defined objective and to which degree the current configuration satisfies the conditions. Optimization seeks to minimize the error. It is simply to choose the system with the lowest error, after exhaustive try for every possible system configuration. The computation time may become more prohibitive for more compli-

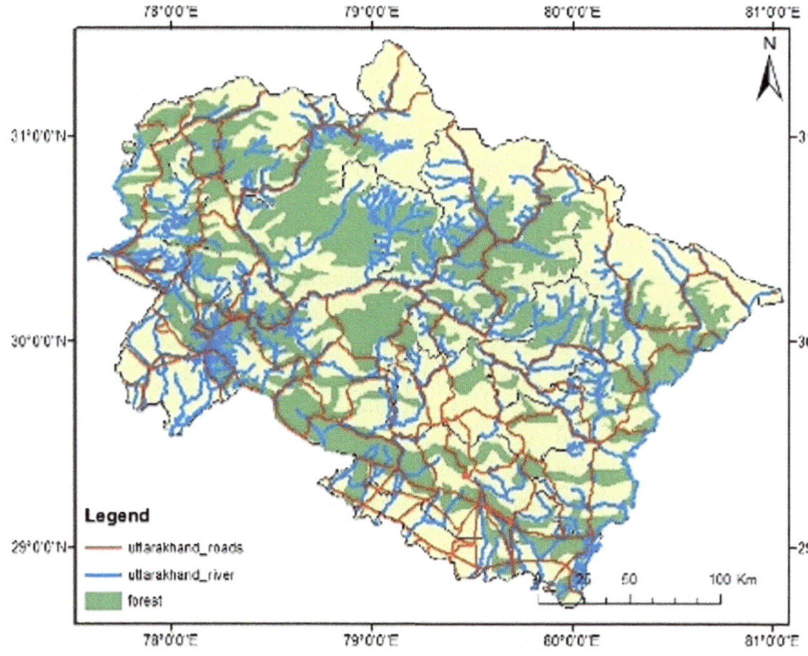

Fig. 1 Demographic map of Uttarakhand State (Study Area)

cated problems. Computation time can be minimized using appropriate optimization models.

This paper is organized into four different sections. Various optimization models based on Meta-heuristic Approaches used for site selection have been discussed in Sect. 2. Section 3 discusses various GIS-based optimization models and also the role of GIS in site suitability. Finally, Sect. 4 concludes by suggesting GIS as a potential tool and by highlighting all the necessary variables required for this optimization problem.

2 Optimization Models

An optimization model always looks for the best solution. It is generally done by generating an initial random solution and then exploiting area in close proximity. If the adjacent solution is better to the current one, then it moves to it. If not, then the algorithm resides at its same position. The commonly used optimization models for base station site selection are based on Meta-heuristic Approaches which includes Simulated Annealing (SA), Tabu Search (TS), Genetic Algorithm (GA), Artificial Bee Colony Optimization (ABC) and Particle Swarm Optimization Technique (PSO).

2.1 *Simulated Annealing Based Models*

Simulated annealing has been shown to be successful in cellular network planning when used for base station placement optimization [1, 2]. While implementing SA, some input parameters, like user distribution and user density have to be determined [2]. Other important parameters which control the algorithm and the methods for choosing their values in an efficient way are addressed in [1]. The suggested SA algorithm allows each BS to make a random move in both "x" and "y" directions. This new location will be accepted if and only if it is a permitted site. A new system "state" is formed when all BSs have their new acceptable locations. The algorithm continues to estimate the cost linked with this state.

Another SA-based framework to enhance previous existing network or to generate completely new networks are presented in [3, 4]. The model recommends not only the base station sites but also base station configuration, such as antenna type (omnidirectional or directional), power control, azimuth, and tilt.

2.2 *Tabu Search Based Models*

Tabu search is capable of base station site planning. In a comparison of local search algorithms for optimal base station location [5], TS provides the most constant final cost value on multiple runs. It maximizes the coverage with least number of base stations.

Another TS approach for cell designing with capacity expansion for mobile communication is presented in [6]. Here the coverage of cellular towers is set constrained in order to satisfy traffic demands.

2.3 Genetic Algorithm Based Models

An application of genetic algorithm (GA) for optimization of cellular network placement provides improved performance in terms of saving time and efficient. The algorithm depends on the selection of parameters such as population size, fitness selection, and reproduction. These parameters were tuned to converge the algorithm optimally [7]. But the algorithm gave its best result only for a flat area. Application of GA with neighborhood constraint [8] when implemented on a small arbitrary urban area, leads to a quantitatively neutral estimated decision-making practice in tower placement.

A study in [9] suggests an efficient algorithm as a solution of large number of tower placement problem. The algorithm based on heuristic approach, propose, and compare three solutions. Greedy approach, ratio heuristics, and GA are allowed to spot the best tower locations from the list of all possible tower locations. While comparing the three solutions, GA proved the best results.

2.4 Hybrid Models

The ability of GA in an area with maximum traffic can be improved using Fuzzy Systems. Fuzzy Systems were used to set the value of the recombination probability and the probability of mutation to avoid premature convergence on GA [10]. In [10, 11], the optimization is done in areas that are planned network, whereas in [12], the optimization of existing towers in an urban area is done using Fuzzy C-Means Clustering (FCM) along with GA. FCM has good ability for clustering existing towers based on the location and distance between the cell tower. The purpose of clustering is to get the point of cluster centers which are evenly distributed throughout the deployment area of BTS tower. Results of clustering were used as input parameters (chromosomes of the initial population) to produce the best fitness of GA. Pattern Swarm Optimization (PSO) [13] is used to minimize the distance between BTS and the furthest Mobile Station MS. By doing this, number of necessary BTS required to cover a particular area is determined. FCM is used to distribute these BTS among the population. Optimization method then determines whether the number of allocated BTS can minimize the distance connecting BTS and farthest MS or not. Optimization output is used to re-cluster. The algorithm stops when the optimal number of base station sites are found. Calculations in PSO are very simple as it has no overlapping or mutation calculation (as in GA), thus taking very less time to evaluate a function.

An effort to design a fuzzy logic-based Decision Support System DSS which is meant to pick most suitable sites for tower placement has been done in [14]. The input parameters such as Cellular interference, End-user density, Ground terrain, etc., are estimated by fuzzy system to get overall suitability as an output parameter. Though the proposed system can be used to select suitable location for setting up new towers but the system can't be used for optimization. Also, accuracy of the system performance is not discussed in paper.

3 Geographic Information Systems (GIS) and Cellular Network Optimization

Most of the models discussed in Sect. 2 were implemented on an arbitrary area. However, site selection is critical while working on a real problem. As towers are to be erected practically on the ground, the authenticity and accuracy of potential tower locations play a major role. Thus, use of high to medium resolution satellite images of an area is vital which has been overlooked by the above papers. GIS is an important tool to manage, analyze and then display large volume of varied data used for local and regional planning activities. GIS approach can solve site selection problems, as presented in papers shown below.

3.1 GIS-Based Optimization Models

GIS helps its users by organizing and combining spatial (like 3D, latitude, longitude), temporal (like time-based changes) and economic (like about trade and industry) information. Paper [15] uses Data Envelopment Analysis (DEA) with GIS frame to control optimal site locations for real land plans. The DEA method builds in the query for selecting locations by maximizing the ratio of outputs to inputs.

Spatial mining with GIS as a tool has been used in [16] to optimize the cell towers distribution. The optimization is fixed by using the Digital Elevation Model (DEM) on the image of the area which was covered with two levels of hierarchy. The study concludes that constructing spatial data as flat data can mark the spatial mining much more effective by reducing the mining to one level only, thus preventing much time and space consumption.

Spatial analyst and proximity analysis tool of ArcGIS was used in the study [17]. The study discusses in what way one can use GIS tool more effectively and efficiently for analysis of sites suitability. It helps to set up more towers in the least time with minimum cost in a certain area. GIS permits the user to conduct a sequence of steps that can visually enable the user to locate best cell tower sites.

In [18], author investigated a structure that can provide more proficient and economical network coverage for dense urban area. The radio propagation–prediction model (ray-tracing) combined with 3D geo-information was used to model the radio signal coverage for one of the cellular service provider. When the suggested method is compared with conventional methods author demonstrates that the spatial analyst methodology can be used to boost signal coverage in any dense urban environment consequently reducing cost for cellular network planning.

Cellular layout design problem (CLDP) can be solved by considering terrain variation and traffic demand [19]. Depending on the type of terrain, user distribution, user demand, and growth rate, distribution of cells was identified. The approach preserves the budget to its minimum using repeaters to meet the traffic demand within the blind spots. However, only flat terrain is considered in the study.

The location of a Mobile Telephone Switching Office (MTSO) can be pretentious by aspects, such as distances between each component, restricted areas, and accessibility. GIS site selection method can be used to identify restricted and potential areas, also in order to find an optimal solution a nonlinear programming model was developed in [20].

By making a buffer zone of individual companies and getting all network companies in a single platform, it can be found that the network coverage in which part of the city is shortfall and in need of towers for the better communications [21].

The planning tools make broad usage of terrain data. Radio-coverage predictions are the basis for planning and designing a proficient and robust cellular network [22].

The type of trees, their densities and heights are some important factors affecting the variation in signal strength. Signal strength projected by Normalized Difference Vegetation Index (NDVI) and viewshed analysis is used in [23]. For an effective telecom planning and development, the author suggests using multi-spectral and stereo-satellite data.

In [24], author discusses all the factors affecting cell tower site selection. It is observed that the site should be adjacent to road for physical access with electricity power and telecommunications network connectivity. To ensure the signal coverage across the terrain, Local zoning ordinances must accommodate tower height requirements. In order to ensure radio signal integrity, sources of electromagnetic interference must be avoided. In addition to this, environmental and wildlife impacts must also be considered. So, while planning mobile tower network, evaluation of population, demographic data, profiles of nearby businesses, the proximity of roads and highways, pedestrian traffic, has to be done.

3.2 GIS-Based Other Site Suitability Models

The procedure of developing a model of GIS application in order to offer a method for supporting location decisions for putting into practice of urban master plans is explained in [25].

To resolve complex issues, the use of GIS is rising rapidly. Paper [26] discussed various prospective of its applications in tourism planning. The author concludes that GIS is a strong and effective tool to serve in planning and decision-making. Recognizing optimum locations of towers, particularly when they function with overlapping regions is a challenging task. Therefore, an innovative tool is needed with three-dimensional analysis and visualization skills which can work for a single environment. Building Information Modeling (BIM) provides a digitally constructed virtual model, which aids to visualize buildings before the actual implementation to be done. GIS is useful to explore already existing objects, while the BIM systems focused on developing objects. The GIS-BIM model is used to identify feasible tower sites [27].

A rule-based spatial model using Arc Macro Language (AML) is developed for planning of infrastructural facilities in the least developed block of Dehradun district [28]. Authors found the GIS approach to be useful and quick for data integration and for spatial modeling required for district-level planning.

While planning cellular network, radiation hazard issues must also be taken into consideration. Because of high radioactivity areas, living beings may suffer from different health problems like infertility, DNA damage, cancer, sleep disorder, gland tumor, etc. So, it is essential to optimally select BTS sites in the design of cellular networks.

Base data, like satellite images, topographic or thematic maps and required field data, must be appropriately used for any suitability analysis [29]. In the study of growth and development of urbanization around a city, it is found that parameters considered for suitability analysis will not have same weightage or usefulness. So, it is essential to set certain score to each category of parameter according to their suitability. Author has given a score of 10 (10-point scale) to vacant land use whereas a score of 4 to forest category of land use. Further, a weighing system is aimed to permit maximum score 550. The area which falls under different suitability zones (like highly suitable, suitable, moderate, etc.) is calculated by dividing suitability score into different ranges.

GIS tools, like overlay and buffer analysis have been used for selecting best alternative site for municipal solid waste land filling, using remote sensing and GPS [30]. Score of respective sites has been estimated using site sensitivity index. Finally, a site with maximum score is selected as suitable.

Multi-criteria analysis and weighted overlay analysis have been done for solid waste disposal site selection by [31]. The GIS-based multi-criteria evaluation method being very simple and flexible, can be useful to a mountainous region [32, 33]. A study to spot appropriate sites of a hilly town, Mussoorie and Nahan municipal areas using GIS-based multi-criteria evaluation is conducted by [33, 34]. Spatial data (Cartosat-1, LISS-IV, and IKONOS satellite data) and Toposheets of the study area were used to generate several thematic information layers; land values, slope, land use, road proximity, soil, and geomorphology. With the generated criteria, maps were standardized by analytical hierarchy process (AHP). In the next stage of analysis, consistency is checked. The final site suitability map can be produced using the formula:

$$\textbf{Site Suitability} = \sum [\textbf{Cn} \times \textbf{Wn}] \tag{1}$$

where

Cn standardized raster cell,
Wn weight derived from AHP pairwise comparisons.

In this paper, different site optimization models have been presented. Summary of all the discussed models is given below in Table 1.

4 Conclusion

- The main objective of site optimization problem is to serve maximum land of a geological region with least number of BTS. GIS may be used as a tool for solving spatial kind of problem.
- Road, River, Dense Forest, Population density, habitation, LULC, DEM, Contour Map, tourist points, and site parameters of existing towers (such as Latitude—Longitude, antenna height, frequencies, antenna type, and azimuth) are important parameters for preparation of suitability map.
- Combining optimization algorithm (FPA) and clustering (FCM) with GIS tool will enhance the accuracy of result.
- Out of various GIS tools, like buffer analysis, Multi-criteria analysis, and weighted overlay analysis, Multi-criteria evaluation method works best in a mountainous region (such as Uttarakhand state).
- A relevant challenge is to develop an integrated and computationally efficient approach to coverage planning for 3G and 4G cellular systems.
- The site should be adjacent to the accessible road with electrical power availability. It increases the probability of telecommunications network connectivity. In order to ensure radio signal integrity, sources of electromagnetic interference must be avoided.
- In addition to this, environmental and wildlife impacts must also be considered.

Table 1 Summary of optimization models

S. No	Models	Input data	Output	Remarks
1	SA-based models	User distribution, user density	Approximate number of required BSs with their position and base station configuration, such as antenna type (omnidirectional or directional), power control, azimuth and tilt	
2	TS-based models	Tabu list	Base station site planning	TS provides the most consistent final cost value of multiple runs
3	GA-based models	Multi-spectral and stereo-satellite data	Effective telecom planning and development	GA with neighborhood constraint leads to a quantitatively unbiased evaluated decision-making process in BTS placement
4	Hybrid model • FCM with GA • FCM with PSO • DSS with Fuzzy Logic	nearest building threshold distance, Interference with other wireless users, Ground terrain, End-user density, Cellular interference, and Electromagnetic radiation level	Overall Site suitability	Combining optimization algorithm with clustering enhances the results
5	GIS-based optimization models GIS tools: • DEA • Spatial mining • Spatial analyst (Viewshed and line of sight) and proximity analysis (buffer) tool	population, demographic data, profiles of nearby businesses, pedestrian traffic, the proximity of roads and highways	Optimal site locations for real estate projects	The signal strength depends upon type of trees, their densities, and their heights. The signal may be diffracted or absorbed on striking on the surface of a building

(continued)

Table 1 (continued)

S. No	Models	Input data	Output	Remarks
6	GIS based other SiteSuitability Models • GIS-BIM • Rule-based spatial model using (AML) • Buffer analysis • Multi-criteria	Base data, like satellite images, topographic or thematic maps and required field data	Tourism planning, site for municipal solid waste land filling, planning of infrastructural facilities	Buffer analysis Multi-criteria analysis and weighted overlay analysis

References

1. Anderson HR, McGeehan JP (1994) Optimizing microcell base station locations using simulated annealing techniques. In: 1994 IEEE 44th vehicular technology conference. IEEE, pp 858–862
2. Yaacoub E, Dawy Z (2014) LTE radio network planning with HetNets: BS placement optimization using simulated annealing. In: 2014 17th IEEE Mediterranean electrotechnical conference (MELECON). IEEE, pp 327–333
3. Prajapati NB, Agravat RR, Hasan MI (2010) Simulated annealing for location area planning in cellular networks. arXiv:1003.3553
4. Mohanty R, Pattanaik S, Bhoi SP (2012) Simulated annealing based placement algorithms and research challenges: a survey. J Global Res Comput Sci 3(6):33–37
5. Krishnamachari B, Wicker SB (2000) Experimental analysis of local search algorithms for optimal base station location. In: Proceedings of international conference on evolutionary computing for computer, communication, control and power, Chennai, India
6. Lee CY, Kang HG (2000) Cell planning with capacity expansion in mobile communications: a tabu search approach. IEEE Trans Veh Technol 49(5):1678–1691
7. Munyaneza J, Kurien A, Van Wyk B (2008) Optimization of antenna placement in 3G networks using genetic algorithms. In: 2008 Third international conference on broadband communications, information technology and biomedical applications. IEEE, pp 30–37
8. Alenoghena CO, Emagbetere JO, Edeko FO (2013) Application of genetic algorithm in radio network coverage optimization—a review. Int J Comput Appl 66(12)
9. Deane JK, Rakes TR, Rees LP (2009) Efficient heuristics for wireless network tower placement. Inf Technol Manage 10(1):55–65
10. Fachrie M, Widowati S, Hanuranto AT (2013) Implementation of fuzzy evolutionary algorithms to determine the position of base transceiver station (BTS). In: National seminar of information technology
11. Alenoghena CO, Emagbetere JO, Aibinu AM (2013) Artificial intelligence based technique for BTS placement. In: IOP Conference series: materials science and engineering, vol 53(1). IOP Publishing, p 012056
12. Rofii F, Toscanni MD, Siswanto D (2016) Optimization of coverage and the number of base transceiver station towers using fuzzy C-means and genetic algorithm. J Theor Appl Inf Technol 93(1):164
13. Onim AO, Kihato PK, Musyoki S (2014) Optimization of base station location in 3G networks using fuzzy clustering and mesh adaptive direct search. In: Proceedings of sustainable research and innovation conference, pp 30–35
14. Singh J, Kaur G, Kaur G (2015) Determining best setup sites for cellular towers using fuzzy logic. In: 2015 International conference on futuristic trends on computational analysis and knowledge management (ABLAZE). IEEE, pp 256–260

15. Li H, Yu L, Cheng EW (2005) A GIS-based site selection system for real estate projects. Constr Innov 5(4):231–241
16. Al-Hamami AH, Hashem SH (2011) Optimal cell towers distribution by using spatial mining and geographic information system. arXiv:1104.2721
17. Kumar BA, Kaur S (2013) GIS based site suitability analysis for setting up of new mobile towers in Neemrana and Behror Tehsil, Rajasthan. In: 14th Esri India user conference
18. Munene EN, Kiema JBK (2014) Optimizing the location of base transceiver stations in mobile communication network planning: case study of the Nairobi Central Business District, Kenya. Kenya Int Interdiscip J Sci Res 1(2):113–127
19. Kamar A, Nawaz SJ, Patwary M, Abdel-Maguid M (2010) Optimized algorithm for cellular network planning based on terrain and demand analysis. In: 2010 2nd International conference on computer technology and development (ICCTD). IEEE, pp 359–364
20. Oh JY, Yu J, Aukerman RA (2007) Decision making tools in cellular telecommunication network design: GIS and NlP. Issues Inf Syst 8(2):97–102
21. Simpi B, Chandrashekarappa KN, Nadan N, Prabhuswamy B (2011) Mobile network system of Bhadravathi Town using remote sending, GIS & GPS, Shimoga District, Karnataka, India. Global J Comput Sci Technol 11(14), 1.0:51–56
22. Wagen JF, Rizk K (2003) Radiowave propagation, building databases, and GIS: anything in common? A radio engineer's viewpoint. Environ Plan 30(5):767–787
23. Naveenchandra B, Lokesh KN, Usha, Bhat HG (2011) Signal strength measurements and coverage estimation of mobile communication network using IRS-IC multispectral and CARTOSAT-1 stereo images. Geospatial world forum
24. Harris M (2011) How cell tower works. Wireless white paper. www.unisonsite.com
25. Al-Shalabi MA, Mansor SB, Ahmed NB, Shiriff R (2006) GIS based multicriteria approaches to housing site suitability assessment. In: XXIII FIG congress, shaping the change, Munich, Germany, Oct, pp 8–13
26. Pareta K (2013) Remote sensing and GIS based site suitability analysis for tourism development. Int J Adv Res Eng Appl Sci 2(5):43–58
27. Irizarry J, Karan EP (2012) Optimizing location of tower cranes on construction sites through GIS and BIM integration. J Inf Technol Constr (ITcon) 17(23):351–366
28. Gupta RD, Garg PK, Arora MK (2000) A GIS based spatial modeling for developmental planning. Proc ICORG 1:265–270
29. Jain K, Subbaiah YV (2007) Site suitability analysis for urban development using GIS. J Appl Sci 7(I8):2576–2583
30. Singh Y, Chauhan MS, Katiyar SK (2012) SWM of Kolar municipality using remote sensing and GIS techniques. Int J Adv Technol Eng Res (IJATER) 2(1):30–33
31. Pradhan J, Samanta K (2015) Site suitability analysis using remote sensing and GIS for proper selection of solid waste disposal ground within Rajarhat Gopalpur municipal area, Kolkata, West Bengal. Int J Geomat Geosci 5(4):640–654
32. Kumar S, Sharma S (2012) Site suitability analysis for urban development: a review. Int J Recent Innov Trends Comput Commun 3(6):3647–3651
33. Kumar M, Shaikh VR (2013) Site suitability analysis for urban development using GIS based multicriteria evaluation technique: a case study of Mussoorie Municipal Area, Dehradun District, Uttarakhand, India. J Indian Soc Remote Sens 41(2):417–424
34. Kumar S, Kumar R (2014) Site suitability analysis for urban development of a Hill Town using GIS based multicriteria evaluation technique: a case study of Nahan Town, Himachal Pradesh, India. Int J Adv Remote Sens GIS 3(1):516

Acquisition of Inaccessible Geospatial Data

Sharwan Ram, Suraj Sawant and Ajay Kumar Patel

Abstract The acquisition of geospatial data from inaccessible areas using conventional methods and man-operated mobile mapping system (MMS) is a critical operation. However, autonomous unmanned vehicles may be used for the acquisition of geospatial data from inaccessible areas rapidly and cost-effectively. Therefore, the objective of this research work is the acquisition of geospatial data using an unmanned mobile mapping system (UMMS) and validating the acquired data using the inexpensive RTK GNSS (Global Navigation Satellite System) system. The data have been collected at the predefined waypoints using UMMS of the study areas. The waypoints and study areas have been selected using Google Earth. The quality of acquired data by the system has been tested with a geodetic GNSS receiver data. It has been observed that the proposed system has the capability to provide data with an accuracy of 2.5 cm. The accuracy of the proposed system degrades as the system navigates under the canopy and buildings.

Keywords Geospatial data · GNSS · UMMS

1 Introduction

The acquisition of geospatial data from inaccessible has been evaluated from the financial and technical point of view [1]. During the acquisition phase of geospatial data the most considerable effort, time, quality of data and cost have been recognized [2]. To acquire the geospatial data conventional methods have been adopted for years. The most widely used instruments are theodolites and total station. However, these

S. Ram (✉) · S. Sawant · A. K. Patel
Geomatics Engineering Group, Indian Institute of Technology Roorkee, Roorkee, Uttarakhand, India
e-mail: sharwandidel@gmail.com

S. Sawant
e-mail: suraj.t.sawant@gmail.com

A. K. Patel
e-mail: ajaypatel.iitb@gmail.com

J. K. Ghosh and I. da Silva (eds.), *Applications of Geomatics in Civil Engineering*, Lecture Notes in Civil Engineering 33, https://doi.org/10.1007/978-981-13-7067-0_53

approaches are labor-intensive, time-consuming, and require inter-visibility between stations. Nowadays, GNSS system has been adopted for the acquisition of geospatial data [3]. This technology provides accurate and reliable data also overcome the limitation arising out of older technology. Henceforth, today Mobile Mapping Systems (MMS) and Unmanned Aerial Vehicles (UAVs) have been used to acquire the geospatial data using GNSS system. Also, these systems are equipped with instruments (such as Lidar, GNSS system, Stereo Cameras) depends on the application and required data type [4–6]. Recently, there is a growing interest in the use of Micro-Unmanned Aerial Vehicles (MUAVs) for the acquisition of geospatial data from inaccessible areas rapidly and cost-effectively [6]. On the other hand, low-cost systems and sensors data have the following limitations such as needed flight permission, expensive and data gets affected due to various factors like a blur, under-or overexposure and sometimes images do not cover the area of interest [7]. Also, the UAV data suffered due to high inclination, poor side overlapping and large distortions than classical aerial data and difficult to photogrammetric processing [8]. The available micro-UAVs equipped with low-cost sensors do not provide qualitative data with inexpensive sensors for surveying and remote sensing applications [9]. The collection of geospatial data with these systems in remote/inaccessible areas is either expensive or time-consuming. The quality of data may improve if acquisition gets coupled to direct geo-referencing with accurate Ground Control Points (GCPs) [9]. But, the establishment of these control points is either difficult or expensive in inaccessible or hazardous areas [10].

The MMS has been frequently used for qualitative, timely and cost-effective data collection. These systems are man operated therefore from these systems acquisition of geospatial data not possible from inaccessible areas such as a frozen lake, mining, desert etc. The problems associated with the acquisition of qualitative geospatial data these areas can be possible through well navigated Unmanned Mobile Mapping System. Thus, auto-navigated UMMS is capable of bringing the dream of the surveyor in reality. In the literature, many auto-navigation systems have been found but these systems are application specific. These systems are not developed for the acquisition of geospatial data. Two unmanned mobile mapping systems have been recognized in the literature. The first one has been developed by Leica and Milrem is an all-terrain system [11]. The second one is a man-portable remotely operated system capable of geospatial data [12].

Therefore, in this research, the emphasis has been given to the acquisition of geospatial data from inaccessible areas using an unmanned mobile mapping system. The system has capabilities of acquired data as well as transferred the obtained data to the ground control station (GCS). Also, the developed system having the ability to establish ground control points. The paper is organized as follows: in Sect. 2 the types of geospatial data have been presented. Section 4 described the unmanned mobile mapping system. Section 4 described the study area from data have been collected. Section 5 described the performance analysis. Section 6 concludes the paper with some final remarks.

2 Geospatial Data

Nowadays, Geospatial data can be defined as "data of earth with proper reference" or "data that identifies the geographic location of the features and boundaries on Earth, such as natural or constructed features, oceans, and more" [13].

2.1 Geospatial Data Types

The geospatial data can be classified into two categories from the prospective acquisition of according to available technology: point data and line or area (see Fig. 1) [14]. The point data provides the location of a point situated on the earth surface. The line data can be described as a combination of continuous/multiple point data along a line. Whereas, area data can be defined as "georeferenced data of the particular area of the earth surface", which can be further classified into three types such as image data, video data, and laser data [13]. The image and video data also required orientation information depends on the application.

2.2 Geospatial Data Representations

After the acquisition, the following actions have been taken such as data gets stored, processed, analyzed, quality assessment and final represent in the digital world. From the perspective of geographic information system (GIS) geospatial data can be represented into two types-vector and raster format (Fig. 2) [16, 17]. Raster data represents an area in the form of a continuous grid of cells/pixel, having each cell with an assigned value. This type of data is used to represents an area in a continuous

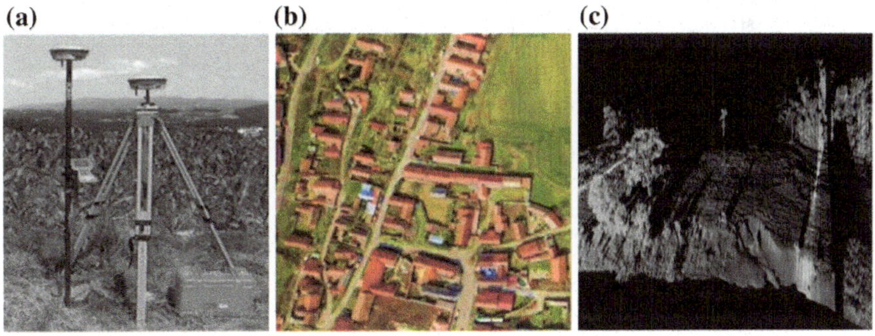

Fig. 1 Examples of **a** point data, **b** image data and **c** point clouds acquired with the mobile laser scanner [15]

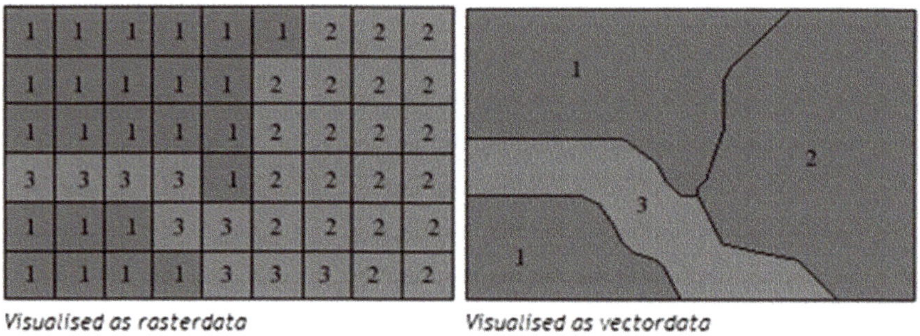

Fig. 2 Geospatial data representations

manner, like a satellite image, an aerial photograph or an elevation surface. Vector data represents an area using points, lines and polygons. Vector data are used to represent any discrete form, such as country borders, land parcels, streets etc. [13].

3 Results and Discussion

The developed UMMS consists of three subunits are navigation and control, data acquisition, and ground control station (GCS) shown in Fig. 4. The architecture and sensors interconnectivity of the developed system is shown in Fig. 3. Navigation and control unit is consists of non-vision sensors such as RTK GNSS receiver, Inertial

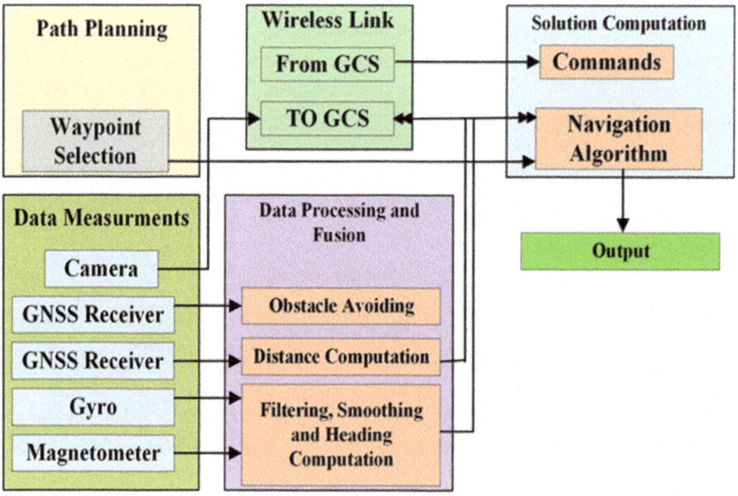

Fig. 3 Architecture framework of the developed system

Fig. 4 The UMMS to acquire the geospatial data from inaccessible areas

Navigation Unit (IMU), processors, ultrasonic distance sensors, and wheel encoders. The navigation unit is responsible for three functions: localization, path planning, and obstacle avoidance. Whereas the control unit is responsible for speed control, sensors data, decode control signals from GCS and for synchronization. The data collection unit consists of active and passive sensors such as RTK GNSS, IMU, cameras and some other sensors (Temperature, Humidity and Pressure etc.). It enables sensors as per given instructions to the system, acquired data, process data, store and transmits and receives the data from sensors to the communication unit.

The GCS is portable hardware/software devices to collect sensor's data remotely, to process and check the quality of the acquired data in real-time, monitor and command. If required, then the sensors onboard the UGV can be remotely controlled from the GCS to collect data from the study area. The obtained data from system real-time is stored at GCS for further real-time or offline processing. In the proposed system GCS consists of a wireless module interface laptop or computer. The wireless module used to send commands to a system and received data from the system.

4 Study Area

The study area is located at latitude 29.865877°–29.865106°N, and longitude 77.897231°–77.895934°E of WGS-84 datum. The study area was chosen in such a way that it involves a plane, sloped and mixed area. To test the accuracy of the system the data have been collected from the selected area. The illustration of the study areas has been shown in Fig. 5. The study area has been chosen using Google Earth, and assumed that area is unknown. Before going to next phase proper analysis is required, that saves the system from any unpredictable damage.

Fig. 5 Study area

5 Field Data Acquisition

The field test was performed at the campus of Indian Institute of Technology Roorkee. The Geodetic GNSS receiver and inexpensive RTK GNSS receiver have been assembled on the UGV. These receiver acquired data at the epoch of 1 s. Two different base stations have been set-up at the known location for the both GNSS receiver. After that system launched into the field and system monitored remotely at the ground control station. The data collected from geodetic GNSS receiver is stored into the memory card. Whereas, the data acquired from RTK GNSS receiver is transmitted over the internet into the real-time. The geodetic GNSS receiver data have been used as reference data to analyze the performance of data collected using RTK GNSS receiver.

5.1 Path Planning

The path planning has been done with the help of Google Earth software (Fig. 6). In this section, the waypoints have selected Google Earth software. From these points, the direction of navigation has computed that used in auto-navigation. The number of waypoints depends on field conditions (such as visibility of GNSS satellites, and variations in elevation profile) and the path to be navigated. If the path is not a straight line, then the numbers of waypoints have been increased. Also, the elevation profile of field area has been analyzed using Google Earth for a rough idea. The elevation variation requires the checking possibility of system navigation. The geodetic

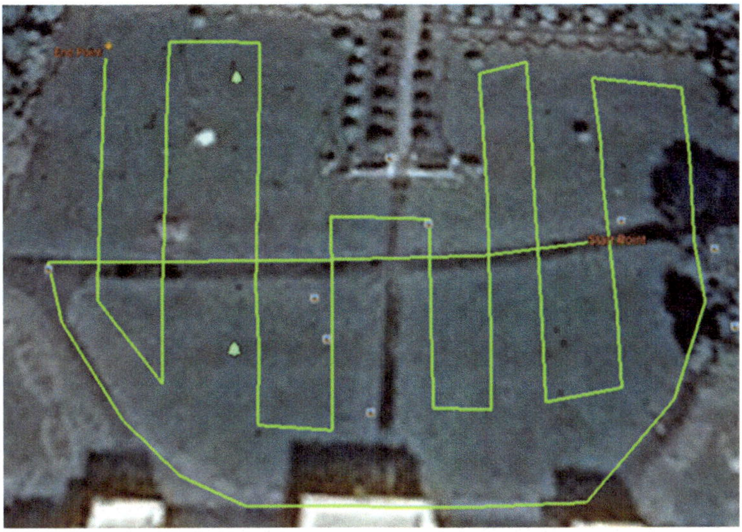

Fig. 6 Planned path to be navigated for the acquisition of data

GNSS receiver data have been used as reference data to analyze the performance of data collected using RTK GNSS receiver. The average elevation of all points has been measured using 216.476 m using geodetic GNSS receiver and 216.39 m using UMMS. This is about 10 cm because UMSS can't navigate and collected data the near to boundary to avoid a collision.

5.2 *Data Acquisition*

Before going into the field or launching system for data collection there is need for preplanning. The pre-planning involves four major steps are checking the availability of satellites over the sites, weather condition, finding the coordinate of the base with the international GNSS service (IGS) stations and the establishment of the base station. After that, the data acquisition phase has been performed using UGV system. Two GNSS receivers were used for data collection, one act as a base station and other as a rover. The real-time corrections from the base station were transmitted to the rover receiver over a wireless channel. The rover receiver gets corrections wirelessly at a rate of 1 Hz and provides centimeter level accuracy positional accuracy. After corrections, rover receiver data have been transmitted to the GCS in real-time over a web server using the internet. The GCS station consists of a computer where real-time data from the vehicle is monitored . The real-time monitoring (Fix Mode, Dilution

Table 1 Minimum, average and maximum elevation observed using both systems

Elevation measured using	Minimum (m)	Maximum (m)	Average (m)
Geodetic GNSS receiver	215.052	218.082	216.476
UMMS	215.09	218.04	216.39

of Precision (DOP) and tracked satellite information) help in judgment that rover data are meeting the required quality or not. The RTK fix position data can be easily acquired where the minimum seven satellites in view and signal to noise ratio above 39 dB.

5.3 Performance Testing

To test the performance reference data have been collected from the study area as shown in Fig. 7. Total 130 points have collected using geodetic GNSS receiver. The data have taken at least 5 min at every point. Similarly using UMMS data have collected along the predefined path. Data have been collected in RTK mode of operation. After data acquisition gets over, the data have been downloaded and processed. To process geodetic GNSS receiver data spectrum survey office software has been used. In performance testing the elevation profile observed using both receivers have been compared is shown in Figs. 8 and 9. From observed data, it has been found that acquisition of geospatial data using low-cost RTK GNSS receiver provides identical results. The minimum value of elevation has been observed 215.09 m using UMMS and 215.052 using geodetic GNSS receiver as shown in Table 1. The maximum value of elevation has been observed 218.04 m using UMMS and 218.082 m using geodetic GNSS receiver as shown in Table 1. At the static point, the maximum difference has been observed 2.5 cm in the elevation. All data have been taken open sky where at least 12 satellites were available. The three points where satellites were not available more than seven, therefore, RTK receiver not work in RTK mode and provides the elevation variation in 1–3 m. It has been found the inexpensive RTK GNSS receiver not reliable where seven satellites do not have signal o noise ratio 39 dB. The average elevation of all points has been measured using 216.476 m using geodetic GNSS receiver and 216.39 m using UMMS. This is about 10 cm because UMSS can't navigate and collected data the near to boundary to avoid a collision.

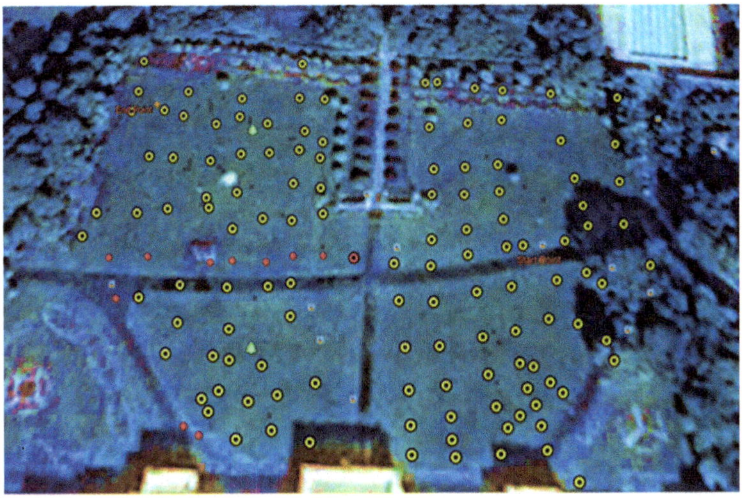

Fig. 7 Reference data collected using geodetic GNSS receiver

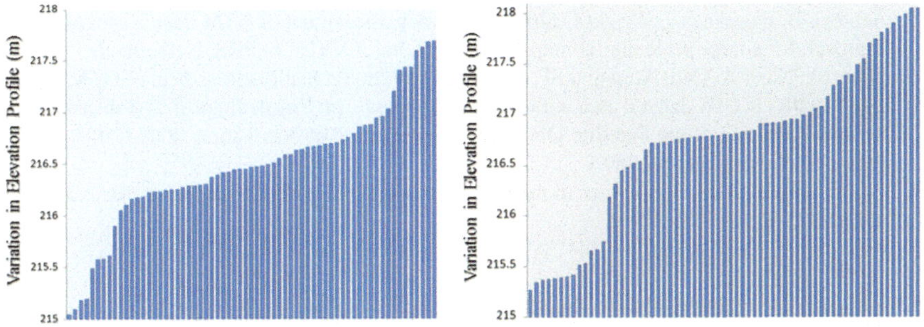

Fig. 8 Study area elevation profile in (m) of study area (left and right sub-areas) observed using geodetic GNSS receiver

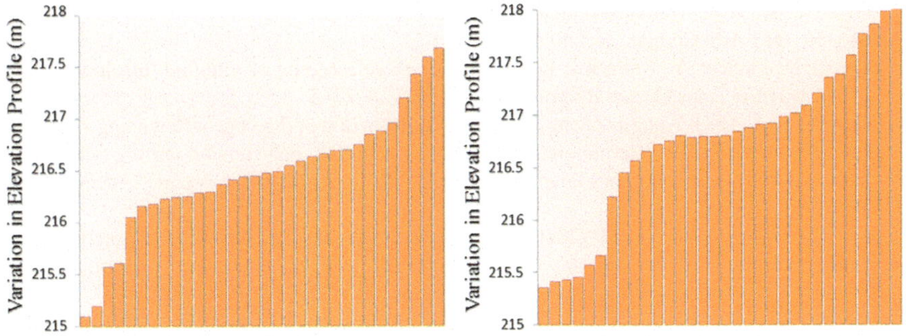

Fig. 9 Elevation profile of study area (left and right sub-areas) observed using UMMS

6 Conclusions

The acquisition of geospatial data from the inaccessible areas is difficult and expensive. The unnamed mobile mapping system able to acquired the geospatial data. The developed having capability of acquired the geospatial data from the inaccessible areas. From the obtained results it has been found that UMMS provides same results as highly accurate GNSS system. Also, the system having the capability to provide data in real-time at the ground control station. The system can be used for other applications such as where real-time geospatial data required, photogrammetry and agriculture. The making use of inexpensive GNSS system is not reliable in dense forest areas and where GNSS satellites get obstructed. The navigation of developed system still a challenge in the highly steep sloped areas.

References

1. Li Z, Zhu Q, Gold C (2004) Digital terrain modeling: principles and methodology. CRC Press
2. Acharya B, Fagerman J, Wright C (2000) Accuracy assessment of DTM data: a cost effective approach for a large scale digital mapping project, vol XXXIII. IAPRS, Netherlands
3. Matori Y, Nasir A, Dedi Atunggal SP, Cahyono BK (2008) Quality assessment of DTM generated from RTK GPS data on area with various sky views. In: Proceedings of 21st international technical meeting of the Satellite Division of the Institute of Navigation (ION GNSS 2008). Savannah, GA, pp 1462–1469
4. Puente I et al (2013) Review of mobile mapping and surveying technologies. Measurement 46(7):2127–2145
5. Francesco N, Remondino F (2014) UAV for 3D mapping applications: a review. Appl Geomatics 6(1):1–15
6. Colomina I, Molina P (2014) Unmanned aerial systems for photogrammetry and remote sensing: a review. ISPRS J Photogramm Remote Sens 92:79–97
7. Andreas F, Kattenborn T, Koch B (2013) UAV-based photogrammetric point clouds-tree stem mapping in open stands in comparison to terrestrial laser scanner point clouds. Int Arch Photogramm Remote Sens Spat Inf Sci 40:141–146
8. QianY et al (2013) Unmanned airship based high resolution images acquisition and the processing. Int J Adv Comput Technol 5(3)
9. Turner D, Lucieer A, Wallace L (2014) Direct georeferencing of ultrahigh-resolution UAV imagery. IEEE Trans Geosci Remote Sens 52(5):2738–2745
10. Guo D, Wu L, Li Q, Wang J, Zheng X (2006) A comparison of direct georeferencing and GPS-supported AT. In: IEEE International conference on geoscience and remote sensing symposium
11. https://www.spar3d.com/news/lidar/leicas-terrain-unmanned-mobile-system/. Accessed 06 Dec 2017
12. Li-Chee-Ming J, Armenakis C (2010) Towards the development of a low-cost remotely-piloted land mobile mapping system. In: Annual conference of the American Society for photogrammetry and remote sensing
13. Neteler M, Mitasova H (2013) Open source GIS: a GRASS GIS approach, vol 689. Springer Science & Business Media
14. Lewis P, Fotheringham S, Winstanley A (2011) Spatial video and GIS. Int J Geogr Inf Sci 25(5):697–716
15. Tommaselli AM et al (2014) Monitoring marginal erosion in hydroelectric reservoirs with terrestrial mobile laser scanner. Int Arch Photogramm Remote Sens Spat Inf Sci 40(5):589

16. Mills JP, Newton I (1996) A new approach to the verification and revision of large-scale mapping. ISPRS J Photogramm Remote Sens 51(1):17–27
17. Winter S, Frank AU (2000) Topology in raster and vector representation. GeoInformatica 4(1):35–65

Printed by Printforce, the Netherlands